The Standard Model and Beyond

The Standard Model and Beyond

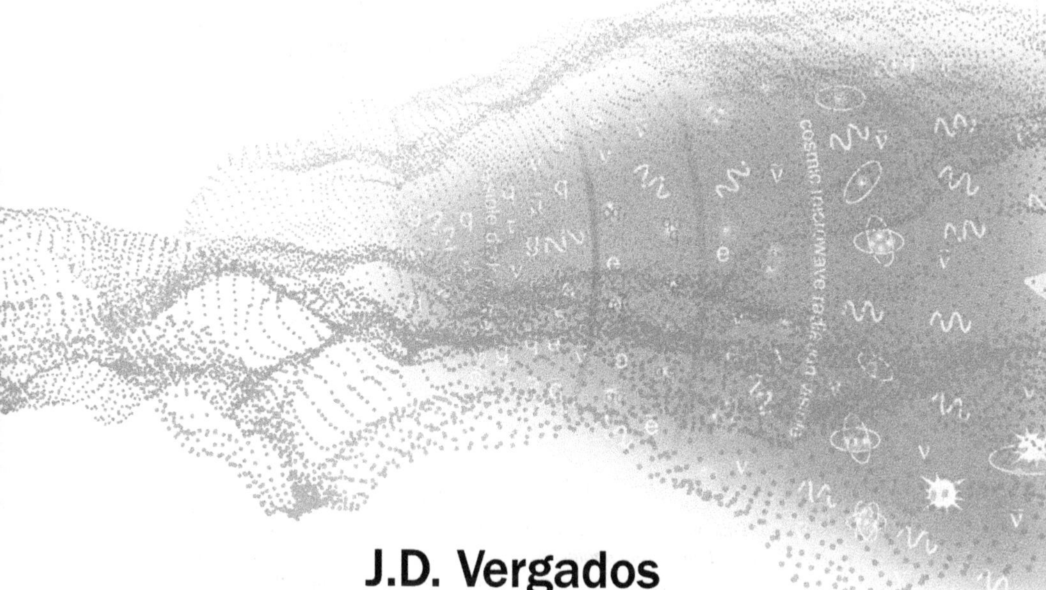

J.D. Vergados
University of Ioannina, Greece

NEW JERSEY • LONDON • SINGAPORE • BEIJING • SHANGHAI • HONG KONG • TAIPEI • CHENNAI • TOKYO

Published by

World Scientific Publishing Co. Pte. Ltd.
5 Toh Tuck Link, Singapore 596224
USA office: 27 Warren Street, Suite 401-402, Hackensack, NJ 07601
UK office: 57 Shelton Street, Covent Garden, London WC2H 9HE

Library of Congress Cataloging-in-Publication Data
Names: Vergados, J. D., author.
Title: The standard model and beyond / Ioannis (J.D.) Vergados (University of Ioannina, Greece).
Description: Singapore ; Hackensack, NJ : World Scientific, [2017] |
Includes bibliographical references and index.
Identifiers: LCCN 2017026743| ISBN 9789813228559 (hardcover ; alk. paper) |
ISBN 9813228555 (hardcover ; alk. paper)
Subjects: LCSH: Particles (Nuclear physics) | Standard model (Nuclear physics) |
Group theory. | Symmetry (Physics)
Classification: LCC QC793.2 .V48 2017 | DDC 539.7/2--dc23
LC record available at https://lccn.loc.gov/2017026743

British Library Cataloguing-in-Publication Data
A catalogue record for this book is available from the British Library.

Typeset by Stallion Press
Email: enquiries@stallionpress.com

To the Students

Who helped in preparing these Lecture Notes

Preface

This book contains the material of a set of Lectures on The Standard Model and Beyond, which were initiated for the first year graduate students of the Department of Physics of Nanjing University during the Fall of 2012. They were originally based on my sketchy notes, which were put together and transformed into LaTex form by the graduate students. I edited these notes, so I am responsible for any mistakes and omissions.

This material has appeared as "Lecture Notes in Particle Physics: The Standard Model and Beyond, Nanjing University Press, Nanjing, China, 2013"

The author is indebted to Professor Yeukkwan Edna Cheung and Dr Konstantin G. Savvidy for their hospitality and useful discussions and to Professor Zhenlin Wang, Dean of the School of Physics, for his kind invitation to visit Nanjing.

The above lectures have been extended and improved taking into account the suggestions of the students of KAIST, who enrolled in and attended a course of Nuclear and Particle Physics during the Fall Semester of 2014.

On this occasion I like to extend my appreciation to the Department of Physics at KAIST and Professor Y. Semetzidis for their hospitality and support.

Finally the applications considered in chapter VII were extended, while on a visit to the University of S. Carolina during the spring term of 2015. This part was supported by a USC Provost's "Visiting Scholars Grant Program". The author is indebted to Professors Frank Avignone and R. Creswick for their support and hospitality.

KAIST, DAEJEON, S. Korea, November 2016

Contents

Preface vii

1. Mathematical Prerequisite A: Elements of Group Theory 1

1.1 Definitions 1
1.2 Matrix groups 2
1.2.1 The exponential of a matrix 6
1.2.2 Determination of the independent parameters . 9
1.3 The structure constants 12
1.4 The metric tensor and Casimir operators 13
1.4.1 The metric tensor 14
1.4.2 The Casimir operator 14
1.5 Representations 16
1.5.1 The fundamental representation 17
1.5.2 The adjoined representation 23
1.6 More representations 25
1.7 Irreducible representations 26
1.8 Homomorphism and isomorphism 32
1.9 Some further examples 32
1.10 The proper orthogonal groups O(3) and O(4) 35
1.11 Realization of $SO(4) = SO(3) \times SO(3)$ and some applications 38
1.11.1 Application to classical mechanics 39
1.11.2 Application to quantum mechanics 41
1.12 The proper Lorentz transformations; the group $SO(3,1)$ 43

1.13 Symmetries and conservation laws; Noether's theorem . . . 44
1.13.1 Global transformations in classical physics . . . 44
1.13.2 Global transformations in quantum mechanics (in the Heisenberg picture) . . . 45
1.13.3 Field theories; scalar fields . . . 45
1.13.4 Field theory; Fermion fields . . . 49
1.14 Problems . . . 50

2. Mathematical Prerequisite B: The Dirac Theory 63

2.1 Preliminaries; the Klein–Gordon equation . . . 63
2.2 The Dirac equation . . . 65
2.2.1 The spinors u and v . . . 68
2.2.2 Projection operators . . . 72
2.2.3 Various representations of the Dirac matrices . . . 72
2.3 Interpretation of negative energy solutions . . . 73
2.4 Transformation of spinors under space inversion . . . 74
2.5 The notion of helicity . . . 76
2.6 Charge conjugation for 4-spinors . . . 79
2.7 Types of currents . . . 81
2.7.1 Problems . . . 83

3. The Standard Model; Particle Content and Symmetry 87

3.1 Brief history . . . 87
3.2 The essential ingredients of the Standard Model . . . 89
3.3 The notion of a local symmetry . . . 91
3.4 Maxwell's equations; an example of Abelian gauge symmetry . . . 94
3.4.1 Maxwell's equations in terms of the potentials . . . 95
3.4.2 Maxwell's equations involving the electromagnetic tensor . . . 99
3.5 Non Abelian gauge transformations . . . 103
3.5.1 Transformations associated with a non Abelian group . . . 104
3.5.2 Some examples . . . 106
3.6 Gauge invariant Lagrangians . . . 107
3.6.1 Only gauge fields . . . 107

3.6.2 The Lagrangian in the presence of scalar fields 110
3.6.3 The Lagrangian in the presence of Fermion fields 110
3.7 The form of the SM Lagrangian 110
3.7.1 The electroweak part 111
3.7.2 The strong part 112

4. The Higgs Mechanism 113

4.1 The Higgs mechanism in global gauge transformations 113
4.1.1 A complex scalar field 113
4.1.2 Two real scalar fields 117
4.1.3 A scalar transforming as a doublet under $SU_I(2)$ 118
4.1.4 Symmetry restoration 122
4.2 The Higgs mechanism in gauge theories 123
4.2.1 The spontaneous symmetry braking (SSB); the unitary gauge 123
4.2.2 How the gauge bosons acquire a mass 124
4.2.3 The electroweak Lagrangian after the SSB . . . 127
4.2.4 Summary 129
4.3 What happened to the Goldstone bosons? 130
4.4 The ρ parameter 132
4.5 Fundamental theorem of Higgs mechanism 133
4.6 Problems 134

5. Fermion Masses and Currents 137

5.1 Fermion masses 137
5.1.1 Hermitian mass matrices 140
5.1.2 Non Hermitian mass matrices; left-right similarity transformations 142
5.2 The currents 144
5.2.1 The left handed currents 144
5.2.2 The right handed currents 146
5.2.3 Both left and right handed currents combined 146
5.2.4 The EM current 147

5.2.5 The charged currents; the KM matrix 147
5.2.6 The neutral currents 148
5.3 The contact interaction 150
5.4 Determination of the Standard Model parameters 151
5.5 The vector boson self-couplings 153
5.6 Summary 155

6. The SM SU(3) Group; Quantum Chromodynamics 157

6.1 Quantum chromodynamics QCD 157
6.2 The color structure of the one gluon exchange potential involving quarks 159
6.3 Approximations at low energies; interaction potentials between quarks 164
6.3.1 One gluon exchange potential in a process involving only baryons 165
6.3.2 One gluon exchange potential in processes involving the creation of a $q\bar{q}$ pair 166
6.3.3 One gluon exchange potential in processes involving meson spectra 167
6.4 Low energy formalism 168
6.4.1 The orbital part at the quark level 168
6.4.2 The kinetic energy part 169
6.4.3 The confining potential 169
6.4.4 Fitting the strength of the confining potential . 170
6.5 Matrix elements involving two quarks 170
6.5.1 The confining potential 171
6.5.2 The one gluon exchange potential 171

7. Rates and Cross Sections in Electroweak Theory 173

7.1 The Feynman diagrams and rules 173
7.1.1 Propagators 174
7.1.2 The simplest diagrams in quantum electrodynamics; the photon-charge interaction 175
7.1.3 Quantum electrodynamics; the interaction between two charges 175
7.1.4 The Feynman rules 177
7.2 Further elementary considerations 178
7.3 Decay widths and cross sections 179

7.3.1 Some useful expressions 179
7.3.2 The phase space integrals 180
7.4 Rutherford scattering 182
7.5 Brief review of the trace technique 185
7.6 The decay of a vector boson 186
7.6.1 The decay widths of vector bosons 186
7.6.2 The discovery of vector bosons 191
7.7 Muon-antimuon production in electron-positron colliders 193
7.8 Electron proton scattering 198
7.8.1 The proton as elementary particle 199
7.8.2 The proton form factors 202
7.8.3 Inelastic electron-proton scattering 202
7.9 Neutrino electron scattering 205
7.10 Compton scattering 208

8. Supersymmetry for Pedestrians 211

8.1 Introduction 211
8.2 The particle content of MSSM 212
8.2.1 Gauge particles 213
8.2.2 Fermions and s-fermions 213
8.2.3 The Higgs content 213
8.3 The Higgs mechanism 214
8.4 The Fermion masses 218
8.5 Supersymmetry breaking 220
8.6 Some remarks about the particle spectrum 220

9. Grand Unification; the SU(5) Example 225

9.1 Mathematical Introduction 225
9.2 The structure of the GUT $SU(5)$ 226
9.3 The particle content 227
9.3.1 The Fermions 228
9.3.2 The gauge bosons 228
9.3.3 The Higgs content 231
9.4 The Higgs mechanism 231
9.5 The gauge boson masses 234
9.6 Gauge couplings 235
9.7 Baryon asymmetry 235

10. A Brief Introduction to Cosmology 243

10.1 Cosmological principles 243
10.2 The expanding universe; the Big Bang scenario 244
10.2.1 The receding of galaxies 245
10.2.2 The background microwave radiation 248
10.2.3 The abundance of primordial ${}^4_2$He and other light elements in the universe 250
10.3 Evolution of the universe 256
10.3.1 A classical prelude 257
10.3.2 The Robertson-Walker metric in the context of GTR 259
10.3.3 Presence of matter and curvature 263
10.3.4 Presence of radiation and curvature (the early years) 264
10.4 The cosmological constant; dark energy 266
10.5 The standard cosmological model 272
10.6 The proper length and the horizon 272
10.6.1 The proper length entering the prototype candles 274
10.6.2 The horizon 275
10.7 Evolution of temperature 277
10.7.1 The relic neutrinos 278
10.7.2 The re-ionization era 283
10.8 Some problems with the Standard Cosmological Model . 284
10.9 The period of inflation 288
10.10 Particle physics and the inflationary scenario 291
10.11 The role of dark matter 297
10.11.1 The rotational velocities 298
10.11.2 Gravitational lensing 301
10.11.3 The observation of bullet cluster 302
10.12 Dark energy 307
10.12.1 Observations with standard candles 307
10.12.2 The microwave background radiation; the earliest picture of the universe 312
10.13 The WMAP and PLANCK observations 315

11. Aspects of Neutrino Physics; Neutrino Oscillations 319

11.1 The elusive neutrino 319
11.2 Lepton flavor conservation 320

11.3 History of neutrino oscillations 321
11.4 Aspects of neutrino masses and mixing 322
11.4.1 The Majorana neutrino mass 323
11.4.2 Transformation properties under C 324
11.4.3 The see-saw mechanism 325
11.5 Neutrino oscillations in the case of two generations . . . 327
11.6 The three generation formalism 329
11.7 Neutrino oscillation experiments 330
11.8 The elusive absolute scale of the neutrino mass 341
11.9 Neutrinos as probes 346
11.10 Problems 348

12. Discrete Symmetries: C, P, T and All That 353

12.1 Space inversion P 353
12.1.1 Parity conservation; parity violation 357
12.2 The helicity of the neutrinos 361
12.2.1 The determination of the helicity of neutrino . . 363
12.2.2 Weak interaction and handedness 364
12.3 Charge conjugation 366
12.3.1 Charge conjugation for 4-spinors 368
12.3.2 The charge conjugation of composite systems . . 369
12.3.3 How good a symmetry is the charge conjugation? 373
12.4 Time reversal 374
12.4.1 Time reversal in classical mechanics 374
12.4.2 Time reversal in quantum mechanics 376
12.4.3 Premature evidence for T violation 380
12.5 CP, the combined action of P and C symmetries 381
12.5.1 The neutral kaon system 383
12.5.2 Strangeness oscillations 387
12.5.3 Kaon regeneration 388
12.5.4 CP violation 391
12.5.5 Other CP violating processes involving quarks . 396
12.5.6 Evidence for T-violation 397
12.6 The combined symmetry CPT 400

13. Appendix: Some Elementary Aspects of Particle Physics 401

13.1 The natural system of units 401
13.2 The Planck natural system of units 404

13.3 Invariants in kinematics 406
13.4 Cross sections and luminosities 407
13.4.1 Luminosity in fixed target experiments 408
13.4.2 Luminosity in colliding beam experiments 409
13.5 Kinematics of particle decay 411
13.6 Invariants in the scattering of two particles 412
13.7 Detection of resonances 416
13.7.1 The signature of a resonance 416
13.7.2 The notion of a resonance 418
13.8 Transformation properties of non invariant quantities . . 421
13.9 Problems 422

Bibliography 425

Index 429

Chapter 1

Mathematical Prerequisite A: Elements of Group Theory

In this chapter we make a very simple exposition of the elements of group theory that one needs to know in order to understand the essentials of the Standard Model (SM) of particle physics. For a more complete treatment the reader is referred to standard group theory texts, see e.g. [Vergados (2016)] and references therein.

1.1 Definitions

A group is a set of elements

$$G : X \equiv \{X_1, X_2 \cdots X_n\}$$

with properties:

(1) $X_1 \in G,\ X_2 \in G \Rightarrow X_1X_2 = X_3,\ X_3 \in G.$
(2) $X_1(X_2X_3) = (X_1X_2)X_3.$
(3) $\exists\ I \in G$ Such that $X_iI = IX_i,\ i = 1, 2, \cdots, n.$
(4) For every $X_i \in G$, there is one element $Y_i \in G$, such that

$$X_iY_i = I.$$

Then the element Y_i is called inverse of X_i, and we write $Y_i = X_i^{-1}$.

Note that the multiplication in general is not commutative, i.e. often;

$$X_iX_j \neq X_jX_i$$

A subset of $G' \subset G$, such that:

- $X_i \in G',\ Y_j \in G' \rightarrow X_iY_j \in G',$
- the elements of G' satisfy all the above properties, forms a subgroup.

Furthermore, if for any two elements of G' we have:

$$X_i Y_i = Y_i X_i$$

then G' is called Abelian.

A group is simple, if does not contain any invariant subgroups. A group is semisimple, if it does not contain any continuous invariant Abelian subgroups.

Example 1:

The set $\{i, -i, 1, -1\}$ forms a group with a multiplication table given in 1.1. This is an Abelian group.

Table 1.1: The multiplication table of the discrete group $G = \{1, -1, i, -i\}$

	i	-i	1	-1
-i	1	-1	-i	i
i	-1	1	i	-i
1	i	-i	1	-1
-1	-i	i	-1	1

The set$\{-1, 1\}$ forms a subgroup of the above group. The subset$\{i, -i\}$ does not form a subgroup.

1.2 Matrix groups

The matrices are well known entities. We will briefly summarize here some aspects relevant to group theory.

$$(\alpha) \Leftrightarrow (\alpha_{ij}) = \begin{pmatrix} \alpha_{11} & \alpha_{12} & \cdots & \alpha_{1n} \\ \alpha_{21} & \alpha_{22} & \cdots & \alpha_{2n} \\ \cdots & \cdots & \cdots & \cdots \\ \alpha_{n1} & \alpha_{n2} & \cdots & \alpha_{nn}, \end{pmatrix} \tag{1.1}$$

$$(\alpha)(\beta) = \gamma \leftrightarrow \ (\alpha\beta)_{ij} = \gamma_{ij}, (\alpha\beta)_{ij} = \sum_k \alpha_{ik}\beta_{kj}.$$

If the determinant Δ of $n \times n$ matrix (α) is not zero $\exists$ (α^{-1}): $(\alpha\alpha^{-1}) = I$, with (α^{-1}) given[1] by:

$$\alpha^{-1} = \frac{1}{\Delta}\begin{vmatrix} \Delta_{11} & \Delta_{12} & \cdots & \Delta_{1n} \\ \Delta_{21} & \Delta_{22} & \cdots & \Delta_{2n} \\ \cdots & \cdots & \cdots & \cdots \\ \Delta_{n1} & \Delta_{n2} & \cdots & \Delta_{nn} \end{vmatrix}. \tag{1.2}$$

with

$$\Delta_{ij} = (-1)^{i+l} C_{ji}$$

where C_{ij} is the co-factor determinant of the matrix obtained after stripping the row i and the column j. Furthermore

$$(\alpha)I = \alpha$$

where I is the identity matrix, i.e. the matrix with 1 along the diagonal and zeros everywhere else. Furthermore, since $det(\gamma) = \det(\alpha)\det(\beta)$,

$$\text{If } \det(\alpha) \neq 0 \text{ and } \det(\beta) \neq 0 \Rightarrow \det(\gamma) \neq 0.$$

Thus the set of $n \times n$ non singular matrices form a group.

For any given matrix (α) one defines the following matrices obtained from it:

$$(\alpha) \rightarrow (\alpha^T),\ (\alpha^T)_{ij} = (\alpha)_{ji} \text{ transpose of } (\alpha)$$

$$(\alpha) \rightarrow (\alpha^+),\ (\alpha^+)_{ij} = ((\alpha)_{ji})^* \text{ Hermitian conjugate of } (\alpha)$$

In other words the Hermitian conjugate is the complex conjugate of the transpose matrix.

A matrix is symmetric if $(\alpha^T) = (\alpha)$ and Hermitian if $(\alpha^+) = (\alpha)$.

Possible subgroups:

(1) The set of unitary matrices. A matrix is said to be unitary if:

$$(U^+) = U^{-1} \Leftrightarrow (U)(U^+) = (U^+)(U) = I \Rightarrow$$

$$(U^+)(U) = I \Rightarrow \sum_k (U^+)_{ik}(U)_{kj} = \delta_{ij} \Rightarrow \sum_k (U^*)_{ki}(U)_{kj} = \delta_{ij}$$

$$(U)(U^+) = I \Rightarrow \sum_k (U)_{ik}(U^+)_{kj} = \delta_{ij} \Rightarrow \sum_k (U)_{ki}(U^*)_{jk} = \delta_{ij}$$

[1] Often when there seems to be no ambiguity we will omit the parenthesis in the notation for the matrix

This shows the orthogonality of the rows and the orthogonality of the columns. In particular:

$$\sum_k U_{ik}^* U_{ik} = 1, \quad \sum_k |U_{ik}|^2 = 1$$

The number of possible parameters $2n^2$. The number of constraints:

$$i) \sum_k |U_{ik}|^2 = 1, \quad i = 1, 2, \cdots, n \to \ n \text{ constraints},$$

$$ii) \sum_k U_{ik} U_{jk}^* = 0 \ \ i \neq j \to 2 \cdot \frac{1}{2} n(n-1) \text{constraints}.$$

Thus the number of independent parameters is:

$$2n^2 - n - 2 \cdot \frac{1}{2}(n-1)n = 2n^2 - n - n^2 + n = n^2$$

Recall now some facts about the scalar product in complex spaces:

- For any two vectors $|\alpha\rangle = (\alpha_1, \alpha_2, \cdots, \alpha_n)$ and $|\beta\rangle = (\beta_1, \beta_2, \cdots, \beta_n)$ the scalar product is defined as:

  $$\langle\beta|\alpha\rangle = \sum_i^n \beta_i^* \alpha_i$$

  i.e. $\langle\beta|$ is the Hermitian conjugate (row vector) $|\beta\rangle$ (column vector), if viewed as a matrix. The thus defined scalar product has the property:

  $$\langle\beta|\alpha\rangle = \langle\alpha|\beta\rangle^*$$

- From a given vector $|\alpha\rangle$ we can obtain another vector after acting on it with a unitary transformation (matrix):

  $$|\alpha\rangle \to |U\alpha\rangle \Leftrightarrow |U\alpha\rangle_i = U_{ij}\alpha_j.$$

  Assuming that operators act on the right we write $U|\alpha\rangle = |U\alpha\rangle$, i.e.

  $$\langle\beta|U\alpha\rangle = \langle\beta|U|\alpha\rangle$$

- $\langle\beta U|\alpha\rangle = \langle\beta|U^+|\alpha\rangle$.
  Indeed:

  $$\langle\beta U|\alpha\rangle \left(\langle\alpha|U\beta\rangle\right)^* = \sum_{i,k} (\alpha_k^* U_{ki} \beta_i)^* =$$

  $$\sum_{i,k} \alpha_k U_{ki}^* \beta_i^* = \sum_{i,k} \beta_i^* (U^+)_{ik} \alpha_k = \langle\beta|U^+|\alpha\rangle$$

Consequently:

$$\langle U\alpha|U|\beta\rangle = \langle\alpha|U^{+}U|\beta\rangle = \langle\alpha|\beta\rangle.$$

That is, the unitary matrices have the nice property of preserving the scalar product of two vectors defined in a complex space. Furthermore if U_1 unitary and U_2 unitary

$$\begin{aligned}(U_1U_2)^{+} &= U_2^{+}U_1^{+},\\ (U_1U_2)^{-1} &= U_2^{-1}U_1^{-1} = U_2^{+}U_1^{+} \Rightarrow\\ (U_1U_2)^{+} &= (U_1U_2)^{-1} \Rightarrow \text{the product } U_1U_2 \text{ is unitary}\end{aligned} \tag{1.3}$$

$$\det(U_1U_2) = \det(U_1)\det(U_2) \neq 0 \Rightarrow \text{the product } U_1U_2 \text{ is non singular}$$

Thus the unitary matrices form a group.

(2) The special unitary group $SU(n)$ is a subset of $U(n)$ such that

$$\det(U) = 1\ .$$

Thus now the number of independent parameters is: $n^2 - 1$.

(3) The set of orthogonal matrices.

For a given matrix (α) its transpose $(\alpha)^T$ is obtained by interchanging rows and columns:$(\alpha)^T_{ij} = (\alpha)_{ji}$. A matrix is said to be orthogonal if:

$$O^T\,O = I \text{ that is} \Leftrightarrow \sum_k O_{ik}O_{jk} = \delta_{ij} \Leftrightarrow \sum_k O_{ki}O_{kj} = \delta_{ji}. \tag{1.4}$$

The number of independent parameters of such a matrix in real space is:

$$d = n^2 - n - \frac{n}{2}(n-1) = \frac{1}{2}n(n-1). \tag{1.5}$$

Note that, since $\det(O^T) = \det(O)$, it follows:

$$\det(O^T)\det(O) = 1 \rightarrow (\det(O))^2 = 1\ . \tag{1.6}$$

$$\Rightarrow \det(O) = \begin{cases} 1 & \text{proper } O \\ -1 & \text{improper } O. \end{cases} \tag{1.7}$$

The elements with $\det(O) = 1$ form a subgroup, indicated by $SO(n)$, while the elements with with $\det(O) = -1$ do not. The two sets cannot be connected by a continuous transformation.

Note for a scalar product defined in real space:

$$\langle OX|OY\rangle = \langle X|O^TO|Y\rangle = \langle X|Y\rangle, \tag{1.8}$$

that is the orthogonal matrices in real space preserve the scalar product. This is not true in complex space. The number of independent parameters in complex space is

$$d(O(n)) = n(n-1)\ . \tag{1.9}$$

1.2.1 *The exponential of a matrix*

One can generalize the usual exponential function over the domain of matrices:

$$e^{(\alpha)} \equiv \sum_{k=0}^{\infty} \frac{(\alpha)^k}{k!},\ \alpha^0 = I\ . \tag{1.10}$$

Theorem 1: The series converge if $|\alpha_{ij}| < \infty$ (bounded)
Theorem 2: If $[\alpha, \beta] = (\alpha)(\beta) - (\beta)(\alpha) = 0$, i.e. they commute. Then

$$e^{(\alpha)} e^{(\beta)} = e^{(\alpha+\beta)}. \tag{1.11}$$

This relation need not hold if $[(\alpha), (\beta)] \neq 0$.
Theorem 3: For a regular matrix β

$$\beta^{-1} e^{\alpha} \beta = e^{(\beta)^{-1}\alpha(\beta)}. \tag{1.12}$$

Theorem 4: For any $n \times n$ matrix

$$e^{(\alpha)} = \sum_{k=0}^{n-1} \gamma_k (\alpha)^k \Rightarrow \alpha^n = c_0 + c_1\alpha + \cdots + c_{n-1}\alpha^{n-1} \tag{1.13}$$

In other words the infinite series collapses to a finite sum.
Theorem 5: Let $\lambda_1,\ \lambda_2 \cdots \lambda_n$ be the eigenvalues of the matrix (α), then $e^{\lambda_1},\ e^{\lambda_2},\ \cdots\ e^{\lambda_n}$ are the eigenvalues of $e^{(\alpha)}$.
Theorem 6: The following are true:

$$e^{\alpha^*} = (e^{\alpha})^* \tag{1.14}$$

$$e^{(\alpha^+)} = (e^{(\alpha)})^+ \tag{1.15}$$

$$(e^{(\alpha)})^{-1} = e^{-(\alpha)} \tag{1.16}$$

$$\det(e^{\alpha}) = e^{tr(\alpha)} \tag{1.17}$$

with

$$tr(\alpha) = \sum_i \alpha_{ii} \tag{1.18}$$

The following hold true:

(1)

$$(\alpha)^{+} = (\alpha) \text{ (Hermitian)} \Rightarrow e^{i\alpha} \text{ is Unitary.}$$

Furthermore:

$(\alpha)^{+} = (\alpha)$, $(\beta)^{+} = (\beta) \Rightarrow (\alpha\beta)^{+} = (\beta)^{+}(\alpha)^{+} = \beta\alpha \Rightarrow \alpha\beta$ is Hermitian

(2)

$$(\alpha)^{+} = -\alpha \text{ (anti-Hermitian matrix) } \Rightarrow e^{\alpha} \text{ is unitary}$$

(3)

$$\alpha^{T} = -\alpha \text{ (anti-symmetric in real apace) } \Rightarrow e^{\alpha} \text{ is orthogonal}$$

Theorem 8: The Cambell-Hausdorff formula holds:

$$e^{-\alpha}\beta e^{\alpha} = \beta + [\beta,\alpha] + \frac{1}{2!}[[\beta,\alpha],\alpha]] + \frac{1}{3!}[[[\beta,\alpha],\alpha],\alpha] + \cdots, \tag{1.19}$$

$$e^{-\alpha}\beta e^{\alpha} = \beta \quad if\ [\beta,\alpha] = 0\,. \tag{1.20}$$

Example 2: Consider the special case of 2×2 Hermitian matrix:

$$(\alpha) = \begin{pmatrix} a & b-ic \\ b+ic & d \end{pmatrix} \tag{1.21}$$

a, b, c, d real. In the space of these matrices one can choose the basis:

$$I = e = \begin{pmatrix} 1\,0 \\ 0\,1 \end{pmatrix}, \quad \sigma_1 = \begin{pmatrix} 0\,1 \\ 1\,0 \end{pmatrix}, \quad \sigma_2 = \begin{pmatrix} 0 & -i \\ i & 0 \end{pmatrix}, \quad \sigma_3 = \begin{pmatrix} 1 & 0 \\ 0 & -1 \end{pmatrix}. \tag{1.22}$$

The last three are the Pauli matrices. One can easily verify that:

$$[\sigma_1,\sigma_2] = 2i\sigma_3,\ [\sigma_2,\sigma_3] = 2i\sigma_1,\ [\sigma_3,\sigma_1] = 2i\sigma_2 \Leftrightarrow [\sigma_j,\sigma_k] = 2\epsilon_{jk\ell}\sigma_\ell. \tag{1.23}$$

Since they constitute a basis we have:

$$\begin{aligned}(\alpha) &= \zeta_0(e) + \zeta_1\sigma_1 + \zeta_2\sigma_2 + \zeta_3\sigma_3 \\ &= \begin{pmatrix} \zeta_0 & 0 \\ 0 & \zeta_0 \end{pmatrix} + \begin{pmatrix} 0 & \zeta_1 \\ \zeta_1 & 0 \end{pmatrix} + \begin{pmatrix} 0 & -i\zeta_2 \\ i\zeta_2 & 0 \end{pmatrix} + \begin{pmatrix} \zeta_3 & 0 \\ 0 & -\zeta_3 \end{pmatrix} \\ &= \begin{pmatrix} \zeta_0+\zeta_3 & \zeta_1 - i\zeta_2 \\ \zeta_1 + i\zeta_2 & \zeta_0 - \zeta_3 \end{pmatrix},\end{aligned} \tag{1.24}$$

Thus

$$a = \zeta_0 + \zeta_3,\ b = \zeta_1,\ c = \zeta_2,\ d = \zeta_0 - \zeta_3 \Rightarrow$$

$$\zeta_0 = \frac{a+d}{2}, \zeta_3 = \frac{a-d}{2}, \zeta_1 = b, \zeta_2 = c.$$

If (α) is traceless, $\zeta_0 = 0$ and

$$a = \zeta_3,\ d = -\zeta_3,\ b = \zeta_1,\ c = \zeta_2,\ (d = -a). \tag{1.25}$$

Thus we encounter only three parameters a, b, c or $\zeta_1,\ \zeta_2,\ \zeta_3$.

The corresponding elements of SU(2) are

$$U_i = e^{i\theta_l \sigma_l/2}. \tag{1.26}$$

The Pauli matrices do not form a group. Note however that:

$$\sigma_1^2 = \sigma_2^2 = \sigma_3^2 = I. \tag{1.27}$$

Thus one can form the following three groups:

$$\{\sigma_1,\ I\}, \{\sigma_2,\ I\}, \{\sigma_3 I\}.$$

One sometimes writes: $T_l = \frac{1}{2}\sigma_i$ with algebra:

$$[T_k, T_l] = i\epsilon_{klm} T_m. \tag{1.28}$$

Summation over repeated indices understood.

Example 3:

Consider the special case of $n = 3$ Hermitian matrix α, associated with $U(3)$. Then

$$\alpha = \text{Hermitian} \Rightarrow U = e^{i\alpha} = \text{unitary}.$$

If the matrix α is traceless, it depends on $n^2 - 1 = 8$ parameters and is a associated with the unimodular $SU(3)$.

$$\alpha = \text{traceless} \rightarrow \det U = e^{itr(\alpha)} = e^0 = 1$$

Proceeding as above we select basis set of traceless 3×3 hermitian matrices:

$$\lambda_1 = \begin{pmatrix} 0 & 1 & 0 \\ 1 & 0 & 0 \\ 0 & 0 & 0 \end{pmatrix}, \lambda_2 = \begin{pmatrix} 0 & -i & 0 \\ i & 0 & 0 \\ 0 & 0 & 0 \end{pmatrix}, \lambda_3 = \begin{pmatrix} 1 & 0 & 0 \\ 0 & -1 & 0 \\ 0 & 0 & 0 \end{pmatrix} \lambda_4 = \begin{pmatrix} 0 & 0 & 1 \\ 0 & 0 & 0 \\ 1 & 0 & 0 \end{pmatrix},$$

$$\lambda_5 = \begin{pmatrix} 0 & 0 & -i \\ 0 & 0 & 0 \\ i & 0 & 0 \end{pmatrix}, \lambda_6 = \begin{pmatrix} 0 & 0 & 0 \\ 0 & 0 & 1 \\ 0 & 1 & 0 \end{pmatrix}, \lambda_7 = \begin{pmatrix} 0 & 0 & 0 \\ 0 & 0 & -i \\ 0 & i & 0 \end{pmatrix}, \lambda_8 = \frac{1}{\sqrt{3}} \begin{pmatrix} 1 & 0 & 0 \\ 0 & 1 & 0 \\ 0 & 0 & -2 \end{pmatrix}. \tag{1.29}$$

With this selection:

$$tr(\lambda_i \lambda_j) = 2\delta_{ij}$$

Table 1.2: The SU(3) structure constants (functions)

μ	ν	ρ	$f_{\mu\nu\rho}$
1	2	3	1
1	3	2	-1
1	4	7	$\frac{1}{2}$
1	5	6	$-\frac{1}{2}$
1	6	5	$\frac{1}{2}$
1	7	4	$-\frac{1}{2}$
2	3	1	1
2	4	6	$\frac{1}{2}$
2	5	7	$\frac{1}{2}$
2	6	4	$-\frac{1}{2}$
2	7	5	$\frac{1}{2}$
3	4	5	$\frac{1}{2}$
3	5	4	$-\frac{1}{2}$
3	6	7	$-\frac{1}{2}$

;

μ	ν	ρ	$f_{\mu\nu\rho}$
3	7	6	$\frac{1}{2}$
4	5	3	$\frac{1}{2}$
		8	$\frac{\sqrt{3}}{2}$
4	6	2	$\frac{1}{2}$
4	7	1	$\frac{1}{2}$
4	8	5	$-\frac{\sqrt{3}}{2}$
5	6	1	$-\frac{1}{2}$
5	7	2	$\frac{1}{2}$
5	8	4	$\frac{\sqrt{3}}{2}$
6	7	3	$-\frac{1}{2}$
		8	$\frac{\sqrt{3}}{2}$
6	8	7	$-\frac{\sqrt{3}}{2}$
7	8	6	$\frac{\sqrt{3}}{2}$

$$[\lambda_\alpha, \lambda_\beta] = 2if_{\alpha\beta\gamma}\lambda_\gamma.$$

Summation over repeated indices understood. Sometimes we will use $T_k = \frac{1}{2}\lambda_k$. Then

$$T_k = \frac{1}{2}\lambda_k \Rightarrow [T_\alpha, T_\beta] = if_{\alpha\beta\gamma}T_\gamma. \tag{1.30}$$

It is adequate to consider $\alpha \leq \beta$. Then the non zero values of the parameters $f_{\alpha\beta\gamma}$ are given in table 1.2: Note that in the special case of $\alpha = 4$, $\beta = 5$ two terms appear in the commutator with the indicated structure constants. Similarly for $\alpha = 7$, $\beta = 8$.

To each generator λ_k, there corresponds a group element, $U_k = e^{i\theta_k \frac{\lambda_k}{2}}$.

1.2.2 *Determination of the independent parameters*

We have seen that a unitary $n \times n$ matrix is characterized by n^2 parameters. These can be given in terms of $\frac{1}{2}n(n-1)$ angles, like in the orthogonal matrices, and $\frac{1}{2}n(n+1)$ phases e^{δ_i}. In the special case that we deal with fermions it is possible for the left (ket) and the right (bra) spaces to be different. In this case we can make appropriate separate rotations in each of the spaces so that $(2n-1)$ phases can be eliminated. Then we are left with:

$$\frac{1}{2}n(n-1) \text{ angles,}$$

$$\frac{1}{2}n(n+1)-(2n-1)=\frac{n^2+n-4n+2}{2}=\frac{(n^2-3n+2)}{2}\ \text{phases}\, e^{i\delta_j}$$

Thus

$$n=2 \Leftrightarrow \ \text{one angle}$$

$$n=3 \Leftrightarrow (1/2)\cdot 3\cdot 2=3\ \text{angles and}\ (n^2-3n+2)/2=1\ \text{phase}\, e^{i\delta}$$

in the case of $SU(n)$ we have n^2-1 parameters. Thus

$$\begin{cases} \frac{1}{2}n(n-1)\ \text{angles.} \\ \frac{1}{2}(n+2)(n-1)\ \text{phases}\, e^{i\delta_j}. \\ \text{we may remove}\ (2n-1)-1=2(n-1)\ \text{phases.} \\ \text{Thus we are left with}\ \frac{1}{2}(n-1)(n-2)\ \text{phases}\, e^{i\delta_j}. \end{cases}$$

Let us illustrate this in the case of $n=2$ and $n=3$.

(1) The SU(2) case.

$$\alpha=\begin{pmatrix} a\ b \\ c\ d \end{pmatrix},\quad \det(\alpha)=1,\quad ad-bc=1\ .$$

$$I=\begin{pmatrix} a^*\ c^* \\ b^*\ d^* \end{pmatrix}\begin{pmatrix} a\ b \\ c\ d \end{pmatrix}=\begin{pmatrix} |a|^2+|c|^2\ a^*b+c^*d \\ b^*a+d^*c\ |b|^2+|d|^2 \end{pmatrix}=\begin{pmatrix} 1\,0 \\ 0\,1 \end{pmatrix}\Rightarrow$$

$$|a|^2+|c|^2=1,\ |b|^2+|d|^2=1,\ a^*b+c*d=0$$

$$\Rightarrow d=a^*,\quad c=-b^*,\ |a|^2+|b|^2=1\ .$$

Thus

$$(\alpha)=\begin{pmatrix} a & b \\ -b^* & a \end{pmatrix}\quad |a|^2+|b|^2=1, a=\cos\theta e^{i\alpha},\ b=\sin\theta e^{i\beta},\ \alpha,\beta\ \text{real,}$$

$$(\alpha)=\begin{pmatrix} \cos\theta e^{i\alpha} & \sin\theta e^{i\beta} \\ -\sin\theta e^{-i\beta} & \cos\theta e^{-i\alpha} \end{pmatrix}.$$

Not all phases are physical, however, if the left vector (bra) and the right vector (ket) happen to be different. To see this, consider the expression:

$$L=(x_L,y_L)\begin{pmatrix} \cos\theta e^{i\alpha} & \sin\theta e^{i\beta} \\ -\sin\theta e^{-i\beta} & \cos\theta e^{-i\alpha} \end{pmatrix}\begin{pmatrix} x_R \\ y_R \end{pmatrix} \tag{1.31}$$

and suppose:

$$x_R \to x_R,\, y_R \to e^{i(\alpha-\beta)} y_R, \quad x_L \to x_L,\, y_L \to e^{i(\alpha+\beta)} y_R,$$

$$\begin{aligned} L = (x_L, y_L) \begin{pmatrix} 1 & 0 \\ 0 & e^{i(\alpha+\beta)} \end{pmatrix} \begin{pmatrix} \cos\theta e^{i\alpha} & \sin\theta e^{i\beta} \\ -\sin\theta e^{-i\beta} & \cos\theta e^{-i\alpha} \end{pmatrix} \begin{pmatrix} 1 & 0 \\ 0 & e^{i(\alpha-\beta)} \end{pmatrix} \begin{pmatrix} x_R \\ y_R \end{pmatrix} \\ = (x_L, y_L) \begin{pmatrix} \cos\theta e^{i\alpha} & \sin\theta e^{i\alpha} \\ -\sin\theta e^{i\alpha} & \cos\theta e^{i\alpha} \end{pmatrix} \begin{pmatrix} x_R \\ y_R \end{pmatrix} \\ = (x_L, y_L) \begin{pmatrix} \cos\theta & \sin\theta \\ -\sin\theta & \cos\theta \end{pmatrix} \begin{pmatrix} x_R \\ y_R \end{pmatrix}. \end{aligned} \tag{1.32}$$

The last step follows by a simple phase change of one of the vectors. Thus we removed all phases.

(2) The SU(3) case. We have:

$$\frac{1}{2} n(n-1) = \frac{1}{2} 3 \cdot 2 = 3 \text{ angles and}$$

$$n^2 - 1 - \frac{1}{2} n(n-1) = \frac{1}{2} n(n+1) - 1 = \frac{1}{2} \cdot 2 \cdot 5 = 5 \text{ phases}\, e^{i\delta_j}.$$

We can remove $2(n-1) = 2 \cdot 2 = 4$ phases and, thus, we are left with 1 phase $e^{i\delta}$.

To see this we proceed as above and consider:

$$L = (x_1, \cdots, x_n)(U) \begin{pmatrix} y_1 \\ y_2 \\ \vdots \\ y_n \end{pmatrix}. \tag{1.33}$$

We then perform $n-1$ left rotations $R_L(1)R_L(2)\cdots R_L(n-1)$ and $n-1$ right rotations $R_R(1)R_R(2)\cdots R_R(n-1)$ and find

$$(U) = R_L(1)R_L(2)\cdots R_L(n-1)(U)R_R(1)R_R(2)\cdots R_R(n-1) \tag{1.34}$$

These rotations can be judiciously chosen to eliminate the redundant phase. After that for $n = 3$ we find the Cabibbo-Kobayashi-Maskawa (CKM) matrix:

$$U_{CKM} = \begin{pmatrix} c_1 & -c_3 s_1 & -s_1 s_3 \\ c_2 s_1 & c_1 c_2 c_3 - s_2 s_3 e^{i\delta} & c_1 c_2 s_3 + c_3 s_2 e^{i\delta} \\ s_1 s_2 & c_1 c_3 s_2 + c_2 s_3 e^{i\delta} & c_1 s_2 s_3 - e^{i\delta} c_2 c_3 \end{pmatrix} \tag{1.35}$$

Table 1.3: The best fit to the KCM matrix

	u	c	t
d	0.97419± 0.00022	0.2257 ± 0.0010	0.00359 ± 0.00016
s	0.2256 ± 0.0010	0.97334 ± 0.00023	$0.0415^{+0.0010}_{-0.0011}$
b	$0.00874^{+.00026}_{-0.00037}$	0.0407 ± 0.0010	$0.999133^{+0.000044}_{-0.000043}$

with $c_i = \cos\theta_i$, $s_i = \sin\theta_i$.

This form is traditionally employed in the hadronic sector. The best fit to experiment is given in table 1.3.

Another possibility, usually employed in the leptonic sector, is the Pontecorvo-Maki-Nakagawa-Sakata (PMNS) matrix:

$$U_{PMNS} = \begin{pmatrix} 1 & 0 & 0 \\ 0 & c_{23} & s_{23} \\ 0 & -s_{23} & c_{23} \end{pmatrix} \begin{pmatrix} c_{13} & 0 & s_{13}e^{i\delta_{13}} \\ 0 & 1 & 0 \\ -s_{13}e^{i\delta_{13}} & 0 & c_{13} \end{pmatrix} \begin{pmatrix} c_{12} & s_{12} & 0 \\ -s_{12} & c_{12} & 0 \\ 0 & 0 & 1 \end{pmatrix}$$

$$= \begin{pmatrix} c_{12}c_{13} & s_{12}c_{13} & s_{13}e^{-i\delta 13} \\ -c_{23}s_{12} - s_{23}s_{13}c_{12}e^{i\delta_{13}} & c_{12}c_{23} - s_{12}s_{23}s_{13}e^{i\delta_{13}} & s_{23}c_{13} \\ -c_{12}c_{23}s_{13}e^{i\delta_{13}} + s_{23}s_{12} & -c_{12}s_{23} - s_{12}s_{13}c_{23}e^{i\delta_{13}} & c_{23}c_{13} \end{pmatrix} \tag{1.36}$$

with

$$c_{ij} = \cos(\theta_{ij}),\ s_{ij} = \sin(\theta_{ij}) \tag{1.37}$$

1.3 The structure constants

We begin by considering a set of operators T_k. Then we generalize the concept of the exponential we encountered in the case of matrices. Thus

$$A = e^{i\epsilon_k T_k} \simeq 1 + i\epsilon_k T_k,$$

$$A^{-1} = e^{-i\epsilon_k T_k} \simeq 1 - i\epsilon_k T_k,$$

$$X_\ell = A^{-1}T_\ell A \simeq (1 - i\epsilon_k T_k)T_\ell(1 + i\epsilon_k T_k)$$

$$\simeq T_\ell + i\epsilon_k(T_\ell T_k - T_k T_\ell) + O(\epsilon^2),$$

$$X_\ell - T_\ell \simeq i\epsilon_k(T_\ell T_k - T_k T_\ell)$$

that is

$$T_\ell T_k - T_k T_\ell = \text{linear combination of the operators } T_m.$$

Thus by defining:

$$[T_k, T_\ell] \equiv T_k T_\ell - T_\ell T_k, \tag{1.38}$$

we get

$$[T_k, T_\ell] = C^m_{k\ell} T_m \tag{1.39}$$

with summation over m understood. The quantities C^m_{kl} are called structure constants and completely determine the local structure of the group. They have the following properties:

(1) $C^m_{k\ell} = -C^m_{\ell k}$.
(2) They satisfy the Jacobi identity. Indeed from the operator identity:

$$[[T_k, T_\ell], T_m] + [[T_\ell, T_m], T_k] + [[T_m, T_k], T_\ell] = 0\ , \tag{1.40}$$

it follows that

$$\sum_n (C^n_{k\ell} C^\rho_{nm} + C^n_{\ell n} C^\rho_{nk} + C^n_{mk} C^\rho_{n\ell}) = 0 \tag{1.41}$$

(3) the above operation defines the product of two elements of the set: $A \times B \equiv [AB - BA] \equiv [A, B]$.

In this sense the above operators constitute an algebra, a Lie algebra.

Note one can show that the A set of $n \times n$ matrices E_{ij} are given such that its elements are given by

$$(E_{ij})_{\alpha,\beta} = \delta_{i\alpha}\delta_{j\beta}. \tag{1.42}$$

The E_{ij} constitute an algebra with the structure constants given by:

$$[E_{ij}, E_{k\ell}] = \delta_{jk} E_{i\ell} - \delta_{i\ell} E_{kj} \tag{1.43}$$

(see problems 10 and 11). For another representation see problems 13 and 14.

1.4 The metric tensor and Casimir operators

Once the structure constants we can obtain two useful concepts for the group, namely the metric tensor and the Casimir operator. The latter is an operator, which commutes with all the members of the algebra characterizing the group.

1.4.1 *The metric tensor*

The metric tensor is defined as follows:

$$g_{\lambda\mu} = C^{\alpha}_{\lambda\beta} C^{\beta}_{\mu\alpha} \tag{1.44}$$

From this, if it is non-singular, we can obtain the tensor $g^{\lambda\mu}$, which is the inverse of the previous. Both are symmetric.
Example 4: Find the metric tensor associated with the algebra $[T_\mu, T_\nu] = i\epsilon_{\mu,\nu,\rho}$, i.e. if $C^{\rho}_{\mu\nu} = i\epsilon_{\mu\nu\rho}$.

$$g_{\mu\nu} = i^2 \epsilon_{\mu\rho\beta}\epsilon_{\nu\beta\rho} = \epsilon_{\mu\beta\rho}\epsilon_{\nu\beta\rho} = 2\delta_{\mu\nu}, \quad g^{\mu\nu} = \frac{1}{2}\delta^{\mu\nu} \tag{1.45}$$

i.e. $g_{\mu\nu}$ is diagonal, in fact twice the identity matrix.

With the use of the metric tensor, among other things, we can see whether an algebra is semisimple, whether or not it has an Abelian subalgebra. If it not semisimple[2] one can find at least one element X_ρ, which commutes with all the other elements of the algebra, i.e.

$$[X_\lambda, X_\rho] = 0, \text{ all } \lambda \to C^{\sigma}_{\lambda,\rho} = 0 \text{ all } \rho, \sigma.$$

This means that an entire row (column) of the matrix has elements, which are zeros, i.e. the metric tensor is singular.

1.4.2 *The Casimir operator*

A Casimir operator $K^{(n)}$ is one that commutes with all the elements of the algebra, i.e.

$$[K^{(n)}, X_\rho] = 0, \text{ for all } \rho.$$

It is a polynomial in the generators of degree n defined as:

$$K^{(n)} = C^{\beta_2}_{\alpha_1\beta_1} C^{\beta_3}_{\alpha_2\beta_2} C^{\beta_4}_{\alpha_3\beta_3} \cdots C^{\beta_1}_{\alpha_n\beta_{n+1}} X^{\alpha_1} X^{\alpha_2} \cdots X^{\alpha_n}, \text{ where } X^{\alpha} = g^{\alpha\beta} X_\beta \tag{1.46}$$

Not all such operators are independent. For $SU(2)$ and $SO(3)$, e.g., we have only one quadratic Casimir operator.

$$K^{(2)} = C^{\sigma}_{\lambda\rho} C^{\rho}_{\mu\sigma} X^{\lambda} X^{\mu} = g_{\lambda,\mu} X^{\lambda} X^{\mu} \tag{1.47}$$

Example 5: The quadratic operator for $SO(3)$. In this case we find:

$$K^{(2)} = \frac{1}{2} \sum_{\rho} (X_\rho)^2$$

[2]The algebra $G \supset G_s = SU(3) \otimes SU(2) \otimes U(1)$, encountered in the standard model and its extensions, is not semisimple, since it contains the abelian group $U(1)$

The operators X_ρ are related to the angular momentum operators of quantum mechanics

$$X_\rho = \frac{1}{\hbar} L_\rho \to K^{(2)} = \frac{1}{2\hbar^2}\left(L_x^2 + L_y^2 + L_z^2\right) = \frac{1}{2\hbar^2}\mathbf{L}^2.$$

For reasons of convenience one uses in this case the quadratic operator $\mathbf{L}^2$ instead of $K^{(2)}$.

In the usual quantum mechanical case, if the Hamiltonian is only rotationally invariant, one has only one Casimir operator and one member of the Lie algebra, namely $\mathbf{L}^2$, L_z. Then one diagonalizes these operators:

$$H|\ell,\, m\rangle = g(\ell)|\ell,\, m\rangle,\; \mathbf{L}^2|\ell,\, m\rangle = \ell(\ell+1)\hbar^2|\ell,\, m\rangle,\; L_z|\ell,\, m\rangle = m\hbar|\ell,\, m\rangle$$

If the symmetry of H is higher than $SO(3)$, then one considers in addition the Hamiltonian operator H. The eigenvalues of these operators specify the state $|n,\, \ell,\, m\rangle$ such that:

$$H|n,\, \ell,\, m\rangle = g(n)|n,\, \ell,\, m\rangle,\; \mathbf{L}^2|n,\, \ell,\, m\rangle = \ell(\ell+1)\hbar^2|n,\, \ell,\, m\rangle,$$

$$L_z|n,\, \ell,\, m\rangle = m\hbar|n,\, \ell,\, m\rangle$$

where $g(n)$ depends on the system. In the hydrogenic atom one finds that the higher symmetry is $SO(4)$ and in the case of the harmonic oscillator it is the $SU(3)$, with $g(n) \propto 1/n^2$ and $g(n) \propto (n + 3/2)$ respectively.

The significance of the Casimir operators is that they commute among themselves and all the elements of the algebra. Suppose that the commuting elements of the algebra are $h_1,\, h_2, \cdots,\, h_\ell$ and the independent Casimir operators are $K^{(2)},\, K^{(3)}, \cdots,\, K^{(n)}$. Suppose, further, that the Hamiltonian H describing a system is invariant under the symmetry, i.e. it commutes with all the elements of the algebra, then the system of operators:

$$H,\, K^{(2)},\, K^{(3)}, \cdots,\, K^{(n)},\, h_1,\, h_2, \cdots,\, h_\ell$$

can be simultaneously diagonalized. This way one obtains the needed quantum numbers and the associated eigenfunctions describing the system.

Strategy: Find (hopefully) the maximum symmetry characterizing a given system. This must contain some sub-symmetries. All these symmetries furnish a set of commuting operators. If this set of operators is complete, it supplies sufficient quantum numbers to fully describe the system.

In the Standard Model one uses the invariance of the Lagrangian and the symmetries of interest are internal, i.e. in addition to the usual ones involving ordinary space. In this case the internal symmetry is $SU(3) \otimes SU(2) \otimes U(1)$.

1.5 Representations

Consider a group G and any element $X \in G$. Consider a mapping of this element to a matrix $T(X)$ such that:

$$XY = Z \Rightarrow T(X)T(Y) = T(Z), \quad X, Y, Z \in G.$$

Then we call $T(X)$ a representation of the group[3].

From the properties of the matrices we easily see that given a representation $T(X)$ can obtain additional representations as follows:

- The complex conjugate representation $T^*(X)$.
  Indeed $T^*(XY) = (T(X)T(Y))^* = T^*(X)T^*(Y)$. This representation is different from the previous one if they contain complex numbers. Even if it contains complex numbers the may be equivalent in the sense there exists a non singular matrix S such that $T^*(X) = S^{-1}T(X)S$. Then a basis can be found that the representation becomes real.
- The contragredient representation T^c.
  This defined by: $T^c(X) = (T^+(X))^{-1}$, i.e. it is the inverse of the Hermitian conjugate of $T(X)$. Then:

  $$T^c(XY) = \left((T(XY))^+\right)^{-1} = \left((T(X)(Y))^+\right)^{-1} =$$

  $$\left((T^+(Y)(T^+(X))\right))^{-1} = \left((T^+(X)\right)^{-1}\left(T^+(Y)\right)^{-1} = T^c(X)T^c(Y)$$

  For unitary representations, which is the case for most of the representations we encounter in physics[4], we have

  $$T^c(X) = T(X).$$

- The representation $\bar{T}(X)$, defined by

  $$\bar{T}(X) = \left((T(X))^T\right)^{-1}, (T(X))^T \text{ is the transpose of } T(X).$$

  The proof proceeds as in the previous case. Again if the representation is orthogonal over a real space

  $$\bar{T}(X) = T(X)$$

---

[3]We are mostly interested in faithful representation, i.e. those for which the mapping involved is 1-1.

[4]The elements of the group are bounded and the space of the group parameters is simply connected.

The problem is then of finding one representation, which will be addressed shortly.

In every Lie group there corresponds a set of infinitesimal operators. By choosing a basis in the Hilbert space on which these operators act, we get a matrix representation γ_k for each operator J_k. Then the corresponding representation for the group element A_k, associated with J_k, is the matrix $T(A_k) = e^{i\alpha_k \gamma_k}$, where α_k is a number. We stress that no summation over k is implied in the exponent!.

1.5.1 *The fundamental representation*

This representation is fundamental in the sense that its dimension equals the dimension of the space the group operators act on. Before proceeding further let us then make a detour:

1.5.1.1 *Representations in function spaces*

Consider a, transformation A, which has an inverse, acting on the coordinates, i.e. $\mathbf{r}' = A\mathbf{r}$. What transformation T_A this induces on the space of functions ψ which depend on the coordinates? We demand that:

$$\mathbf{r} \to \mathbf{r}', \psi \to \psi' \text{ such that } \psi'(\mathbf{r}') = \psi(\mathbf{r}) \tag{1.48}$$

i.e.

$$T(A)\psi(\mathbf{r}') = \psi(\mathbf{r}) \Rightarrow T(A)\psi(A\mathbf{r}) = \psi(\mathbf{r}) \Rightarrow \tag{1.49}$$

or

$$T(A)\psi(\mathbf{r}) = \psi(A^{-1}\mathbf{r}), \tag{1.50}$$

which defines the operator. We will apply this in some special cases.
Example 6: Consider the transformation of coordinates:

$$L: \quad x \to x - a, \, a = \text{ a constant} \tag{1.51}$$

Find the transformation T_a it induces for action on the space of differential functions $f(x)$.

The desired relation is

$$T_a f(x) = f(L^{-1}x) = f(x+a).$$

$$f(x+a) = f(x) + \frac{a}{1!}\frac{df(x)}{dx} + \frac{a^2}{2!}\frac{d^2 f(x)}{dx^2} + \cdots = \left(1 + \frac{a}{1!}\frac{d}{dx} + \frac{a^2}{2!}\frac{d^2}{dx^2} \cdots\right) f(x)$$

$$= e^{a\frac{d}{dx}} f$$

Thus

$$T_a = e^{a\frac{d}{dx}}$$

We verify

$$T_a f = e^{a\frac{d}{dx}} f = \sum_n^{\infty} a^n \frac{d^n}{dx^n} f = f(x+a)$$

Usually instead of $\frac{d}{dx}$ we use the momentum operator $p_x = \frac{\hbar}{i}\frac{d}{dx}$, which is a Hermitian operator. Therefore

$$T_a = e^{ia\frac{p_x}{\hbar}} \tag{1.52}$$

We can generalize Eq. (1.51) in three dimensions:

$$L: \quad \mathbf{r} \to \mathbf{r} - \mathbf{a},\ \mathbf{a} = \text{ constant} \tag{1.53}$$

Then proceeding as above for each component we find the analog of Eq. (1.52):

$$T_{\mathbf{a}} = e^{i\mathbf{a}.\mathbf{p}/\hbar},\ \mathbf{p} = \frac{\hbar}{i}\nabla \tag{1.54}$$

Example 7: Consider the transformation of coordinates $R(-\theta)$ in two dimensions, e.g. in the (x,y) plane[5]:

$$\begin{pmatrix} x' \\ y' \end{pmatrix} = \begin{pmatrix} \cos\theta & -\sin\theta \\ \sin\theta & \cos\theta \end{pmatrix} \begin{pmatrix} x \\ y \end{pmatrix} \tag{1.55}$$

Find the transformation it induces on the space of functions $f(x,y)$, namely

$$T_\theta f(x,y) = f(\tilde{x},\tilde{y}), \begin{pmatrix} \tilde{x} \\ \tilde{y} \end{pmatrix} = R_\theta \begin{pmatrix} x \\ y \end{pmatrix},\ R_\theta = \begin{pmatrix} \cos\theta & \sin\theta \\ -\sin\theta & \cos\theta \end{pmatrix}$$

(note the change in sign of θ) or

$$T_\theta = f\left(x\cos\theta + y\sin\theta, x(-)\sin\theta + y\cos\theta\right)$$

The infinitesimal transformation is:

$$T_\epsilon f(x,y) = f(x,y) + \epsilon\left(\frac{\partial f}{\partial\tilde{x}}\frac{d\tilde{x}}{d\theta} + \frac{\partial f}{\partial\tilde{y}}\frac{d\tilde{y}}{d\theta}\right)_{\theta=0} = f(x,y) + \epsilon\left(y\frac{\partial f}{\partial x} + (-x)\frac{\partial f}{\partial y}\right)$$

[5]We intentionally write rotations in the plane rather than around an axis. In three dimensions, of course, it makes no difference, since we have three planes and three axes. So it is immaterial whether we talk about a rotation in the plane (x,y) or around the $z$axis, which is defined to be perpendicular to it. This is not true in higher dimensions. In the four dimensions, e.g., we have 4 axes, but 6 planes. So the number of rotations is 6, not 4.

or

$$T_\epsilon f(x,y) = (1 - \epsilon X) f(x,y) = \left(1 - i\epsilon\frac{1}{\hbar}L_{xy}\right) f(x,y),$$

where

$$X = x\frac{\partial}{\partial y} - y\frac{\partial}{\partial x},\ L_{x,y} = \frac{\hbar}{i}\left(x\frac{\partial}{\partial y} - y\frac{\partial}{\partial x}\right)$$

The finite transformation becomes:

$$T_\theta = e^{-i\theta L_{xy}/\hbar}$$

The quantities L_{xy} are the familiar angular momentum operators in coordinate space for rotations in the plane x, y. This can be generalized in three dimensions. The angular momentum operators become:

$$L_{jk} = \frac{\hbar}{i}\left(x_j\frac{\partial}{\partial x_k} - x_k\frac{\partial}{\partial x_j}\right) \tag{1.56}$$

The corresponding group rotation operators in the corresponding plane are

$$R_{jk} = T_{\theta_{jk}} = e^{-i\theta_{jk}L_{jk}/\hbar}\ \text{(no summation)} \tag{1.57}$$

Since the angular momentum operators do not commute, one has to be careful. We have to specify what is the specific rotation (which axes and what rotation angles and in which order they are performed). It is common to specify the rotation in terms of the Euler angles α, β and γ as we have done above.

After this detour we will consider some interesting cases.

1.5.1.2 *The orthogonal groups*

Consider the operator given by Eq. (1.56) acting on the three dimensional space (x_1, x_2, x_3). Choose a basis $|i\rangle$ in this space: $|i\rangle \Leftrightarrow x_i$. Then

$$L_{12}|1\rangle = L_{12}x_1 = ix_2 = i|2\rangle, L_{12}|2\rangle = L_{12}x_2 = -ix_1 = -i|1\rangle,$$

$$L_{12}|3\rangle = L_{12}x_3 = 0\ .$$

Thus the corresponding matrix is:

$$\gamma_{12} = \begin{pmatrix} 0 & -i & 0 \\ i & 0 & 0 \\ 0 & 0 & 0 \end{pmatrix}.$$

In an analogous fashion we get:

$$\gamma_{13} = \begin{pmatrix} 0 & 0 & -i \\ 0 & 0 & 0 \\ i & 0 & 0 \end{pmatrix}, \quad \gamma_{23} = \begin{pmatrix} 0 & 0 & 0 \\ 0 & 0 & -i \\ 0 & i & 0 \end{pmatrix}.$$

The corresponding elements of the rotation group are:

$$R(\gamma_{jk}) = e^{i\theta_{jk}\gamma_{jk}}, \quad \theta_{jk} \text{ a set of numbers.}$$

Now we notice that

$$\gamma_{jk}^2 = I_{jk},\, I_{12} = \begin{pmatrix} 1\,0\,0 \\ 0\,1\,0 \\ 0\,0\,0 \end{pmatrix},\, I_{13} = \begin{pmatrix} 1\,0\,0 \\ 0\,0\,0 \\ 0\,0\,1 \end{pmatrix},\, I_{23} = \begin{pmatrix} 0\,0\,0 \\ 0\,0\,1 \\ 0\,1\,0 \end{pmatrix} \Rightarrow$$

$$\gamma_{jk}^{2\ell} = I_{jk},\, \gamma_{jk}^{2\ell+1} = \gamma_{jk},\, \ell = 1, 2, \cdots, \ell$$

Thus

$$R(\gamma_{jk}) = e^{i\theta_{jk}\gamma_{jk}} = I + \sum_{\ell=1} \left(\frac{i^{2\ell}}{(2\ell)!}\theta_{jk}^{2\ell}\gamma_{jk}^{2\ell} + \frac{i^{2\ell+1}}{(2\ell+1)!}\theta_{jk}^{2\ell+1}\gamma_{jk}^{2\ell+1} \right) =$$

$$I + I_{jk}\sum_{\ell=1} \left((-1)^{\ell}\frac{\theta_{jk}^{2\ell}}{(2\ell)!} + i\gamma_{jk} + (-1)^{\ell}\frac{\theta_{jk}^{2\ell+1}}{(2\ell+1)!} \right) =$$

$$I + I_{jk}(\cos\theta_{jk} - 1) + i\gamma_{jk}\sin\theta_{jk}$$

We thus find:

$$R_z \leftrightarrow R_{12} = \begin{bmatrix} \cos\theta_{12} & \sin\theta_{12} & 0 \\ -\sin\theta_{12} & \cos\theta_{12} & 0 \\ 0 & 0 & 1 \end{bmatrix} \text{ rotation in the plane (1,2),} \quad (1.58)$$

$$R_y \Longleftrightarrow R_{13} = \begin{bmatrix} \cos\theta_{13} & 0 & \sin\theta_{13} \\ 0 & 1 & 0 \\ -\sin\theta_{13} & 0 & \cos\theta_{13} \end{bmatrix} \text{ rotation in the plane (1,3),} \quad (1.59)$$

$$R_x \Longleftrightarrow R_{23} = \begin{bmatrix} 1 & 0 & 0 \\ 0 & \cos\theta_{23} & \sin\theta_{23} \\ 0 & -\sin\theta_{23} & \cos\theta_{23} \end{bmatrix} \text{ rotation in the plane (2,3).} \quad (1.60)$$

This is the fundamental (basic) representation of the group $SO(3) \equiv R(3)$, with $R(3)$ the group of rotations in 3-dimensional space. It is fundamental in the sense that its dimension is the same with the dimension of the space, i.e. the non trivial representation with the lowest dimension. This can trivially be extended to SO(n)

The above matrices can be found directly in the case of $SO(3)$ without going through the algebra. For more complicated groups one has no choice but to start from the algebra and proceed as above. The opposite, i.e. to get the generators of the algebra from those of the group, is, however, simpler as is exhibited in the next example.

Example 8: Consider the above group generators of SO(3). Obtain the generators of the corresponding algebra.
We can do this by proceeding in the opposite direction by considering the expansion:

$$R_{ij} = R_{ij}(0) + i\theta_{ij}J_{ij}, \quad J_{ij} = -i\left.\frac{dR}{d\theta_{ij}}\right|_{\theta_{ij}=0}. \tag{1.61}$$

Then

$$T_3 \Longleftrightarrow J_{12} = -i\left.\frac{dR}{d\theta_{12}}\right|_{\theta_{12}=0} = \begin{bmatrix} 0 & -i & 0 \\ i & 0 & 0 \\ 0 & 0 & 0 \end{bmatrix},$$

$$T_2 \Longleftrightarrow J_{13} = -i\left.\frac{dR}{d\theta_{13}}\right|_{\theta_{13}} = \begin{bmatrix} 0 & 0 & -i \\ 0 & 0 & 0 \\ i & 0 & 0 \end{bmatrix},$$

$$T_1 \Longleftrightarrow J_{23} = -i\left.\frac{dR}{d\theta_{12}}\right|_{\theta 23=0} = \begin{bmatrix} 0 & 0 & 0 \\ 0 & 0 & -i \\ 0 & i & 0 \end{bmatrix}.$$

These are the same with the matrices γ_{ij} above. We can verify that:

$$[T_k, T_\ell] = i\epsilon_{k\ell m}T_m. \tag{1.62}$$

We can also define the operators:

$$K_{ij} = \left.\frac{dR}{d\theta_{ij}}\right|_{\theta_{ij}=0}. \tag{1.63}$$

Then

$$K_1 \equiv K_{23} = \left.\frac{dR}{d\theta_{23}}\right|_{\theta_{23}=0} = \begin{bmatrix} 0 & 0 & 0 \\ 0 & 0 & 1 \\ 0 & -1 & 0 \end{bmatrix},$$

$$K_2 \equiv K_{13} = \left.\frac{dR}{d\theta_{13}}\right|_{\theta_{13}=0} = \begin{bmatrix} 0 & 0 & 1 \\ 0 & 0 & 0 \\ -1 & 0 & 0 \end{bmatrix},$$

$$K_3 \equiv K_{12} = \left.\frac{dR}{d\theta_{12}}\right|_{\theta 12=0} = \begin{bmatrix} 0 & 1 & 0 \\ -1 & 0 & 0 \\ 0 & 0 & 0 \end{bmatrix},$$

which are antisymmetric.We can verify that

$$[K_k, K_\ell] = \epsilon_{k\ell m}K_m. \tag{1.64}$$

1.5.1.3 *The unimodular groups $SU(n)$*

We have already done in the case $n = 2$ and $n = 3$.
Example 9: The SU(2) case. This has already been done accomplished by the Pauli matrices:

$$\sigma_1 = \begin{pmatrix} 0 & 1 \\ 1 & 0 \end{pmatrix}, \sigma_2 = \begin{pmatrix} 0 & -i \\ i & 0 \end{pmatrix}, \sigma_3 = \begin{pmatrix} 1 & 0 \\ 0 & -1 \end{pmatrix}. \tag{1.65}$$

One can define the matrices

$$T_1 = s_1 = \frac{1}{2}\sigma_1, \, T_2 = s_2 = \frac{1}{2}\sigma_2, \, T_3 = s_3 = \frac{1}{2}\sigma_3. \tag{1.66}$$

The Pauli matrices σ_i represent the fundamental representation of $SU(2)$.
Example 10: The $SU(3)$ group.
The fundamental representation is furnished by the matrices λ_i, $i = 1, 2, \cdots, 8$ (see Eq. (1.29) and table 1.2).

This can be generalized to $U(n)$ drawing from our experience with the quantized harmonic oscillator. Recall that we start with sates $|i\rangle$ and indexcreation operators creation and indexdestruction operators destruction operators a_i^+ and a_i respectively such that

$$a_i^+|0\rangle = |i\rangle, \, a_i|0\rangle = 0$$

where $|0\rangle$ is the vacuum state. Furthermore they obey the commutation rules

$$[a_i, a_j^+] = \delta_{i,j}, \, [a_i, a_j] = 0, \, [a_i^+, a_j^+] = 0$$

Then the generators of $U(n)$ are defined as

$$A_{k\ell} = a_k^+ a_\ell \tag{1.67}$$

with the commutator rules:

$$[A_{jk}, A_{mn}] = A_{jn}\delta_{km} - A_{mk}\delta_{jn} \tag{1.68}$$

Indeed

$$[A_{jk}, A_{mn}] = [a_j^+ a_k, a_m^+ a_n] = [a_j^+, a_m^+ a_n]a_k + a_j^+[a_k, a_m^+ a_n] =$$

$$a_m^+[a_j^+, a_n]a_k + a_j^+[a_k, a_m^+]a_n = -\delta_{jn}a_m^+ a_k + \delta_{km}a_j^+ a_n = A_{jn}\delta_{km} - A_{mk}\delta_{jn}$$

The elements A_{ii}, $i = 1, 2, \cdots, n$ form an Abelian subalgebra.
The result is that the fundamental representation is obtained by acting with the generators on the basis $|i\rangle = a_i^+|0\rangle$, i.e.

$$A_{jk}|m\rangle = \delta_{km}|j\rangle \Leftrightarrow \text{ representation } (e_{jk}) \Leftrightarrow \text{ a matrix } (e_{jk})_{\alpha,\beta} = \delta_{j\alpha}\delta_{k\beta} \tag{1.69}$$

that is the matrix (e_{jk}) has 1 in row j and column k and zero everywhere else.

For $SU(n)$ a slight complication arises, since the matrices have to be reaceless. The mathematicians choose the elements :

$$h_k = e_{k,k} - e_{k+1,k+1},\ k = 1, 2, \cdots, n-1 \tag{1.70}$$

As we have seen the physicists prefer a different basis. In the case of $SU(3)$

$$\lambda_1 = e_{12} + e_{21},\ \lambda_2 = -i(e_{12} - e_{21}),\ \lambda_4 = e_{13} + e_{31},\ \lambda_5 = -i(e_{13} - e_{31}),\ \lambda_6 = e_{23} + e_{32},$$

$$\lambda_7 = -i(e_{23} - e_{32}), \lambda_3 = e_{11} - e_{22}, \lambda_8 = \frac{1}{\sqrt{3}}(e_{11} + e_{22} - 2e_{33})$$

1.5.2 *The adjoined representation*

The abstract algebra of n elements, in a given basis λ_k, is defined by the structure constants:

$$[\lambda_k, \lambda_\ell] = c^m_{k\ell}\lambda_m. \tag{1.71}$$

Then a $n \times n$ basis $\{Adj(\lambda_k)\}$ for the algebra can be defined by:

$$Adj(\lambda_k)|\ell\rangle = c^m_{k\ell}|m\rangle. \tag{1.72}$$

The set of matrices $Adj(\lambda_k)$ defines the **adjoined** or **adjoint representation**. Suppose, e.g., some abstract algebra is given:

$$[\lambda_1, \lambda_2] = i\lambda_3, \quad [\lambda_2, \lambda_3] = i\lambda_1, \quad [\lambda_3, \lambda_1] = i\lambda_2, \tag{1.73}$$

then the independent structure constants are:

$$C^3_{12} = i, \quad C^1_{23} = i, \quad C^2_{31} = i.$$

Writing for simplicity $T_k = Adj(\lambda_k)$ we find:

$$T_1|1\rangle = 0, \tag{1.74}$$

$$T_1|2\rangle = C^3_{12}|3\rangle = i|3\rangle, \tag{1.75}$$

$$T_1|3\rangle = C^2_{13}|2\rangle = -i|2\rangle. \tag{1.76}$$

Thus the obtained matrix (column vectors) is:

$$T_1 = \begin{bmatrix} 0 & 0 & 0 \\ 0 & 0 & -i \\ 0 & i & 0 \end{bmatrix}, \tag{1.77}$$

$$T_2|1\rangle = C^3_{21}|3\rangle = -i|3\rangle, \tag{1.78}$$

$$T_2|2\rangle = 0, \tag{1.79}$$

$$T_2|3\rangle = C^1_{23}|1\rangle = i|1\rangle, \tag{1.80}$$

$$T_2 = \begin{bmatrix} 0 & 0 & i \\ 0 & 0 & 0 \\ -i & 0 & 0 \end{bmatrix}, \tag{1.81}$$

Note the sign of T_2. Furthermore:

$$T_3|1\rangle = C^2_{31}|2\rangle = i|2\rangle, \tag{1.82}$$

$$T_3|2\rangle = C^1_{32}|1\rangle = -i|1\rangle, \tag{1.83}$$

$$T_3|3\rangle = 0, \tag{1.84}$$

$$T_3 = \begin{bmatrix} 0 & -i & 0 \\ i & 0 & 0 \\ 0 & 0 & 0 \end{bmatrix}. \tag{1.85}$$

One can verify that this set obeys the same set of commutators as those of Eq. (1.73), namely

$$[T_1, T_2] = iT_3,\ [T_2, T_3] = iT_1,\ [T_3, T_1] = iT_2. \tag{1.86}$$

This is the adjoined representation corresponding to the algebra of of the operators λ_i with the above structure constants. We have two possibilities:

- $\lambda_i = (1/2)\sigma_i$. Then talk about the adjoined representation of $SU(2)$, which is three dimensional.
- If, on the other hand, the λ_i are identified with the fundamental representation γ_{ij} of $SO(3)$ found above, then we talk about the adjoined representation of $SO(3)$. Both are 3-dimensional.

In other words in the case of the $SO(3)$ the fundamental and the adjoined are equivalent, since the commutation relations are the same.

Note now that since $T_k^2 = I_k$, with I_k the identity matrix without the 1 in the kth column and row, we have

$$R_k = e^{i\theta_k T_k} = I + \gamma_0 I_k + \gamma_1 T_k. \tag{1.87}$$

Since the non zero eigenvalues of T_k are (1,-1), Eq. (1.87) yields:

$$e^{i\theta_k} = 1 + \gamma_0 + \gamma_1, \quad e^{-i\theta_k} = 1 + \gamma_0 - \gamma_1.$$

From these we get $\gamma_0 = \cos\theta_k - 1$, $\gamma_1 = i\sin\theta_k$. Thus Eq. (1.87) becomes:

$$e^{i\theta_k T_k} = I + (\cos\theta_k - 1)I_k + i\sin\theta_i T_k,$$

that is

$$R_1 = e^{i\theta_1 T_1} = \begin{bmatrix} 1 & 0 & 0 \\ 0 & \cos\theta_1 & \sin\theta_1 \\ 0 & -\sin\theta_1 & \cos\theta_1 \end{bmatrix}, \tag{1.88}$$

$$R_2 = e^{i\theta_2 T_2} = \begin{bmatrix} \cos\theta_2 & 0 & \sin\theta_2 \\ 0 & 1 & 0 \\ -\sin\theta_2 & 0 & \cos\theta_2 \end{bmatrix}, \tag{1.89}$$

$$R_3 = e^{i\theta_3 T_3} = \begin{bmatrix} \cos\theta_3 & \sin\theta_3 & 0 \\ -\sin\theta_3 & \cos\theta_3 & 0 \\ 0 & 0 & 1 \end{bmatrix}. \tag{1.90}$$

These are a set proper orthogonal transformations (rotations) on the planes around the Cartesian axes. It is preferred to use the plane notation and rewrite the above equations by the substitution:

$$1 \to 23,\ 2 \to 3,\ 3 \to 12, \text{e.g.} T_1 \to T_{23},\ \theta_1 \to \theta_{23|}, \text{etc}$$

The plane notation is the one that can be generalized to more than 3 dimensions.

Exercise 1: Consider the following basis for the Pauli matrices:

$$\tau_+ = \begin{pmatrix} 0 & 1 \\ 0 & 0 \end{pmatrix}, \quad \tau_- = \begin{pmatrix} 0 & 0 \\ 1 & 0 \end{pmatrix}, \quad \tau_0 = \begin{pmatrix} 1 & 0 \\ 0 & -1 \end{pmatrix}. \tag{1.91}$$

Show that:

$$[\tau_+, \tau_-] = \tau_0, \quad [\tau_0, \tau_+] = 2\tau_+, \quad [\tau_0, \tau_-] = -2\tau_- \tag{1.92}$$

From these, by making the correspondence $+ \Leftrightarrow 1$, $- \Leftrightarrow 2$ and $\Leftrightarrow 3$, we construct the matrices corresponding to the Adjoined representation:

$$I_+ = \begin{pmatrix} 0 & 0 & -2 \\ 0 & 0 & 0 \\ 0 & 1 & 0 \end{pmatrix}, \quad I_- = \begin{pmatrix} 0 & 0 & 0 \\ 0 & 0 & 2 \\ -1 & 0 & 0 \end{pmatrix}, \quad I_0 = \begin{pmatrix} 2 & 0 & 0 \\ 0 & -2 & 0 \\ 0 & 0 & 0 \end{pmatrix}.$$

Verify that these 3×3 matrices obey the same commutation relations as the original Pauli matrices.

1.6 More representations

From the fundamental and the adjoined representations discussed above we can construct more larger representations.

Theorem: Suppose that we know two representations of an algebra $\mathcal{A}$ and $\mathcal{B}$. Let us now suppose $A \in \mathcal{A}$ and $B \in \mathcal{B}$. Then we obtain a new representation

$\mathcal{A} \otimes \mathcal{B}$, known as Kronecker product, as follows: Extend the matrix A by substituting $A_{ij} \to A_{ij}B$. Thus if A is an $m \times m$ matrix and B is an $n \times n$ matrix, the new representation is $nm \times nm$ matrix.

Consider, e.g., $\mathcal{A} = \sigma_1, \sigma_2, \sigma_3$, see Eq. (1.65, and $\mathcal{B}$ the identity 2×2 matrix. Then we obtain the representation:

$$\Sigma_1 = \begin{pmatrix} (0) & I \\ I & (0) \end{pmatrix} = \begin{pmatrix} 0 & 0 & 1 & 0 \\ 0 & 0 & 0 & 1 \\ 1 & 0 & 0 & 0 \\ 0 & 1 & 0 & 0 \end{pmatrix},$$

$$\Sigma_2 = \begin{pmatrix} (0) & -iI \\ iI & (0) \end{pmatrix} = \begin{pmatrix} 0 & 0 & -i & 0 \\ 0 & 0 & 0 & -i \\ i & 0 & 0 & 0 \\ 0 & i & 0 & 0 \end{pmatrix},$$

$$\Sigma_3 = \begin{pmatrix} (I) & (0) \\ (0) & -I \end{pmatrix} = \begin{pmatrix} 1 & 0 & 0 & 0 \\ 0 & 1 & 0 & 0 \\ 0 & 0 & -1 & 0 \\ 0 & 0 & 0 & -1 \end{pmatrix}.$$

One can verify that:

$$[\Sigma_j, \Sigma_k] = 2\epsilon_{jkm}\Sigma_m,$$

i.e. it is the same as that involving the Pauli matrices σ_i (see Eq. (1.65)). Sometimes the group elements are given and one proceeds to construct the Algebra. Then, from the structure constants of the algebra, a new representation can be obtained.

1.7 Irreducible representations

Not all representations of the algebra are necessary. It is sufficient to know and study the irreducible representations. This is one of the main goals of group theory but it is beyond our goals. There exist many excellent books on this subject [Wybourne (1974)],[Vergados (2016)]. We will proceed here using rather primitive techniques, which are adequate for our purposes in this introductory chapter.

An $n \times n$ representation $\Gamma = (\Gamma_i, i = 1, 2, \cdots r)$ is reducible if a non singular matrix can be found such that:

$$S^{-1}\Gamma_i S = \text{ block diagonal form } = \begin{pmatrix} (\gamma_i^{(1)}) & 0 & 0 & 0 & 0 \\ 0 & (\gamma_i^{(2)}) & 0 & 0 & 0 \\ 0 & 0 & (\gamma_i^{(3)}) & 0 & 0 \\ 0 & 0 & 0 & \cdots & 0 \\ 0 & 0 & 0 & 0 & (\gamma_i^{(\ell)}) \end{pmatrix} \tag{1.93}$$

S is the same for all Γ_i, $\gamma^k_{(i)}$ is a $n_k \times n_k$ matrix such $\sum_{k=1}^{\ell} n_k = n$. If such a transformation does not exist the representation is **irreducible.**

One can find whether or nor a representation is irreducible by calculating the corresponding Casimir operators associated with this representation. If they are diagonal the representation is perhaps irreducible. If it is reducible the diagonal elements of the Abelian sub-algebra will give a hint about the transformation required to reduce it. We will illustrate this by some examples.

Example 11: Let us consider the matrices Γ_1, Γ_2, Γ_3 as follows:

$$\Gamma_1 = \begin{pmatrix} 0 & \frac{\sqrt{3}}{2} & 0 & 0 \\ \frac{\sqrt{3}}{2} & 0 & 1 & 0 \\ 0 & 1 & 0 & \frac{\sqrt{3}}{2} \\ 0 & 0 & \frac{\sqrt{3}}{2} & 0 \end{pmatrix}, \Gamma_2 = \begin{pmatrix} 0 & -\frac{i\sqrt{3}}{2} & 0 & 0 \\ \frac{i\sqrt{3}}{2} & 0 & -i & 0 \\ 0 & i & 0 & -\frac{i\sqrt{3}}{2} \\ 0 & 0 & \frac{i\sqrt{3}}{2} & 0 \end{pmatrix}, \Gamma_3 = \begin{pmatrix} \frac{3}{2} & 0 & 0 & 0 \\ 0 & \frac{1}{2} & 0 & 0 \\ 0 & 0 & -\frac{1}{2} & 0 \\ 0 & 0 & 0 & -\frac{3}{2} \end{pmatrix} \tag{1.94}$$

One can verify that

$$[\Gamma_k, \Gamma_m] = i\epsilon_{kmn}\Gamma_n$$

that is they form a representation of $SU(2)$. This representation is irreducible since:

$$\Gamma_1^2 + \Gamma_2^2 + \Gamma_3^2 = 3/2 \times 5/2 \text{ and } \Gamma_3 \text{ is diagonal with } \Leftrightarrow s = 3/2$$

Example 12: Let us again consider the 4×4 matrices Γ_1, Γ_2, Γ_3 as follows:

$$\Gamma_1 = \begin{pmatrix} 0 & \frac{1}{2} & \frac{1}{2} & 0 \\ \frac{1}{2} & 0 & 0 & \frac{1}{2} \\ \frac{1}{2} & 0 & 0 & \frac{1}{2} \\ 0 & \frac{1}{2} & \frac{1}{2} & 0 \end{pmatrix}, \Gamma_2 = \begin{pmatrix} 0 & -\frac{i}{2} & -\frac{i}{2} & 0 \\ \frac{i}{2} & 0 & 0 & -\frac{i}{2} \\ \frac{i}{2} & 0 & 0 & -\frac{i}{2} \\ 0 & \frac{i}{2} & \frac{i}{2} & 0 \end{pmatrix}, \Gamma_3 = \begin{pmatrix} 1 & 0 & 0 & 0 \\ 0 & 0 & 0 & 0 \\ 0 & 0 & 0 & 0 \\ 0 & 0 & 0 & -1 \end{pmatrix} \tag{1.95}$$

One sees immediately that this is reducible, since the Casimir operator constructed from it is not a multiple of the identity. The form of the value of the Casimir operator as well as the diagonal values of Γ_3 suggest that it contains a representation associated with spin one, containing $m = 1, 0, -1$ and a trivial representation with spin zero. So we identify

$$|s=1, m=1\rangle \Leftrightarrow |1\rangle,\ |s=1, m=0\rangle \Leftrightarrow \frac{1}{\sqrt{2}}|2\rangle + \frac{1}{\sqrt{2}}|3\rangle,$$

$$|s=1, m=-1\rangle \Leftrightarrow |4\rangle,\ |s=0, m=0\rangle \Leftrightarrow \frac{1}{\sqrt{2}}|2\rangle - \frac{1}{\sqrt{2}}|3\rangle$$

The desired unitary transformation is:

$$S = \begin{pmatrix} 1 & 0 & 0 & 0 \\ 0 & \frac{1}{\sqrt{2}} & 0 & \frac{1}{\sqrt{2}} \\ 0 & \frac{1}{\sqrt{2}} & 0 & -\frac{1}{\sqrt{2}} \\ 0 & 0 & 1 & 0 \end{pmatrix}$$

and thus we find $\gamma_1^{(2)} = \gamma_2^{(2)} = \gamma_3^{(2)} = (0)$, the last row and column of the reduced matrix contain only 0 and the upper left block is:

$$\gamma_1^{(1)} = \begin{pmatrix} 0 & \frac{1}{\sqrt{2}} & 0 \\ \frac{1}{\sqrt{2}} & 0 & \frac{1}{\sqrt{2}} \\ 0 & \frac{1}{\sqrt{2}} & 0 \end{pmatrix},\ \gamma_2^{(1)} = \begin{pmatrix} 0 & -\frac{i}{\sqrt{2}} & 0 \\ \frac{i}{\sqrt{2}} & 0 & -\frac{i}{\sqrt{2}} \\ 0 & \frac{i}{\sqrt{2}} & 0 \end{pmatrix},\ \gamma_3^{(1)} = \begin{pmatrix} 1 & 0 & 0 \\ 0 & 0 & 0 \\ 0 & 0 & -1 \end{pmatrix} \tag{1.96}$$

which is the familiar adjoined representation of $SU(2)$. The remaining one is just the scalar representation associated with $s = 0$.

Example 13: Let us now consider the 6×6 matrices Γ_1, Γ_2, Γ_3 as follows

$$\Gamma_1 = \begin{pmatrix} 0 & \frac{1}{2} & \frac{1}{\sqrt{2}} & 0 & 0 & 0 \\ \frac{1}{2} & 0 & 0 & \frac{1}{\sqrt{2}} & 0 & 0 \\ \frac{1}{\sqrt{2}} & 0 & 0 & \frac{1}{2} & \frac{1}{\sqrt{2}} & 0 \\ 0 & \frac{1}{\sqrt{2}} & \frac{1}{2} & 0 & 0 & \frac{1}{\sqrt{2}} \\ 0 & 0 & \frac{1}{\sqrt{2}} & 0 & 0 & \frac{1}{2} \\ 0 & 0 & 0 & \frac{1}{\sqrt{2}} & \frac{1}{2} & 0 \end{pmatrix},\ \Gamma_2 = \begin{pmatrix} 0 & -\frac{i}{2} & -\frac{i}{\sqrt{2}} & 0 & 0 & 0 \\ \frac{i}{2} & 0 & 0 & -\frac{i}{\sqrt{2}} & 0 & 0 \\ \frac{i}{\sqrt{2}} & 0 & 0 & -\frac{i}{2} & -\frac{i}{\sqrt{2}} & 0 \\ 0 & \frac{i}{\sqrt{2}} & \frac{i}{2} & 0 & 0 & -\frac{i}{\sqrt{2}} \\ 0 & 0 & \frac{i}{\sqrt{2}} & 0 & 0 & -\frac{i}{2} \\ 0 & 0 & 0 & \frac{i}{\sqrt{2}} & \frac{i}{2} & 0 \end{pmatrix} \tag{1.97}$$

together with the diagonal matrix Γ_3 with elements (3/2, 1/2, 1/2, −1/2, −1/2, −3/2) along the diagonal. One can verify that:

$$[\Gamma_k, \Gamma_\ell] = i\epsilon_{k\ell m}\Gamma_m. \tag{1.98}$$

In other words it is a six dimensional representation of $SU(2)$.

One can immediately see that this representation is reducible, since the operator $\Gamma_1^2 + \Gamma_2^2 + \Gamma_2^3$ is not a multiple of the identity. Another basis, however, can be found via a unitary transformation S. To get a hint about how to proceed we look at the diagonal elements of Γ_3. One sees that there are exist two sets of degenerate m values, associated with $m = 1/2$ and $m = -1/2$. The presence of $m = \pm 3/2$ suggests that it must contain a representation $S = 3/2$ involving the $m = 3/2, 1/2, -1/2, -3/2$. Then the remaining $m = \pm 1/2$ are absorbed by the $s = 1/2$ representation. So we can take any linear combination of the degenerate m-values. In hindsight I choose:

$$|3/2,3/2\rangle \Leftrightarrow |1\rangle, |3/2,1/2\rangle \Leftrightarrow \sqrt{1/3}|\rangle|2\rangle + \sqrt{2/3}|\rangle|3\rangle,$$

$$|3/2,-1/2\rangle \Leftrightarrow \sqrt{2/3}|\rangle|4\rangle + \sqrt{1/3}\rangle|5\rangle, |1/2,1/2\rangle \Leftrightarrow -\sqrt{2/3}|2\rangle + \sqrt{1/3}|\rangle|3\rangle,$$

$$|1/2,-1/2\rangle \Leftrightarrow -\sqrt{1/3}|4\rangle + \sqrt{2/3}|\rangle|5\rangle, |3/2,3/2\rangle \Leftrightarrow |6\rangle$$

This leads to the unitary matrix:

$$\begin{pmatrix} 1 & 0 & 0 & 0 & 0 & 0 \\ 0 & \frac{1}{\sqrt{3}} & 0 & 0 & -\sqrt{\frac{2}{3}} & 0 \\ 0 & \sqrt{\frac{2}{3}} & 0 & 0 & \frac{1}{\sqrt{3}} & 0 \\ 0 & 0 & \sqrt{\frac{2}{3}} & 0 & 0 & -\frac{1}{\sqrt{3}} \\ 0 & 0 & \frac{1}{\sqrt{3}} & 0 & 0 & \sqrt{\frac{2}{3}} \\ 0 & 0 & 0 & 1 & 0 & 0 \end{pmatrix} \tag{1.99}$$

so that the above representation can be cast in the form of Eq. (1.93) with:

$$\gamma_1^{(1)} = \begin{pmatrix} 0 & \frac{\sqrt{3}}{2} & 0 & 0 \\ \frac{\sqrt{3}}{2} & 0 & 1 & 0 \\ 0 & 1 & 0 & \frac{\sqrt{3}}{2} \\ 0 & 0 & \frac{\sqrt{3}}{2} & 0 \end{pmatrix}, \gamma_2^{(1)} = \begin{pmatrix} 0 & -\frac{i\sqrt{3}}{2} & 0 & 0 \\ \frac{i\sqrt{3}}{2} & 0 & -i & 0 \\ 0 & i & 0 & -\frac{i\sqrt{3}}{2} \\ 0 & 0 & \frac{i\sqrt{3}}{2} & 0 \end{pmatrix},$$

$$\gamma_3^{(1)} = \begin{pmatrix} \frac{3}{2} & 0 & 0 & 0 \\ 0 & \frac{1}{2} & 0 & 0 \\ 0 & 0 & -\frac{1}{2} & 0 \\ 0 & 0 & 0 & -\frac{3}{2} \end{pmatrix} \tag{1.100}$$

and

$$\gamma_1^{(2)} = \begin{pmatrix} 0 & \frac{1}{2} \\ \frac{1}{2} & 0 \end{pmatrix}, \gamma_2^{(2)} = \begin{pmatrix} 0 & -i\frac{1}{2} \\ i\frac{1}{2} & 0 \end{pmatrix}, \gamma_3^{(2)} = \begin{pmatrix} \frac{1}{2} & 0 \\ 0 & -\frac{1}{2} \end{pmatrix} \tag{1.101}$$

These matrices satisfy commutation relations analogous to those of Eq. (1.98), namely

$$[\gamma_m^{(k)}, \gamma_n^{(k)}] = i\epsilon_{mn\ell}\gamma_\ell^{(k)},\ k = 1,2$$

Table 1.4: Some simple C-G coefficients

$m_2 \rightarrow$	$\frac{1}{2}$	$-\frac{1}{2}$
$j \downarrow$		
$j_1+\frac{1}{2}$	$\sqrt{\frac{j_1+1/2+m}{2j_1+1}}$	$\sqrt{\frac{j_1+1/2-m}{2j_1+1}}$
$j_1-\frac{1}{2}$	$-\sqrt{\frac{j_1+1/2-m}{2j_1+1}}$	$\sqrt{\frac{j_1+1/2+m}{2j_1+1}}$

Thus the Γ_i are reducible, while the $\gamma_k^{(i)}$ are irreducible. As we have already seen that those with $k=1$ are associated with spin $3/2$, while those with $k=2$ with spin $1/2$.

Conclusion: The Kronecker product of two representations of the same algebra is, in general, reducible. Of special interest to particle physics are:

The SU(2) case.

Here we encounter the reduction of two angular momenta, known from quantum mechanics. We know

$$|j_1,m_1\rangle \otimes |j_2,m_2\rangle = \sum_{J=|j_1-j_2|}^{j_1+j_2} C^J |J, m_1+m_2\rangle \tag{1.102}$$

We even know the Clebsch-Gordan (C-G) coefficients entering this reduction, which have been tabulated. They can also be obtained using the package Mathematica. In the above example this reducible representation arises out of the Kronecker product $|1/2,m_1\rangle \otimes |1,m_2\rangle$. Clearly

$$|\frac{1}{2}\frac{1}{2}\rangle|1,1\rangle = |\frac{3}{2}\frac{3}{2}, \quad |\frac{1}{2}-\frac{1}{2}\rangle|1,-1\rangle = |\frac{3}{2}-\frac{3}{2}\rangle$$

For the cases $m_1+m_2=\pm 1/2$ using the coefficients used in the above example we get the results given in table 1.5. The general formula $J_1 \otimes 1/2$ is also simple, see table 1.4. Another useful case is the $j=1 \otimes j=1$ (see table 1.6).

The SU(3) case.

This is more complicated [Vergados (2016)]. We will give here some simple

Table 1.5: The coefficients for the reduction of the product $|1/2, m_1\rangle \otimes |1, m_2\rangle$ of $SO(3)$ for $m_1 + m_2 = \pm\frac{1}{2}$

j_1	m_1	j_2	m_2	$J=\frac{1}{2}$	$J=\frac{3}{2}$
$\frac{1}{2}$	$\frac{1}{2}$	1	0	$\frac{1}{\sqrt{3}}$	$\frac{\sqrt{2}}{\sqrt{3}}$
$\frac{1}{2}$	$-\frac{1}{2}$	1	1	$-\frac{\sqrt{2}}{\sqrt{3}}$	$\frac{1}{\sqrt{3}}$
$\frac{1}{2}$	$-\frac{1}{2}$	1	0	$-\frac{1}{\sqrt{3}}$	$\frac{\sqrt{2}}{\sqrt{3}}$
$\frac{1}{2}$	$\frac{1}{2}$	1	-1	$\frac{\sqrt{2}}{\sqrt{3}}$	$\frac{1}{\sqrt{3}}$

Table 1.6: C-G coefficients for the reduction $j_1 = 1 \otimes j_2 = 1 \Rightarrow J,\ J = 0, 1, 2$. The coefficients for $M = \pm 2$ are not listed, since they admit a unique solution and the C-G coefficient is unity.

			$M=0$				
m_1	m_2	J	$C-G$	J	$C-G$	J	$C-G$
-1	1	0	$\frac{1}{\sqrt{3}}$	1	$-\frac{1}{\sqrt{2}}$	2	$\frac{1}{\sqrt{6}}$
0	0	0	$-\frac{1}{\sqrt{3}}$	1	0	2	$\frac{\sqrt{2}}{\sqrt{6}}$
1	-1	0	$\frac{1}{\sqrt{3}}$	1	$\frac{1}{\sqrt{2}}$	2	$\frac{1}{\sqrt{6}}$

			$M=\pm1$		
m_1	m_2	J	$C-G$	J	$C-G$
±1	0	1	$-\frac{1}{\sqrt{2}}$	2	$\frac{1}{\sqrt{2}}$
0	±1	1	$\frac{1}{\sqrt{2}}$	2	$\frac{1}{\sqrt{2}}$

reductions staring with the fundamental representation $\underline{3}$.

$$\begin{aligned}
&\underline{3} \otimes \underline{3} = \underline{6} + \underline{\bar{3}} \\
&\underline{3} \otimes \underline{\bar{3}} = \underline{8} + \underline{1} \\
&\underline{6} \otimes \underline{3} = \underline{8} + \underline{10} \\
&\underline{\bar{3}} \otimes \underline{\bar{3}} = \underline{\bar{6}} + \underline{3} \\
&\underline{\bar{6}} \otimes \underline{\bar{3}} = \underline{8} + \underline{\bar{10}}
\end{aligned} \tag{1.103}$$

With a bar over the representation we indicate the complex conjugate representation The unitary groups admit complex representation. The representation $\underline{\bar{3}}$, the complex conjugate of $\underline{3}$, is antisymmetric (see problem 14). $\underline{6}$ is completely symmetric and its conjugate is the representation $\underline{\bar{6}}$. The representation $\underline{8}$ is the regular (adjoined) representation, which is real (self adjoined).

1.8 Homomorphism and isomorphism

Consider two algebras A and A' and a mapping P such that:

$$P : X \in A \to X' \in A' \text{or } X' = P(X)$$

We say that the mapping is a homomorphism if:

- for any set of numbers α and β: $P(\alpha X + \beta Y) = \alpha P(X) + \beta P(Y)$.
- $P([X, Y]) = [P(X), P(Y)]$, that is the corresponding structure constants are the same.

Then the algebra A' is homomorphic to the algebra A

The homomorphism is an isomporphism if it is 1 to 1 (1-1). Thus the algebra of the operators T_i (see Eq. (1.86)) and $s_i = (1/2)\sigma_i$ (see Eq. (1.65)) are isomorphic. On the other hand the algebra σ_i is not isomorphic to the previous since:

$$[\sigma_j, \sigma_\ell] = 2i\epsilon_{j\ell k}\sigma_k.$$

Thus a proper basis for the algebra must be chosen to establish isomorphism.

We remark that two algebras maybe isomorphic, but the associated groups need not be. Such, e.g., is the case for the groups $SO(3)$ and $SU(2)$.

1.9 Some further examples

Example 14: We will find the adjoined(3×3) representation associated with the Pauli matrices: Recall the adjoined representation $Adj(J_i)$ for a set of operators J_i, $i = 1, \cdots n$ with algebra

$$[J_i, J_j] = C^k_{ij} J_k$$

with structure constants C^k_{ij} is defined by:

$$Adj(J_i)|j\rangle = C^k_{ij}|k\rangle,$$

where $|\alpha\rangle$ is a basis in the n-dimensional space.

We choose the following basis for the Pauli matrices:

$$\tau_+ = \begin{pmatrix} 0 & 1 \\ 0 & 0 \end{pmatrix}, \qquad \tau_- = \begin{pmatrix} 0 & 0 \\ 1 & 0 \end{pmatrix}, \quad \tau_0 = \begin{pmatrix} 1 & 0 \\ 0 & -1 \end{pmatrix}.$$

Then

$$[\tau_+, \tau_-] = \begin{pmatrix} 0 & 1 \\ 0 & 0 \end{pmatrix}\begin{pmatrix} 0 & 0 \\ 1 & 0 \end{pmatrix} - \begin{pmatrix} 0 & 0 \\ 1 & 0 \end{pmatrix}\begin{pmatrix} 0 & 1 \\ 0 & 0 \end{pmatrix} = \begin{pmatrix} 1 & 0 \\ 0 & 0 \end{pmatrix} - \begin{pmatrix} 0 & 0 \\ 0 & 1 \end{pmatrix} = \tau_3,$$

$$[\tau_0, \tau_+] = \begin{pmatrix} 1 & 0 \\ 0 & -1 \end{pmatrix}\begin{pmatrix} 0 & 1 \\ 0 & 0 \end{pmatrix} - \begin{pmatrix} 0 & 1 \\ 0 & 0 \end{pmatrix}\begin{pmatrix} 1 & 0 \\ 0 & -1 \end{pmatrix} = 2\tau_+,$$

$$[\tau_0, \tau_-] = \begin{pmatrix} 1 & 0 \\ 0 & -1 \end{pmatrix}\begin{pmatrix} 0 & 0 \\ 1 & 0 \end{pmatrix} - \begin{pmatrix} 0 & 0 \\ 1 & 0 \end{pmatrix}\begin{pmatrix} 1 & 0 \\ 0 & -1 \end{pmatrix} = -2\tau_-.$$

The structure constants are given by the matrix:

$$\begin{pmatrix} & \tau_+ & \tau_- & \tau_0 \\ \tau_+ & 0 & 2 & -1 \\ \tau_- & -2 & 0 & 1 \\ \tau_0 & 1 & -1 & 0 \end{pmatrix}.$$

For simplicity of notation we write $Adj(\tau_\alpha) = I_\alpha$. Then from the table of the structure constants we find:

$$I_+|1\rangle = 0, \quad I_+ = C^k_{+,-}|k\rangle = 2|3\rangle, \quad I_+|3\rangle = C^k_{+,0}|k\rangle = -|1\rangle,$$

$$I_-|1\rangle = C^k_{-,+}|0\rangle = -2|3\rangle, \quad I_-|2\rangle = 0, \quad I_-|3\rangle = C^k_{-,0}|k\rangle = |1\rangle,$$

$$I_0|1\rangle = C^k_{0,+}|k\rangle = |1\rangle, \quad I_0|2\rangle = C^k_{0,-}|k\rangle = -|2\rangle, \quad I_0|3\rangle = 0.$$

From these we construct (column-wise) the matrices:

$$I_+ = \begin{pmatrix} 0 & 0 & -2 \\ 0 & 0 & 0 \\ 0 & 1 & 0 \end{pmatrix}, \qquad I_- = \begin{pmatrix} 0 & 0 & 0 \\ 0 & 0 & 2 \\ -1 & 0 & 0 \end{pmatrix}, \qquad I_0 = \begin{pmatrix} 2 & 0 & 0 \\ 0 & -2 & 0 \\ 0 & 0 & 0 \end{pmatrix}.$$

One can now verify that these 3×3 matrices obey the same commutation relations as the original Pauli matrices as given in the above table. It is a 1-1 mapping. This is the adjoined representation. We see that what really enters in its construction is the structure constants. In other words this representation can always be constructed once an acceptable set of structure constants is given.

Another Cartesian basis is:

$$J_1 = \begin{pmatrix} 0 & -1 & 0 \\ -1 & 0 & 1 \\ 0 & 1 & 0 \end{pmatrix}, \quad J_2 = \begin{pmatrix} 0 & i & 0 \\ -i & 0 & -i \\ 0 & i & 0 \end{pmatrix}, \quad J_3 = \begin{pmatrix} 1 & 0 & 0 \\ 0 & 0 & 0 \\ 0 & 0 & -1 \end{pmatrix}$$

or normalized differently:

$$\lambda_1 = \begin{pmatrix} 0 & -\frac{1}{\sqrt{2}} & 0 \\ -\frac{1}{\sqrt{2}} & 0 & \frac{1}{\sqrt{2}} \\ 0 & \frac{1}{\sqrt{2}} & 0 \end{pmatrix}, \quad \lambda_2 = \begin{pmatrix} 0 & \frac{i}{\sqrt{2}} & 0 \\ -\frac{i}{\sqrt{2}} & 0 & -\frac{i}{\sqrt{2}} \\ 0 & \frac{i}{\sqrt{2}} & 0 \end{pmatrix}, \quad \lambda_3 = \begin{pmatrix} 1 & 0 & 0 \\ 0 & 0 & 0 \\ 0 & 0 & -1 \end{pmatrix},$$

$$\begin{aligned} \mathrm{tr}(\lambda_i \lambda_j) &= 2\delta_{ij}, \\ [\lambda_i, \lambda_j] &= i\epsilon_{ijk}\lambda_k \end{aligned}$$

or another different normalization

$$I_1 = \frac{1}{2}\begin{pmatrix} 0 & -1 & 0 \\ -1 & 0 & 1 \\ 0 & 1 & 0 \end{pmatrix}, \quad I_2 = \frac{1}{2}\begin{pmatrix} 0 & i & 0 \\ -i & 0 & -i \\ 0 & i & 0 \end{pmatrix}, \quad I_3 = \frac{1}{\sqrt{2}}\lambda_3,$$

$$\boxed{\mathrm{tr}\,(I_i I_j) = \delta_{ij},}$$

$$[2I_i, 2I_j] = i\epsilon_{ijk}2I_k \Rightarrow \boxed{[I_i, I_j] = \tfrac{i}{2}\epsilon_{ijk}I_k.}$$

Example 15:

Consider the isotriplet of Scalar fields ϕ^+, ϕ^0, ϕ^-, i.e. scalars transforming like $I = 1$ under the $SU(2)$

$$\begin{aligned} T_+ |1, m\rangle &= \sqrt{2}\,|1, m+1\rangle\,, \\ T_- |1, m\rangle &= \sqrt{2}\,|1, m-1\rangle\,, \\ T_0 |1, m\rangle &= m\,|1, m\rangle\,, \end{aligned}$$

$$\begin{aligned} T_+ |1, 1\rangle &= 0, \\ T_+ |1, 0\rangle &= \sqrt{2}\,|1, 1\rangle\,, \\ T_- |1, 1\rangle &= \sqrt{2}\,|1, 0\rangle \end{aligned}$$

$$T_+ = \begin{pmatrix} 0 & \sqrt{2} & 0 \\ 0 & 0 & \sqrt{2} \\ 0 & 0 & 0 \end{pmatrix}, \quad T_- = \begin{pmatrix} 0 & 0 & 0 \\ \sqrt{2} & 0 & 0 \\ 0 & \sqrt{2} & 0 \end{pmatrix}, \quad T_0 = \begin{pmatrix} 1 & 0 & 0 \\ 0 & 0 & 0 \\ 0 & 0 & -1 \end{pmatrix},$$

$$\begin{aligned} T_+ &= T_1 + iT_2, \\ T_- &= T_1 - iT_2, \end{aligned}$$

$$2T_1 = \begin{pmatrix} 0 & \sqrt{2} & 0 \\ 0 & 0 & \sqrt{2} \\ 0 & 0 & 0 \end{pmatrix} + \begin{pmatrix} 0 & 0 & 0 \\ \sqrt{2} & 0 & 0 \\ 0 & \sqrt{2} & 0 \end{pmatrix} = \begin{pmatrix} 0 & \sqrt{2} & 0 \\ \sqrt{2} & 0 & \sqrt{2} \\ 0 & \sqrt{2} & 0 \end{pmatrix},$$

$$\lambda_1 = T_1 = \begin{pmatrix} 0 & 1/\sqrt{2} & 0 \\ 1/\sqrt{2} & 0 & 1/\sqrt{2} \\ 0 & 1/\sqrt{2} & 0 \end{pmatrix} = \frac{1}{\sqrt{2}} \begin{pmatrix} 0 & 1 & 0 \\ 1 & 0 & 1 \\ 0 & 1 & 0 \end{pmatrix},$$

$$\lambda_2 = T_2 = -\frac{i}{2}(T_+ - T_-)$$

$$= -\frac{i}{2}\left\{ \begin{pmatrix} 0 & \sqrt{2} & 0 \\ 0 & 0 & \sqrt{2} \\ 0 & 0 & 0 \end{pmatrix} - \begin{pmatrix} 0 & 0 & 0 \\ \sqrt{2} & 0 & 0 \\ 0 & \sqrt{2} & 0 \end{pmatrix} \right\} = \frac{1}{\sqrt{2}} \begin{pmatrix} 0 & -i & 0 \\ i & 0 & -i \\ 0 & i & 0 \end{pmatrix}$$

$$T_3 = \frac{1}{\sqrt{2}} \begin{pmatrix} 1 & 0 & 0 \\ 0 & 0 & 0 \\ 0 & 0 & -1 \end{pmatrix}.$$

In summary:

- The $I = 1/2$ representation of the $SU(2)$ is described by the Pauli matrices

$$\tau_1 = \begin{pmatrix} 0 & 1 \\ 1 & 0 \end{pmatrix} \quad \tau_2 = \begin{pmatrix} 0 & -i \\ i & 0 \end{pmatrix} \quad \tau_3 = \begin{pmatrix} 1 & 0 \\ 0 & -1 \end{pmatrix}.$$

- the $I = 1$ representation of the $SU(2)$ group is described by the

$$\lambda_1 = \frac{1}{\sqrt{2}} \begin{pmatrix} 0 & 1 & 0 \\ 1 & 0 & 1 \\ 0 & 1 & 0 \end{pmatrix}, \quad \lambda_2 = \frac{1}{\sqrt{2}} \begin{pmatrix} 0 & -i & 0 \\ i & 0 & -i \\ 0 & i & 0 \end{pmatrix} \quad \lambda_3 = \begin{pmatrix} 1 & 0 & 0 \\ 0 & 0 & 0 \\ 0 & 0 & -1 \end{pmatrix}.$$

$$[\lambda_k, \lambda_l] = i\epsilon_{klm}\lambda_m,$$
$$\mathrm{tr}\,(\lambda_k\lambda_l) = 2\delta_{kl}.$$

1.10 The proper orthogonal groups O(3) and O(4)

Before generalizing the results we have obtained in three dimensions, we note that in three dimensions $n = 3$ the number of planes $(1/2)n(n-1)$ coincides with the number of axes n. This of course is not true in higher dimensions. We have seen that the number of parameters in the case of a real orthogonal group is $(1/2)n(n-1)$, i.e. it coincides with the number of planes. Thus we can talk about a rotation in a given plane. A rotation in the (1,2) plane also leaves the fourth axis unchanged. Thus in going from $n = 3$ to $n = 4$ we have:

$$R_3 = \begin{bmatrix} \cos\theta_3 & \sin\theta_3 & 0 \\ -\sin\theta_3 & \cos\theta_3 & 0 \\ 0 & 0 & 1 \end{bmatrix} \to R_{1,2} = \begin{bmatrix} \cos\theta_{1,2} & \sin\theta_{1,2} & 0 & 0 \\ -\sin\theta_{1,2} & \cos\theta_{1,2} & 0 & 0 \\ 0 & 0 & 1 & 0 \\ 0 & 0 & 0 & 1 \end{bmatrix}, \tag{1.104}$$

$$R_2 = \begin{bmatrix} \cos\theta_2 & 0 & \sin\theta_2 \\ 0 & 1 & 0 \\ -\sin\theta_2 & 0 & \cos\theta_2 \end{bmatrix} \to R_{1,3} = \begin{bmatrix} \cos\theta_{13} & 0 & \sin\theta_{1,3} & 0 \\ 0 & 1 & 0 & 0 \\ -\sin\theta_{1,3} & 0 & \cos\theta_{1,3} & 0 \\ 0 & 0 & 0 & 1 \end{bmatrix}, \tag{1.105}$$

$$R_1 = \begin{bmatrix} 1 & 0 & 0 \\ 0 & \cos\theta_1 & \sin\theta_1 \\ 0 & -\sin\theta_1 & \cos\theta_1 \end{bmatrix} \to R_{2,3} = \begin{bmatrix} 1 & 0 & 0 & 0 \\ 0 & \cos\theta_{2,3} & \sin\theta_{2,3} & 0 \\ 0 & -\sin\theta_{2,3} & \cos\theta_{2,3} & 0 \\ 0 & 0 & 0 & 1 \end{bmatrix}. \tag{1.106}$$

One has, of course, three additional rotations involving the 4th component:

$$R_{1,4} = \begin{bmatrix} \cos\theta_{1,4} & 0 & 0 & \sin\theta_{1,4} \\ 0 & 1 & 0 & 0 \\ 0 & 0 & 1 & 0 \\ -\sin\theta_{1,4} & 0 & 0 & \cos\theta_{1,4} \end{bmatrix}, R_{2,4} = \begin{bmatrix} 1 & 0 & 0 & 0 \\ 0 & \cos\theta_{2,4} & 0 & \sin\theta_{2,4} \\ 0 & 0 & 1 & 0 \\ 0 & -\sin\theta_{2,4} & 0 & \cos\theta_{2,4} \end{bmatrix}, \tag{1.107}$$

$$R_{3,4} = \begin{bmatrix} 1 & 0 & 0 & 0 \\ 0 & 1 & 0 & 0 \\ 0 & 0 & \cos\theta_{3,4} & \sin\theta_{3,4} \\ 0 & 0 & -\sin\theta_{3,4} & \cos\theta_{3,4} \end{bmatrix}. \tag{1.108}$$

Using Eq. (1.61) we obtain the generators:

$$J_{1,2} = \begin{bmatrix} 0 & -i & 0 & 0 \\ i & 0 & 0 & 0 \\ 0 & 0 & 0 & 0 \\ 0 & 0 & 0 & 0 \end{bmatrix}, J_{1,3} = \begin{bmatrix} 0 & 0 & -i & 0 \\ 0 & 0 & 0 & 0 \\ i & 0 & 0 & 0 \\ 0 & 0 & 0 & 0 \end{bmatrix}, J_{1,4} = \begin{bmatrix} 0 & 0 & 0 & -i \\ 0 & 0 & 0 & 0 \\ 0 & 0 & 0 & 0 \\ i & 0 & 0 & 0 \end{bmatrix},$$

$$J_{2,3} = \begin{bmatrix} 0 & 0 & 0 & 0 \\ 0 & 0 & -i & 0 \\ 0 & i & 0 & 0 \\ 0 & 0 & 0 & 0 \end{bmatrix}, J_{2,4} = \begin{bmatrix} 0 & 0 & 0 & 0 \\ 0 & 0 & 0 & -i \\ 0 & 0 & 0 & 0 \\ 0 & i & 0 & 0 \end{bmatrix}, J_{3,4} = \begin{bmatrix} 0 & 0 & 0 & 0 \\ 0 & 0 & 0 & 0 \\ 0 & 0 & 0 & -i \\ 0 & 0 & i & 0 \end{bmatrix}.$$

These generators obey the commutators

$$[J_{ij}, J_{k\ell}] = i\left(\delta_{ik}J_{j\ell} + \delta_{j\ell}J_{ik} - \delta_{i\ell}J_{jk} - \delta_{jk}J_{i\ell}\right). \tag{1.109}$$

Sometimes in the literature one finds the notation:

$$M_1 = J_{2,3},\ M_2 = J_{3,1},\ M_3 = J_{1,2},\ N_1 = J_{1,4},\ N_2 = J_{2,4},\ N_3 = J_{3,4}.$$

Then one can show:

$$[M_j, M_\ell] = i\epsilon_{j\ell k}M_k, \quad [N_j, N_\ell] = i\epsilon_{j\ell k}M_k, \quad [M_j, N_\ell] = i\epsilon_{j\ell k}N_k. \tag{1.110}$$

The group $O(4)$ is semisimple, since it does not have continuous Abelian invariant subgroups. It is not simple, however, since it has non Abelian invariant subgroups. Indeed Define:

$$A_i = \frac{M_i + N_i}{2}, \quad B_i = \frac{M_i - N_i}{2}.$$

Then

$$[A_j, A_\ell] = i\epsilon_{j\ell k}A_k, \quad [B_j, B_\ell] = i\epsilon_{j\ell k}B_k, \quad [A_j, B_\ell] = 0. \tag{1.111}$$

The two sets of generators are disjoint. Thus we write:

$$O(4) = O(3) \otimes O(3).$$

Its properties are essentially those of $O(3)$.

The reader must show that for O(4) the following matrices constitute a basis:

$$A_1 = \begin{pmatrix} 0 & 0 & \frac{1}{2} & 0 \\ 0 & 0 & 0 & \frac{1}{2} \\ \frac{1}{2} & 0 & 0 & 0 \\ 0 & \frac{1}{2} & 0 & 0 \end{pmatrix}, \quad A_2 = \begin{pmatrix} 0 & 0 & -\frac{i}{2} & 0 \\ 0 & 0 & 0 & -\frac{i}{2} \\ \frac{i}{2} & 0 & 0 & 0 \\ 0 & \frac{i}{2} & 0 & 0 \end{pmatrix}, \quad A_3 = \begin{pmatrix} \frac{1}{2} & 0 & 0 & 0 \\ 0 & \frac{1}{2} & 0 & 0 \\ 0 & 0 & -\frac{1}{2} & 0 \\ 0 & 0 & 0 & -\frac{1}{2} \end{pmatrix}, \tag{1.112}$$

$$B_1 = \begin{pmatrix} 0 & \frac{1}{2} & 0 & 0 \\ \frac{1}{2} & 0 & 0 & 0 \\ 0 & 0 & 0 & \frac{1}{2} \\ 0 & 0 & \frac{1}{2} & 0 \end{pmatrix}, \quad B_2 = \begin{pmatrix} 0 & -\frac{i}{2} & 0 & 0 \\ \frac{i}{2} & 0 & 0 & 0 \\ 0 & 0 & 0 & -\frac{i}{2} \\ 0 & 0 & \frac{i}{2} & 0 \end{pmatrix}, \quad B_3 = \begin{pmatrix} \frac{1}{2} & 0 & 0 & 0 \\ 0 & -\frac{1}{2} & 0 & 0 \\ 0 & 0 & \frac{1}{2} & 0 \\ 0 & 0 & 0 & -\frac{1}{2} \end{pmatrix}. \tag{1.113}$$

Then verify that

$$[A_k, A_l] = i\epsilon_{klm}A_m, \quad [B_k, B_l] = i\epsilon_{klm}B_m, \quad [A_k, B_l] = 0. \tag{1.114}$$

Hint: consider the Kronecker product $\mathcal{A} \otimes \mathcal{B}$ when

i) $\mathcal{A} = (1/2)(\sigma_1, \sigma_2, \sigma_3)$, $\mathcal{B} = I$,
i) $\mathcal{A} = I$, $\mathcal{B} = (1/2)(\sigma_1, \sigma_2, \sigma_3)$.

The O(4) symmetry appears both in classical and quantum physics, when the system is described by a potential $\propto 1/r$. It is due to this symmetry that in the Kepler's problem the line connected the two focuses of the ellipse is not moving (the orbit is plane due to the conservation of angular momentum, valid for any central potential. The shape of the orbit is a consequence of both angular momentum and energy conservation). A small perturbation destroys this symmetry (the perihelion moves! Similarly in hydrogenic atoms we have a high degree of degeneracy, the energy depends n, but it does not depend on the orbital angular momentum momentum quantum number ℓ, as is the case for any other central potential. A small perturbation destroys this symmetry, the extra degeneracy is removed.

1.11 Realization of $SO(4) = SO(3) \times SO(3)$ and some applications

This section may be omitted at a first reading, since it does not affect the Standard Model physics. The mathematically oriented reader may, instead, attempt to do problem 16 at the end of this chapter.

We consider the standard angular momentum operator $\mathbf{L} = \mathbf{r} \times \mathbf{p}$ and the *Rurïge*-Lenz vector is defined by

$$\mathbf{R} = \frac{1}{2\mu}\left(\mathbf{p} \times \mathbf{L} - \mathbf{L} \times \mathbf{p}\right) - k\hat{r},\ F = -\frac{k}{r^2}\hat{r}, \tag{1.115}$$

where μ is the reduced mass of the particle

$$\frac{1}{\mu} = \frac{1}{m} + \frac{1}{M}$$

The system is specified by the Hamiltonian

$$H = \frac{\mathbf{p}^2}{2\mu} - \frac{k}{r} \tag{1.116}$$

Then it is easy to verify that

$$[\mathbf{L}, H] = 0,\ [\mathbf{R}, H] = 0 \tag{1.117}$$

and

$$[L_j, L_k] = i\hbar\epsilon_{jk\ell}L_{jk\ell}. \tag{1.118}$$

One can also show that

$$[L_j, R_k] = i\hbar\epsilon_{jk\ell}R_{jk\ell},\ [R_j, R_k] = -2i\frac{\hbar}{\mu}\epsilon_{jk\ell}HR_{jk\ell} \tag{1.119}$$

Define now

$$\tilde{R}_j = Rj \begin{cases} \sqrt{-\frac{\mu}{2H}} \; E < 0 \\ \\ \sqrt{\frac{\mu}{2H}} \;\; E > 0 \end{cases} \tag{1.120}$$

Then

$$[\tilde{R}_j, \tilde{R}_k] = \begin{cases} i\hbar\epsilon_{jk\ell}\tilde{R}_\ell \;\; E < 0 \\ -i\hbar\epsilon_{jk\ell}\tilde{R}_\ell \; E > 0 \end{cases} \tag{1.121}$$

Defining now

$$M_i = \frac{L_i}{\hbar}, \; N_i = \frac{\tilde{R}_i}{\hbar}, \tag{1.122}$$

we recover Eqs 1.110 for $E < 0$ and proceed as above. For $E > 0$ we get $[N_j, N_k] = -i\epsilon_{jk\ell}N_\ell$, but we will not concern ourselves in this case.

1.11.1 *Application to classical mechanics*

The vector **R**, sometimes indicated by **A**, is perpendicular to **L**,

$$\mathbf{R}.\mathbf{L} = 0. \tag{1.123}$$

The energy is given by

$$E = \frac{1}{2\mu}p^2 - \frac{k}{r}$$

The magnitude of **R** is given by

$$R = k\sqrt{1 + \frac{L^2}{2\mu k^2}E}$$

and the angular momentum is $\mathbf{L} = \mathbf{r} \times \mathbf{p}$. The proof of conservation of **R** is quite simple.

$$\frac{d\mathbf{R}}{dt} = \frac{d\mathbf{p}}{dt} \times \mathbf{L} - \mu k \left(\frac{1}{r}\frac{d\mathbf{r}}{dt} - \frac{\mathbf{r}}{r^2}\frac{dr}{dt} \right)$$

but

$$\frac{d\mathbf{p}}{dt} = -k\frac{\mathbf{r}}{r^3} \Rightarrow$$

$$\frac{d\mathbf{R}}{dt} = -\frac{k}{\mu}\left(\frac{\mathbf{r} \times \mathbf{L} - (\mathbf{r}.\mathbf{p})\mathbf{r}}{r^3} + \frac{\mathbf{p}}{r} \right)$$

But

$$\mathbf{r} \times \mathbf{L} = \mathbf{r} \times (\mathbf{r} \times \mathbf{p}) = (\mathbf{r}.\mathbf{p})\mathbf{r} - r^2\mathbf{p}$$

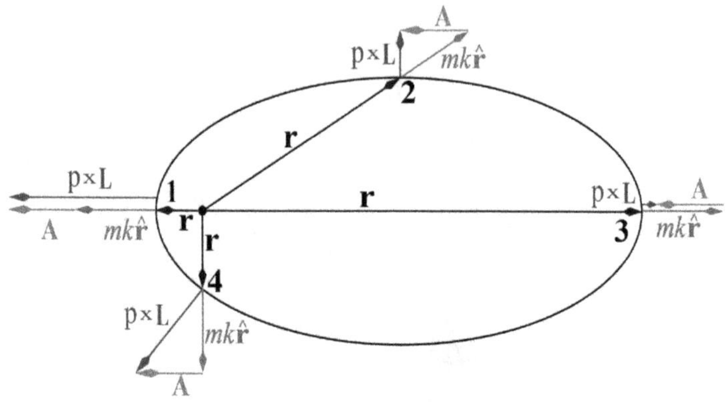

Fig. 1.1: The role of the Rünge-Lenz vector, here indicated by **A** in Keplerian motion.

Thus

$$\frac{d\mathbf{R}}{dt} = 0 \tag{1.124}$$

Its relevance to the Kepler problem is given in Fig. 1.1. Since the vectors **L** and **R** are conserved and perpendicular we can define a natural coordinate system:

$$\hat{e}_3 = \frac{\mathbf{L}}{L}, \; \hat{e}_1 = \frac{\mathbf{R}}{R}, \; \hat{e}_2 = \hat{e}_3 \times e_1 \tag{1.125}$$

The orbit is perpendicular to **L**. So

$$\mathbf{r} = r(\hat{e}_1 \cos\theta + \hat{e}_2 \sin\theta) \tag{1.126}$$

$$\mathbf{p} = \mu\frac{d\mathbf{r}}{dt} = \mu\left(\hat{e}_1\left(\cos\theta\frac{dr}{dt} - r\sin\theta\frac{d\theta}{dt}\right) + \hat{e}_2\left(\sin\theta\frac{dr}{dt} + r\cos\theta\frac{d\theta}{dt}\right)\right)$$

but

$$\mathbf{L} = \mu r^2\frac{d\theta}{dt}\hat{e}_3 \Rightarrow \frac{d\theta}{dt} = \frac{mur^2}{L}.$$

Thus

$$\mathbf{R} = \hat{e}_1\left(L\frac{dr}{dt}\sin\theta + \left(\frac{L^2}{\mu r} - k\right)\cos\theta\right) - \hat{e}_2\left(L\frac{dr}{dt}\cos\theta - \left(\frac{L^2}{\mu r} - k\right)\sin\theta\right)$$

But by definition $\mathbf{R} = r\hat{e}_1$, which implies

$$L\frac{dr}{dt}\cos\theta = \left(\frac{L^2}{\mu r} - k\right)\sin\theta \Rightarrow \frac{dr}{dt} = \frac{1}{L}\left(\frac{L^2}{\mu r} - k\right)\tan\theta$$

$$\mathbf{R} = \hat{e}_1 \left(\frac{L}{\mu} r - k \right) \sin\theta$$

Thus

$$R = k\sqrt{1 + \frac{L^2}{2\mu k^2} E} = \left(\frac{L}{\mu r} - k \right) \frac{1}{\cos\theta}$$

or

$$r = \frac{\ell}{1 + e\cos\theta} \text{ equation of the orbit} \tag{1.127}$$

with

$$\ell = \frac{L^2}{\mu k}, \; e = \sqrt{1} + \frac{2L^2}{\mu} \frac{E}{k^2}$$

The eccentricity is defined by

$$e^2 = 1 + \frac{2L^2}{\mu k^2} E$$

$$a = \frac{L^2}{\mu k} \frac{1}{1 - e^2} = -\frac{k}{2E} b = a\sqrt{1 - e^2}$$

1.11.2 *Application to quantum mechanics*

Since the operators A and B commute we can simultaneously diagonalize them and get a wave function:

$$\psi = |a, m_a; b, m_b\rangle \tag{1.128}$$

in the notation of two angular momenta, such that

$$A_0\psi = m_a\psi, \; A^2\psi = a(a+1)\psi, \; B_0\psi = m_b\psi, \; B^2\psi = b(b+1)\psi. \tag{1.129}$$

$$\tag{1.130}$$

Then in the special case of a hydrogenic atom we find:

$$C^{(2)} = \sqrt{\frac{\mu}{-2H}} \mathbf{L}.\mathbf{R} = 0$$

This implies the condition that $a(a+1) - b(b+1) = 0$, which leads to $a = b$ (the solution $a = -(b+1)$ is not acceptable since both a and b should be positive). Thus

$$C^{(1)}\psi = 2a(a+1)\psi. \tag{1.131}$$

But

$$C^{(1)} = \frac{1}{2\hbar^2}\left(\mathbf{L}^2 - \frac{\mu}{2E}\mathbf{R}^2\right)$$

while after some effort one can show that:

$$\mathbf{R}^2 = \frac{2H}{\mu}\left(\mathbf{L}^2 + \hbar^2 + k^2\right)$$

that is

$$C^{(1)} = -\frac{1}{2\hbar^2}\left(\hbar^2 + \frac{k^3\mu}{2E}\right)$$

$$-\frac{1}{2\hbar^2}\left(\hbar^2 + \frac{k^3\mu}{2E}\right)\psi = 2a(a+1)\psi.$$

Finally

$$-\frac{\mu k^2}{4\hbar^2 E} = \frac{1}{2}\left(2a+1\right)^2,$$

$$E_n = -\frac{1}{2}mc^2(Ze^2)^2\frac{1}{n^2}\frac{M}{m+M},\ (n = (2a+1)),\ a = 0, 1/2, 1, 3/2\cdots \tag{1.132}$$

The above wave functions $|a, m_a, b, m_b\rangle$ do not have a good angular momentum. Since, however,

$$\mathbf{L} = \hbar(\mathbf{A} + \mathbf{B}),$$

we know that the allowed ℓ values are constrained to be

$$|a - b| \leq \ell \leq a + b.$$

Thus the wave functions of good angular momentum are going to be of the form:

$$|n\ell m\rangle = \sum_{m_a, m_b} \langle m_a, m_b|\ell, m\rangle \delta_{m_a+m_b-m}|a, m_a, b, m_b\rangle \tag{1.133}$$

where $\langle m_a, m_b|\ell, m\rangle$ are the Clebsch-Gordan coefficients familiar from the angular momentum theory, see e.g. [Vergados (2016)]. It is adequate to construct the maximum weight state. The radial part is very hard to construct. To this end ladder operators constructed in terms of the higher symmetry of the hydrogenic atoms, which is $SO(4, 2$. see e.g. [Wybourne (1974); Baru and Bronzin (1971)].

1.12 The proper Lorentz transformations; the group $SO(3,1)$

At this point we should briefly discuss the algebra generated by the group of the proper Lorentz transformations $L(1,3)$. It is customary to change the index $4 \rightarrow 0$. Then the elements $R_{i,j}$ and $J_{i,j}$, $i,j = 1,2,3$ remain the same, describing rotations in ordinary space. Those that contain the time component are modified:

$$R_{i,4} \text{ (rotations)} \rightarrow \Lambda_{0,i} \text{ (boosts)},$$

$$\Lambda_{0,1} = \begin{bmatrix} \cosh\xi_1 & \sinh\xi_1 & 0 & 0 \\ \sinh\xi_1 & \cosh\xi_1 & 0 & 0 \\ 0 & 0 & 1 & 0 \\ 0 & 0 & 0 & 1 \end{bmatrix}, \Lambda_{0,2} = \begin{bmatrix} \cosh\xi_2 & 0 & \sinh\xi_2 & 0 \\ 0 & 1 & 0 & 0 \\ \sinh\xi_2 & 0 & \cosh\xi_2 & 0 \\ 0 & 0 & 0 & 1 \end{bmatrix}, \tag{1.134}$$

$$\Lambda_{0,3} = \begin{bmatrix} \cosh\xi_3 & 0 & 0 & \sinh\xi_3 \\ 0 & 1 & 0 & 0 \\ 0 & 0 & 1 & 0 \\ \sinh\xi_3 & 0 & 0 & \cosh\xi_3 \end{bmatrix}, \tag{1.135}$$

where $\tanh\xi_i = -\upsilon_i/c$, υ the relative velocity of the two frames.

The infinitesimal generators are

$$L_{1,2} = \begin{bmatrix} 0 & 0 & 0 & 0 \\ 0 & 0 & -i & 0 \\ 0 & i & 0 & 0 \\ 0 & 0 & 0 & 0 \end{bmatrix}, L_{1,3} = \begin{bmatrix} 0 & 0 & 0 & 0 \\ 0 & 0 & 0 & -i \\ 0 & 0 & 0 & 0 \\ 0 & i & 0 & 0 \end{bmatrix}, L_{2,3} = \begin{bmatrix} 0 & 0 & 0 & 0 \\ 0 & 0 & 0 & 0 \\ 0 & 0 & 0 & -i \\ 0 & 0 & i & 0 \end{bmatrix},$$

$$L_{0,1} = \begin{bmatrix} 0 & -i & 0 & 0 \\ -i & 0 & 0 & 0 \\ 0 & 0 & 0 & 0 \\ 0 & 0 & 0 & 0 \end{bmatrix}, L_{0,2} = \begin{bmatrix} 0 & 0 & -i & 0 \\ 0 & 0 & 0 & 0 \\ -i & 0 & 0 & 0 \\ 0 & 0 & 0 & 0 \end{bmatrix}, L_{0,3} = \begin{bmatrix} 0 & 0 & 0 & -i \\ 0 & 0 & 0 & 0 \\ 0 & 0 & 0 & 0 \\ -i & 0 & 0 & 0 \end{bmatrix}.$$

Then we find the commutators:

$$[L_{0,j}, L_{0,k}] = -iL_{j,k}, \quad [L_{0,j}, L_{k,n}] = i\left(\delta_{jn}L_{0,k} - \delta_{jk}L_{0,n}\right), \tag{1.136}$$

$$[L_{k,j}, L_{m,n}] = i\left(\delta_{km}L_{j,n} + \delta_{nj}L_{k,m} - \delta_{kn}L_{j,m} - \delta_{jm}L_{k,n}\right). \tag{1.137}$$

Note that if we define:$J_{0,k} = -iL_{0,k}$, $J_{m,n} = L_{m,n}$ we obtain the same commutators as in the case of $O(4)$, namely:

$$[J_{ij}, J_{k\ell}] = i\left(\delta_{ik}J_{j\ell} + \delta_{j\ell}J_{ik} - \delta_{i\ell}J_{jk} - \delta_{jk}J_{i\ell}\right), \tag{1.138}$$

We say that the two groups are locally isomorphic. They are not, of course, isomorphic in the large. The elements corresponding to boosts are not bounded.

At this point we should mention that it is common in physics to define the generators in terms of infinitesimal operators containing derivatives. Thus

- for SO(n):

$$J_{j,k} = \frac{\hbar}{i}\left(x_j \frac{\partial}{\partial x_k} - x_k \frac{\partial}{\partial x_j}\right), \quad j,k = 1,2,\cdots n. \tag{1.139}$$

  They obey the rule:

$$[J_{ij}, J_{k\ell}] = i\hbar\left(\delta_{ik}J_{j\ell} + \delta_{j\ell}J_{ik} - \delta_{i\ell}J_{jk} - \delta_{jk}J_{i\ell}\right). \tag{1.140}$$

- For the Lorentz group L(1,3):
  construct $L_{\mu,\nu}$, $\mu,\nu = 0,1,2,3$, as follows:

$$L_{j,k} = \frac{\hbar}{i}\left(x_j \frac{\partial}{\partial x_k} - x_k \frac{\partial}{\partial x_j}\right), \quad L_{0,j} = \frac{\hbar}{i}\left(x_0 \frac{\partial}{\partial x_j} + x_j \frac{\partial}{\partial x_0}\right),$$
$$x_0 = ct,\ j,k = 1,2,3. \tag{1.141}$$

  These obey the commutator rules

$$[L_{0,j}, L_{0,k}] = -i\hbar L_{j,k}, \quad [L_{0,j}, L_{k,n}] = i\hbar\left(\delta_{jn}L_{0,k} - \delta_{jk}L_{0,n}\right),$$
$$[L_{k,j}, L_{m,n}] = i\hbar\left(\delta_{km}L_{j,n} + \delta_{nj}L_{k,m} - \delta_{kn}L_{j,m} - \delta_{jm}L_{k,n}\right). \tag{1.142}$$

1.13 Symmetries and conservation laws; Noether's theorem

A symmetry implies a conservation law and conversely if a conservation law is found experimentally, we are not fully satisfied till we find the symmetry behind it.

We will examine here the following cases:

1.13.1 *Global transformations in classical physics*

We have seen in section 1.5.1 the connection between space translations and momentum as well as between rotations in space and angular momentum. One can derive a similar formal relation between energy and translations in time. Hence the connection between invariance under such transformations and conservation laws:

$$\begin{array}{c} t \to t + \alpha \\ \mathbf{r} \to \mathbf{r} + \mathbf{b} \end{array} \Rightarrow \begin{array}{c} \frac{dE}{dt} = 0 \\ \frac{d\mathbf{p}}{dt} = 0, \end{array}$$

$$\mathbf{r}' = R\mathbf{r} \Leftrightarrow r_i' = \sum R_{ij} r_j \Leftrightarrow \frac{d\mathbf{L}}{dt} = 0,$$

$$\mathbf{L} = \mathbf{r} \times \mathbf{p}.$$

1.13.2 *Global transformations in quantum mechanics (in the Heisenberg picture)*

$$\frac{dO}{dt} = i\,[H, O] \qquad \text{any operator } O,$$

$$\frac{dO}{dt} = 0 \Leftrightarrow [H, O] = 0,$$

$$\frac{d\mathbf{J}}{dt} = i\,[H, \mathbf{J}] \text{ (e.g. angular momentum)},$$

$$\frac{d\mathbf{J}}{dt} = 0 \Leftrightarrow [H, \mathbf{J}] = 0.$$

We have seen that:

$$[J_k, J_l] = i\epsilon_{klm} J_m \qquad k, l = 1, 2, 3.$$

Conservation of angular momentum implies energy degeneracy:

$$\phi_{jm} \Leftrightarrow \epsilon_j \qquad \text{independent of } m \Leftrightarrow 2j+1 \qquad \text{degenerate states}$$

1.13.3 *Field theories; scalar fields*

In field theories the invariance of the Lagrangian under a symmetry transformation implies the existence of a conserved current J_μ^a, i.e.

$$\delta\mathcal{L} = 0 \Leftrightarrow \partial^\mu J_\mu^a = 0$$

Consider first only scalar fields. This system is described by a Lagrangian density of the form:

$$\mathcal{L}\left(\phi_i, \partial^\mu \phi_i\right), \quad i = 1, 2, \cdots n,$$

with Lagrangian:

$$L = \int \mathcal{L}\left(\phi_i, \partial^\mu \phi_i\right) d^3x.$$

The action is defined by

$$S = \int L dt$$

The variation of the action leads to:

$$\delta S = \int d^4x \left[\partial_\mu \frac{\partial \mathcal{L}}{\partial \left(\partial_\mu \phi_i \right)} - \frac{\partial \mathcal{L}}{\partial \phi_i} \right] \delta \phi_i + \int d^4x \frac{\partial}{\partial x_\alpha} \left(\frac{\partial \mathcal{L}}{\partial \phi_i} \delta \phi_i \right) \tag{1.143}$$

By setting the variation equal to zero and ignoring for the moment the surface term, which can vanish as we shall see below under some symmetry of the Lagrangian density, we get the equations of motion (Euler equations)

$$\partial_\mu \frac{\partial \mathcal{L}}{\partial \left(\partial_\mu \phi_i \right)} - \frac{\partial \mathcal{L}}{\partial \phi_i} = 0.$$

Let us now consider a case such that the Lagrangian density $\mathcal{L}$ is invariant under a under a set of transformations. This leads to a conservation law. Indeed let:

$$\begin{aligned} \phi_i(x) \to \phi_i'(x) &= \phi_i(x) + \delta\phi_i(x) \\ \delta\phi_i &= \phi_i' - \phi_i = (O\phi_i - \phi_i) = i\epsilon_\alpha t^\alpha_{ij} \phi_j(x), \end{aligned}$$

$$\left[t^a, t^b \right] = i C^{ab}_c t^c$$

$\hookrightarrow$ structure constants,

$$\delta \mathcal{L} = \frac{\partial \mathcal{L}}{\partial \phi_i} \delta \phi_i + \frac{\partial \mathcal{L}}{\partial \left(\partial_\mu \phi_i \right)} \delta \left(\partial_\mu \phi_i \right).$$

Note $\delta \left(\partial_\mu \phi_i \right) = \partial_\mu \phi_i' - \partial_\mu \phi_i = \partial_\mu \delta \phi_i \Rightarrow$

$$\begin{aligned} \partial_\mu \frac{\partial \mathcal{L}}{\partial \left(\partial_\mu \phi_i \right)} - \frac{\partial \mathcal{L}}{\partial \phi_i} = 0 \Rightarrow \delta \mathcal{L} &= \partial_\mu \left(\frac{\partial \mathcal{L}}{\partial \left(\partial_\mu \phi_i \right)} \right) \delta \phi_i + \frac{\partial \mathcal{L}}{\partial \left(\partial_\mu \phi_i \right)} \partial_\mu \left(\delta \phi_i \right) \\ &= \partial_\mu \left(\frac{\partial \mathcal{L}}{\partial \left(\partial_\mu \phi_i \right)} \delta \phi_i \right) = \epsilon^\alpha \partial_\mu \left[i \frac{\partial \mathcal{L}}{\partial \left(\partial_\mu \phi_i \right)} t^\alpha_{ij} \phi_j \right]. \end{aligned}$$

Define

$$J^\alpha_\mu = -i \frac{\partial \mathcal{L}}{\partial \left(\partial_\mu \phi_i \right)} t^\alpha_{ij} \phi_j.$$

Then

$$\delta \mathcal{L} = 0 \Leftrightarrow \partial^\mu J^a_\mu = 0$$

We call this a conserved current. Thus the observed charge

$$Q^{(a)} = \int d^3x J^a_0(x)$$

is a generator of the group.

Example 16: Consider the Lagrangian density describing two real fields

$$\mathcal{L} = \frac{1}{2}\left[(\partial_\mu \phi_1)^2 + (\partial_\mu \phi_2)^2\right] - \frac{\mu}{2}\left(\phi_1^2 + \phi_2^2\right) + \frac{\lambda}{4}\left(\phi_1^2 + \phi_2^2\right)^2 \tag{1.144}$$

This is invariant under a rotation:

$$\phi_1' = \phi_1 \cos\alpha - \phi_2 \sin\alpha, \phi_2' = \phi_1 \sin\alpha + \phi_2 \cos\alpha \tag{1.145}$$

or

$$\begin{pmatrix} \phi_1' \\ \phi_2' \end{pmatrix} = \begin{pmatrix} \cos\alpha & -\sin\alpha \\ \sin\alpha & \cos\alpha \end{pmatrix} \begin{pmatrix} \phi_1 \\ \phi_2 \end{pmatrix} \tag{1.146}$$

This is an Abelian transformation with a generator:

$$t = -i\frac{d}{d\alpha}\begin{pmatrix} \cos\alpha & -\sin\alpha \\ \sin\alpha & \cos\alpha \end{pmatrix}\Bigg|_{\alpha=0} = \begin{pmatrix} 0 & i \\ -i & 0 \end{pmatrix} \Rightarrow t_{12} = i,\, t_{21} = -i$$

Thus

$$J_\mu = -i\partial_\mu \phi_1 (i\phi_2) - i\partial_\mu \phi_2 (-i\phi_1) = (\partial_\mu \phi_1)\phi_2 - (\partial_\mu \phi_2)\phi_1$$

and

$$\partial_\mu J^\mu = (\partial_\mu \phi_1)\partial^\mu \phi_2 - (\partial_\mu \phi_2)\partial^\mu \phi_1 + \overbrace{\underbrace{(\partial^\mu \partial_\mu \phi_1)\phi_2}_{-m^2\phi_1\phi_2} \underbrace{-(\partial^\mu \partial_\mu \phi_2)\phi_1}_{+m^2\phi_1\phi_2}}^{\text{use eq. of motion}}$$

$$= -m^2\phi_1\phi_2 + m^2\phi_1\phi_2 = 0.$$

It is helpful to go to ϕ, ϕ^* basis. Then

$$\phi = \frac{1}{\sqrt{2}}(\phi_1 + i\phi_2) \quad \phi^* = \frac{1}{\sqrt{2}}(\phi_1 - i\phi_2), \;\; \phi\phi^* = \frac{1}{2}\left(\phi_1^2 + \phi_2^2\right)$$

$$\begin{aligned}
\Rightarrow (\partial_\mu \phi \partial^\mu \phi^*) &= \frac{1}{2}(\partial_\mu \phi_1 + i\partial_\mu \phi_2)(\partial^\mu \phi_1 - i\partial^\mu \phi_2) \\
&= \frac{1}{2}(\partial_\mu \phi_1 \partial^\mu \phi_1 + i\partial^\mu \phi_1 \partial_\mu \phi_2 - i\partial_\mu \phi_1 \partial^\mu \phi_2 + \partial_\mu \phi_2 \partial^\mu \phi_2) \\
&= \frac{1}{2}(\partial_\mu \phi_1 \partial^\mu \phi_1 + \partial_\mu \phi_2 \partial^\mu \phi_2) \Rightarrow
\end{aligned}$$

$$\mathcal{L} = (\partial_\mu \phi^* \partial^\mu \phi) - \mu^2 \phi\phi^* + \lambda\left|\phi\phi^*\right|^2$$

Now

$$\phi \to \phi' = \frac{1}{\sqrt{2}}(\phi_1' + i\phi_2') = \frac{1}{\sqrt{2}}\left[\cos\alpha\phi_1 - \sin\alpha\phi_2 + i(\sin\alpha\phi_1 + \cos\alpha\phi_2)\right]$$

$$= \frac{1}{\sqrt{2}}\left[(\cos\alpha + i\sin\alpha)\phi_1 + i(\cos\alpha + i\sin\alpha)\phi_2\right]$$

$$= \frac{1}{\sqrt{2}}\left[e^{i\alpha}\phi_1 + ie^{i\alpha}\phi_2\right] = e^{i\alpha}\frac{(\phi_1 + i\phi_2)}{\sqrt{2}} = e^{i\alpha}\phi$$

Thus

$$\phi' = e^{i\alpha}\phi,\ \phi^{*\prime} = e^{-i\alpha}\phi^* \tag{1.147}$$

This an Abelian $U(1)$ with $t_{ij} = i\delta_{ij}$. Thus

$$J_\mu = -i\partial_\mu\phi^*\phi + i\partial_\mu\phi\phi^* \tag{1.148}$$

Furthermore

$$\begin{aligned}\partial_\mu J^\mu &= -i\left(\partial_\mu\partial^\mu\phi^*\right)\phi - i\partial_\mu\phi^*\partial^\mu\phi + i\partial_\mu\phi^*\partial^\mu\phi + i\left(\partial_\mu\partial^\mu\phi\right)\phi^* \\ &= im^2\phi^*\phi - im^2\phi\phi^* = 0,\end{aligned}$$

that is

$$\partial_\mu J^\mu = 0 \tag{1.149}$$

Example 17: SU (2) Symmetry

$$\phi = \begin{pmatrix}\phi_1 \\ \phi_2\end{pmatrix} \qquad \phi^\dagger = \left(\phi_1^*, \phi_2^*\right),$$

$$\mathcal{L} = \frac{1}{2}\left(\partial_\mu\phi^\dagger\right)\left(\partial^\mu\phi\right) - \frac{\mu^2}{2}\phi^\dagger\phi + \frac{\lambda}{4}\left(\phi^\dagger\phi\right)^2 \phi_i \to \phi_i' = i\epsilon^a\frac{\tau_{ij}^a}{2}\phi_j$$

$$\tau^1 = \begin{pmatrix}0 & 1 \\ 1 & 0\end{pmatrix} \qquad \tau^2 = \begin{pmatrix}0 & -i \\ i & 0\end{pmatrix} \qquad \tau^3 = \begin{pmatrix}1 & 0 \\ 0 & -1\end{pmatrix},$$

$$\begin{aligned}J_\mu^a &= -\frac{i}{2}\left(\partial_\mu\phi_i^\dagger\tau_{ij}^a\phi_j - \phi_i^\dagger\tau_{ij}^a\partial_\mu\phi_j\right), \\ J_0^a &= -\frac{i}{2}\left(\partial_0\phi_i^\dagger\tau_{ij}^a\phi_j - \phi_i^\dagger\tau_{ij}^a\partial_0\phi_j\right) \\ &= -\frac{i}{2}\left(n_i\tau_{ij}^a\phi_j - \phi_i^\dagger\tau_{ij}^a n_j^\dagger\right),\end{aligned}$$

with $\pi_i = \partial_0\phi_i^\dagger$. Then

$$[n_i(\mathbf{x},t), \phi_j(\mathbf{x}',t)] = \delta^3(\mathbf{x}-\mathbf{x}')(-e)\,\delta_{ij} \text{ and}$$

$$Q^a = \int d^3x J_0^a(x),$$

$$[Q_a, Q_b] = i\epsilon_{abc}Q_c.$$

Furthermore

$$\left[J_0^a, J_0^b\right] = \left(-\frac{i}{2}\tau_{ij}^a\right)\left(-\frac{i}{2}\tau_{kl}^b\right)\left[n_i\phi_j - \phi_i^\dagger n_j^\dagger, n_k\phi_l - \phi_k^\dagger n_l^\dagger\right]$$

$$= \left(-\frac{i}{2}\right)^2 \left(\tau_{ij}^a \tau_{kl}^b\right)\left\{\left[n_i\phi_j, n_k\phi_l\right] - \left[n_i\phi_j, \phi_k^\dagger n_l^\dagger\right] - \left[\phi_i^\dagger n_j^\dagger, n_k\phi_l\right] + \left[\phi_i^\dagger n_j^\dagger, \phi_k^\dagger n_l^\dagger\right]\right\}$$

$$= \left(-\frac{i}{2}\right)^2 \left(\tau_{ij}^a \tau_{kl}^b\right)\left\{\left[n_i, n_k\phi_l\right]\phi_j + n_i\left[\phi_j, n_k\phi_l\right] + \left[\phi_i^\dagger, \phi_k^\dagger n_l^\dagger\right] n_j^\dagger + \phi_i^\dagger\left[n_j^\dagger, \phi_k^\dagger n_l^\dagger\right]\right\}$$

$$= \left(-\frac{i}{2}\right)^2 \left(\tau_{ij}^a \tau_{kl}^b\right)\left(\delta_{il} n_k\phi_j - \delta_{jk} n_i\phi_l - \delta_{il}\phi_k^\dagger n_j^\dagger + \delta_{jk}\phi_i^\dagger n_l^\dagger\right)$$

$$= \left(-\frac{i}{2}\right)^2 \left(\tau_{ij}^a \tau_{ki}^b n_k\phi_j - \tau_{ij}^a \tau_{jl}^b n_i\phi_l - \tau_{ij}^a \tau_{ki}^b \phi_k^\dagger n_j^\dagger + \tau_{ij}^a \tau_{jl}^b \phi_i^\dagger n_l^\dagger\right)$$

$$= \left(-\frac{i}{2}\right)^2 \left[\left(\tau^b\tau^a\right)_{kj} n_k\phi_j - \left(\tau^a\tau^b\right)_{il} n_i\phi_l - \left(\tau^b\tau^a\right)_{kj}\phi_k^\dagger n_j^\dagger + \left(\tau^a\tau^b\right)_{il}\phi_i^\dagger n_l^\dagger\right]$$

$$= \left(-\frac{i}{2}\right)^2 \left[\left(\tau^b\tau^a - \tau^a\tau^b\right)_{il} n_i\phi_l - \left(\tau^b\tau^a - \tau^a\tau^b\right)_{il}\phi_i^\dagger n_l^\dagger\right]$$

$$= \left(-\frac{i}{2}\right)^2 2\left[i\epsilon_{bac}\tau_{il}^c n_i\phi_l - i\epsilon_{bac}\tau_{il}^c\phi_i^\dagger n_l^\dagger\right]$$

$$= \frac{i}{2}\epsilon_{abc}\tau_{il}^c\left[n_i\phi_l - \phi_i^\dagger n_l^\dagger\right].$$

That is

$$\left[J_0^a, J_0^b\right] = i\epsilon_{abc}J_0^c, \quad \left[Q_a, Q_b\right] = i\epsilon_{abc}Q_c \tag{1.150}$$

and

$$\left[Q_a, Q_b\right] = \int d^3x' d^3x \left[J_0^a(x), J_0^b(x')\right] = i\epsilon_{abc}\int d^3x J_0^c(x) = i\epsilon_{abc}Q_c \tag{1.151}$$

1.13.4 *Field theory; Fermion fields*

Described, e.g., by the Dirac equation:

$$i\gamma^\mu\partial_\mu\psi - m\psi = 0. \tag{1.152}$$

The relevant current is given by:

$$J_\mu = \bar{\psi}\gamma_\mu\psi \tag{1.153}$$

Such that:

$$\partial_\mu J^\mu = \left(\partial_\mu\bar{\psi}\right)\gamma^\mu\psi + \bar{\psi}\gamma^\mu\partial_\mu\psi = im\bar{\psi}\psi + \bar{\psi}\left(-im\psi\right) = 0, \tag{1.154}$$

$$
\begin{aligned}
Q &= \int d^3x \left(\bar{\psi}\gamma_0\psi\right), \\
J_0 &= \rho c, \qquad \mathbf{J} = \rho \mathbf{v}, \\
Q &= \int \rho(\mathbf{x})\, d^3x = \int \frac{J_0(\mathbf{x})}{c} d^3x, \\
Q &= \int d^3x J_0(\mathbf{x}) .
\end{aligned} \tag{1.155}
$$

Furthermore

$$\mathcal{L}_{\text{int}}^{EM} = eJ_\mu^{EM} A^\mu(x), \tag{1.156}$$

$$J_\mu(x) = \bar{e}\gamma_\mu e + \bar{\mu}\gamma_\mu\mu + \bar{\tau}\gamma_\mu\tau \qquad \text{in } SU(2). \tag{1.157}$$

For quarks in flavor $SU(3)$

$$J_\mu = \frac{2}{3}\left(\bar{u}\gamma_\mu u\right) - \frac{1}{3}\bar{d}\gamma_\mu d - \frac{1}{3}\bar{s}\gamma_\mu s, \tag{1.158}$$

$$J_\mu^a = \bar{q}(x)\gamma_\mu \frac{\lambda_a}{2} q(x), \tag{1.159}$$

$$J_{\mu,s}^a = \bar{q}\gamma_\mu\gamma_5 \frac{\lambda_a}{2} q(x), \tag{1.160}$$

$$
\begin{aligned}
J_\mu = J_\mu^3 + \frac{1}{\sqrt{3}} J_\mu^8 &\simeq \frac{1}{2}\bar{q}\gamma_\mu \begin{pmatrix} 1 & & \\ & -1 & \\ & & 0 \end{pmatrix} q + \frac{1}{2}\frac{1}{3}\bar{q}\begin{pmatrix} 1 & & \\ & 1 & \\ & & -2 \end{pmatrix}\gamma_\mu q \\
&= \bar{q}\gamma_\mu \left\{ \begin{pmatrix} \frac{1}{2} & & \\ & -\frac{1}{2} & \\ & & 0 \end{pmatrix} + \begin{pmatrix} \frac{1}{6} & & \\ & \frac{1}{6} & \\ & & -\frac{1}{3} \end{pmatrix} \right\} q = \bar{q}\gamma_\mu \begin{pmatrix} \frac{2}{3} & & \\ & -\frac{1}{3} & \\ & & -\frac{1}{3} \end{pmatrix} q.
\end{aligned}
$$

1.14 Problems

(1) In class we used the Pauli matrices

$$\tau_1 = \begin{pmatrix} 0 & 1 \\ 1 & 0 \end{pmatrix}, \quad \tau_2 = \begin{pmatrix} 0 & -i \\ i & 0 \end{pmatrix}, \quad \tau_3 = \begin{pmatrix} 1 & 0 \\ 0 & -1 \end{pmatrix}, \tag{1.161}$$

("Cartesian" basis).

- By taking appropriate linear combinations of the above matrices, construct a new basis of 2x2 matrices satisfying the commutator rules:

  $$[\tau_0, \tau_+] = \tau_+, [\tau_0, \tau_-] = -\tau_-, [\tau_+, \tau_-] = \tau_0, (\text{"spherical" basis}) \tag{1.162}$$

  Then express the gauge bosons in the new basis.

  Note: It is customary to define the new bosons so that they remain normalized.

- Using the commutators of Eq. (1.162), construct both a "spherical" and a "Cartesian" matrix 3x3 basis. Select the normalization in the Cartesian basis i) to be 2 and ii) to be 1 and supply the commutator relations in each case. Comment in comparison to those of Eq. (1.65).

(2) Imagine that, instead of the isospin 1/2 representation of $SU(2)$ we used in class, we had used the isovector one, $I = 1$.

- Construct the corresponding set of "Cartesian" generators I_1, I_2 and I_3 and make a suitable decomposition of the gauge fields.
- Do the same for the "spherical" set I_+, I_- and I_0
- Does your representation agree with the 3x3 one of the previous problem? Explain.

Note: you are allowed to use the standard angular momentum experience.

(3) Consider the group of orthogonal transformations in 4 dimensions. In three dimensions one has three axes and three planes. So rotation around an axes is equivalent with rotations in a plane perpendicular to it. In a space with more dimensions the number of axes is less than the number of planes. Thus it makes no sense to talk about rotation around an axis. In $n > 3$ dimensions we have n axes but $(1/2)n(n-1)$ planes. And we talk about rotations in a given plane.

i Show that there exist six elementary rotations, each one of them corresponding to a rotation on a plane.

ii Find the corresponding generators a) represented by real antisymmetric matrices and b) Represented by Hermitian matrices.

iii obtain all commutation rules and construct a table giving the non vanishing commutators.

iv Using these generators construct the rotation matrix representing the product of any two different elementary rotations of your choice. Does your answer agree with that obtained by using i) above?

v Show that a basis can be found: So that

$$[A_k, A_l] = i\epsilon_{klm}A_m, \quad [B_k, B_l] = i\epsilon_{klm}B_m, \quad [A_k, B_l] = 0 \tag{1.163}$$

What does this signify? Is the group $O[4]$ simple?
Hint: If you cannot find such a basis consider:

$$A_1 = \begin{pmatrix} 0 & 0 & 0 & -\frac{i}{2} \\ 0 & 0 & -\frac{i}{2} & 0 \\ 0 & \frac{i}{2} & 0 & 0 \\ \frac{i}{2} & 0 & 0 & 0 \end{pmatrix}, A_2 = \begin{pmatrix} 0 & 0 & \frac{i}{2} & 0 \\ 0 & 0 & 0 & -\frac{i}{2} \\ -\frac{i}{2} & 0 & 0 & 0 \\ 0 & \frac{i}{2} & 0 & 0 \end{pmatrix},$$

$$A_3 = \begin{pmatrix} 0 & -\frac{i}{2} & 0 & 0 \\ \frac{i}{2} & 0 & 0 & 0 \\ 0 & 0 & 0 & -\frac{i}{2} \\ 0 & 0 & \frac{i}{2} & 0 \end{pmatrix} B_1 = \begin{pmatrix} 0 & 0 & 0 & \frac{i}{2} \\ 0 & 0 & -\frac{i}{2} & 0 \\ 0 & \frac{i}{2} & 0 & 0 \\ -\frac{i}{2} & 0 & 0 & 0 \end{pmatrix},$$

$$B_2 = \begin{pmatrix} 0 & 0 & \frac{i}{2} & 0 \\ 0 & 0 & 0 & \frac{i}{2} \\ -\frac{i}{2} & 0 & 0 & 0 \\ 0 & -\frac{i}{2} & 0 & 0 \end{pmatrix}, B_3 = \begin{pmatrix} 0 & -\frac{i}{2} & 0 & 0 \\ \frac{i}{2} & 0 & 0 & 0 \\ 0 & 0 & 0 & \frac{i}{2} \\ 0 & 0 & -\frac{i}{2} & 0 \end{pmatrix}$$

(4) Repeat the step iii of the previous problem in the case of the eight generators λ_i, $i = 1, ..., 8$ of the group $SU(3)$.

(5) In the case of the previous problem find the matrix corresponding to the product of any two different elementary "rotations" of the type $U_k = e^{i\omega_k \lambda_k}$, which are non diagonal.

Hint: You may find it useful to proceed by finding the eigenvalues of the corresponding matrices.

(6) If you can afford the time and you have the energy:
Try to understand from the literature the role of $SO(4)$ in the case of potentials proportional to $1/r$, e.g. the *unge*-Lenz vector in the Keplerian motion and in understanding the degeneracy and obtaining of the spectrum of the hydrogen atom in Quantum Mechanics.

(7) Try to generalize the results of the first problem in the case of the Lorentz group.

Hint: Now the rotations in the three planes $(i, 4)$, $i = 1, 2, 3$ are actually boosts, $\pm \sin\theta \to \sinh(\xi)$, with ξ real. Now the generators associated with boosts are different and the commutators involving them have different signs than in the case of $O(4)$. The matrices A_i, however, can be chosen to be:

$$A_1 = \begin{pmatrix} 0 & 0 & 0 & -\frac{1}{2} \\ 0 & 0 & -\frac{i}{2} & 0 \\ 0 & \frac{i}{2} & 0 & 0 \\ -\frac{1}{2} & 0 & 0 & 0 \end{pmatrix}, A_2 = \begin{pmatrix} 0 & 0 & \frac{i}{2} & 0 \\ 0 & 0 & 0 & -\frac{1}{2} \\ -\frac{i}{2} & 0 & 0 & 0 \\ 0 & -\frac{1}{2} & 0 & 0 \end{pmatrix},$$

$$A_3 = \begin{pmatrix} 0 & -\frac{i}{2} & 0 & 0 \\ \frac{i}{2} & 0 & 0 & 0 \\ 0 & 0 & 0 & -\frac{1}{2} \\ 0 & 0 & -\frac{1}{2} & 0 \end{pmatrix}$$

$$B_1 = \begin{pmatrix} 0 & 0 & 0 & \frac{1}{2} \\ 0 & 0 & -\frac{i}{2} & 0 \\ 0 & \frac{i}{2} & 0 & 0 \\ \frac{1}{2} & 0 & 0 & 0 \end{pmatrix}, B_2 = \begin{pmatrix} 0 & 0 & \frac{i}{2} & 0 \\ 0 & 0 & 0 & \frac{1}{2} \\ -\frac{i}{2} & 0 & 0 & 0 \\ 0 & \frac{1}{2} & 0 & 0 \end{pmatrix}, B_3 = \begin{pmatrix} 0 & -\frac{i}{2} & 0 & 0 \\ \frac{i}{2} & 0 & 0 & 0 \\ 0 & 0 & 0 & \frac{1}{2} \\ 0 & 0 & \frac{1}{2} & 0 \end{pmatrix}$$

(8) Show that if a representation $A(J_i) = e^{i\theta_i J_i}$ of a group is given, its complex conjugate$A^*(J_i) = e^{-i\theta_i J_i^*}$ is also a representation. This is, of course, significant only if some of the J_i contain complex numbers.

Regardless of whether you proved the above or not, show that in the case of $SU(2)$, $J_i = \sigma_i$, the complex representation is obtained by:

$$\sigma_1 \to \sigma_1^{'} = -\sigma_1, \sigma_2 \to \sigma_2^{'} = \sigma_2, \sigma_3 \to \sigma_3^{'} = -\sigma_3$$

Find the structure constants in the case of the $\sigma_i^{'}$'s. Furthermore show that, unlike the $SU(3)$ the complex conjugate is equivalent to the original one. Construct the transformation matrix that achieves this.

(9) Show that the set of operators

$$L_{m,n} = \frac{\hbar}{i}\left(x_m \frac{\partial}{\partial x_n} - x_n \frac{\partial}{\partial x_m}\right), m, n = 1, 2 \cdots N$$

constitute a Lie algebra and obtain its structure constants. Specialize this in the case of $N = 3$ and give such operators a physical meaning in the context of quantum mechanics.

(10) Consider the set of $n \times n$ matrices.

- A set of $n \times n$ matrices E_{ij} are given such that its elements are given by $(E_{ij})_{\alpha,\beta} = \delta_{i\alpha}\delta_{j\beta}$.
- Show that this set constitutes a basis in the space of $n \times n$ matrices.
- Show that it constitutes a Lie Algebra and obtain its structure constants.

Hint: show that:

$$[E_{ij}, E_{k\ell}] = \delta_{jk}E_{i\ell} - \delta_{i\ell}E_{kj}.$$

- Is the algebra semisimple? If it not, find an Abelian subalgebra.
- What is the number of the elements of the algebra? The maximum number of the Abelian elements?
- Discuss in particular the case of a traceless set.

(11) Consider the set of $n \times n$ matrices E_{ij} discussed in the previous problem.

Regardless of whether you proved the relation

$$[E_{ij}, E_{k\ell}] = \delta_{jk} E_{i\ell} - \delta_{i\ell} E_{kj}.$$

you can use these matrices as a basis to study the groups $U(2)$, $SU(2)$ and $SU(3)$ we have discussed in class. In particular i) find the cummutators (structure constants) for the group $U(2)$. ii) Do the same for the group SU(2) and write down the Pauli matrices in this basis. iii) Go as far as you can in doing the same for the group $SU(3)$ using the basis $h_1 = E_{11} - E_{22}$, $h_2 = E_{22} - E_{33}$ and E_{ij}, $i \neq j$. Comment on the form of these structure constants compared to those obtained with the λ_i.

(12) Consider the algebra of SU(2) or SO(3) (they are isomorphic). Then

- Obtain the basis $|jm\rangle$ by considering two suitable commuting operators. The essential ingredients should be i)the commutation relations of the algebra, in particular in the basis J_+, j_- and J_0, the unitarity of the representation and the fact that $\langle\psi|\psi\rangle \geq 0$, $|\psi = J_+|jm\rangle$
- Show in particular that j is integral or half integral and m has values, between j and $-j$. There exist two representation D^j and $D^{-(j+1)}$, which, however are equivalent.

(13) Consider the quantum harmonic oscillator and the operators

$$\boldsymbol{\xi} = \sqrt{(m\omega)/2\hbar}\mathbf{r}, \mathbf{p} = \sqrt{2m\hbar\omega}\boldsymbol{\eta},\ \boldsymbol{\eta} = \frac{1}{i}\frac{\partial}{\partial \boldsymbol{\xi}}$$

- Show that the harmonic oscillator Hamiltonian takes the form

$$H = \frac{1}{2}\hbar\omega(\boldsymbol{\xi}^2 + \boldsymbol{\eta}^2)$$

- Define the operators $a_k^+ = \boldsymbol{\xi}_k - i\boldsymbol{\eta}_k$, $a_k = \boldsymbol{\xi}_k + \boldsymbol{\eta}_k$, $k = 1, 2, 3$ and show that:

$$[a_k, a_m] = 0,\ [a_k^+, a_m^+] = 0,\ [a_m, a_k^+] = \delta_{km}$$

$$H = \hbar\omega\left(N + \frac{3}{2}\right), N = \sum_{i=k}^{3} a_k^+ a_k = \text{ the number operator}$$

- The operators $A_{k\ell} = a_k^+ a_\ell$ constitute a basis for $SU(3)$ and its subgroup $SU(2)$
- In the chain $SU(3) \supset SU(2)$ a set of commuting operators can be found:

  $$K^{(2)}, K^{(3)}, Q_0, \Lambda^2, \Lambda_0$$

  where $K^{(2)}$, $K^{(3)}$ the Casimir operators of $SU(3)$ and

  $$Q_0 = A_{11} + A_{22} - 2A_{33}, \Lambda_0 = \frac{1}{2}(A_{11} - A_{22}),$$

  $$\Lambda^2 = A_{21}A_{12} + \Lambda_0(\Lambda_0 + 1)$$

- In the chain $SU(3) \supset SO(3)$ one considers instead the operators:

  $$L_1 = L_{23} = -i(A_{23} - A_{32}),\ L_2 = L_{31} = -i(A_{31} - A_{13}),$$

  $$L_3 = L_{12} = -i(A_{12} - A_{21})$$

  Find the structure constants of these operators and show that they can be identified with the angular momentum operators. Show further that in this case a commuting set of operators, in addition to the Casimir operators of $SU(3)$, contains L_{12} and $\mathbf{L}^2$. There is a missing operator, which cannot be easily constructed.
- In both chains the operator N can be simultaneously diagonalized (and hence the Hamiltonian itself).

(14) Consider the operators defined in the previous problem.

- In the case of many particles $i = 1, 2, \cdots, A$ one defines the operators:

  $$A_{\mu\nu} = \sum_i^A a_\mu^+(i)a_\nu(i),\ h = \hbar\omega \sum_i^A \left(a_\mu^+(i)a_\nu(i) + \frac{3}{2}\right).$$

  These operators shift quanta of the type ν to the type μ. They satisfy the commutation rules:

  $$[A_{\mu\rho}, A_{\nu\sigma}] = \delta_{\mu\sigma}A_{\rho\nu} - \delta_{\nu\rho}A_{\mu\sigma}, \quad [h, A_{\mu\nu|}] = 0 \qquad (1.164)$$

  The first of these is the basic algebra of $SU(3)$, while the last equation guarantees the invariance of the harmonic oscillator Hamiltonian under the $SU(3)$ transformations.

- One can now construct various representations, e.g. the adjoined representation which depends only on the structure constants as we discussed in class. Another one is the fundamental representation, based on the basis:

$$|\rangle = x \text{ (one quantum in the x-direction)}, \; |2\rangle = y, \; |3\rangle = z \tag{1.165}$$

Thus

$$A_{xy} \Rightarrow (\alpha)_{xy} = \begin{pmatrix} 0 & 1 & 0 \\ 0 & 0 & 0 \\ 0 & 0 & 0 \end{pmatrix} \tag{1.166}$$

Proceed similarly for the other elements of the algebra and their adjoined. Verify that, by combining these, you obtain the λ_i matrices we discussed in class.

- Another 3-dimensional basis can be constructed from the antisymmetric vectors:

$$|\bar{1}\rangle = \frac{1}{\sqrt{2}}(yz - zy)\,,\; |\bar{2}\rangle = \frac{1}{\sqrt{2}}(zx - xz)\,,\; |\bar{3}\rangle = \frac{1}{\sqrt{2}}(xy - yx) \tag{1.167}$$

where in this writing x, y and z do not commute. Repeat the above procedure in this case to obtain the matrices $(\bar{\alpha})_{ij}$ and $(\bar{\lambda})_i$. e.g.:

$$A_{xy} \Rightarrow (\bar{\alpha})_{xy} = \begin{pmatrix} 0 & -1 & 0 \\ 0 & 0 & 0 \\ 0 & 0 & 0 \end{pmatrix} \tag{1.168}$$

This is sometimes called conjugate (not complex conjugate) of the previous. These new matrices are not the same with the previous. Can they be obtained from the previous by a unitary transformation? If so they are equivalent and we say that the symmetry does not admit conjugate representations[6]. If not, are the two algebras isomorphic? For $SU(n)$, $n > 2$ no!

- From the two basis construct a new basis $|j\rangle|\bar{j}\rangle$, in some convenient order, and obtain the corresponding 9-dimensional

[6]There exist symmetries that they do not, but the symmetry $SU(n)$ does.

representation of $SU(3)$, e.g.[7]:

$$\Lambda_1 = \begin{pmatrix} 0 & -1 & 0 & 1 & 0 & 0 & 0 & 0 & 0 \\ -1 & 0 & 0 & 0 & 1 & 0 & 0 & 0 & 0 \\ 0 & 0 & 0 & 0 & 0 & 1 & 0 & 0 & 0 \\ 1 & 0 & 0 & 0 & -1 & 0 & 0 & 0 & 0 \\ 0 & 1 & 0 & -1 & 0 & 0 & 0 & 0 & 0 \\ 0 & 0 & 1 & 0 & 0 & 0 & 0 & 0 & 0 \\ 0 & 0 & 0 & 0 & 0 & 0 & 0 & -1 & 0 \\ 0 & 0 & 0 & 0 & 0 & 0 & -1 & 0 & 0 \\ 0 & 0 & 0 & 0 & 0 & 0 & 0 & 0 & 0 \end{pmatrix},$$

$$\Lambda_2 = \begin{pmatrix} 0 & -i & 0 & -i & 0 & 0 & 0 & 0 & 0 \\ i & 0 & 0 & 0 & -i & 0 & 0 & 0 & 0 \\ 0 & 0 & 0 & 0 & 0 & -i & 0 & 0 & 0 \\ i & 0 & 0 & 0 & -i & 0 & 0 & 0 & 0 \\ 0 & i & 0 & i & 0 & 0 & 0 & 0 & 0 \\ 0 & 0 & i & 0 & 0 & 0 & 0 & 0 & 0 \\ 0 & 0 & 0 & 0 & 0 & 0 & 0 & -i & 0 \\ 0 & 0 & 0 & 0 & 0 & 0 & i & 0 & 0 \\ 0 & 0 & 0 & 0 & 0 & 0 & 0 & 0 & 0 \end{pmatrix},$$

$$\Lambda_3 = \begin{pmatrix} 0 & 0 & 0 & 0 & 0 & 0 & 0 & 0 & 0 \\ 0 & 2 & 0 & 0 & 0 & 0 & 0 & 0 & 0 \\ 0 & 0 & 1 & 0 & 0 & 0 & 0 & 0 & 0 \\ 0 & 0 & 0 & -2 & 0 & 0 & 0 & 0 & 0 \\ 0 & 0 & 0 & 0 & 0 & 0 & 0 & 0 & 0 \\ 0 & 0 & 0 & 0 & 0 & -1 & 0 & 0 & 0 \\ 0 & 0 & 0 & 0 & 0 & 0 & -1 & 0 & 0 \\ 0 & 0 & 0 & 0 & 0 & 0 & 0 & 1 & 0 \\ 0 & 0 & 0 & 0 & 0 & 0 & 0 & 0 & 0 \end{pmatrix} \qquad (1.169)$$

etc.

- Show that the representation you obtained is reducible, in fact:

$$\underline{3} \otimes \underline{\bar{3}} = \underline{9} = \underline{8} + \underline{1}$$

where $\underline{8}$ is the regular (adjoined) representation. In fact the resulting representations had to be self adjoined. Can you see why?

[7] It took me a whole day to devise a code to put the zeros in the right places. So there exists a solution, if you hate to do it by hand for so many matrices. You, being more familiar with modern technology, must be more efficient.

This is a general result, which can be obtained by much more powerful methods, holding for $SU(n)$ and other groups. It also holds if you take successive products of the fundamental representations. In fact one can obtain the conjugate of the fundamental this way. This is why the fundamental representation has the name that it does!

- Find a suitable unitary matrix that reduces the representation you have obtained and show that the scalar $\underline{1}$ is obtained by a change of basis:

$$\underline{1} = \frac{1}{\sqrt{3}}\left(|1\rangle|\bar{1}\rangle + |2\rangle|\bar{2}\rangle + |3\rangle|\bar{3}\rangle\right)$$

(Democracy at work!)

(15) Consider the generators of the algebra SO(2,1):

- The relevant generators are:

$$J_1 \equiv J_{01} = \frac{1}{i}\left(x_0\frac{\partial}{\partial x_1} + x_1\frac{\partial}{\partial x_0}\right),$$
$$J_2 \equiv J_{02} = \frac{1}{i}\left(x_0\frac{\partial}{\partial x_2} + x_2\frac{\partial}{\partial x_0}\right),$$
$$J_3 \equiv_{12} = \frac{1}{i}\left(x_1\frac{\partial}{\partial x_2} - x_2\frac{\partial}{\partial x_1}\right) \quad (1.170)$$

The operator J_{12} is the same with that of $SO(2) \subset SO(3)$.

- Show that these operators satisfy the commutation rules:

$$[J_1, J_2] = -iJ_3,\ [J_2, J_3] = iJ_1,\ [J_3, J_1] = -iJ_2 \quad (1.171)$$

They resemble the commutations of the SO(3), except for one sign.

- The corresponding group elements are:

$$\Lambda_{01} = \begin{pmatrix} \cosh\xi & \sinh\xi & 0 \\ \sinh\xi & \cosh\xi & 0 \\ 0 & 0 & 1 \end{pmatrix},\ \Lambda_{02} = \begin{pmatrix} \cosh\eta & 0 & \sinh\eta \\ 0 & 1 & 0 \\ \sinh\eta & 0 & \cosh\eta \\ 0 & 0 & 1 \end{pmatrix},$$

$$\Lambda_{12} = \begin{pmatrix} 1 & 0 & 0 \\ 0 & \cos\theta & \sin\theta \\ 0 & \sin\theta & \cos\theta \end{pmatrix} \quad (1.172)$$

Only the last is a bona fide rotation, the other two are "pseudorotations". Furthermore the first two are not bounded.

- The above operators leave invariant the expression:

$$x_0^2 - x_1^2 - x_2^2 = \text{invariant} \tag{1.173}$$

- One can show that the Casimir operator, which commutes with all three is:

$$J^2 \equiv J_1^2 + J_2^2 - J_3^2$$

One can also define:

$$J_+ = iJ_1 - J_2,\; J_- = iJ_1 + J_2,\; J_0 = J_3$$

with commutation rules:

$$[J_0, J_+] = J_+,\; [J_0, J_-] = -J_-,\; [J_+, J_-] = 2J_0$$

- Then show that:

$$J^2 = \frac{1}{2}(J_+J_- + J_-J_+) - J_0^2 = -J_-J_+ - J_0(J_0 + 1)$$

- Since J^2 and J_0 commute they can be simultaneously diagonalized.
- Go as far as you can following the procedure adopted in problem i) for $SU(2)$ and $SO(3)$. Show that the unitary representations are infinitely dimensional. Why the analogy breaks down?

(16) Study the extra symmetry of the potential energy $U(r) = -k/r$ by observing that the Rünge-Lenz vector is conserved.

The Rünge-Lenz vector is defined by

$$\mathbf{R} = \mathbf{p} \times \mathbf{L} - \mu k\hat{r}, \text{ in a cental field given by } \mathbf{F} = -\frac{k}{r^2}\hat{r} \tag{1.174}$$

This vector is sometimes indicated by **A**.

R is perpendicular to **L**,

$$\mathbf{R.L} = 0. \tag{1.175}$$

The energy is given by

$$E = \frac{1}{2\mu}p^2 - \frac{k}{r}$$

The magnitude of **R** is given by

$$R = k\sqrt{1 + \frac{L^2}{2\mu k^2}E}$$

and the angular momentum is $\mathbf{L} = \mathbf{r} \times \mathbf{p}$. The proof of conservation of $\mathbf{R}$ is quite simple.

$$\frac{d\mathbf{R}}{dt} = \frac{d\mathbf{p}}{dt} \times \mathbf{L} - \mu k \left(\frac{1}{r}\frac{d\mathbf{r}}{dt} - \frac{\mathbf{r}}{r^2}\frac{dr}{dt} \right)$$

but

$$\frac{d\mathbf{p}}{dt} = -k\frac{\mathbf{r}}{r^3} \Rightarrow$$

$$\frac{d\mathbf{R}}{dt} = -\frac{k}{\mu} \left(\frac{\mathbf{r} \times \mathbf{L} - (\mathbf{r}.\mathbf{p})\mathbf{r}}{r^3} + \frac{\mathbf{p}}{r} \right)$$

But

$$\mathbf{r} \times \mathbf{L} = \mathbf{r} \times (\mathbf{r} \times \mathbf{p}) = (\mathbf{r}.\mathbf{p})\mathbf{r} - r^2\mathbf{p}$$

Thus

$$\frac{d\mathbf{R}}{dt} = 0 \tag{1.176}$$

Its relevance to the Kepler problem is given in Fig. 1.1. Since the vectors $\mathbf{L}$ and $\mathbf{R}$ are conserved and perpendicular we can define a natural coordinate system:

$$\hat{e}_3 = \frac{\mathbf{L}}{L},\ \hat{e}_1 = \frac{\mathbf{R}}{R},\ \hat{e}_2 = \hat{e}_3 \times e_1 \tag{1.177}$$

The orbit is perpendicular to $\mathbf{L}$. So

$$\mathbf{r} = r(\hat{e}_1 \cos\theta + \hat{e}_2 \sin\theta) \tag{1.178}$$

$$\mathbf{p} = \mu\frac{d\mathbf{r}}{dt} = \mu \left(\hat{e}_1 \left(\cos\theta\frac{dr}{dt} - r\sin\theta\frac{d\theta}{dt} \right) + \hat{e}_2 \left(\sin\theta\frac{dr}{dt} + r\cos\theta\frac{d\theta}{dt} \right) \right)$$

but

$$\mathbf{L} = \mu r^2\frac{d\theta}{dt}\hat{e}_3 \Rightarrow \frac{d\theta}{dt} = \frac{mur^2}{L}.$$

Thus

$$\mathbf{R} = \hat{e}_1 \left(L\frac{dr}{dt}\sin\theta + \left(\frac{L^2}{\mu r} - k \right) \cos\theta \right)$$

$$-\hat{e}_2 \left(L\frac{dr}{dt}\cos\theta - \left(\frac{L^2}{\mu r} - k \right) \sin\theta \right)$$

But, by definition, $\mathbf{R} = r\hat{e}_1$, which implies

$$L\frac{dr}{dt}\cos\theta = \left(\frac{L^2}{\mu r} - k \right) \sin\theta \Rightarrow \frac{dr}{dt} = \frac{1}{L}\left(\frac{L^2}{\mu r} - k \right) \tan\theta$$

$$\mathbf{R} = \hat{e}_1 \left(\frac{L}{\mu} r - k \right) \sin\theta$$

Thus

$$R = k\sqrt{1 + \frac{L^2}{2\mu k^2} E} = \left(\frac{L}{\mu r} - k \right) \frac{1}{\cos\theta}$$

or

$$r = \frac{\ell}{1 + e\cos\theta} \text{ equation of the orbit} \tag{1.179}$$

with

$$\ell = \frac{L^2}{\mu k}, \; e = \sqrt{1 + \frac{2L^2}{\mu} \frac{E}{k^2}}$$

The eccentricity is defined by

$$e^2 = 1 + \frac{2L^2}{\mu k^2} E$$

$$a = \frac{L^2}{\mu k} \frac{1}{1 - e^2} = -\frac{k}{2E} \, b = a\sqrt{1 - e^2}$$

Furthermore

$$R^2 = \mu^2 k^2 + 2mEL^2 \tag{1.180}$$

The eccentricity can be written as

$$a(1 \pm e^2) = \ell, \; \ell = \frac{L^2}{\mu k}, \; + \text{ for ellipse }, \; - \text{ for hyperbola}$$

Since the vector **R** is conserved and it is always along the symmetry axes, pointing from the center of the force to the perihelium, its conservation implies that the perihelium does not move. The R*ü*nge vector gives one additional constant of motion on top of the four resulting conservation of energy and angular momentum, since we have the two constraints given by equations 1.175 and 1.180.

(17) Quantum mechanical description of the R*ü*nge-Lenz vector.

One has to be a bit careful in the definition of **R**, since the momentum and angular momentum operators do not commute.

$$R_s = -\mu k \hat{r}_s + \frac{1}{2} \epsilon_{sjk} (p_j L_k + L_j p_k) \tag{1.181}$$

Then one can consider the operators

$$R_0 = R_3, \; R_{\pm 1} = \frac{1}{\sqrt{2}} (R_1 \mp R_2) \tag{1.182}$$

These connect different eigenstates of angular momentum The Casimir invariant is written as

$$K_1 = -I - \frac{\mu k^2}{2h^2} H^{-1} \tag{1.183}$$

where H is the Hamiltonian operator and I the identity operator. The operator C_1 is quantized to $n^2 - 1$ yielding the well known eigenvalues of the energy

$$E_n = -\frac{\mu k^2}{2\hbar^2}\frac{1}{n^2} \tag{1.184}$$

and they are independent of the angular momentum quantum number.

Chapter 2

Mathematical Prerequisite B: The Dirac Theory

2.1 Preliminaries; the Klein–Gordon equation

Recall from ordinary quantum mechanics for a free particle

$$\begin{aligned}
E &= \frac{p^2}{2m}, \\
E &\to \hbar i\frac{\partial}{\partial t}, \quad \vec{p} \to -\hbar i\vec{\nabla}, \\
i\frac{\partial\psi}{\partial t} &= -\frac{\nabla^2}{2m}\psi.
\end{aligned}$$

In the presence of a potential:

$$H = \frac{p^2}{2m} + V \Rightarrow H = -\hbar^2\frac{\nabla^2}{2m} + V.$$

This leads to the Schrödinger equation

$$\Rightarrow i\hbar\frac{\partial\psi}{\partial t} = H\psi. \tag{2.1}$$

Then one defines the probability and current density

$$\begin{aligned}
\rho &= \psi^*\psi \geq 0, \\
\vec{J}_\rho &= \frac{\hbar i}{2m}(\psi\nabla\psi^* - \psi^*\nabla\psi).
\end{aligned} \tag{2.2}$$

with the conservation law:

$$\frac{\partial\rho}{\partial t} + \vec{\nabla}\cdot\vec{J} = 0. \tag{2.3}$$

In going to the relativistic case, we found it convenient to use the natural system of units ($\hbar = 1$, $c = 1$), see chapter 13. Thus the energy momentum relation becomes:

$$p_\mu p^\mu = m^2 \Rightarrow E^2 = \vec{p}^2 + m^2$$

or setting $x^0 = ct$:

$$p_\mu = i\frac{\partial}{\partial x^\mu} \equiv i\partial_\mu, \quad p^2 = -\partial_\mu\partial^\mu, \quad \partial_\mu\partial^\mu = \frac{\partial^2}{c^2\partial t^2} - \nabla^2.$$

Thus for a scalar particle:

$$\left(\frac{\partial^2}{\partial t^2} - \nabla^2\right)\phi + m^2\phi^2 = 0 \text{ (Klein-Gordon equation).} \tag{2.4}$$

From the Klein-Gordon equation we get:

$$\phi^*\left(\frac{\partial^2\phi}{\partial t^2} - \nabla^2\phi + m^2\phi\right) = 0,$$

$$\phi\left(\frac{\partial^2\phi^*}{\partial t^2} - \nabla^2\phi^* + m^2\phi^*\right) = 0.$$

Subtracting the two we get:

$$\left(\phi^*\frac{\partial^2\phi}{\partial t^2} - \phi\frac{\partial^2\phi^*}{\partial t^2}\right) - (\phi^*\nabla^2\phi - \phi\nabla^2\phi^*) = 0 \Rightarrow$$

$$\frac{\partial}{\partial t}\left(\phi^*\frac{\partial\phi}{\partial t} - \phi\frac{\partial\phi^*}{\partial t}\right) = \nabla.\left(\phi^*\vec{\nabla}\phi - \phi\vec{\nabla}\phi^*\right).$$

Defining:

$$\rho = \frac{i}{2}\left(\phi^*\frac{\partial\phi}{c\partial t} - \phi\frac{\partial\phi^*}{c\partial t}\right), \quad \vec{J} = \frac{i}{2}\left(\phi^*\vec{\nabla}\phi - \phi\vec{\nabla}\phi^*\right),$$

we get the conservation law:

$$\frac{\partial\rho}{c\partial t} + \vec{\nabla}\cdot\vec{J} = 0 \tag{2.5}$$

We note that the space component for the current has the proper form. The time component, however, is unusual. It contains a time derivative. This happens, because unlike the Schrödinger equation, which is first order in the time derivative, the Klein-Gordon equation is of second order. This, of course, presents a problem, since ρ is not always positive, and cannot be interpreted as a probability.

Even in the case of a plane wave:

$$\phi = Ne^{-i(Et-\vec{p}\cdot\vec{x})}, \; E = \pm\sqrt{p^2+m^2}$$

(for the normalization N see problem 1 at the end of this chapter). Thus in the case of a free particle we have:

$$\rho = |N|^2E, \quad E = \pm\sqrt{p^2+m^2}$$

i.e. ρ can become negative for negative energy. The presence of negative energies lead to the abandonment of the Klein-Gordon equation, quite prematurely as we will see later.

We can write the above conservation equation in covariant form:

$$\partial_\mu j^\mu = 0.$$

with

$$j_\mu = \frac{i}{2}\left(\phi^* \partial_\mu \phi - i\phi \partial_\mu \phi^*\right) = (\rho, J),\ j^\mu = (\rho, -J).$$

We thus find in the case of a free particle:

$$J^\mu = |N|^2 (E, \vec{p}) = |N|^2 p_\mu.$$

2.2 The Dirac equation

Motivated by the non relativistic case, which is first order in time, Dirac attempted to set up an equation of the type:

$$i\frac{\partial \psi(x,t)}{\partial t} = H\psi(x),$$
$$H = -i\left(\alpha_1 \frac{\partial}{\partial x^1} + \alpha_2 \frac{\partial}{\partial x^2} + \alpha_3 \frac{\partial}{\partial x^3}\right) + \beta m \tag{2.6}$$

with α_i, $i = 1, 3, 3$, β to be determined. In operator form:

$$i\frac{\partial}{\partial t} = -i\vec{\alpha}\cdot\vec{\nabla} + \beta m,$$
$$-\frac{\partial^2}{\partial t^2} = \left(-i\vec{\alpha}\cdot\vec{\nabla} + \beta m\right)\left(-i\vec{\alpha}\cdot\vec{\nabla} + \beta m\right),$$

that is

$$-\frac{\partial^2}{\partial t^2}\psi = -\alpha_i^2 \frac{\partial^2 \psi}{\partial x_i^2} - \sum_{i<j} \left(\alpha_i \alpha_i + \alpha_j \alpha_i\right) \frac{\partial^2 \psi}{\partial x_i \partial x_i} - im\left(\alpha_i \beta + \beta \alpha_i\right) \frac{\partial \psi}{\partial x_i}$$
$$+ \beta^2 m^2 \psi. \tag{2.7}$$

This should be brought to the form:

$$-\frac{\partial^2}{\partial t^2} = -\nabla^2 + m^2.$$

So we demand:

$$\alpha_1^2 = \alpha_2^2 = \alpha_3^2 = \beta^2 = 1,$$
$$\alpha_i \alpha_j + \alpha_j \alpha_i = 0, \quad i \neq j,$$
$$\alpha_i \beta + \beta \alpha_i = 0. \tag{2.8}$$

These conditions cannot be satisfied by ordinary numbers. Dirac realized that it is possible to have them satisfied if α_i, β are matrices. He then proceeded to construct these matrices starting from the Pauli matrices σ_i, which satisfy

$$\sigma_i\sigma_j + \sigma_j\sigma_i = 2I\delta_{ij}.$$

His choice was:

$$\alpha_i = \begin{pmatrix} 0 & \sigma_i \\ \sigma_i & 0 \end{pmatrix}, \quad \beta = \begin{pmatrix} 1 & 0 \\ 0 & -1 \end{pmatrix}, \tag{2.9}$$

i.e.

$$\alpha_1 = \begin{pmatrix} 0 & \sigma_1 \\ \sigma_1 & 0 \end{pmatrix} = \begin{pmatrix} 0&0&0&1 \\ 0&0&1&0 \\ 0&1&0&0 \\ 1&0&0&0 \end{pmatrix}, \quad \alpha_2 = \begin{pmatrix} 0&0&0&-i \\ 0&0&i&0 \\ 0&-i&0&0 \\ i&0&0&0 \end{pmatrix},$$

$$\alpha_3 = \begin{pmatrix} 0&0&1&0 \\ 0&0&0&-1 \\ 1&0&0&0 \\ 0&-1&0&0 \end{pmatrix}, \quad \beta = \begin{pmatrix} 1&0&0&0 \\ 0&1&0&0 \\ 0&0&-1&0 \\ 0&0&0&-1 \end{pmatrix}, \tag{2.10}$$

$$\{\alpha_i, \alpha_i\} = 2\delta_{ij}I, \; \{\alpha, \beta\} = 0, \; \beta^2 = 1.$$

In practice it is more convenient to use a different set of matrices

$$\gamma^\mu = (\gamma_0, \gamma_i), \quad \gamma_0 = \beta, \; \gamma_i = \beta\alpha_i$$

or more explicitly the Dirac matrices:

$$\gamma^0 = \begin{pmatrix} 1 & 0 \\ 0 & -1 \end{pmatrix}, \gamma^k = \begin{pmatrix} 0 & \sigma_k \\ -\sigma_k & 0 \end{pmatrix}. \tag{2.11}$$

These satisfy:

$$\{\gamma_\mu, \gamma_\nu\} = (\gamma_\mu\gamma_\nu + \gamma_\nu\gamma_\mu) = 2g_{\mu\nu}. \tag{2.12}$$

One then can show:

$$(\gamma_0)^2 = 1, \; (\gamma_i)^2 = -1, \; (\gamma^0)^\dagger = \gamma^0, \; (\gamma^k)^\dagger = -\gamma^k.$$

Then the Dirac equation reads:

$$\begin{aligned} i\beta\frac{\partial\psi}{\partial t} &= (-i\beta\vec{\alpha}\cdot\vec{\nabla} + m)\psi = 0 \Rightarrow \\ i\gamma^0\frac{\partial\psi}{\partial t} &= -i\vec{\gamma}\cdot\vec{\nabla} + m \text{ or} \\ i\gamma^0\frac{\partial\psi}{\partial t} &+ i\vec{\gamma}\cdot\vec{\nabla}\psi - m\psi = 0, \end{aligned}$$

i.e. more compactly in covariant form

$$(i\gamma^\mu \partial_\mu - m)\psi = 0 \text{ (The Dirac equation)}, \tag{2.13}$$

with

$$\partial_\mu = \frac{\partial}{\partial x^\mu} = \left(\frac{\partial}{\partial x^0}, \frac{\partial}{\partial x^1}, \frac{\partial}{\partial x^2}, \frac{\partial}{\partial x^3}\right) = \left(\frac{\partial}{\partial x_0}, -\frac{\partial}{\partial x_1}, -\frac{\partial}{\partial x_2}, -\frac{\partial}{\partial x_3}\right).$$

Taking the Hermitian conjugate of (2.13) we get:

$$-i\partial_\mu \psi^+ (\gamma^\mu)^+ - m\psi^+ = 0$$

or

$$-i\partial_\mu \psi^+ (\gamma^\mu)^+ \gamma_0 - m\psi^+ \gamma_0 = 0.$$

Defining now $\bar{\psi} = \psi^+ \gamma_0$ and noting that $\gamma_0 \gamma_\mu^+ \gamma_0 = \gamma_\mu$ we get:

$$\bar{\psi}(i\gamma^\mu \partial_\mu + m) = 0 \text{ (conjugate Dirac equation)}, \tag{2.14}$$

with the understanding that ∂_μ acts on the left. Eq. (2.14) is called the conjugate of Eq. (2.13).

Multiplying (2.13) from the left with $\bar{\psi}$ and (2.14) from the right with ψ and adding the two we find:

$$\bar{\psi}\gamma^\mu (\partial_\mu)\psi + (\partial_\mu \bar{\psi})\gamma^\mu \psi = 0.$$

This can be written as:

$$\partial_\mu \left(\bar{\psi}\gamma^\mu \psi\right) = 0$$

or finally

$$\partial_\mu j^\mu = 0, \quad j^\mu = \bar{\psi}\gamma^\mu \psi. \tag{2.15}$$

This defines the (probability) current which is conserved. In particular we find:

$$j^0 = \bar{\psi}\gamma^0 \psi = \psi^+ \psi = |\psi_1|^2 + |\psi_2|^2 + |\psi_3|^2 + |\psi_4|^2 \geq 0, \tag{2.16}$$

which is positive definite. Thus the Dirac equation, being of first order in time, avoids the problem of negative probability, a problem that plagued the Klein-Gordon equation.

2.2.1 *The spinors u and v*

For a free particle, $\psi = ue^{-i(Et-\vec{p}\cdot\vec{x})}$, $i\partial_\mu = (i)(-ip_\mu = p_\mu \Rightarrow$

$$(i\gamma^\mu\partial_\mu - m)\psi = 0 \Leftrightarrow (p_\mu\gamma^\mu - m)u = 0 \tag{2.17}$$

and $\bar{\psi}= \bar{u}e^{i(Et-\vec{p}\cdot\vec{x})}$, $i\partial_\mu = (i)(ip_\mu = -p_\mu \Rightarrow$

$$\bar{\psi}(i\gamma^\mu\partial_\mu + m) = 0 \Leftrightarrow \bar{u}(\gamma^\mu p_\mu - m) = 0. \tag{2.18}$$

Note that in momentum space both expressions contain $-m$.

Multiplying Eq. (2.17) from the left with $\bar{\psi}\gamma^\nu$ and Eq. (2.18) from the right with $\gamma^\nu\psi$ and adding the two we get:

$$\bar{u}\left(\gamma^\mu\gamma^\nu + \gamma^\nu\gamma^\mu\right)p_\mu u = 2m\bar{u}\gamma^\nu u.$$

Then using the fundamental Eq. (2.12) we get:

$$\bar{u}\gamma^\nu u = \bar{u}u\frac{p^\nu}{m}. \tag{2.19}$$

Returning to the plane wave solution we write :

$$\psi \sim e^{ipx}u \text{ (outgoing)}, \psi \sim e^{-ipx}u \text{ (incoming)}, px = Et - \vec{x}\cdot\vec{p}. \tag{2.20}$$

In this context we choose:

$$\text{positive energy: } \psi^{(+)} = e^{-ipx}u, \text{ negative energy: } \psi^{(-)} = e^{ipx}\upsilon$$

we find respectively

$$(\not{p} - m)u = 0, \quad (\not{p} + m)\upsilon = 0 \tag{2.21}$$

(note the sign difference). The quantity $\not{p}$ defined by:

$$\not{p} = p_\mu\gamma^\mu = E\beta - \mathbf{p}_k\gamma^k = \begin{pmatrix} E & -\mathbf{p}.\boldsymbol{\sigma} \\ \mathbf{p}.\boldsymbol{\sigma} & -E \end{pmatrix}. \tag{2.22}$$

We will consider each case separately.

- the positive energy solution:

$$(\not{p} - m)u = 0,$$

$$u = \begin{pmatrix} \phi \\ \chi \end{pmatrix}, \ \phi, \chi \text{ two component spinors :}$$

$$\begin{pmatrix} 1 \\ 0 \end{pmatrix} \Leftrightarrow S_z = \frac{1}{2}, \begin{pmatrix} 0 \\ 1 \end{pmatrix} \Leftrightarrow S_z = -\frac{1}{2}. \tag{2.23}$$

Then noting that

$$\not{p} = \begin{pmatrix} E & -\vec{p}\cdot\sigma \\ \vec{p}\cdot\sigma & -E \end{pmatrix},$$

we find

$$\begin{pmatrix} E-m & -\vec{p}\cdot\sigma \\ \vec{p}\cdot\sigma & -E-m \end{pmatrix}\begin{pmatrix} \phi \\ \chi \end{pmatrix} = 0. \tag{2.24}$$

That is

$$(E-m)\phi - (\vec{p}\cdot\vec{\sigma})\chi = 0,$$

$$(E+m)\chi = \vec{\sigma}\cdot\vec{p}\phi \Rightarrow \tag{2.25}$$

$$\chi = \frac{\vec{\sigma}\cdot\vec{p}}{E+m}\phi, \text{ or } u = N\begin{pmatrix} \phi \\ \frac{\vec{\sigma}\cdot\vec{p}}{E+m}\phi \end{pmatrix}. \tag{2.26}$$

Since the two component spinors are normalized, $|\phi|^2 = 1$, the normalization condition $\bar{u}u = 1$ yields:

$$|N|^2(1 - \frac{(\vec{\sigma}\cdot\vec{p})(\vec{\sigma}\cdot\vec{p})}{(E+m)^2}) = 1, |N|^2(1 - \frac{p^2}{(E+m)^2})$$

$$= 1 \Rightarrow |N|^2(1 - \frac{E-m}{E+m}) = 1$$

or

$$|N|^2\frac{2m}{E+m} \Rightarrow N = \sqrt{\frac{E+m}{2m}}.$$

Thus

$$u = \sqrt{\frac{E+m}{2m}}\begin{pmatrix} \phi \\ \frac{\vec{\sigma}\cdot\vec{p}}{E+m}\phi \end{pmatrix}. \tag{2.27}$$

We note that, with this normalization:

$$u^\dagger u = \frac{E+m}{2m}\left[1 + \frac{E^2-m^2}{(E+m)^2}\right] = \frac{E+m}{2m}\left(\frac{E+m+E-m}{E+m}\right) = \frac{E}{m}.$$

Some authors use the normalization:

$$\bar{u}u = 2m \Rightarrow u = \sqrt{(E+m)}\begin{pmatrix} \phi \\ \frac{\vec{\sigma}\cdot\vec{p}}{E+m}\phi \end{pmatrix},$$

in which case:

$$u^\dagger u = (E+m)\frac{2E}{E+m} = 2E.$$

- Negative energy solution:

$$(\not{p} + m)v = 0,$$

$$\begin{pmatrix} E+m & -\vec{p}\cdot\sigma \\ \vec{p}\cdot\sigma & -E+m \end{pmatrix} \begin{pmatrix} \phi \\ \chi \end{pmatrix} = 0$$

$$(E+m)\phi = (\vec{\sigma}\cdot\vec{p})\chi, (\vec{\sigma}\cdot\vec{p})\phi + = E - m)\chi$$

Solving again for the coefficient of $E+m$, in this case ϕ, we get

$$\phi = \frac{\vec{\sigma}\cdot\vec{p}}{E+m}\chi \Rightarrow v = N \begin{pmatrix} \frac{\vec{\sigma}\cdot\vec{p}}{E+m}\chi \\ \chi \end{pmatrix}. \tag{2.28}$$

If we choose again:

$$N = \sqrt{\frac{E+m}{2m}} \Rightarrow v = \sqrt{\frac{E+m}{2m}} \begin{pmatrix} \frac{\vec{\sigma}\cdot\vec{p}}{E+m}\chi \\ \chi \end{pmatrix},$$

We can, of course, rewrite it as:

$$v = \sqrt{\frac{E+m}{2m}} \begin{pmatrix} \frac{\vec{\sigma}\cdot\vec{p}}{E+m}\phi \\ \phi \end{pmatrix}. \tag{2.29}$$

Then

$$\bar{v}v = \frac{E+m}{2m}\left[\frac{p^2}{(E+m)^2} - 1\right] = \frac{E+m}{2m}\left[\frac{(E+m)(E-m)}{(E+m)^2} - 1\right] =$$

$$\frac{E+m}{2m}\left[\frac{E-m}{E+m} - 1\right] = \frac{E+m}{2m}\frac{-2m}{E+m} = -1.$$

i.e.

$$v^\dagger v = E/m.$$

Sometimes the choice $N = \sqrt{E+m}$ is made. Then

$$v^\dagger v = (E+m)\left[\frac{(E+m)(E-m)}{(E+m)^2} + 1\right] = (E+m)\frac{2E}{E+m} = 2E.$$

Summarizing for plane wave solutions, by choosing the momentum as the quantization axis, we have:

$$u_1 = \sqrt{\frac{E+m}{2m}} \begin{pmatrix} 1 \\ 0 \\ \frac{p}{E+m} \\ 0 \end{pmatrix}, \quad v_1 = \sqrt{\frac{E+m}{2m}} \begin{pmatrix} \frac{p}{E+m} \\ 0 \\ 1 \\ 0 \end{pmatrix},$$

$$u_2 = \sqrt{\frac{E+m}{2m}} \begin{pmatrix} 0 \\ 1 \\ 0 \\ -\frac{p}{E+m} \end{pmatrix}, \quad v_2 = \sqrt{\frac{E+m}{2m}} \begin{pmatrix} 0 \\ -\frac{p}{E+m} \\ 0 \\ 1 \end{pmatrix}. \tag{2.30}$$

In the static limit:

$$u_1 = \begin{pmatrix} 1 \\ 0 \\ 0 \\ 0 \end{pmatrix}, u_2 = \begin{pmatrix} 0 \\ 1 \\ 0 \\ 0 \end{pmatrix}, v_1 = \begin{pmatrix} 0 \\ 0 \\ 1 \\ 0 \end{pmatrix}, v_2 = \begin{pmatrix} 0 \\ 0 \\ 0 \\ 1 \end{pmatrix}. \tag{2.31}$$

In this presentation we can see that one cannot throw away the negative energy solutions, since, then, one does not have a complete set.

With the above ingredients one can summarize the results as follows:

$$\bar{u}(\mathbf{p}, s)u(\mathbf{p}, s) = 1 \ , \ \bar{v}(\mathbf{p}, s)v(\mathbf{p}, s) = -1$$

$$\bar{u}(\mathbf{p}, s)\gamma^\mu u(\mathbf{p}, s) = \frac{1}{m}p^\mu \ , \ \bar{v}(\mathbf{p}, s)\gamma^\mu v(\mathbf{p}, s) = -\frac{1}{m}p^\mu$$

$$\sum_{s=\pm 1/2} u(\mathbf{p}, s)\bar{u}(\mathbf{p}, s) = \frac{1}{2m}(\not{p} + m) \ , \ v(\mathbf{p}, s)\bar{v}(\mathbf{p}, s) = \frac{1}{2m}(\not{p} - m), \ \not{p} = p_\mu \gamma^\mu \tag{2.32}$$

Note: To prove the second of the above equations you observe that the right hand side must be proportional to p^μ, the only available 4-vector. You write it as Λp^μ. Then you multiply the resulting equation with p_μ and you make a Lorentz contraction. After that, using the Dirac equation, you find $\Lambda = 1/m$ and $\Lambda = -1/m$ for the spinors u and v respectively.
The third equation is, of course, a matrix equation.

Returning back to the free Dirac equation and considering fermion spinor

$$u = \begin{pmatrix} \phi \\ \chi \end{pmatrix},$$

we get:

$$(p_\mu \gamma^\mu - m) \begin{pmatrix} \phi \\ \chi \end{pmatrix} = 0, \quad p_\mu \gamma^\mu = E_0 \gamma^0 - \vec{p} \cdot \vec{\gamma},$$

$$\Rightarrow \begin{pmatrix} E_0 - m & -\vec{p} \cdot \vec{\sigma} \\ \vec{p} \cdot \vec{\sigma} & -E - m \end{pmatrix} \begin{pmatrix} \phi \\ \chi \end{pmatrix} = 0 \tag{2.33}$$

or

$$\begin{aligned} &(E - m)\phi = (\vec{p} \cdot \vec{\sigma})\chi, \\ &(\vec{p} \cdot \vec{\sigma})\phi = (E + m)\chi \Rightarrow \\ &(E - m)\phi = \frac{\vec{\sigma} \cdot \vec{p}}{E + m} \vec{\sigma} \cdot \vec{p}\phi \Rightarrow \\ &((E - m) - \frac{p^2}{E + m})\phi = 0. \end{aligned} \tag{2.34}$$

This is an eigenvalue problem, leading to the solution:

$$E - m = \frac{p^2}{E+m} \Rightarrow E^2 - m^2 = p^2 \Rightarrow E = \pm\sqrt{m^2+p^2}.$$

In other words the energy can be both positive and negative.

2.2.2 *Projection operators*

The notion of projection operators is useful in computation. A projection operator projects a vector defined into a space into a subspace. Consider, e.g., a complete set of states $\mathcal{N} = \{[|i\rangle, i = 1, 2, \cdots, N\}$ and a complete set of states $\mathcal{N}' = \{|\alpha\rangle, \alpha = 1, 2, \cdots, N'\}$, $\mathcal{N}' \subset \mathcal{N}$. Then a projection operator P is defined by:

$$P|\psi\rangle = \sum_{\alpha=1}^{N'} c_\alpha |\alpha\rangle,\ c_\alpha = \text{constants},\ \psi \in \mathcal{N}.$$

A special case is if:

$$P|\psi\rangle = |\psi\rangle.$$

Now we notice the following:

- $(\not p + m)/2m$ is a projection operator for the spinor u.
  Indeed using the Dirac equation, we get:

$$\frac{\not p + m}{2m} u = \frac{\not p - m}{2m} u + \frac{2m}{2m} u = u. \tag{2.35}$$

- $(-\not p + m)/2m$ is a projection operator for the spinor v.
  Again

$$\frac{-\not p + m}{2m} v = \frac{-\not p - m}{2m} v + \frac{2m}{2m} v = v. \tag{2.36}$$

2.2.3 *Various representations of the Dirac matrices*

There exist many equivalent representations of the Dirac matrices. The most common are:

- The above introduced Dirac-Pauli representation γ^μ_{DP}:

$$\gamma^0 = \begin{pmatrix} 1 & 0 \\ 0 & -1 \end{pmatrix}, \vec{\gamma} = \begin{pmatrix} 0 & \vec{\sigma} \\ -\vec{\sigma} & 0 \end{pmatrix}, \gamma^5 = \begin{pmatrix} 0 & 1 \\ 1 & 0 \end{pmatrix}.$$

- The Weyl representation γ^μ_W, to be discussed below in connection with the helicity:

$$\gamma^5 = \begin{pmatrix} -1 & 0 \\ 0 & 1 \end{pmatrix}, \vec{\gamma} = \begin{pmatrix} 0 & \vec{\sigma} \\ -\vec{\sigma} & 0 \end{pmatrix}, \gamma^0 = \begin{pmatrix} 0 & 1 \\ 1 & 0 \end{pmatrix}.$$

This is accomplished by a unitary matrix U, which diagonalizes the Dirac matrix γ_5, e.g.

$$U = \frac{1}{\sqrt{2}} \begin{pmatrix} 1 & 1 \\ -1 & 1 \end{pmatrix}, \quad U^\dagger = \frac{1}{\sqrt{2}} \begin{pmatrix} 1 & -1 \\ 1 & 1 \end{pmatrix},$$

$$U^\dagger U = \frac{1}{2} \begin{pmatrix} 1 & 1 \\ -1 & 1 \end{pmatrix} \begin{pmatrix} 1 & -1 \\ 1 & 1 \end{pmatrix} = \frac{1}{2} \begin{pmatrix} 2 & 0 \\ 0 & 2 \end{pmatrix} = \mathbf{1}. \tag{2.37}$$

Indeed this is equivalent to the D-P representation, since:

$$\boxed{\gamma^\mu_{DP} = U^\dagger \gamma^\mu_W U.} \tag{2.38}$$

- The Majorana representation γ^μ_M

$$\gamma^0_M = \begin{pmatrix} 1 & 0 \\ 0 & -1 \end{pmatrix}, \gamma^1_M = \begin{pmatrix} i\sigma_3 & 0 \\ 0 & i\sigma_3 \end{pmatrix},$$
$$\gamma^2_M = \begin{pmatrix} 0 & -\sigma_2 \\ \sigma_2 & 0 \end{pmatrix}, \gamma^3_M = \begin{pmatrix} -i\sigma_1 & 0 \\ 0 & -i\sigma_1 \end{pmatrix}, \gamma^5_M = \begin{pmatrix} \sigma_2 & 0 \\ 0 & -\sigma_2 \end{pmatrix}. \tag{2.39}$$

This is equivalent to the Dirac-Pauli representation since:

$$\gamma^\mu_{DP} = U^\dagger \gamma^\mu_M U,$$

$$U = \frac{1}{\sqrt{2}} \gamma^0_{DP} \left(1 + \gamma^2_{DP}\right).$$

2.3 Interpretation of negative energy solutions

What is the interpretation of the negative energy solutions? Dirac gave the following interpretation:

(1) Fermions obey the Pauli principle
(2) In the ground state of the system all negative states are fully occupied.
(3) Given enough energy $> 2mc^2$, a negative energy particle, e.g. an electron, can be excited. Leaving behind an electron hole $\Leftrightarrow$ positron. Thus an electron positron pair is created out of the vacuum.

The above interpretation is fine with fermions, which obey the Pauli Principle and the Dirac equation.

In general consider the phase:

$$e^{-i(-E)(-t)} = e^{-iEt}.$$

noting that $Et < 0$, either if $E < 0$, $t > 0$ or $E > 0$, $t < 0$, we may make the correspondence:

A particle with $E < 0$, $t < 0 \Leftrightarrow$ antiparticle with $E > 0$, $t > 0$

Thus suppose the time t runs upward, $t \uparrow$, we have:

e^+ momentum p$\downarrow$ $(E < 0) \Leftrightarrow e^-$ momentum $p \uparrow (E > 0)$

It can be generalized to any kind of particles if one interprets the negative energy states as associated with antiparticles. The negative energy particles can be thought as traveling backwards in time, a picture first proposed by Stueckelberg and Feynman. The situation is exhibited in Figs. 2.1 and 2.2. Note that in Fig. 2.2 all actual particles move forward in time, the intermediate ones, called "virtual", can move both ways.

Given enough energy particle-antiparticle pairs can be crated out of the vacuum. So the negative value of the probability density encountered in th he Klein-Gordon equation is simply a manifestation of the fact that in relativity the number of particles plus antiparticles is not conserved. One can produce particles and antiparticles out of the vacuum. The only restriction is that conservation laws have to be obeyed (energy, momentum, angular momentum, all kinds of "charges", including, of course, the electric charge). Conservation of charge implies that the number of charged particles must be the same with the number of their anti antiparticles. This restriction, of course, does not apply if the particles are absolutely neutral.

The Klein-Gordon equation was prematurely abandoned. It gave us the benefit of the invention of the Dirac theory. It is valid for scalar particles in the same way as the Dirac theory governs the behavior of spin 1/2 particles.

2.4 Transformation of spinors under space inversion

The space inversion is defined

$$P : \mathbf{x} \to -\mathbf{x},\, t \to t.$$

Under such a transformation:

$$\mathbf{p} \to -\mathbf{p}, \mathbf{L} \to \mathbf{L}\,(\mathbf{L} = \mathbf{x} \times \mathbf{p}), \mathbf{s} \to \mathbf{s}\,(\mathbf{s} = \text{spin operator}), \mathbf{E} \to -\mathbf{E}, \mathbf{B} \to \mathbf{B} \text{ etc.} \tag{2.40}$$

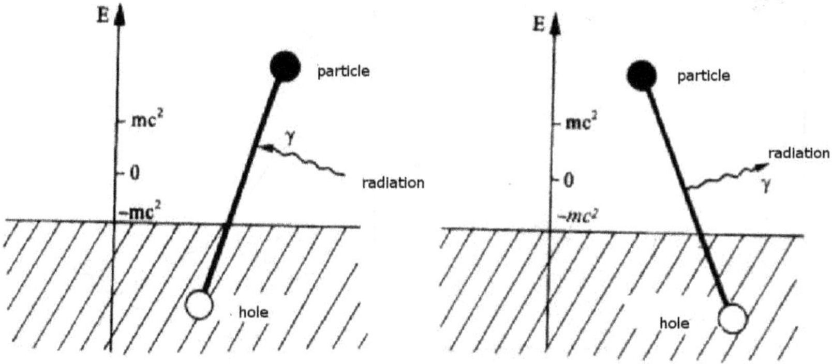

Fig. 2.1: (a): Pair production. A particle of negative energy absorbs radiation and it is transformed into a state of positive energy. (b) Pair annihilation. A particle with positive energy falls in a hole, an entity with negative energy, producing radiation with positive energy.

In the case of the Dirac spinors this operator is generalized to be the product of two operators, P followed by multiplication with γ_5, i.e.

$$\mathcal{P} = \gamma_0 P \tag{2.41}$$

Then

$$\psi(\mathbf{x}, t) \rightarrow \psi'(\mathbf{x}, t) = \mathcal{P}\psi(\mathbf{x}, t)$$

In the special case of a free particle:

$$\psi(\mathbf{x}, t) = e^{i(Et - \mathbf{p}.\mathbf{x})}\sqrt{\frac{E+m}{2m}}\begin{pmatrix} \phi \\ \frac{\vec{\sigma}\cdot\vec{p}}{E+m}\phi \end{pmatrix}. \tag{2.42}$$

We find

$$\psi'(\mathbf{x}, t) = e^{i(Et + \mathbf{p}.\mathbf{x})}\sqrt{\frac{E+m}{2m}}\begin{pmatrix} \phi \\ \frac{-\vec{\sigma}\cdot\vec{p}}{E+m}\phi \end{pmatrix}, \tag{2.43}$$

as expected. Also

$$u(\mathbf{p}, s) = \sqrt{\frac{E+m}{2m}}\begin{pmatrix} \phi \\ \frac{\vec{\sigma}\cdot\vec{p}}{E+m}\phi \end{pmatrix} \rightarrow u'(\mathbf{p}', s) = \gamma_0 u(\mathbf{p}, s) = \sqrt{\frac{E+m}{2m}}\begin{pmatrix} \phi \\ \frac{-\vec{\sigma}\cdot\vec{p}}{E+m}\phi \end{pmatrix} \tag{2.44}$$

Associated with this symmetry we associate a multiplicative quantum number π, called parity, which takes values ± 1. Parity is conserved in electromagnetic and strong interactions, but, to the surprise of everyone, it was found to be violated in weak interactions (see section 12.1.1).

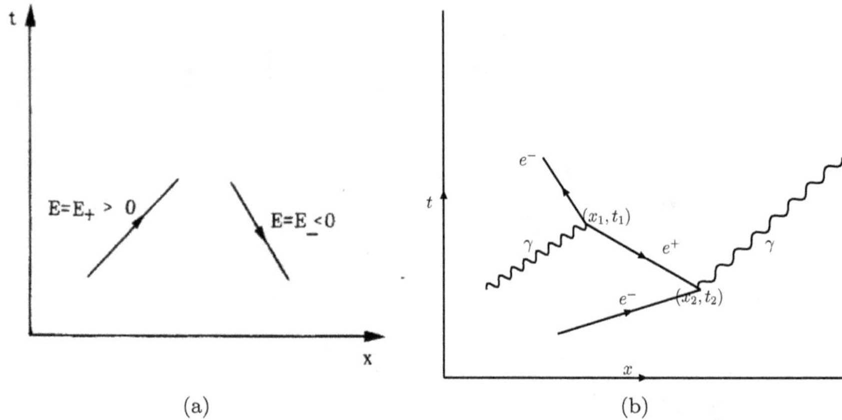

Fig. 2.2: The Stueckelberg-Feynman picture of negative energy interpreted as motion backwards in time. (a) A particle with positive energy + moves as usual. The antiparticle with negative energy − can be thought of as a particle with negative energy moving backwards in time. Both of them are moving to the right, x increases. (b) At the space-time point (x_1, t_1) one sees a photon producing an electron positron pair. The electron is moving forward in time, while the anti-particle (positron) moves backward in time. It the space time point (x_2, t_2) this positron meets another electron and they annihilate into a photon. In this process the positron is virtual.

2.5 The notion of helicity

(1) Ordinary quantum mechanics.
The helicity operator is defined by

$$h = \vec{\sigma} \cdot \hat{p}.$$

It has two eigenvectors:

$$\psi_1 = \frac{1}{2}(1 - \vec{\sigma} \cdot \hat{p})\phi, \quad \psi_2 = \frac{1}{2}(1 + \vec{\sigma} \cdot \hat{p})\phi.$$

Indeed

$$h\psi_1 = \frac{1}{2}(\vec{\sigma} \cdot \hat{p} - \vec{\sigma} \cdot \hat{p}\vec{\sigma} \cdot \hat{p})\psi = -\frac{1}{2}(1 - \vec{\sigma} \cdot \hat{p})\phi = -\psi_1,$$
$$h\psi_2 = \frac{1}{2}(1 + \vec{\sigma} \cdot \hat{p})\phi = \psi_2. \tag{2.45}$$

Thus, taking the direction of momentum as the quantization axis, we see that ψ_2 corresponds to the spin parallel to the momentum and ψ_1 to the case the spin is in the opposite to the momentum direction.

(2) Massless particles in the Dirac theory.
In the context of the special theory of relativity, the helicity has a meaning only for massless particles. For massive particles it is not an invariant quantity. A particle's helicity can be positive in one frame, but it may become negative in another frame, since the particle's momentum may become opposite.

In particle physics we deal with chiral fields defined by:

$$\psi_L = \frac{1}{2}(1-\gamma_5)\psi \text{ (left)}, \quad \psi_R = \frac{1}{2}(1+\gamma_5)\psi \text{ (right)}, \quad \gamma_5 = \begin{pmatrix} 0 & 1 \\ 1 & 0 \end{pmatrix} \tag{2.46}$$

Note that since $\gamma_5^2 = 1$ it is trivial to shoe

$$\gamma_5\psi_L = -\psi_L, \quad \gamma_5\psi_R = \psi_L R$$

that is to say thi chiral fields are eigenfields of γ_5. One also defines the helicity operator:

$$\Sigma = \begin{pmatrix} \vec{\sigma}\cdot\hat{p} & 0 \\ 0 & \vec{\sigma}\cdot\hat{p} \end{pmatrix}. \tag{2.47}$$

We consider the following cases:

α) Positive energy:

$$\psi = \begin{pmatrix} 1 \\ \vec{\sigma}\cdot\hat{p} \end{pmatrix}\phi, \tag{2.48}$$

$$\begin{aligned}\psi_L &= \frac{1}{2}(1-\gamma_5)\psi = \frac{1}{2}(1-\gamma_5)\begin{pmatrix} 1 \\ \vec{\sigma}\cdot\hat{p} \end{pmatrix}\phi \\ &= \frac{1}{2}\begin{pmatrix} 1 & -1 \\ -1 & 1 \end{pmatrix}\begin{pmatrix} 1 \\ \vec{\sigma}\cdot\hat{p} \end{pmatrix}\phi = \begin{pmatrix} \frac{1}{2}(1-\vec{\sigma}\cdot\hat{p})\phi \\ -\frac{1}{2}(1-\vec{\sigma}\cdot\hat{p})\phi \end{pmatrix}.\end{aligned}$$

Similarly

$$\psi_R = \frac{1}{2}(1+\gamma_5)\psi = \begin{pmatrix} \frac{1}{2}(1+\vec{\sigma}\cdot\hat{p})\phi \\ \frac{1}{2}(1+\vec{\sigma}\cdot\hat{p})\phi \end{pmatrix}. \tag{2.49}$$

Then it is trivial to show that

$$\Sigma\psi_L = -\psi_L, \quad \Sigma\psi_R = \psi_R$$

For positive energy solutions the $+(-)$ helicity coincides with the right (left) chirality.

β) Negative energy:

$$\psi = \begin{pmatrix} -\vec{\sigma}\cdot\hat{p} \\ 1 \end{pmatrix}\phi, \tag{2.50}$$

$$\psi_L = \frac{1}{2}(1-\gamma_5)\psi = \frac{1}{2}(1-\gamma_5)\begin{pmatrix} -\vec{\sigma}\cdot\hat{p} \\ 1 \end{pmatrix}\phi$$
$$= \frac{1}{2}\begin{pmatrix} 1 & -1 \\ -1 & 1 \end{pmatrix}\begin{pmatrix} -\vec{\sigma}\cdot\hat{p} \\ 1 \end{pmatrix}\phi = \begin{pmatrix} -\frac{1}{2}(1+\vec{\sigma}\cdot\hat{p}) \\ \frac{1}{2}(1+\vec{\sigma}\cdot\hat{p}) \end{pmatrix}\phi,$$

$$\psi_R = \frac{1}{2}(1+\gamma_5)\psi = \frac{1}{2}(1+\gamma_5)\begin{pmatrix} -\vec{\sigma}\cdot\hat{p} \\ 1 \end{pmatrix}\phi = \frac{1}{2}\begin{pmatrix} 1 & 1 \\ 1 & 1 \end{pmatrix}\begin{pmatrix} -\vec{\sigma}\cdot\hat{p} \\ 1 \end{pmatrix}\phi$$
$$= \begin{pmatrix} \frac{1}{2}(1-\vec{\sigma}\cdot\hat{p})\phi \\ \frac{1}{2}(1-\vec{\sigma}\cdot\hat{p})\phi \end{pmatrix}, \tag{2.51}$$

$$\Sigma\psi_L = \psi_L, \quad \Sigma\psi_R = -\psi_R.$$

For negative energy solutions the $-(+)$ helicity coincides with the right (left) chirality.

We stress that the chiral fields are not eigenfields of the helicity operator, if they have non zero mass.

(3) The Weyl representation of spinors

We choose the Weyl representation of the Dirac matrices.

$$\vec{\gamma} = \begin{pmatrix} 0 & \vec{\sigma} \\ -\vec{\sigma} & 0 \end{pmatrix}, \gamma^0 = \begin{pmatrix} 0 & 1 \\ 1 & 0 \end{pmatrix}, \gamma^5 = \begin{pmatrix} -1 & 0 \\ 0 & 1 \end{pmatrix}.$$

Then

$$\psi = \begin{pmatrix} \phi \\ \chi \end{pmatrix},$$
$$\psi_L = \begin{pmatrix} 1 & 0 \\ 0 & 0 \end{pmatrix}\begin{pmatrix} \phi \\ \chi \end{pmatrix} = \begin{pmatrix} \phi \\ 0 \end{pmatrix},$$
$$\psi_R = \begin{pmatrix} 0 & 0 \\ 0 & 1 \end{pmatrix}\begin{pmatrix} \phi \\ \chi \end{pmatrix} = \begin{pmatrix} 0 \\ \chi \end{pmatrix},$$
$$\Leftrightarrow \phi = \psi_L, \chi = \psi_R. \tag{2.52}$$

The Dirac equation becomes:

$$\begin{pmatrix} 0 & E-\vec{\sigma}\cdot\vec{p} \\ E+\vec{\sigma}\cdot\vec{p} & 0 \end{pmatrix}\begin{pmatrix} \phi \\ \chi \end{pmatrix} = m\begin{pmatrix} \phi \\ \chi \end{pmatrix} \to$$

$$(E - \vec{\sigma} \cdot \vec{p})\chi = m\phi, \quad (E + \vec{\sigma} \cdot \vec{p})\phi = m\chi \rightarrow$$

$$(E - \vec{\sigma} \cdot \vec{p})\psi_R = m\psi_L, \quad (E + \vec{\sigma} \cdot \vec{p})\psi_L = m\psi_R \text{ (Weyl equations)} \tag{2.53}$$

For massless particles ($m = 0$) The Weyl equations become:

$$(E - \vec{\sigma} \cdot \vec{p})\psi_R = 0, \quad (E + \vec{\sigma} \cdot \vec{p})\psi_L = 0. \tag{2.54}$$

Thus:

- $E > 0$

$$(1 - \vec{\sigma} \cdot \hat{p})\psi_R = 0 \Rightarrow \vec{\sigma} \cdot \vec{p}\psi_R = \psi_R, \quad (1 + \vec{\sigma} \cdot \hat{p})\psi_L = 0 \Rightarrow \vec{\sigma} \cdot \vec{p}\psi_L = -\psi_L, \tag{2.55}$$

$$\psi_R \leftrightarrow \text{ positive helicity}, \quad \psi_L \leftrightarrow \text{ negative helicity}.$$

- $E < 0$

$$(1 + \vec{\sigma} \cdot \hat{p})\psi_R = 0 \Rightarrow \vec{\sigma} \cdot \vec{p}\psi_R = -\psi_R, \quad (1 - \vec{\sigma} \cdot \hat{p})\psi_L = 0 \Rightarrow \vec{\sigma} \cdot \vec{p}\psi_L = \psi_L, \tag{2.56}$$

$$\psi_R \leftrightarrow \text{ negative helicity}, \quad \psi_L \leftrightarrow \text{ positive helicity}.$$

The only particles believed to be massless is the photon and the neutrino. While the photon is known to exist in both helicities, the neutrino has been found to only left handed (helicity -1) (see section 12.2.1).

2.6 Charge conjugation for 4-spinors

Charge conjugation is an operation which changes the sign of the electric charge of any particle The charge conjugation operator will be denoted by C. Under C

$$q \rightarrow -q, \, \mathbf{J} \rightarrow -\mathbf{J}, \, \mathbf{E} \rightarrow -\mathbf{E}, \, \mathbf{B} \rightarrow -\mathbf{B} \tag{2.57}$$

(see section 12.3). The charge conjugation in Dirac theory is in addition defined by:

$$\psi \rightarrow \psi^c = C(\bar{\psi})^T = C\gamma_0\psi^*, \text{ (under charge conjugation transformation).}$$

Then ψ^c satisfies the Dirac equation

$$(i\partial\gamma^\mu - m)\,\psi^c = 0, \tag{2.58}$$

provided that C is non singular and satisfies the condition:

$$C^{-1}\gamma_\mu C = -\left(\gamma_\mu\right)^T . \tag{2.59}$$

Indeed then

$$C^{-1}\psi^c = \bar{\psi}^T \to \left(C^{-1}\psi^c\right)^T = \bar{\psi} \to \bar{\psi} = \left(\psi^c\right)^T \left(C^{-1}\right)^T .$$

Thus the Dirac equation:

$$\bar{\psi}\left(i\partial_\mu \gamma^\mu + m\right) = 0$$

implies

$$\left(\psi^c\right)^T \left(C^{-1}\right)^T \left(i\partial_\mu \gamma^\mu + m\right) = 0 \Rightarrow \left(i\partial_\mu \left(\gamma^\mu\right)^T + m\right) C^{-1}\psi^c$$

$$\Rightarrow \left(i\partial_\mu C \left(\gamma^\mu\right)^T C^{-1} + m\right)\psi^c$$

That is ψ^c satisfies the Dirac equation, (2.58), provided that C is non singular and it can be chosen to meet the condition of Eq. (2.59). In fact such solutions exist and one may impose the additional conditions:

$$C^{-1} = -C = C^+ = C^T .$$

A possible choice:

$$C = i\gamma_2\gamma_0$$

One can show that under the charge conjugation transformation:

$$\bar{\psi} \to \left(\bar{\psi}\right)^c = C\gamma_0 \left(\bar{\psi}\right)^* = C\gamma_0 \left(\psi^+\gamma_0\right)^* = C\gamma_0\gamma_0\psi^T = C\psi^T .$$

Another useful relation is

$$\bar{\psi^c} = \left(\psi^c\right)^+ \gamma_0 = \left(C\gamma_0\psi^*\right)^+ \gamma_0 = \psi^T \gamma_0 C^+ \gamma 0 = -\psi^T \gamma_0 C \gamma 0 = \psi^T C,$$

Since $\gamma_0 C\gamma 0 = \gamma_0(i\gamma_2\gamma_0)\gamma_0 = (i\gamma_0\gamma_2) = -(i\gamma_2\gamma_0) = -C$

With this formalism Dirac showed that if a fermion exists, its charge conjugate also should exist, which was identified as its antiparticle. This proved to be one of the triumphs of physics.

Chiral states transform analogously:

$$\psi_{L,R} \to \psi^c_{L,R} = P_{L,R}\psi^c, \quad P_L = \frac{1}{2}(1-\gamma_5), \quad P_R = \frac{1}{2}(1+\gamma_5).$$

Note, however, that:

$$\left(\psi_R\right)^c = C\left(\bar{\psi}_R\right)^T = C\left(\bar{\psi}\frac{(1-\gamma_5)}{2}\right)^T = C\frac{(1-\gamma_5)}{2}\left(\bar{\psi}\right)^T =$$

$$\frac{(1-\gamma_5)}{2} C \left(\bar{\psi}\right)^T = (\psi^c)_L = \psi^c_L.$$

Similarly:

$$(\psi_L)^c = (\psi^c)_R = \psi^c_R.$$

We have only two independent chiral combinations. Usually these are taken to be:

$$(\psi_L, \psi_R) \text{ or } (\psi_L, (\psi_L)^c).$$

Charge conjugation is found to s a good symmetry in the case of the electromagnetic and strong interactions, but it is violated in weak interactions (see 12.3.3).

By combining the charge conjugation (C) and space inversion symmetry $\mathcal{P}$ we obtain a new symmetry CP. This symmetry has also been found to be violated in weak interactions (see section 12.5.4). For many years the neutral Kaon system was the only one, which exhibited such violation, but in recent years it has been seen in few other systems as well.

2.7 Types of currents

We consider the following forms:

- Scalar form:

$$\bar{u}u \Leftrightarrow \text{transforms like a scalar; one component.} \quad (2.60)$$

- 4-vector form:

$$\bar{u}\gamma_\mu u \Leftrightarrow \text{transforms like a vector; 4 components.} \quad (2.61)$$

- Tensor form:

$$\bar{u}\frac{\sigma_{\alpha\beta}}{\sqrt{2}}u, \ \sigma_{\alpha\beta} = \frac{1}{2}\left(\gamma_\alpha\gamma_\beta - \gamma_\beta\gamma_\alpha\right), \quad (2.62)$$

  i.e. it transforms like a tensor (6 components).

- Pseudo-scalar[1] form:

$$\bar{u}\gamma_5 u \Leftrightarrow \text{transforms like a pseudoscalar; one component.} \quad (2.63)$$

- Pseudo-vector form:

$$\bar{u}\gamma_\mu\gamma_5 u \Leftrightarrow \text{transforms like a pseudovector; 4 components.} \quad (2.64)$$

[1]Show that the matrix γ_5 given by Eq. 2.46 satisfies the equations

$$\gamma_5 = i\gamma^0\gamma^1\gamma^2\gamma^3, \quad [\gamma_5, \gamma^\mu] = 0, \ \mu = 0, 1, 2, 3$$

We thus have a total of 16 components. The matrices

$$O_i = \{1,\ \gamma_\mu,\ \sigma_{\alpha\beta},\ \gamma_5,\ \gamma_\mu\gamma_5\}$$

form a complete set in the space formed by any combination of products of γ matrices.

Indeed

$$tr(O_\alpha O_\beta) = 0, \alpha \neq \beta.$$

That is they form an orthogonal set with respect to the scalar product defined by the trace. Furthermore, with the exception of the tensor:

$$tr(O_\ell O_\ell) = 4$$

For the tensor part:

$$\begin{aligned}
\sigma_{\alpha\beta}\sigma_{\alpha\beta} &= \frac{1}{4}(\gamma_\alpha\gamma_\beta - \gamma_\beta\gamma_\alpha)(\gamma_\alpha\gamma_\beta - \gamma_\beta\gamma_\alpha) \\
&= \frac{1}{4}[\gamma_\alpha\gamma_\beta\gamma_\alpha\gamma_\beta - \gamma_\alpha\gamma_\beta\gamma_\beta\gamma_\alpha - \gamma_\beta\gamma_\alpha\gamma_\alpha\gamma_\beta + \gamma_\beta\gamma_\alpha\gamma_\beta\gamma_\alpha], \\
tr(\sigma_{\alpha\beta}\sigma_{\alpha\beta}) &= \frac{4}{4}[-2g_{\alpha\alpha}g_{\beta\beta} - 2g_{\alpha\alpha}g_{\beta\beta}] = -4\ . \qquad (2.65)
\end{aligned}$$

To this end we note that the O_i are linearly independent. Indeed

$$\sum_k C_k O_k = 0 \Rightarrow \sum_k C_k O_j O_k = 0, \text{ some } j \Rightarrow$$

$$0 = tr\sum_k O_j O_k C_k = tr\sum \delta_{jk} C_k = C_j.$$

That is

$$\sum C_k O_k = 0 \Rightarrow C_k = 0.$$

Examples:

(1) Operator type: $\gamma_\mu\gamma_\nu\gamma_5$.

We will show that there exists coefficients such that:

$$\gamma_\mu\gamma_\nu\gamma_5 = \sum_k C_k O_k.$$

Then

$$C_j = tr(O_j\gamma_\mu\gamma_\nu\gamma_5).$$

From the trace properties, see chapter 7, we note the for the scalar and all operators with an odd number of γ_μ matrices the trace is zero. For the pseudoscalar part we find that

$$C_{ps} = \frac{1}{4}tr(\gamma_\mu\gamma_\nu\gamma_5)\gamma_5 = tr(\gamma_\mu\gamma_\nu) = g_{\mu\nu},$$

while for the tensor part:

$$C_T = \frac{1}{tr(\sigma_{\alpha\beta}\sigma_{\alpha\beta})} tr(\gamma_\mu\gamma_\nu\gamma_5\sigma_{\alpha\beta}).$$

That is:

$$\begin{aligned} C_T &= -\frac{1}{8}[tr\gamma_\mu\gamma_\nu\gamma_5\gamma_\alpha\gamma_\beta - tr(\gamma_\nu\gamma_\mu\gamma_5\gamma_\beta\gamma_\alpha)] \\ &= -\frac{1}{8}[tr\gamma_\mu\gamma_\nu\gamma_\alpha\gamma_\beta\gamma_5 - tr\gamma_\mu\gamma_\nu\gamma_\beta\gamma_\alpha\gamma_5] \\ &= -\frac{1}{8}i(\epsilon_{\mu\nu\alpha\beta} - \epsilon_{\mu\nu\beta\alpha})4 \\ &= -i\epsilon_{\mu\nu\alpha\beta}. \end{aligned}$$

Thus

$$\gamma_\mu\gamma_\nu\gamma_5 = \gamma_5 g_{\mu\nu} - \sum_{\alpha\beta}(i\epsilon_{\mu\nu\alpha\beta})\sigma_{\alpha\beta}.$$

(2) Operator type: $\gamma_\lambda\gamma_\mu\gamma_\nu$

Now only the operators with an odd number of matrices γ_μ contribute.

$$C^V_{\lambda\mu\nu\alpha} = \frac{1}{4}tr(\underbrace{\gamma_\lambda\gamma_\mu\gamma_\nu\gamma_\alpha}\ ) = g_{\mu\lambda}g_{\nu\alpha} + g_{\mu\nu}g_{\alpha\lambda} - g_{\mu\alpha}g_{\lambda\nu},$$

$$C^A_{\lambda\mu\nu\alpha} = \frac{1}{4}tr(\gamma_\lambda\gamma_\mu\gamma_\nu\gamma_\alpha\gamma_5),$$

$$C^A_{\lambda\mu\nu\alpha} = \frac{1}{4}tr\gamma_\lambda\gamma_\mu\gamma_\nu\gamma_\alpha\gamma_5 = i\epsilon_{\lambda\mu\nu\alpha}.$$

Thus

$$\gamma_\lambda\gamma_\mu\gamma_\nu = \sum_\alpha \left\{(g_{\lambda\mu}g_{\nu\alpha} + g_{\mu\nu}g_{\alpha\lambda} - g_{\alpha\mu}g_{\lambda\nu})\gamma_\alpha + i\epsilon_{\lambda\mu\nu\alpha}\gamma_\alpha\gamma_5\right\}.$$

2.7.1 Problems

(1) In the case of the Klein-Gordon the normalized free particle solutions are label by their momentum and are indicated by

$$\phi^{(+)}_{\mathbf{p}}(x) = \frac{1}{(2\pi)^{3/2}}\frac{1}{\sqrt{2\omega_p}}e^{-i(\omega_p t - \mathbf{px})},$$

$$\phi^{(-)}_{\mathbf{p}}(x) = \frac{1}{(2\pi)^{3/2}}\frac{1}{\sqrt{2\omega_p}}e^{i(\omega_p t + \mathbf{px})}$$

with $\omega_p = \sqrt{p^2 + m^2}$. The scalar product is defined by

$$\langle\psi_1|\psi_2\rangle = i\int d^3\mathbf{x}\,(\psi_1^*(\mathbf{x})\partial_0\psi_2(\mathbf{x}) - \psi_2(\mathbf{x})\partial_0\psi_1^*(\mathbf{x}))$$

Show that:

$$\langle\phi^{(+)}_{\mathbf{p}}(x)|\phi^{(+)}_{\mathbf{p}}(x)\rangle = 1,\ \langle\phi^{(-)}_{\mathbf{p}}(x)|\phi^{(-)}_{\mathbf{p}}(x)\rangle = -1,\ \langle\phi^{(+)}_{\mathbf{p}}(x)|\phi^{(-)}_{\mathbf{p}}(x)\rangle = 0$$

(2) Consider the the Dirac equation

$$(i\partial_\mu \gamma^\mu - m)\psi = 0 \tag{2.66}$$

- show that it is invariant under a set global Abelian transformations

$$S = e^{ig\theta}, \quad \theta = \text{constant real function} \tag{2.67}$$

but it is not invariant under a set of local (gauge) transformations $\theta = \theta(\mathbf{x}, x_0)$.
- Modify it suitably so that it is invariant under gauge transformations with $g = e$. Do you recognize the result?

(3) Repeat the previous procedure in the case of Klein-Gordon equation.

$$(\partial_\mu \partial^\mu + m^2)\phi = 0 \tag{2.68}$$

(4) Imagine that in the world there exit three spin 1/2 fermions with the same mass obeying the above Dirac equation with ψ representing three 4 component spinors, i.e. the column $\psi = (\psi_1, \psi_2, \psi_3)^T$. Repeat the analysis of problem 1, if the group of transformations is the familiar group $SU(3)$. Do you notice anything unusual?

(5) Repeat the analysis of the previous problem if instead of spin 1/2 particles one has three scalar particles $\phi = (\phi_1, \phi_2, \phi_3)^T$.

(6) Using the Dirac equation show that the current

$$J_\mu = \bar{\psi}\gamma_\mu\gamma_5\psi \tag{2.69}$$

is not conserved so long as the fermions ψ have a non zero mass.

(7) A chiral global transformation on the fermion fields is defined as

$$\psi_L \to \psi' L = e^{i\alpha_L \gamma_5}\psi_L, \psi_R \to \psi' R = e^{i\alpha_L \gamma_5}\psi_R,$$

$$\psi_L = \frac{1}{2}(1 - \gamma_5)\psi, \ \psi_R = \frac{1}{2}(1 + \gamma_5)\psi.$$

Noting that the the eigenvalues of γ_5 are ± 1 with ψ_L and ψ_R show that:

$$\psi' L = e^{-i\alpha_L}\psi_L, \ \psi' R = e^{i\alpha_L}\psi_R$$

Consider now the Lagrangian

$$\mathcal{L} = Y_f \bar{\psi}_L \psi_R \phi + \text{HC}, Y_f \text{ Yukawa coupling constant}$$

where HC stands for Hermitian Conjugate and ϕ is a scalar field transforming as $\phi \to \phi' = e^{-i\alpha}\phi$. Show that the above Lagrangian is invariant under global chiral transformations provided that the corresponding charges obey $\alpha = \alpha_R + \alpha_L$

(8) For the Dirac matrices:

i Show that the members of the set:

$$1, \gamma_\mu, \sigma_{\mu\nu}, \gamma_5, \gamma_\mu\gamma_5$$

are linearly independent and constitute a basis in the 16 dimensional space of the spinors.

ii Using the properties of traces express each of the expressions

$$\sigma_{\mu\nu}\gamma_5, \ \gamma_\lambda\gamma_\mu\gamma_\nu \text{ and } \gamma_\lambda\gamma_\mu\gamma_\nu\gamma_5$$

in terms of the above set.

(9) The electron magnetic moment. Suppose that the interaction of a photon with an electron is given by:

$$\mathcal{L} = -eJ_\lambda A^\lambda, \ J_\lambda = \bar{u}\gamma_\lambda u \tag{2.70}$$

where e is the unit of charge and A^λ describes the photon field (vector potential). See, e.g., section 1.13.4.

- Concentrate on the space part of this interaction $\lambda = k \neq 0$. Show that:

$$\mathcal{L} = e\bar{u}(\mathbf{p}_1', s')\gamma.\mathbf{A}u(\mathbf{p}_1, s). \tag{2.71}$$

Thus

$$\mathcal{L} = e\frac{E+m_e}{2m_e}\left(\phi^+(1, \frac{\boldsymbol{\sigma}.p_1'}{E+m_e})\begin{pmatrix} 0 & \boldsymbol{\sigma}.\mathbf{A} \\ \boldsymbol{\sigma}.\mathbf{A} & 0 \end{pmatrix}\begin{pmatrix} \phi \\ \phi\frac{\boldsymbol{\sigma}.p_1}{E+m_e} \end{pmatrix}\right) \tag{2.72}$$

- Show further that in the space of two component spinors one can write:

$$\mathcal{L} = \phi^+\Omega\phi, \ \Omega = e\left(\frac{\boldsymbol{\sigma}.p_1'}{2m_e}\boldsymbol{\sigma}.\mathbf{A} + \boldsymbol{\sigma}.\mathbf{A}\frac{\boldsymbol{\sigma}.p_1}{2m_e}\right) \tag{2.73}$$

- Using the fact that the Pauli matrices obey the rule:

$$\boldsymbol{\sigma}_\alpha\boldsymbol{\sigma}_\beta = \delta_{\alpha\beta} + i\epsilon_{\alpha\beta\gamma}\boldsymbol{\sigma}_\gamma$$

with $\epsilon_{\alpha\beta\gamma}$ the completely antisymmetric symbol. Show that

$$(\boldsymbol{\sigma}.\mathbf{B})(\boldsymbol{\sigma}.\mathbf{C}) = \mathbf{B}.\mathbf{C} + i\boldsymbol{\sigma}.(\mathbf{B}\times\mathbf{C})$$

As a result

$$\Omega = \frac{e}{2m_e}\left(\mathbf{P}.\mathbf{A} - i\boldsymbol{\sigma}.(\mathbf{q}\times\mathbf{A})\right), \ \mathbf{P} = \mathbf{p}_1 + \mathbf{p}_1', \ \mathbf{q} = \mathbf{p}_1 - \mathbf{p}_1' \tag{2.74}$$

where $\mathbf{q}$ is the momentum transfer to the photon.

- Noting that

$$\mathbf{q} \times A = \frac{\hbar}{i} \nabla \times A = \frac{\hbar}{i} \mathbf{B} \tag{2.75}$$

where $\mathbf{B}$ is the magnetic field, show further that

$$\Omega = \frac{e}{2m_e} (\mathbf{P}.\mathbf{A} - \hbar\boldsymbol{\sigma}.\mathbf{B}) = \frac{e}{2m_e} \mathbf{P}.\mathbf{A} + \boldsymbol{\mu}_e.\mathbf{B} \tag{2.76}$$

the second term yields the interaction of the electron spin with the magnetic field, i.e.

$$\boldsymbol{\mu}_e = -\frac{e\hbar}{2m_e} \boldsymbol{\sigma} = -\mu_B \boldsymbol{\sigma} = -g_s \mu_B \mathbf{s} \tag{2.77}$$

where $\mathbf{s}$ is the spin of the electron and

$$g_s = 2,\ \mu_B = \frac{e\hbar}{2m_e} = \text{Bohr's magneton}, e > 0 \tag{2.78}$$

$\boldsymbol{\mu}_e$ is the magnetic moment of the electron.
- Similarly for other elementary fermions, e.g. for the muon:

$$\boldsymbol{\mu}_\mu = \frac{m_e}{m_\mu} \boldsymbol{\mu}_e \tag{2.79}$$

i.e. the muon magnetic moment is about 210 times smaller than that of the electron.

So in the Dirac theory g_s takes twice the classical value, in agreement with experiment. In practice, of course, one should consider other higher order (loop) corrections, yielding the $g_s - 2$ corrections.

Chapter 3

The Standard Model; Particle Content and Symmetry

3.1 Brief history

The first step towards the Standard Model was Sheldon Glashow's discovery in 1961 of a way to combine the electromagnetic and weak interactions. In 1967 Steven Weinberg and Abdus Salam incorporated the Higgs mechanism into Glashow's electroweak interaction, giving it its modern form.

It was designed to explain all the then known physics, including the following key ingredients.

- The known interactions relevant to particle physics were three, named electromagnetic, strong and weak interactions.
- All these interactions conserve energy, momentum and angular momentum.
- The EM interaction is long range, while the strong and weak interactions are of short range. The first is mediated by a massless particle, the photon, while the weak interaction is expected to be mediated by very heavy charged bosons, not known. The strong interaction was not understood along the lines of the Yukawa theory after the pions did not seem to be elementary.
- The known elementary particles were classified as Fermions with spin 1/2 and bosons with spin 1. The lightest of these are the neutrinos and the photon respectively. They have zero mass.
- The zero mass particles are characterized by a quantum number called helicity[1], which is the orientation of the spin with respect to the momentum. The number of helicity states (polarizations) of the photon are two (two transverse polarizations), but the neutrino has

[1]The helicity is not a good quantum number for massive particles, since it depends on the frame of reference it is observed, see section 12.2.

only one polarization. More specifically it can only be left handed (and the antineutrino right handed). In both cases less than $(2s+1)$ allowed by their spin.

- The particles which feel the strong interaction are called hadrons. The spin 1/2 hadrons were are also called **baryons**. Those that do not feel the strong interaction are called **leptons**. The first are characterized by a **baryon number** and the second by a **lepton number**.
- All the known interactions conserve charge, baryon and lepton number.
- Parity, associated with the symmetry under space inversion, seemed conserved in strong and electromagnetic interactions, but it is maximally violated in weak interactions (see section 12.1.1).
- The only interaction the neutrinos feel is the weak interaction.
- The known elementary Fermions are the **quarks** and leptons and they come into families. The hadrons the (u, d), (c, s) and perhaps others and the leptons as $(\nu_e, e^-), (\nu_\mu, \mu^-)$ and perhaps others. Lepton flavor seems to be conserved in all three interactions, but the quark flavor is a little bit violated in weak interactions.
- The quarks carry an additional quantum number called **color**, which is responsible for the the strong interaction.
- In condensed matter physics the phenomenon of a **spontaneous symmetry breaking** was known, e.g. the spontaneous magnetization. In other words, even if the symmetry of the interaction between the physical entities has a symmetry, e.g. it is rotationally invariant, the ground state of the system does not respect this symmetry. Thus even in the absence of an external magnetic field, all the elementary dipoles come out aligned at least in a sufficiently small region.
- The notion of spontaneous symmetry breaking was elucidated by the existence of spin zero particles (scalars) and suggested that the true symmetries of nature, involving the internal degrees of freedom, are not global, holding for all space time ponts simultaneously, but **local or gauge symmetries**.
- All particles are initially characterized by zero mass, but some of them may attain a mass after the spontaneous break down of the symmetry by the scalars called **Higgs**.

3.2 The essential ingredients of the Standard Model

Those not familiar with introductory particle physics may want to look at the appendix at the end of the book before proceeding further. The standard model (SM) of Elementary particles consists of the following ingredients:

i) A symmetry.
Its proposers guessed that the smallest possible symmetry could correspond to a simple group but it must contain 3 group factors, i.e. it is of the form

$$G_s = SU_c(3) \otimes SU_I(2) \otimes U_Y(1) \tag{3.1}$$

Each of them being a special unitary group in dimensions 3, 2 and 1 respectively (see chapter 1 on group theory). The first refers to the color group (c), the second is the weak isospin (I) and the last behaves just a like number whose value is indicated by Y, called hypercharge. They supposed that this is a gauge symmetry in a manner we will shortly discuss.

ii) A particle content.
All particles are initially massless. We distinguish two possibilities:

1) The gauge bosons.
These are spin one particles. Once the symmetry is known there is no freedom. They must transform like the adjoined representation of the group. Their structure and interactions are fixed by the symmetry. With the above symmetry the number of gauge bosons is the same as the number of generators of the group, i.e. $(3^2 - 1) + (2^2 - 1) + 1 = 12$. They are indicated as follows:
$G^i_\mu, i = 1, \cdots, 8$ for $SU_c(3)$, A^k_μ, $k, 1, 2, 3$ for $SU_I(2)$, $B\mu$ for $U(1)$

2) All other particles are put in by hand in some judiciously chosen representation of the symmetry group.

These other particles form a minimal set including:

i) Matter particles. The Fermions are put ad hoc fields were put in chiral multiplets of some representation of the above symmetry groups in chiral multiplets. The chiral components of a Fermion field f are given as follows:

$$f_L = \frac{1}{2}(1 - \gamma_5)f,\ f_R = \frac{1}{2}(1 + \gamma_5)f \text{ such that } \gamma_5 f_L = -f_L,\ \gamma_5 f_R = f_R \tag{3.2}$$

(See section 2.5). The left handed in $SU(2)$ doublets ($I = 1/2$) and the right handed in $SU(2)$ singlets ($I = 0$). The quantum number Y called **hypercharge** is defined so that

$$Q = I_3 + \frac{Y}{2}, \; Q = \text{ the charge in units of } e \qquad (3.3)$$

where by convention the upper component of isodoublet ($I_3 = 1/2$) is that with the highest charge. They belong to the two classes.

(1) Leptons (charged like the electron etc and their neutrinos). They are appear in three flavors (generations) designated as follows:

$$\begin{pmatrix} \nu_e \\ e^- \end{pmatrix}_L, \; e^-_R, \; \begin{pmatrix} \nu_\mu \\ \mu^- \end{pmatrix}_L, \; \mu^-_R, \; \begin{pmatrix} \nu_\tau \\ \tau^- \end{pmatrix}_L, \; \mu^-_R,$$

$$\ell_L \leftrightarrow (I, Y) = (1/2, -1), \; \ell_R \leftrightarrow (I, Y) = (0, -2), \qquad (3.4)$$

with $\ell \leftrightarrow$ lepton family (e, μ, τ) where the hypercharge Y was determined by Eq. (3.3). Note the conspicuous absence of the right handed neutrino. Also note that at that time not all generations were known. The number of flavors is immaterial for the standard model (SM).

(2) Hadrons i.e the quarks.

$$\begin{pmatrix} u^\alpha \\ d^\alpha \end{pmatrix}_L, \; u^\alpha_R, \; d^\alpha_R, \; \begin{pmatrix} c^\alpha \\ s^\alpha \end{pmatrix}_L, \; c^\alpha_R, \; s^\alpha_R, \; \begin{pmatrix} t^\alpha \\ b^\alpha \end{pmatrix}_L, \; c^\alpha_R, \; b^\alpha_R \qquad (3.5)$$

with $\alpha = r, g, b$ color indices. Furthermore we make the correspondence:

$$\begin{aligned} &q^\alpha_L \leftrightarrow (I, Y) = (1/2, 1/3), q^\alpha \leftrightarrow \text{ quark generation} \\ &(u^\alpha, c^\alpha, t^\alpha)_R \Leftrightarrow (I, Y) = (0, 4/3), \\ &(d^\alpha, s^\alpha, b^\alpha)_R \Leftrightarrow (I, Y) = (0, -2/3). \end{aligned} \qquad (3.6)$$

Note that quarks of both charges, i.e. $2/3$ as well as $(-1/3)$, have right handed components. Note also that not all quarks where known at that time. The weak interaction acts vertically between members of the isodoublet, while the strong interaction acts horizontally between quarks of the same flavor with different color indices. From this point of view the quarks are color triplets under the group $SU_c(3)$, indicated as $(r, g, b) \Leftrightarrow$ (red, green, blue). More explicitly we may write

$$\begin{pmatrix} u^r & u^g & u^b \\ d^r & d^g & d^b \end{pmatrix}_L, \; (u^r, u^g, u^b)_R, \; (d^r, d^g, d^b)_R.$$

Similarly for the other generations. The right handed isosinglet quarks do not participate in the weak interaction, but they participate in other interactions, e.g. the strong interaction.

ii) A complex scalar particle (spin zero particle) with zero color also put arbitrarily in an isodoublet:

$$\begin{pmatrix} \phi^0 \\ \phi^- \end{pmatrix}, (I, Y) = \left(\frac{1}{2}, -1\right), \quad \text{(Higgs doublet)} \tag{3.7}$$

Such an elementary particle did not exist till 2012, when a particle has been found at LHC, CERN. But it was very crucial for the success of the SM to cause the spontaneous symmetry breaking as we will see later in chapter 4.

A chart of the elementary particles is presented in Fig. 3.1.

Of course the antiparticles of the above exist, but we do not bother writing them down. For the Fermions they are as provided by the Dirac theory (see chapter 2), for the gauge bosons they can be chosen so that the particle coincides with its own antiparticle. For the Higgs its antiparticle is

$$\begin{pmatrix} \phi^+ \\ \phi^* \end{pmatrix}, (I, Y) = \left(\frac{1}{2}, 1\right)$$

The standard model

i) using field theory provides interactions consistent with the symmetry and thus methods of calculating physical observables.
ii) explained all the available data at the time it was proposed. It predicted new phenomena that were discovered later. It is at present consistent with the experimental results (almost all of them).

3.3 The notion of a local symmetry

Consider the transformation:

$$\psi(\vec{r}, t) \longrightarrow \psi'(\vec{r}, t) = S(\vec{r}, t)\psi(\vec{r}, t),$$

with $S(\vec{r}, t) = e^{ig\Theta(\vec{r},t)}$, $\Theta(\vec{r}, t)$ a real function of $\vec{r}$ and t and g a constant. Then

$$\frac{\partial \psi'}{\partial x} = \frac{\partial}{\partial x}(S\psi) = S\frac{\partial \psi}{\partial x} + \frac{\partial S}{\partial x}\psi \neq S\frac{\partial \psi}{\partial x} \Longrightarrow \frac{\partial \psi'}{\partial x} \neq S\frac{\partial \psi}{\partial x}.$$

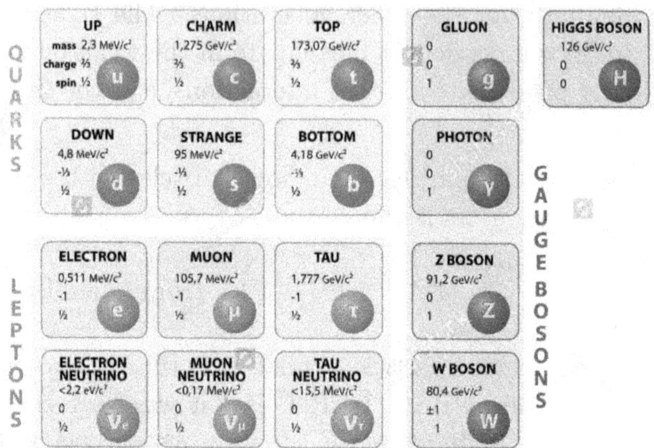

Fig. 3.1: A poster like view of elementary particles with their measured masses. The color quantum number is not indicated. The gauge boson particles γ, Z and $W^{\pm}$ are not the original ones of the SM, but a linear combination of them obtained after the spontaneous symmetry breaking.

i.e. the derivative does not transform he same way. So the theory need not be invariant under such transformations, if it contains space derivatives. How can we make it invariant? We try to generalize the notion of derivative $\partial/\partial x \to D_x$:

$$D_x = \frac{\partial}{\partial x} - igA_x(\vec{r}, t),$$

with $A_x(\vec{r}, t)$ to be specified. We demand $D'_x\psi' = e^{ig\theta(\vec{r},t)}D_x\psi$, i.e. the generalized derivative to transform like ψ itself. Then we have:

$$\begin{aligned} D'_x\psi' &= \left(\frac{\partial}{\partial x} - igA'_x\right) S\psi(\vec{r}, t) \\ &= S\left(\frac{\partial}{\partial x} - igA'_x\right)\psi(\vec{r}, t) + ig\frac{\partial\Theta}{\partial x}S\psi(\vec{r}, t) \\ &= S\left(\left(\frac{\partial}{\partial x} - igA_x\right) + ig(A_x - A'_x) + ig\frac{\partial\Theta}{\partial x}\right)\psi(\vec{r}, t), \quad (3.8) \\ D'_x\psi' &= SD_x\psi(\vec{r}, t) + ig\left((A_x - A'_x) + \frac{\partial\Theta}{\partial x}\right)S\psi(\vec{r}, t) \quad (3.9) \end{aligned}$$

$$\Longrightarrow D'_x\psi' = SD_x\psi(\vec{r}, t) \Longleftrightarrow ig(A_x - A'_x) + ig\frac{\partial\Theta}{\partial x} = 0\ . \quad (3.10)$$

Thus one must expand the theory to include another particle described by the function A_x which must satisfy the transformation property:

$$A'_x = A_x + \frac{\partial\Theta}{\partial x}. \quad (3.11)$$

We can easily extend this to a 3-dimensional space with coordinates $\mathbf{x} = (x^1, x^2, x^3)$:

$$\nabla \to \vec{D} = \nabla + ig\vec{A},\ \vec{A}' = \vec{A} + \nabla\Theta \tag{3.12}$$

and to the time component

$$\frac{\partial}{x_0} \to D_0 = \frac{\partial}{x_0} + igA_0,\ A_0' = A_0 + \frac{\partial\Theta}{\partial x_0} \tag{3.13}$$

Instead of Eq. (3.12) we could easily have chosen:

$$\nabla \to \vec{D} = -\nabla + ig\vec{A},\ \vec{A}' = \vec{A} - \nabla\Theta \tag{3.14}$$

The last equation is appropriate in the context the special theory of relativity. It is then preferred to use $\mu = 0, 1, 2, 3$ and define:

$$\partial_\mu = \left(\frac{\partial}{\partial x^0}, \frac{\partial}{\partial x^1}, \frac{\partial}{\partial x^2}, \frac{\partial}{\partial x^3}\right) = \left(\frac{\partial}{\partial x^0}, -\frac{\partial}{\partial x_1}, -\frac{\partial}{\partial x_2}, -\frac{\partial}{\partial x_3}\right) = \left(\frac{\partial}{\partial x_0}, -\nabla\right).$$

Then we can combine Eqs. (3.14) and (3.13) and write down the generalized derivative in covariant form:

$$D_\mu = \partial_\mu - igA_\mu \tag{3.15}$$

$$A'_\mu = A_\mu + \partial_\mu\Theta \tag{3.16}$$

$$D'_\mu\psi' = SD_\mu\psi. \tag{3.17}$$

Sometimes it is beneficial to write the middle of the previous equations in a form that contains explicitly the group element S, i.e.

$$A'_\mu = SA_\mu S^{-1} - \frac{i}{g}\left(\partial_\mu S\right)S^{-1} \tag{3.18}$$

For Abelian symmetries the two forms are the same. For non Abelian theories the latter form is preferred.

Example 1: Consider the kinetic energy of a scalar field:

The Lagrangian involving a scalar field is given by:

$$\mathcal{L} = \frac{1}{2}(i\partial_\mu\phi)^+(i\partial^\mu\phi) + \frac{1}{2}m_\phi^2|\phi|^2 - V(|\phi|)$$

This is not gauge invariant. Gauge invariance requires the existence of a gauge boson A_μ. Furthermore

$$(i\partial_\mu\phi)^+(i\partial^\mu\phi) \Rightarrow (iD_\mu\phi)^+(iD^\mu\phi) = (i\partial_\mu + gA_\mu\phi)^+(i\partial_\mu + gA^\mu\phi)^+ =$$

$$(i\partial_\mu\phi)^+(i\partial^\mu\phi) + igA^\mu(\phi^*\partial_\mu\phi - \phi\partial_\mu\phi^*) + g^2A_\mu A^\mu|\phi|^2$$

The second term vanishes if the scalar field is real. The last term is an interaction between the scalar field and the gauge boson.

Example 2: The kinetic term of a Fermion
In the pregauge era the kinetic energy of the Dirac spinor ψ was given by:

$$\mathcal{L} = \bar{\psi}\gamma^{\mu} i\partial_{\mu}\psi).$$

Gauge invariance requires:

$$\bar{\psi}\gamma^{\mu} i\partial_{\mu}\psi \Rightarrow \bar{\psi}\gamma^{\mu}(i\partial_{\mu} + gA_{\mu})\psi = \bar{\psi}\gamma^{\mu} i\partial_{\mu}\psi + gA_{\mu}\bar{\psi}\gamma^{\mu}\psi$$

Thus if the electron exists the photon must exist and electromagnetism should exist! The last term is the interaction of the Fermion with the photon (vector potential) with $g = e = \sqrt{4\pi\alpha}$ (see section 13.1).

The precise form of these interactions in the context of the standard model will be given later.

3.4 Maxwell's equations; an example of Abelian gauge symmetry

Recall Maxwell's equations whereby the coupling is $g = e = \sqrt{4\pi\alpha}$. We define a vector potential $\vec{A}$ as follows:

$$\vec{\nabla} \cdot \vec{B} = 0 \Longrightarrow \text{ there exists an } \vec{A} \text{ such that } \vec{B} = \vec{\nabla} \times \vec{A} \tag{3.19}$$

The the original equation is no longer needed. Let us now consider the other homogeneous equation:

$$\vec{\nabla} \times \vec{E} = -\frac{\partial \vec{B}}{\partial t} \tag{3.20}$$

Replacing $\vec{B}$ as above we get

$$\vec{\nabla} \times \vec{E} = -\vec{\nabla} \times \frac{\partial \vec{A}}{\partial t} \Longrightarrow$$

$$\vec{\nabla} \times \left(\vec{E} + \frac{\partial \vec{A}}{\partial t}\right) = 0. \tag{3.21}$$

This means that there exists a scalar potential ψ, such that:

$$\vec{E} + \frac{\partial \vec{A}}{\partial t} = -\vec{\nabla}\psi \Longrightarrow$$

$$\vec{E} = -\frac{\partial \vec{A}}{\partial t} - \vec{\nabla}\psi.$$

Setting $\psi = A^0 c$ we get:

$$\vec{E} = -c\left(\frac{\partial \vec{A}}{\partial x_0} + \vec{\nabla} A_0\right). \tag{3.22}$$

In another gauge, using $\vec{A}' = \vec{A} - \nabla\Theta$ and $A_0' = A_0 + \frac{\partial\Theta}{\partial x_0}$ we get:

$$\vec{B}' = \vec{\nabla} \times A' = \vec{\nabla} \times (\vec{A} - \vec{\nabla}\Theta) = \vec{\nabla} \times \vec{A} - \vec{\nabla} \times \vec{\nabla}\Theta = \vec{B},$$

$$\vec{E}' = -c\left(\frac{\partial \vec{A}'}{\partial x_0'} + \vec{\nabla}A_0'\right) = -c\left(\frac{\partial \vec{A}}{\partial x_0} - \frac{\partial}{\partial x_0}\vec{\nabla}\Theta + \vec{\nabla}A_0 + \vec{\nabla}\frac{\partial\Theta}{\partial x_0}\right) = \vec{E}.$$

So Maxwell's equations are invariant under a gauge transformation. In fact the name gauge for the transformation came from electromagnetism.

In the above formulas the function Θ was assumed to be a scalar function. So various such transformations commute. i.e the transformations constitute an Abelian group.

3.4.1 *Maxwell's equations in terms of the potentials*

Let us now examine the other two non homogeneous Maxwell's equations. The first equation is written as :

$$\nabla.\vec{E} = \frac{\rho}{\epsilon_0} \tag{3.23}$$

and it now becomes

$$-\nabla.\left(\frac{\vec{A}}{\partial t} + \nabla\psi\right) = \frac{\rho}{\epsilon_0} \Rightarrow -\nabla^2\psi - \frac{\partial}{\partial t}\nabla.\vec{A} = \frac{\rho}{\epsilon_0}, \tag{3.24}$$

i.e. both potentials appear, they are coupled.

The second inhomogenious equation is:

$$\nabla \times \vec{B} = \mu_0\vec{J} + \frac{1}{c^2}\frac{\vec{E}}{\partial t},\ \mu_0 = \frac{1}{\epsilon_0 c^2} \tag{3.25}$$

and it can now be written as

$$\nabla \times (\nabla \times \vec{A}) = \mu_0\vec{J} - \frac{1}{c^2}\frac{1}{\partial t}\left(\frac{\partial\vec{A}}{\partial t} + \nabla\psi\right) \tag{3.26}$$

It can be simplified via an identity (which holds in Cartesian field components), namely

$$\nabla \times (\nabla \times \vec{A} = \nabla(\nabla.\vec{A}) - \nabla^2\vec{A}$$

to get

$$\nabla(\nabla.\vec{A}) - \nabla^2\vec{A} = \mu_0\vec{J} - \frac{1}{c^2}\frac{\partial^2\vec{A}}{\partial t^2} - \frac{1}{c^2}\nabla\frac{\partial\psi}{\partial t} \tag{3.27}$$

This equation is still complicated and, again, the two potentials are mixed.

Since the fields $\vec{A}$ and ψ are not uniquely determined we can select them, i.e. we can make a convenient choice. The following possibilities are common in classical electrodynamics:

- The Lorentz gauge

$$\partial_\mu A^\mu = 0 \Rightarrow \nabla.\vec{A} = -\frac{1}{c^2}\frac{\partial \psi}{\partial t} \tag{3.28}$$

We notice that it is a Lorentz invariant, i.e. if it holds in a given inertial frame it holds to any other inertial frame.
- The Coulomb gauge

$$\nabla.\vec{A} = 0 \tag{3.29}$$

- One axial gauge:

$$A_1 = 0, \text{ or } A_2 = 0, \text{ or } A_3 = 0, \text{ or } \psi = 0. \tag{3.30}$$

Yes, the scalar potential ψ can be chosen to be zero! The students should answer the question: How does electrostatics work in this case?

We will now simplify the inhomogeneous equations by considering the Lorentz and the Coulomb gauges.

- The Lorentz gauge.
Now we can eliminate $\nabla.\vec{A}$ in the first of Maxwell's equations to get:

$$\frac{1}{c^2}\frac{\partial^2 \psi}{\partial t^2} - \nabla^2 \psi = \frac{\rho}{\epsilon_0}$$

which is the wave equation for the scalar potential.
We can also eliminate $\partial\psi/\partial t$ in the other inhomogeneous equation to get:

$$\nabla \times (\nabla \times \vec{A}) = \mu_0 \vec{J} - \frac{1}{c^2}\frac{1}{\partial t}\left(\frac{\partial \vec{A}}{\partial t}\right) + \nabla(\nabla.\vec{A}) \tag{3.31}$$

In Cartesian components of the field it can again be simplified to yield:

$$\frac{1}{c^2}\frac{\partial^2 \vec{A}}{\partial t^2} - \nabla^2 \vec{A} = \mu_o \vec{J}, \tag{3.32}$$

which is again the wave equation for each of the components of the vector potential.

- The Coulomb gauge $\nabla.\vec{A} = 0$.

  Now the first equation becomes:
  $$\nabla^2\psi = -\frac{\rho}{\epsilon_0},$$
  which is the same as Poisson's equation of electrostatics. It holds, however, in this gauge even if we have a time dependence of the density, $\rho = \rho(t)$.

  The second equation becomes:
  $$\frac{1}{c^2}\frac{\partial^2\vec{A}}{\partial t^2} - \nabla^2\vec{A} = \mu_0\vec{J} - -\frac{1}{c^2}\nabla\frac{\partial\psi}{\partial t}$$
  It has been simplified, but the two potentials do not decouple. This is not a serious problem, however, if the first equation can be solved. We can then substitute this solution into the second equation, which adds an additional inhomogeneous term (in addition to the current density).

  One may conclude by saying that the Coulomb gauge oversimplifies the equation for ψ, but it makes the equation for $\vec{A}$ a lot harder.

There has been a considerable discussion whether the potentials are only a useful theoretical concept or are in fact measurable, since they are not uniquely defined. We know now, however, that at the quantum level they are physical. This was established by the Aharonov-Bohm effect which was experimentally verified[2]. Let us, therefore proceed to exploit some properties of the photon as described by the vector potential, in the convenient Lorentz gauge
$$\partial_\mu A^\mu = 0.$$
Recall now that under a gauge transformation:
$$A'_\mu = A_\mu + \frac{\partial\Theta}{\partial x^\mu}$$

[2] An ideal solenoid (i.e. infinitely long and with a perfectly uniform current distribution) encloses a magnetic field B, but does not produce any magnetic field outside of its cylinder, and thus the charged particle (e.g. an electron) passing outside experiences no magnetic field B. Continuity of the vector potential demands, however, that there must be a (curl-free) vector potential **A** outside the solenoid with an enclosed flux:
$$\oint \mathbf{A}.d\boldsymbol{\ell} = \int \nabla\times\mathbf{A}.\hat{n}dS = \int \mathbf{B}.\hat{n}dS = \Phi$$
and so the relative phase of particles passing through one slit or the other is altered by whether the solenoid current is turned on or off. This corresponds to an observable shift of the interference fringes on the observation plane. The same phase effect is responsible for the quantized-flux requirement in superconducting loops, which has, of course, been observed.

and

$$\partial^\mu A'_\mu = \partial^\mu A_\mu + \partial^\mu \partial_\mu \Theta.$$

Θ can be chosen to satisfy the wave equation:

$$\partial^\mu \partial_\mu \Theta = \left(-\nabla^2 + \frac{1}{c^2}\frac{\partial^2}{\partial t^2}\right)\Theta = 0.$$

Thus

$$\partial^\mu A'_\mu = \partial^\mu A_\mu,$$

i.e. the Lorentz condition can be satisfied in all gauges. As a result the gauge field has two polarizations only.

Indeed the condition

$$\partial_\mu A^\mu = 0$$

for a plane wave yields

$$A^\mu = \epsilon^\mu e^{-ikx} \Longrightarrow k_\mu A^\mu = 0 \Longrightarrow k_\mu \epsilon^\mu = 0$$

One can choose Θ so that the time component of the polarization is zero, i.e. the polarization can be chosen to be space-like, namely,

$$\epsilon_\mu = (0, \vec{\epsilon}).$$

Then writing

$$\vec{\epsilon} = \epsilon_L + \epsilon_T,\ \epsilon_L = \text{ longitudinal i.e. in the direction of momentum,}$$
$$\epsilon_T = \text{ transverse}$$

we find:

$$k^\mu \epsilon_\mu = 0 \Rightarrow \epsilon_L \cdot \vec{k} = 0 \Rightarrow \epsilon_L = 0.$$

Therefore the longitudinal component vanishes, i.e. for a massless photon we have two degrees of freedom, not three as expected for a particle of spin one.

3.4.2 *Maxwell's equations involving the electromagnetic tensor*

Define:

$$F_{\mu\nu}\frac{i}{g}[D_\mu, D_\nu]. \tag{3.33}$$

Then

$$\begin{aligned} F_{\mu\nu} \quad & \frac{i}{g}[\partial_\mu - igA_\mu, \partial_\nu - igA_\nu] \\ &= \frac{i}{g}\{[\partial_\mu, \partial_\nu] - ig[\partial_\mu, A_\nu] - ig[A_\mu, \partial_\nu] - g^2[A_\mu, A_\nu]\}. \end{aligned} \tag{3.34}$$

Since

$$[\partial_\mu, \partial_\nu] = 0, [\partial_\mu, A_\nu] = \partial_\mu A_\nu, [\partial_\nu, A_\mu] = \partial_\nu A_\mu,$$

we get

$$F_{\mu\nu} = \frac{i}{g}\left\{(-ig)[\partial_\mu A_\nu - \partial_\nu A_\mu] - g^2[A_\mu, A_\nu]\right\},$$

namely,

$$F_{\mu\nu} = \partial_\mu A_\nu - \partial_\nu A_\mu - ig[A_\mu, A_\nu]. \tag{3.35}$$

Thus,

$$\begin{aligned} F'_{\mu\nu} &= \frac{i}{g}[D'_\mu, D'_\nu] \\ &= \frac{i}{g}[SD_\mu S^{-1}, SD_\nu S^{-1}] \\ &= \frac{i}{g}[SD_\mu S^{-1}SD_\nu S^{-1} - SD_\nu S^{-1}SD_\mu S^{-1}] \\ &= \frac{i}{g}S[D_\mu D_\nu - D_\nu D_\mu]S^{-1} \end{aligned}$$

or

$$F'_{\mu\nu} = SF_{\mu\nu}S^{-1}. \tag{3.36}$$

In the Abelian case, $[A_\mu, A_\nu] = 0$, so

$$F'_{\mu\nu} = F_{\mu\nu}, F_{\mu\nu} = \partial_\mu A_\nu - \partial_\nu A_\mu, \tag{3.37}$$

which is just the EM field tensor.

In electromagnetism the tensor $F_{\mu\nu}$ is nothing but the electric and the magnetic fields. One can easily see that:

i) the tensors with one time component contain the electric field. Indeed:

$$F_{0k} = \partial_0 A_k - \partial_k A_0 = \frac{\partial}{\partial x_0}A_k + \frac{\partial}{\partial x_k}A_0 = \frac{1}{c}\left(\frac{\partial}{\partial t}A_k + \frac{\partial}{\partial x_k}\psi\right) = -\frac{1}{c}E_k,$$

$$F_{k0} = \frac{1}{c} E_k$$

ii) Tensors with both indices non zero
When both indices are non zero we obtain the magnetic field, e.g.:

$$F_{12} = -F_{21} = \partial_1 A_2 - \partial_2 A_1 = -\frac{\partial}{\partial x_1} A_2 + \frac{\partial}{\partial x_2} A_1 = -(\nabla \times \vec{A})_3 = -\vec{B}_3$$

Similarly for the other components

$$F_{13} = -\frac{\partial}{\partial x_1} A_3 + \frac{\partial}{\partial x_3} A_1 = (\nabla \times \vec{A})_2 = \vec{B}_2,$$

$$F_{23} = -\frac{\partial}{\partial x_2} A_3 + \frac{\partial}{\partial x_3} A_2 = (\nabla \times \vec{A})_1 = -\vec{B}_1.$$

We could also have used

$$F^{\mu\nu} = g^{\mu\alpha} g^{\nu\beta} F_{\alpha\beta},$$

which yields the opposite electric field, but the same magnetic field. We thus find

$$F_{\mu\nu} = \begin{array}{c|cccc} & 0 & 1 & 2 & 3 \\ \hline 0 & 0 & -E_1/c & -E_2/c & -E_3/c \\ 1 & E_1/c & 0 & -B_3 & B_2 \\ 2 & E_2/c & B_3 & 0 & -B_1 \\ 2 & E_3/c & -B_2 & B_1 & 0 \end{array} , \quad F^{\mu\nu} = \begin{array}{c|cccc} & 0 & 1 & 2 & 3 \\ \hline 0 & 0 & E_1/c & E_2/c & E_3/c \\ 1 & -E_1/c & 0 & -B_3 & B_2 \\ 2 & -E_2/c & B_3 & 0 & -B_1 \\ 2 & -E_3/c & -B_2 & B_1 & 0 \end{array}$$

It is amusing that had gauge theories preceded the standard EM treatment, we could live with the six components of this tensor, with no need to use the fields $\vec{E}$ and $\vec{B}$ at all[3].

One can also see that the two inhomogeneous Maxwell equations take the form:

$$\partial^\mu F_{\mu\nu} = \mu_0 J_\nu, \quad \text{where } J_\mu = (\rho c, \vec{J}) \text{ is the current density.} \tag{3.38}$$

e.g. for $\nu = 0$ we have:

$$\partial^\mu F_{\mu 0} = \partial^k F_{k0} = \frac{\partial}{x_k} \frac{E_k}{c} = \frac{1}{c} \nabla . \vec{E}$$

Also

$$J_0 = \mu_0 \rho c = \frac{\rho}{c\epsilon_0}$$

[3] We will see below that, e.g., this happens in the case of the color gauge symmetry $SU_c(3)$, where the components of $F^\alpha_{\mu\nu}$ stand for the fields themselves, called gluons.

in other words

$$\partial^\mu F_{\mu 0} = \mu_0 J_0 \Leftrightarrow \nabla.\vec{E} = \frac{\rho}{\epsilon_0}$$

Analogously starting with $\nu \neq 0$ we obtain the other inhomogeneous equation:

$$\nabla \times \vec{B} = \mu_0 \vec{J} + \frac{1}{c^2}\frac{\partial \vec{E}}{\partial t}$$

The homogeneous equations cannot be obtained this way, but from a third rank tensor or from the dual second rank tensor $\tilde{F}^{\mu\nu}$ written this way:

$$\partial_\mu \tilde{F}^{\mu\nu} = 0, \quad \tilde{F}^{\mu\nu} = \frac{1}{2}\epsilon^{\mu\nu\alpha\beta}F_{\alpha\beta} \tag{3.39}$$

with $\epsilon^{\mu\nu\alpha\beta}$ the completely antisymmetric symbol in all four indices, i.e.:

$$-\epsilon^{\mu\nu\alpha\beta} = \epsilon_{\mu\nu\alpha\beta} = \begin{cases} 1, \text{ if } \mu\nu\alpha\beta \text{ is an even permutation of } 0123 \\ -1, \text{ if } \mu\nu\alpha\beta \text{ is an odd permutation of } 0123 \\ 0, \text{ otherwise.} \end{cases}$$

Thus e.g.

$$\tilde{F}^{12} = \frac{1}{2}\left(\epsilon^{1203}F_{03} + \epsilon^{1203}F_{30}\right) = \epsilon^{1203}F_{03} = \epsilon^{0123}F_{03} = \frac{E_3}{c},$$

$$\tilde{F}^{13} = -\epsilon^{0123}F_{02} = -\frac{E_2}{c},\ \tilde{F}^{23} = \epsilon^{0123}F_{23} = \frac{E_1}{c}$$

Similarly

$$\tilde{F}^{03} = \frac{1}{2}\left(\epsilon^{0312}F_{12} + \epsilon^{0321}F_{21}\right) = \epsilon^{0123}F_{12} = -F_{12} = B_3$$

$$\tilde{F}^{01} = \epsilon^{0123}F_{23} = B_1,\ \tilde{F}^{02} = -\epsilon^{0123}F_{23} = B_2$$

We thus obtain

$$\tilde{F}^{\mu\nu} = \begin{array}{c|cccc} & 0 & 1 & 2 & 3 \\ \hline 0 & 0 & B_1 & B_2 & B_3 \\ 1 & -B_1 & 0 & E_3/c & -E_2/c \\ 2 & -B_2 & -E_3/c & 0 & E_1/c \\ 2 & -B_3 & E_2/c & -E_1/c & 0 \end{array}, \quad \tilde{F}_{\mu\nu} = \begin{array}{c|cccc} & 0 & 1 & 2 & 3 \\ \hline 0 & 0 & B_1 & B_2 & B_3 \\ 1 & -B_1 & 0 & -E_3/c & E_2/c \\ 2 & -B_2 & E_3/c & 0 & -E_1/c \\ 2 & -B_3 & -E_2/c & E_1/c & 0 \end{array},$$

In other words in going from $F_{\mu\nu}$ to $\tilde{F}^{\mu\nu}$ we find that the roles of $\vec{E}$ and $\vec{B}$ are simply interchanged. Now

$$\partial_\mu \tilde{F}^{\mu 3} = \partial_{x_0}\tilde{F}^{03} + \partial_{x_1}\tilde{F}^{13} + \partial_{x_2}\tilde{F}^{23} = \frac{1}{c}\frac{\partial B_3}{\partial t} - \frac{\partial(-E_2/c)}{\partial x_1} - \frac{\partial E_1/c}{\partial x_2} =$$

$$\frac{1}{c}\left(\frac{\partial B_3}{\partial t}+(\nabla\times\vec{E})_3\right)\Rightarrow(\nabla\times\vec{E})_3=-\frac{\partial B_3}{\partial t}$$

Thus for considering the expression $\partial_\mu\tilde{F}^{\mu k}, k\neq 0$, e.g. $k=3$ we find

$$\partial_\mu\tilde{F}^{\mu 3}=0\Rightarrow\left(\nabla\times\vec{E}\right)_3=-\left(\frac{\partial\vec{B}}{\partial t}\right)_3\Rightarrow\nabla\times\vec{E}=-\frac{\partial\vec{B}}{\partial t}.$$

Furthermore

$$0=\partial_\mu\tilde{F}^{\mu 0}=\partial_{x_1}\tilde{F}^{10}+\partial_{x_2}\tilde{F}^{20}+\partial_{x_3}\tilde{F}^{30}=\frac{\partial B_1}{\partial x_1}+\frac{\partial B_2}{\partial x_2}+\frac{\partial B_3}{\partial x_3}\Rightarrow\nabla.\vec{B}=0$$

One can construct two invariants:

$$\mathcal{L}_g=-\frac{1}{4}F^{\mu\nu}F_{\mu\nu}=\frac{1}{2}\vec{E}^2-\frac{1}{2}\vec{B}^2,\ \mathcal{L}_g^{'}=-\lambda_g^{'}\frac{1}{4}\tilde{F}^{\mu\nu}F_{\mu\nu}=\lambda_g^{'}\vec{E}.\vec{B}.$$

The first term can be identified with the electric energy density minus the magnetic energy density. This term vanishes in empty space. In the second term $\lambda_g^{'}$ can be non zero only in the presence CP violation[4]. CP is the product of charge conjugation and space inversion (see section 12.5). Under CP:

$$\vec{E}\rightarrow\vec{E},\ \vec{B}\rightarrow-\vec{B}\Rightarrow\mathcal{L}_g^{'}\rightarrow-\mathcal{L}_g^{'}$$

Conclusion: Had we known that charged particles exist and had a Lagrangian describing them in terms of their wave functions and their space and time derivatives, we would have predicted the existence of the photon by demanding gauge invariance of this Lagrangian. The photon would have been the field A_μ of Maxwell. If, in addition, our model had been complete, in the sense of the Lagrangian containing currents like J_μ, we would have found Maxwell's equations from the equations of motion. History, of course, chose otherwise and Maxwell found the correct equations of classical electrodynamics. He did so by going beyond the prevailing beliefs of his day. It took 20 years for his theory to be accepted, when the EM waves he predicted were finally produced by antennas (the connection with light came later). Gauge invariance was, of course, contained in his equations without even him realizing it.

History is not going to repeat itself, since now we already have the Abelian prototype. In fact gauge theories were invented to discover new theories, following similar steps with this prototype.

---

[4]A term containg these fields can exist, e.g.

$$\tilde{\mathcal{L}}_{gga}=\tilde{\lambda}_g\tilde{F}^{\mu\nu}F_{\mu\nu}\,a.$$

This can induce a two photon ($\gamma\gamma$)-pseudoscalar (e.g. axion a) coupling and is being searched, e.g., in axion detection experiments, in which an axion can be converted to a photon in the presence of a magnetic field. At present only upper limits in the relevant coupling $\tilde{\lambda}_g$ exist.

3.5 Non Abelian gauge transformations

In general the function $\Theta(x_\mu)$ can be an operator. So one has to keep track of the order of the operations. Thus:

$$D'_\mu \psi' = S D_\mu \psi \Rightarrow D'_\mu S\psi = S D_\mu \psi, \; S^{-1} D'_\mu S = D_\mu$$

$$\begin{aligned} D'_\mu \psi' &= \left(\frac{\partial}{\partial x^\mu} - igA'_\mu\right) S\psi = \left(\frac{\partial}{\partial x^\mu} S\right)\psi + S\left(\frac{\partial}{\partial x^\mu}\psi\right) - igA'_\mu S\psi \\ &= S\left(\frac{\partial \psi}{\partial x^\mu} - igA_\mu \psi\right) + (igSA_\mu - gA'_\mu S)\psi + \frac{\partial S}{\partial x^\mu}\psi. \end{aligned} \tag{3.40}$$

That is $D'_\mu \psi' = S D_\mu \psi$, if

$$\begin{aligned} ig(SA_\mu - A'_\mu S) + \frac{\partial S}{\partial x^\mu} &= 0 \text{ or} \\ A'_\mu S = SA_\mu - \frac{i}{g}\frac{\partial S}{\partial x^\mu}&. \end{aligned}$$

Thus the transformation law for the gauge fields is:

$$\begin{aligned} S^{-1} A'_\mu S = A_\mu - \frac{i}{g} S^{-1} \frac{\partial S}{\partial x^\mu} &\Rightarrow \\ A'_\mu = S A_\mu S^{-1} + \frac{i}{g}\frac{\partial S}{\partial x^\mu} S^{-1}& \end{aligned} \tag{3.41}$$

or

$$A_\mu = S^{-1} A'_\mu S + \frac{i}{g} S^{-1} \frac{\partial S}{\partial x^\mu} \tag{3.42}$$

Now we distinguish the following simple cases:

- Θ=constant. Then

  $$S^{-1} A'_\mu S = A_\mu.$$

  This corresponds to case encountered previously, i.e. local (non gauged) transformations.
- $S^{-1} A_\mu = A_\mu S^{-1}$, i.e. they commute. Then

  $$A'_\mu = A_\mu + \frac{\partial \Theta}{\partial x^\mu},$$

i.e. the result is just as before.

3.5.1 *Transformations associated with a non Abelian group*

Consider a group G,which has two elements, $S_1 \in G$ and $S_2 \in G$ such that $S = S_2 S_1 \in G$. How do the gauge fields transform? We have:

$$S_1 : \phi \longrightarrow \phi' = e^{igS_1}\phi,$$

$$S_2 : \phi' \longrightarrow \phi'' = e^{igS_2}\phi' \Longrightarrow$$

$$S_1^{-1} A'_\mu S_1 = A_\mu - \frac{i}{g} S_1^{-1} \frac{\partial S_1}{\partial x^\mu},$$

$$S_2^{-1} A''_\mu S_2 = A'_\mu - \frac{i}{g} S_2^{-1} \frac{\partial S_2}{\partial x^\mu}.$$

Then we will show that:

$$S^{-1} A''_\mu S = A_\mu - \frac{i}{g} S \frac{\partial S}{\partial x^\mu}.$$

Indeed,

$$\begin{aligned}
S^{-1} A''_\mu S &= S_1^{-1} S_2^{-1} A''_\mu S_2 S_1 \\
&= S_1^{-1} \left(A'_\mu - \frac{i}{g} S_2^{-1} \frac{\partial S_2}{\partial x^\mu} \right) S_1 \\
&= S_1^{-1} A'_\mu S_1 - \frac{i}{g} S_1^{-1} S_2^{-1} \frac{\partial S_2}{\partial x^\mu} S_1, \\
&= A_\mu - \frac{i}{g} S_1^{-1} \frac{\partial S_1}{\partial x^\mu} - \frac{i}{g} S_1^{-1} S_2^{-1} \left(\frac{\partial (S_2 S_1)}{\partial x^\mu} - S_2 \frac{\partial S_1}{\partial x^\mu} \right) \\
&= A_\mu - \frac{i}{g} S_1^{-1} \frac{\partial S_1}{\partial x^\mu} - \frac{i}{g} S_1^{-1} S_2^{-1} \frac{\partial (S_2 S_1)}{\partial x^\mu} + \frac{i}{g} S_1^{-1} \frac{\partial S_1}{\partial x^\mu} \Longrightarrow \\
S^{-1} A''_\mu S &= A_\mu - \frac{i}{g} S \frac{\partial S}{\partial x^\mu}. \qquad (3.43)
\end{aligned}$$

q.e.d.

Suppose now that $\Theta = \omega^\alpha T_\alpha$ with T_α a generator of the algebra associated with the group, such that $Tr\,(T_\alpha) = 0$. For sufficiently small ω^α:

$$S \simeq 1 + ig\omega^\alpha T_\alpha$$

and

$$\begin{aligned}
A'_\mu &= (S A_\mu S^{-1}) - \frac{i}{g} (\partial_\mu S) S^{-1} \Rightarrow \\
A'_\mu &\simeq (1 + ig\omega^\alpha T_\alpha) A_\mu (1 - ig\omega^\alpha T_a) - \frac{i}{g} (ig)(\partial_\mu \omega^\alpha T_\alpha), \qquad (3.44)
\end{aligned}$$

or ignoring higher order terms in ω we get:

$$A'_\mu \simeq A_\mu + ig\omega^\alpha \left(T_\alpha A_\mu - A_\mu T_\alpha\right) + \partial_\mu \omega^\alpha T_\alpha \Rightarrow \tag{3.45}$$

$$A'_\mu \simeq A_\mu + ig\omega^\alpha [T_\alpha, A_\mu] + \partial_\mu \omega^\alpha T_\alpha. \tag{3.46}$$

Since the gauge fields transform according to the regular (adjoined) representation, they can be decomposed according to the generators of the algebra T_α, which are normalized in some convenient way, e.g.

$$Tr(T_\alpha T_\beta) = \delta_{\alpha\beta} \text{ or } Tr(T_a T_b) = \frac{1}{2}\delta_{\alpha\beta}. \tag{3.47}$$

The second normalization is more commonly used, see e.g. section 3.5.2. Thus we get:

$$A_\mu = A^\gamma_\mu T_\gamma, \quad A'_\mu = \left(A^\gamma_\mu\right)' T_\gamma. \tag{3.48}$$

$$A'_\mu \simeq A_\mu + ig\omega^\alpha [T_\alpha, T_\beta] A^\beta_\mu + \partial_\mu \omega^\gamma T_a,$$
$$A'_\mu - A_\mu \simeq ig\omega^\alpha C^\gamma_{\alpha\beta} A^\gamma_\mu T_\gamma + \partial_\mu \omega^\gamma T_\gamma,$$

where $C^\gamma_{\alpha\beta}$ are the structure constants of the algebra:

$$[T_\alpha, T_\beta] = C^\gamma_{\alpha\beta} T_\gamma. \tag{3.49}$$

They depend, of course, on the chosen normalization. Thus:

$$\left(A^\gamma\right)' = A^\gamma_\mu + ig\omega^\alpha A^\beta_\mu C^\gamma_{\alpha\beta} + \partial_\mu \omega^\gamma,$$

or rearranging:

$$(A^\alpha_\mu)' = A^\alpha_\mu + ig\omega^\gamma C^\alpha_{\gamma\beta} A^\beta_\mu + \partial_\mu \omega^\alpha. \tag{3.50}$$

This is the celebrated rule of transformation of the gauge fields. In the non abelian case the fields are admixed. In the abelian case the structure constants are zero and there is no mixing.

In all cases, of course, we have:

$$D_\mu = \partial_\mu - igA_\mu \tag{3.51}$$

Where A_μ depends on the symmetry group and it can be decomposed in terms of its generators.

3.5.2 *Some examples*

(1) An Abelian group with g elements :

$$D_\mu = \partial_\mu - i g^\alpha A_\mu^\alpha,\ C^\alpha_{\gamma\beta} = 0, \Longrightarrow (A_\mu^\alpha)' = A_\mu^\alpha + \partial_\mu \omega^\alpha,\ \ \alpha = 1,2,...,g. \tag{3.52}$$

(2) The group SU(2). Its generators are

$$T_i = \frac{1}{2}\tau_i, \quad i = 1,2,3\ , \tag{3.53}$$

with τ_i the Pauli matrices:

$$\tau_1 = \begin{pmatrix} 0 & 1 \\ 1 & 0 \end{pmatrix}, \tau_2 = \begin{pmatrix} 0 & -i \\ i & 0 \end{pmatrix}, \tau_3 = \begin{pmatrix} 1 & 0 \\ 0 & -1 \end{pmatrix}. \tag{3.54}$$

In the case of the operators T_i

$$[T_\gamma, T_\beta] = i\epsilon_{\gamma\beta\alpha} T_\alpha \Rightarrow C^\alpha_{\gamma\beta} = i\epsilon_{\gamma\beta\alpha}$$

Thus

$$D_\mu = \partial_\mu - ig\frac{1}{2}\left(A_\mu^1\tau_1 + A_\mu^2\tau_2 + A_\mu^3\tau_3\right),\ \text{ that is:}$$
$$\left(A_\mu^\alpha\right)' = A_\mu^\alpha - g\omega^\gamma A_\mu^\beta \epsilon_{\gamma\beta\alpha} + \partial_\mu\omega^\alpha, \quad \alpha = 1,2,3. \tag{3.55}$$

One may write the previous equation as:

$$\vec{A}'_\mu = \vec{A}_\mu - g\vec{\omega} \times \vec{A}_\mu + \partial_\mu \vec{\omega}$$

with the understanding that it is a vector equation in isospin space.

(3) The group $SU_I(2) \times U_Y(1)$ of the SM:
There are three isospin generators as before:

$$T_i = \frac{1}{2}\tau_i,\ i = 1,2,3 \text{ with commutators } [T_a, T_b] = i\epsilon_{abc}T_c$$

and one generator Y of the $U_Y(1)$ such that $[T_i, Y] = 0$. As we have mentioned before associated with the generator Y is a quantum number, the eigenvalue of Y, called weak hypercharge Y_W. It is normalized so that for any particle:

$$Q = I_3 + \frac{Y_W}{2}$$

The corresponding gauge fields are:

$$A_\mu^i,\ i = 1,2,3,\ (I = 1, Y_W = 0), \quad B_\mu\ (I = 0, Y_W = 0). \tag{3.56}$$

Furthermore

$$D_\mu = \partial_\mu - i\frac{g}{2}\left(A_\mu^1\tau_1 + A_\mu^2\tau_2 + A_\mu^3\tau_3\right) - i\frac{g'}{2} Y B_\mu,$$

$$\left(A_\mu^\alpha\right)' = A_\mu^\alpha - ig\omega^\gamma A_\mu^\beta \epsilon_{\gamma\beta\alpha} + \partial_\mu \omega^\alpha,$$
$$B'_\mu = B_\mu + \partial_\mu \omega^Y, \tag{3.57}$$

since the structure constants involving the generator Y are zero. Any particle carrying hypercharge interacts with the field B_μ, just like the photon interacts with all charged particles.

One usually writes Y instead of Y_W, but there should not be any confusion with the generator Y of $U_Y(1)$

(4) $SU_c(3)$ color group.
There exist 8 generators $T_\alpha = \frac{1}{2}\lambda_\alpha$, $\alpha = 1, 2, \cdots, 8$ and, hence, 8 independent gauge fields G_μ^α. The generators λ_α are defined as:

$$\lambda_1 = \begin{pmatrix} 0 & 1 & 0 \\ 1 & 0 & 0 \\ 0 & 0 & 0 \end{pmatrix}, \lambda_2 = \begin{pmatrix} 0 & -i & 0 \\ i & 0 & 0 \\ 0 & 0 & 0 \end{pmatrix}, \lambda_3 = \begin{pmatrix} 1 & 0 & 0 \\ 0 & -1 & 0 \\ 0 & 0 & 0 \end{pmatrix}, \lambda_4 = \begin{pmatrix} 0 & 0 & 1 \\ 0 & 0 & 0 \\ 1 & 0 & 0 \end{pmatrix},$$
$$\lambda_5 = \begin{pmatrix} 0 & 0 & -i \\ 0 & 0 & 0 \\ i & 0 & 0 \end{pmatrix}, \lambda_6 = \begin{pmatrix} 0 & 0 & 0 \\ 0 & 0 & 1 \\ 0 & 1 & 0 \end{pmatrix}, \lambda_7 = \begin{pmatrix} 0 & 0 & 0 \\ 0 & 0 & -i \\ 0 & i & 0 \end{pmatrix},$$
$$\lambda_8 = \frac{1}{\sqrt{3}} \begin{pmatrix} 1 & 0 & 0 \\ 0 & 1 & 0 \\ 0 & 0 & -2 \end{pmatrix}, \tag{3.58}$$

with commutators:

$$[T_\alpha, T_\beta] = C_{\alpha\beta}^\gamma T_\gamma, \; C_{\alpha\beta}^\gamma = i f_{\alpha\beta\gamma}. \tag{3.59}$$

Thus

$$D_\mu = \partial_\mu - ig^s \sum_\alpha G_\mu^\alpha T_\alpha, \; (G_\mu^\alpha)' = G_\mu^\alpha - g_s \omega^\gamma G_\mu^\beta f_{\gamma\beta\alpha} + \partial\omega_\mu^\alpha. \tag{3.60}$$

The structure constants $f_{\alpha\beta\gamma}$ can be found in table 1.2 of chapter 1 (group theory). You must have noticed the presence of 3 coupling constants g, g' and g^s, since the group is not simple, but the product of three symmetries.

3.6 Gauge invariant Lagrangians

3.6.1 *Only gauge fields*

In discussing EM we have found it useful to define the tensor $F_{\mu\nu}$ via Eq. (3.33), i.e.

$$F_{\mu\nu} = \frac{i}{g}[D_\mu, D_\nu] = \partial_\mu A_\nu - \partial_\nu A_\mu - ig[A_\mu, A_\nu]. \tag{3.61}$$

Then we found that under a gauge transformation S one gets:

$$F'_{\mu\nu} = \frac{i}{g}[D'_\mu, D'_\nu] = SF_{\mu\nu}S^{-1} \tag{3.62}$$

Now:

- Abelian case,$[A_\mu, A_\nu] = 0$, so

$$F_{\mu\nu} = \partial_\mu A_\nu - \partial_\nu A_\mu, \tag{3.63}$$

which is just the EM field tensor.

- Non Abelian case. Then

$$A_\mu = A^\alpha_\mu T_\alpha, \quad F_{\mu\nu} = F^\alpha_{\mu\nu} T_\alpha.$$

So we have:

$$ig[A_\mu, A_\nu] = igA^\beta_\mu A^\gamma_\nu [T_\beta, T_\gamma] = igA^\beta_\mu A^\gamma_\nu C^\alpha_{\beta\gamma} T_\alpha, \tag{3.64}$$

$$F^\alpha_{\mu\nu} T_\alpha = (\partial_\mu A^\alpha_\nu - \partial_\nu A^\alpha_\mu) T_\alpha - igA^\beta_\mu A^\gamma_\nu C^\alpha_{\beta\gamma} T_\alpha$$

or

$$F^\alpha_{\mu\nu} = (\partial_\mu A^\alpha_\nu - \partial_\nu A^\alpha_\mu) - igA^\beta_\mu A^\gamma_\nu C^\alpha_{\beta\gamma}. \tag{3.65}$$

Thus

$$trF_{\mu\nu}F^{\mu\nu} = F^\alpha_{\mu\nu} F^{\mu\nu\beta} tr(T_\alpha T_\beta), \quad = F^\alpha_{\mu\nu} F^{\mu\nu\beta} \delta_{\alpha\beta}$$

Where we used the normalzation $tr(T_\alpha T_\beta) = \delta_{\alpha\beta}$. Thus

$$trF_{\mu\nu}F^{\mu\nu} = F^\alpha_{\mu\nu} F^{\mu\nu\alpha}. \tag{3.66}$$

Considering only gauge fields the Lagrangian is:

$$\mathcal{L}_g = -\frac{1}{4} tr(F_{\mu\nu}F^{\mu\nu}) = -\frac{1}{4} F^\alpha_{\mu\nu} F^{\mu\nu\alpha}. \tag{3.67}$$

Summarizing: In the case of the standard model $SU(2) \times U(1) \times SU(3)$ we have

(1) Symmetry $SU(2)$: $A^\alpha_\mu, \quad \alpha = 1, 2, 3.$
$C^\alpha_{\beta\gamma} = i\epsilon_{\beta\gamma\alpha} \Rightarrow F^\alpha_{\mu\nu} = \partial_\mu A^\alpha_\nu - \partial_\nu A^\alpha_\nu + g_2 \epsilon_{\beta\gamma\alpha} A^\beta_\nu A^\gamma_\mu$
(2) Symmetry $U(1)$: B_μ
$F_{\mu\nu} = \partial_\mu B_\nu - \partial_\nu B_\mu.$

(3) Symmetry $SU(3)$: The gauge fields are called gluons, G^α_μ, $\alpha = 1, 2, \cdots, 8$.
$C^\alpha_{\beta\gamma} = if_{\beta\gamma\alpha} \Rightarrow F^\alpha_{\mu\nu} = \partial_\mu G^\alpha_\nu - \partial_\nu G^\alpha_\mu + g_3 f_{\beta\gamma\alpha} G^\beta_\mu G^\gamma_\nu$, $\alpha = 1, 2, \cdots, 8$.
In the case of $SU_c(3)$ rewriting explicitly Eq. (3.67) with the aid of Eq. (3.65) we obtain:

$$\begin{aligned}\mathcal{L}_g = &-\frac{1}{4}\left(\partial_\mu G^\alpha_\nu - \partial_\nu G^\alpha_\mu\right)\left(\partial^\mu G^{\nu\alpha} - \partial^\nu G^{\mu\alpha}\right) \\ &- g_3 f_{\beta\gamma\alpha} \partial_\mu G^\alpha_\nu G^{\mu\beta} G^{\nu\gamma} - \frac{1}{4} g_3^2 f_{\beta\gamma\alpha} f_{\delta\epsilon\alpha} G^\beta_\mu G^\gamma_\nu G^{\mu\delta} G^{\nu\epsilon}\end{aligned} \tag{3.68}$$

We see that in in addition to the first term, which contains only derivatives, known as kinetic term, we notice that there exist cubic and quartic self interactions among the gluons (see Fig. 3.2). Note that initially there there appeared four cubic in the fields terms. These, however, have been reduced to one by using symmetry relations under the interchange of the indices.

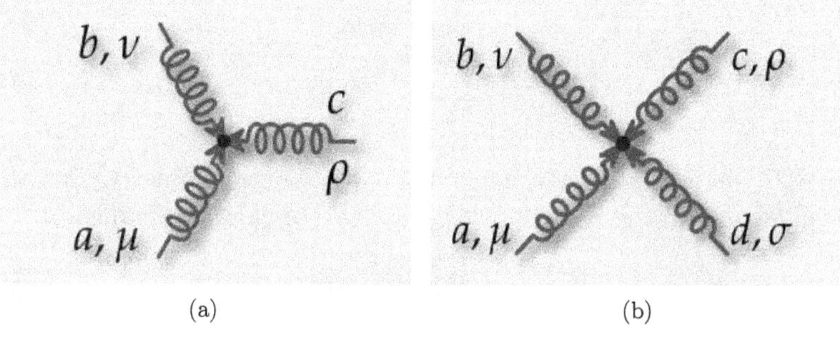

Fig. 3.2: The cubic (a) and quartic (b) interactions of the $SU_c(3)$ gauge bosons (gluons). Analogous interactions do not exist in electrodynamics (no photon self couplings).

This has profound implications in quantum chromodynamics (see chapter 6). Such interactions do not appear in the case of Abelian symmetry.

3.6.2 *The Lagrangian in the presence of scalar fields*

$$\mathcal{L} = -\frac{1}{4} tr(F_{\mu\nu}F^{\mu\nu}) + \mathcal{L}_0(\phi, D_\mu\phi), \quad \mathcal{L}_0(\phi, D_\mu\phi) = (D_\mu\phi)^\dagger(D^\mu\phi) - V(\phi),$$

where $V(\phi)$ is the scalar potential. Since $A_\mu^\dagger = A_\mu$, we have:

$$(\partial_\mu + igA_\mu)\phi^\dagger(\partial_\mu - igA_\mu)\phi = \partial_\mu\phi^\dagger\partial^\mu\phi + g^2|\phi|^2 A_\mu A^\mu - igA^\mu\partial_\mu\phi^\dagger + igA_\mu\partial^\mu\phi.$$

Thus

$$\mathcal{L}_0 = \partial_\mu\phi^\dagger\partial^\mu\phi + ig(A_\mu\partial^\mu\phi - A^\mu\partial_\mu\phi^\dagger) + g^2 A_\mu A^\mu|\phi|^2 - V(\phi). \tag{3.69}$$

3.6.3 *The Lagrangian in the presence of Fermion fields*

It can be shown that the Euler-Lagrange equation leads to the Dirac equation for a Lagrangian density of the form:

$$\mathcal{L}_f = \sum_\psi \bar{\psi}\gamma^\mu\partial_\mu\psi. \tag{3.70}$$

This, however, is not gauge invariant. It can become gauge invariant in the usual way:

$$\mathcal{L}_f = \sum_\psi \bar{\psi}\gamma^\mu D_\mu\psi, \tag{3.71}$$

where D_μ, as we have seen, depends on the gauge symmetry. It can be written, as e.g. in the SM see below, in terms of chiral Fermions:

$$\mathcal{L}_f = \sum_\psi \left(\bar{\psi}_L\gamma^\mu D_\mu\psi_L + \bar{\psi}_R\gamma^\mu D_\mu\psi_R\right) \tag{3.72}$$

which form the basis for the standard model. This will be discussed in some detail below.

3.7 The form of the SM Lagrangian

We will discuss here in some detail the electroweak and the the strong sectors separately.

3.7.1 *The electroweak part*

$$\mathcal{L} = \mathcal{L}_{\text{gauge}} + \mathcal{L}_\phi + \mathcal{L}_f + \mathcal{L}_{\text{Yukawa}},$$

Let us look at each term:

- The $\mathcal{L}_{\text{gauge}}$ term

$$\mathcal{L}_{\text{gauge}} = -\frac{1}{4}F^i_{\mu\nu}F^{i\mu\nu} - \frac{1}{4}B_{\mu\nu}B^{\mu\nu},\ (i = 1,2,3),$$

$$F^i_{\mu\nu} = \partial_\mu A^i_\nu - \partial_\nu A^i_\mu + g\varepsilon_{ijk}A^j_\mu A^k_\nu,\ (i = 1,2,3),$$

$$B_{\mu\nu} = \partial_\mu B_\nu - \partial_\nu B_\mu,$$

- The scalar field term.

$$\mathcal{L}_\phi = (D_\mu\phi)^+ D^\mu\phi - V(\phi),$$

$$D_\mu\phi = \left(\partial_\mu - i\left(g\frac{\tau^i}{2}A^i_\mu + \frac{g'}{2}B_\mu\right)\right)\phi.$$

- The Fermion sector:

$$\mathcal{L}_f = \sum_m \left(\bar{q}^0_{mL} i\gamma^\mu D_\mu q^0_{mL} + \bar{l}^0_{mL} i\gamma^\mu D_\mu l^0_{mL}\right.$$

$$\left. + \bar{u}^0_{mR} i\gamma^\mu D_\mu u^0_{mR} + \bar{d}^0_{mR} i\gamma^\mu D_\mu d^0_{mR} + \bar{e}^0_{mR} i\gamma^\mu D_\mu e^0_{mR}\right),$$

with

$$q^0_{mL} = \begin{pmatrix} u^0_m \\ d^0_m \end{pmatrix}_L,\ l^0_{mL} = \begin{pmatrix} \nu^0_m \\ e^0_m \end{pmatrix}_L,$$

where m is flavor index and, in the case of quarks, the color index is suppressed

$$D_\mu q^0_{mL} = \left(\partial_\mu - i\left(g\frac{\tau^i}{2}A^i_\mu + \frac{g'}{6}B_\mu\right)\right)q^0_{mL},$$

$$D_\mu l^0_{mL} = \left(\partial_\mu - i\left(g\frac{\tau^i}{2}A^i_\mu - \frac{g'}{2}B_\mu\right)\right)l^0_{mL},$$

$$D_\mu u^0_{mR} = \left(\partial_\mu - i\frac{2}{3}g' B_\mu\right)u^0_{mR},$$

$$D_\mu d^0_{mR} = \left(\partial_\mu + i\frac{1}{3}g' B_\mu\right)d^0_{mR},$$

$$D_\mu e^0_{mR} = \left(\partial_\mu + ig' B_\mu\right)e^0_{mR}.$$

- The Yukawa couplings.

  These involve the interactions of Fermions ψ_L and ψ_R with the scalar field (Higgs) is of the form:

$$\mathcal{L} = \sum_f \bar{\psi}_L \psi_R \phi + \text{H.C} \tag{3.73}$$

  with H.C meaning the Hermitian conjugate of the first term. This will be discussed later in 5.1.

For the behavior of the term $(D_\mu \phi)^+ D^\mu \phi$ after the spontaneous breakdown of the symmetry, see chapter 4 on Higgs.

3.7.2 *The strong part*

The relevant Lagrangian is:

$$\mathcal{L}_{SU_c(3)} = -\frac{1}{4} F^a_{\mu\nu} F^{a\mu\nu} + \sum_{a,b} \left(\bar{q}^b_{L\alpha} \gamma^\mu \left(iD^a_\mu \right)^\alpha_\beta q^{b\beta}_L + \bar{q}^b_{R\alpha} \gamma^\mu \left(iD^a_\mu \right)^\alpha_\beta q^{b\beta}_R \right),$$

where the quarks are chiral (L or R), a counts the gauge boson, b is flavor index, α, β are color indices and

$$F^i_{\mu\nu} = \partial_\mu G^i_\nu - \partial_\nu G^i_\mu + g f_{ijk} G^j_\mu G^k_\nu, (i, j, k = 1, 2, \cdots, 8),$$

with λ^i the generators of $SU_c(3)$ with commutators: $[\lambda^i, \lambda^j] = 2 f_{ijk} \lambda^k$ and

$$(D_\mu)^\alpha_\beta = (D_\mu)_{\alpha\beta} = \partial_\mu \delta_{\alpha\beta} - i g_3 G^i_\mu \left(\frac{\lambda^i}{2} \right)_{\alpha\beta}.$$

Before concluding this section we should mention that the group $SU_c(3)$ is not broken. The gluons remain massless. The resulting theory is QCD (see chapter 6 on QCD).

Chapter 4

The Higgs Mechanism

4.1 The Higgs mechanism in global gauge transformations

We will examine the Lagrangian of a scalar in the case that this is invariant under global symmetry transformations

4.1.1 *A complex scalar field*

We begin with the a complex scalar scalar field ϕ and a Lagrangian given by:

$$\mathcal{L} = \frac{1}{2} \left(\partial_\mu \phi^*\right) \partial^\mu \phi - V\left(\phi\right),$$

where $V(\phi)$ is given by:

$$V\left(\phi\right) = V\left(-\phi\right) \qquad \text{even function of } \phi, \tag{4.1}$$

$$V\left(\phi\right) = \kappa \frac{\mu^2}{2} \phi \phi^* + \frac{\lambda}{4} \left(\phi \phi^*\right)^2, \ \kappa = \pm 1. \tag{4.2}$$

We assume that $\lambda > 0$, so that the potential is bounded from below. We rewrite this potential in terms of the real and imaginary parts of the Higgs field:

$$\phi = \frac{1}{\sqrt{2}}(\phi_1 + i\phi_2), \qquad \phi^* = \frac{1}{\sqrt{2}}(\phi_1 - i\phi_2).$$

Then

$$V\left(\phi\right) = \kappa \frac{\mu^2}{4} \left(\phi_1^2 + \phi_2^2\right) + \frac{\lambda}{16} \left(\phi_1^2 + \phi_2^2\right)^2, \ \kappa = \pm 1. \tag{4.3}$$

This potential is exhibited in Fig. 4.1. We see that for $\kappa = 1$ the potential attains the minimum value 0 at the point $\phi = 0$.

We we will consider the case $\kappa = -1$. The condition for an extremum is

$$\frac{\partial V}{\partial \phi_1} = 0 \Rightarrow -\frac{1}{2}\mu^2 \phi_1 + \frac{1}{4}\lambda \left(\phi_1^2 + \phi_2^2\right) \phi_1 = 0,$$

$$\frac{\partial V}{\partial \phi_2} = 0 \Rightarrow -\frac{1}{2}\mu^2 \phi_2 + \frac{1}{4}\lambda \left(\phi_1^2 + \phi_2^2\right) \phi_2 = 0 \ .$$

These are satisfied by:

$$(\phi_1 = 0, \quad \phi_2 = 0) \text{ or } \left(\phi_1^2 + \phi_2^2\right) = \frac{2\mu^2}{\lambda}.$$

The condition for minimum is[1]:

$$\Delta = \frac{\partial^2 V}{\partial \phi_1^2} + 2\frac{\partial^2 V}{\partial \phi_1 \partial \phi_2} + \frac{\partial^2 V}{\partial \phi_2^2} > 0.$$

We find

$$\begin{aligned}
\frac{\partial^2 V}{\partial \phi_1^2} &= -\frac{1}{2}\mu^2 + \frac{1}{4}\lambda\left(3\phi_1^2 + \phi_2^2\right)\\
\frac{\partial^2 V}{\partial \phi_2^2} &= -\frac{1}{2}\mu^2 + \frac{1}{4}\lambda\left(\phi_1^2 + 3\phi_2^2\right)\\
\frac{\partial^2 V}{\partial \phi_1 \partial \phi_2} &= \frac{1}{2}2\lambda\phi_1\phi_2.
\end{aligned}$$

Thus

$$\begin{aligned}
\Delta &= -\mu^2 + \frac{1}{2}\lambda\left[3\left(\phi_1^2 + \phi_2^2\right) + \phi_1^2 + \phi_2^2\right] + 2\lambda\phi_1\phi_2\\
&= -\mu^2 + 2\lambda\left(\phi_1^2 + \phi_2^2\right) + 2\lambda\phi_1\phi_2.
\end{aligned}$$

[1]From calculus we know that the extrema of a function $f(x,y)$ of two variables at a critical point x_0, y_0, that is at a point where the first two derivatives vanish, i.e.

$$f_x(x,y)|_{x=x_0,y=y_0} = 0, \;\; f_y(x,y)|_{x=x_0,y=y_0} = 0$$

is expressed as follows

$$\begin{cases}
D = AB - C^2 > 0, \; A > 0 \text{ or } B > 0, & \text{local minimum}\\
D = AB - C^2 > 0, \; A < 0 \text{ or } B < 0, & \text{local maximum}\\
D = AB - C^2 < 0, & \text{saddle point}\\
D = AB - C^2 = 0, & \text{further examination is needed.}
\end{cases}$$

In the above expression

$$A = \left.\frac{\partial^2 f}{\partial x^2}\right|_{x=x_0,y=y_0}, \; B = \left.\frac{\partial^2 f}{\partial y^2}\right|_{x=x_0,y=y_0}, \; C = \left.\frac{\partial^2 f}{\partial x \partial y}\right|_{x=x_0,y=y_0}$$

In the problems we encounter here it happens that $D = 0$ and, thus, the above criterion is not sufficient. Note, however, that for the scalar potentials we are interested in, the quadratic term near a critical point away from the origin becomes:

$$f(x,y) - f(x_0,y_0) = A(x-x_0)^2 + B(y-y_0)^2 + 2C(x-x_0)(y-y_0)$$

In our case however$A > 0, B > 0$ and $C = \pm\sqrt{AB}$. So we find

$$f(x,y) - f(x_0,y_0) = \left(\sqrt{A}(x-x_0) \pm \sqrt{B}(y-y_0)\right)^2 > 0.$$

For such critical points we have a minimum at the critical point. Furthermore:

$$A + B + 2C = \left(\sqrt{A} \pm \sqrt{B}\right)^2 > 0,$$

i.e. our criterion follows. In the case of potentials with more than two coordinates, the procedure is the same.

For $\phi_1 = 0, \quad \phi_2 = 0 \to \Delta = -\mu^2 < 0$, i.e. we have a maximum. For $\lambda\left(\phi_1^2+\phi_2^2\right) = 2\mu^2$ we find:

$$\begin{aligned}\Delta &= -\frac{1}{2}\lambda\left(\phi_1^2+\phi_2^2\right) + \lambda\left(\phi_1^2+\phi_2^2\right) + \lambda\phi_1\phi_2 \\ &= \frac{1}{2}\lambda\left(\phi_1+\phi_2\right)^2 > 0,\end{aligned}$$

that is we have a minimum. The situation is exhibited in Fig. 4.1b. We say that the Higgs acquires a non zero vacuum expectation value.

In writing Eq. (4.2) we assume μ is real. Had we written $+\mu^2/2$ in the potential we find a minimum only if the higgs field vanishes. We then say that the Higgs does not acquire a vacuum expectation value (see Fig. 4.1a). In the case of Fig. 4.1b we find that the minima lie on a circle with radius υ:

$$\langle\phi_1\rangle^2 + \langle\phi_2\rangle^2 = \upsilon^2, \tag{4.4}$$

with $\langle\phi_1\rangle$ and $\langle\phi_2\rangle$ are the values of ϕ_1 and ϕ_2 at the location of the minimum. We write

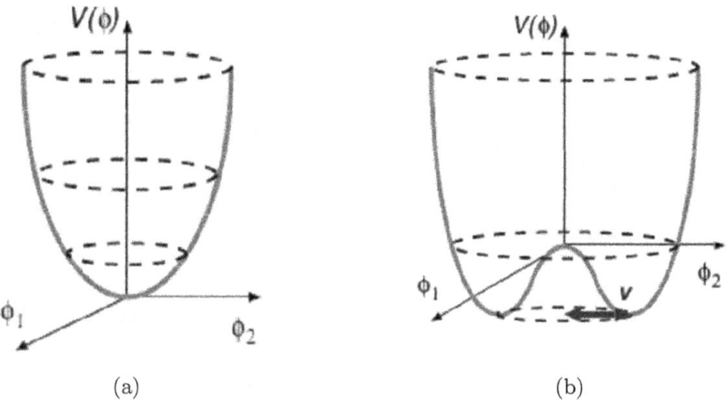

Fig. 4.1: The higgs potential given by Eq. (4.3) in the case of $\kappa = 1$ (a) and $\kappa = -1$ (b).

$$\langle\phi\rangle = \frac{1}{\sqrt{2}}e^{i\beta}\sqrt{2\mu^2/\lambda}.$$

Then away from the minimum by selecting $\beta = 0$ we get:

$$\phi_1 = X_1 + \langle\phi_1\rangle = X_1 + \upsilon,\ \phi_2 = iX_2 \tag{4.5}$$

Furthermore

$$V(\phi) = -\frac{\mu^2}{4}\left((X_1+v)^2 + X_2^2\right) + \frac{\lambda}{16}\left((X_1+v)^2+X_2^2\right)^2, \quad v=\sqrt{2\mu^2/\lambda}$$

$$= -\frac{1}{4}\lambda v^2\left(X_1^2+X_2^2+2X_1 v+v^2\right)$$

$$+\frac{\lambda}{16}\left[\left(X_1^2+X_2^2\right)^2 + 2v^2\left(3X_1^2+X_2^2\right)+4vX_1\left(X_1^2+X_2^2\right)+4v^3X_1+v^4\right]$$

$$= -\frac{1}{16}\lambda v^4 + \frac{1}{2}\lambda v^2 X_1^2 + \frac{\lambda}{16}\left(X_1^2+X_2^2\right)^2 + 4vX_1\left(X_1^2+X_2^2\right)$$

$$= \lambda v^2 X_1^2 - \frac{\mu^4}{4\lambda} + V_{\text{int}}$$

$$\boxed{V_{\text{int}} = \frac{\lambda}{16}\left[\left(X_1^2+X_2^2\right)^2 + 4\left(\frac{2\mu^2}{\lambda}\right)^{1/2} X_1\left(X_1^2+X_2^2\right)\right]}$$

$$V(\phi) = \tfrac{1}{2}\mu^2 X_1^2 + 0X_2^2 - \tfrac{\mu^4}{4\lambda} + V_{\text{int}}(X_1, X_2),$$

$\downarrow$ positive mass; $\downarrow$ vanished X_2 massless

In other words the mass of the scalar is

$$m^2 = \mu^2 = \frac{1}{2}\lambda v^2.$$

We write

$$\langle\phi_1\rangle = v,\ \langle\phi_2\rangle = 0.$$

$\hookrightarrow$ Goldstone boson (remains massless)

We could have chosen $\beta = \pi/2$, then

$$\langle\phi_1\rangle = 0 \qquad \langle\phi_2\rangle = v.$$

Now the ϕ_1 is the Goldstone boson, while ϕ_2 acquires mass m.

In either case the symmetry is broken spontaneously , i.e. the vacuum does not have the symmetry of the original rotationally invariant Lagrangian.

To avoid tedious operations in more complex situations we can obtain the masses by diagonalizing the mass matrix $m^2_{i,j}$ defined by:

$$m^2_{ij} = \left.\frac{\partial^2 V}{\partial\phi_i\partial\phi_j}\right|_{\phi=\langle\phi\rangle}.$$

i.e. evaluated at the location of the minimum. In the above example:

$$\left.\frac{\partial^2 V}{\partial\phi_1^2}\right|_{\phi_1=v,\phi_2=0} = -\frac{1}{2}\mu^2 + \frac{1}{4}\lambda\left(3\phi_1^2+\phi_2^2\right)\Big|_{\phi_1=v,\phi_2=0} = -\frac{1}{2}\mu^2+\frac{3}{4}\lambda v^2 = \frac{1}{2}\lambda v^2,$$

$$\left.\frac{\partial^2 V}{\partial \phi_2^2}\right|_{\phi_1=v,\phi_2=0} = -\frac{1}{2}\mu^2 + \frac{1}{4}\lambda\left(\phi_1^2 + 3\phi_2^2\right)\Big|_{\phi_1=v,\phi_2=0} = -\frac{1}{2}\mu^2 + \frac{1}{4}\lambda v^2 = 0,$$

$$\left.\frac{\partial^2 V}{\partial \phi_1 \partial \phi_2}\right|_{\phi_1=v,\phi_2=0} = \left.\frac{1}{2}\lambda\phi_1\phi_2\right|_{\phi_1=v,\phi_2=0} = 0.$$

The matrix is diagonal given by

$$\begin{pmatrix} \frac{1}{2}\lambda v^2 & 0 \\ 0 & 0 \end{pmatrix},$$

which has one non zero eigenvalue:

$$m_1^2 = \frac{1}{2}\lambda v^2.$$

Exercise: Investigate the case of a real field ϕ with three components $\phi^T = (\phi_1, \phi_2, \phi_3)$ (T stands for transpose). The potential is rotationally invariant.

4.1.2 *Two real scalar fields*

Consider two real scalar fields ϕ_1 and ϕ_2 with a scalar potential given by:

$$V(\phi) = -\frac{\mu^2}{2}\left(\phi_1^2 + \phi_2^2\right) + \frac{\lambda}{4}\left(\phi_1^2 + \phi_2^2\right)^2 \tag{4.6}$$

Proceeding as in the previous case we find that we have a local maximum for $\langle\phi_1\rangle = \langle\phi_2\rangle = 0$ and a minimum for:

$$\langle\phi_1\rangle^2 + \langle\phi_2\rangle^2 = v^2,\ v^2 = \frac{\mu^2}{\lambda}$$

Let us suppose that only ϕ_1 can acquire a vacuum expectation value, i.e. $\langle\phi_1\rangle = v$, $\langle\phi_2\rangle = 0$. Then expanding the potential (4.6) around the expectation value of the fields, i.e. by writing it in terms of two new fields η and π, as $\phi = \eta + v$ and $\phi_2 = \pi$ and using $\mu^2 = v^2\lambda$ we obtain:

$$V(\eta,\pi) = \frac{1}{2}(2\lambda v^2)\eta^2 + \frac{1}{4}\lambda\left(\eta^2 + \pi^2\right)^2 + \lambda v\eta(\eta^2 + \pi^2) - \frac{1}{4}\lambda v^2 \tag{4.7}$$

We see that the particle which acquired a vacuum expectation value attained a mass $m_\eta = v\sqrt{2\lambda}$, while the particle π remained massless.

This model, σ model, played an important role in the history of particle physics, before the emergence of the SM. In those days the strong interaction was supposed to be mediated by mesons, e.g. the η, π mesons etc. It was the first example of symmetry breaking, the idea borrowed from solid state physics, with the mesons identified with quark condensates[2].

[2]The strong interaction was supposed to be invariant under the (strong) isospin group

4.1.3 *A scalar transforming as a doublet under $SU_I(2)$*

We will consider a complex scalar field transforming like a doublet under $SU_I(2)$:

$$\phi = \begin{pmatrix} \phi^+ \\ \phi^0 \end{pmatrix}, \qquad \phi^* = \begin{pmatrix} \phi^- \\ \phi^{0*} \end{pmatrix}, \qquad \tilde{\phi} = i\sigma_2 \phi^* = \begin{pmatrix} \phi^{0*} \\ -\phi^- \end{pmatrix}.$$

Note that when we want to emphasize the $SU(2)$ character of the fields we consider ϕ and $\tilde{\phi}$. Thus

$$\phi^+ \Leftrightarrow I_3 = \frac{1}{2} \qquad \phi^- \Leftrightarrow I_3 = -\frac{1}{2},$$
$$\phi_0 \Leftrightarrow I_3 = -\frac{1}{2} \qquad \phi_0^* \Leftrightarrow I_3 = \frac{1}{2}.$$

Sometimes we find it convenient to write:

$$\phi^+ = \frac{\phi_1 + i\phi_2}{\sqrt{2}}, \qquad \phi^- = \frac{\phi_1 - i\phi_2}{\sqrt{2}}.$$

In any case the potential takes the form:

$$V(\phi) = -\frac{\mu^2}{2}\left(\phi^\dagger \phi\right) + \frac{\lambda}{4}\left(\phi^\dagger \phi\right)^2,$$

with

$$\phi^\dagger \phi = \frac{1}{2}\left(\phi_1^2 + \phi_2^2\right) + |\phi_0|^2$$

$$\boxed{\left(\phi^\dagger \phi\right) = \tfrac{1}{2}\left(\phi_1^2 + \phi_2^2 + R^2 + I^2\right),}$$

with

$$\Re\phi_0 = \frac{R}{\sqrt{2}} \qquad \Im\phi_0 = \frac{I}{\sqrt{2}}.$$

$SU(2)$. Under this $SU(2)$ the η meson was an isoscalar, while the pion was an isovector, appearing in three states π^+, π^- and π_0 with the indicated charges. The analysis proceeds as above with $\pi^2 = \pi_0^2 + \pi^+\pi^- + \pi^-\pi^+$. The pions cannot acquire a vacuum expectation value, since that would violate isospin symmetry and remained massless. The Lagrangian describing the strong interaction is still invariant under chiral transformations of the quark fields, see problem 7, chapter 2:

$$Y_q \bar{q}_L \phi_1 q_R \rightarrow e^{i(-\alpha+\alpha_{q_L}\alpha_{q_R})} \bar{q}_L \phi_1 q_R,$$

if $\alpha = \alpha_{q_L} + \alpha_{q_R}$, but it is broken spontaneously after the η acquires a vacuum expectation value,

$$Y_q \bar{q}_L \phi_1 q_R \rightarrow Y_q e^{i(-\alpha+\alpha_{q_L}+\alpha_{q_R})} \bar{q}_L \eta q_R + Y_q e^{i(\alpha_{q_L}+\alpha_{q_R})} \bar{q}_L v q_R$$

via the second term. We will see later that in such cases the quarks acquire a mass $m_q = Y_q v$ (see ch. 5). Once this happens the pions viewed as a bound state of a quark and antiquark can attain their relatively small mass of about 140 MeV.

Thus

$$\boxed{V(\phi) = -\tfrac{\mu^2}{4}\left(\phi_1^2+\phi_2^2+R^2+I^2\right)+\tfrac{\lambda}{16}\left(\phi_1^2+\phi_2^2+R^2+I^2\right)^2,}$$

Note that:

$$\frac{\partial V}{\partial z} = z\left(-\frac{\mu^2}{2}+\frac{\lambda}{4}\left(\phi_1^2+\phi_2^2+R^2+I^2\right)\right), \quad z=\phi_1,\,\phi_2,\,R,\,I.$$

The extremum occurs when:

$$R=I=\phi_1=\phi_2=0 \text{ or } \left(-\frac{\mu^2}{2}+\frac{\lambda}{4}\left(\phi_1^2+\phi_2^2+R^2+I^2\right)\right)=0.$$

Proceeding as above we find that the the condition for a minimum is $\Delta > 0$ where

$$\Delta = \frac{\partial^2 V}{\partial R^2}+\frac{\partial^2 V}{\partial I^2}+\frac{\partial^2 V}{\partial X_1^2}+$$

$$\frac{\partial^2 V}{\partial X_2^2}+2\frac{\partial^2 V}{\partial R\partial I}+2\frac{\partial^2 V}{\partial R\partial X_1}+2\frac{\partial^2 V}{\partial R\partial X_2}+2\frac{\partial^2 V}{\partial I\partial X_1}+2\frac{\partial^2 V}{\partial I\partial X_2}+2\frac{\partial^2 V}{\partial X_1\partial X_2}$$

$$= -2\mu^2+\frac{3}{2}\lambda\left(X_1^2+X_2^2+R^2+I^2\right)+\lambda\left(RI+RX_1+RX_2+IX_1+IX_2+X_1X_2\right)$$

$$= \frac{\lambda}{2}\left(X_1+X_2+R+I\right)^2 > 0.$$

The minimum is achieved if:

$$\boxed{\left(\langle\phi_1\rangle^2+\langle\phi_2\rangle^2+\langle R\rangle^2+\langle I\rangle^2\right)=\upsilon^2, \quad \upsilon^2=\tfrac{2\mu^2}{\lambda}.}$$

Due to charge conservation only the neutral components can develop a vacuum expectation value, i.e.

$$\langle\phi\rangle = \begin{pmatrix} 0 \\ e^{i\beta}\upsilon/\sqrt{2}\end{pmatrix}. \tag{4.8}$$

We can choose $\beta = 0$, i.e.:

$$\langle\phi_1\rangle = \langle\phi_2\rangle = I = 0,\ \langle R\rangle = \upsilon,$$

in which case:

$$\langle\phi\rangle = \begin{pmatrix} 0 \\ \upsilon/\sqrt{2}\end{pmatrix}. \tag{4.9}$$

The situation is exhibited in Fig. 4.2.

Expressing the fields ϕ^0, ϕ_1 and ϕ_2 around the vacuum expectation value, i.e. by writing:

$$\phi^0 = \frac{1}{\sqrt{2}}\left(\eta+\upsilon+iI\right),\ \phi_1 = \frac{1}{\sqrt{2}}X_1,\ \phi_2 = \frac{1}{\sqrt{2}}X_2,$$

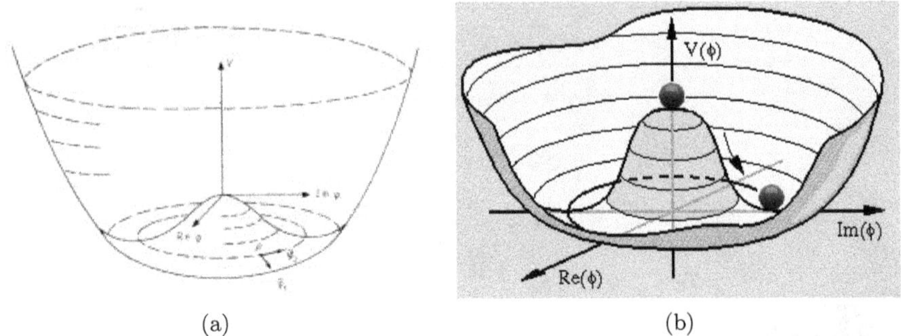

Fig. 4.2: The Higgs potential in the case of the SM (the Higgs is a complex isodoublet field, acquiring a vacuum expecation value) (a). If originally found at the origin (false vacuum), it eventually rolls down to the true vacuum (b).

we find:

$$V = -\frac{\mu^2}{4}\left(X_1^2 + X_2^2 + \eta^2 + I^2 + 2\eta v + v^2\right) + \frac{\lambda}{16} V_Q,$$

with

$$\begin{aligned} V_Q &= \left(X_1^2 + X_2^2 + \eta^2 + v^2 + 2\eta v + I^2\right)^2 \\ &= \left(X_1^2 + X_2^2 + R^2 + I^2\right)^2 + 4\eta^2 v^2 + v^4 + 4\eta v\left(X_1^2 + X_2^2 + \eta^2 + I^2\right) \\ &\quad +2v^2\left(X_1^2 + X_2^2 + \eta^2 + I^2\right) + 4\eta v^3 \\ &= \left(X_1^2 + X_2^2 + \eta^2 + I^2\right)^2 + 4\eta v\left(X_1^2 + X_2^2 + \eta^2 + I^2\right) \\ &\quad +2v^2\left(X_1^2 + X_2^2 + 3\eta^2 + I^2\right) + 4\eta v^3 + v^4. \end{aligned}$$

Now we see that:

(1) Constant term

$$-\frac{\mu^2}{4} v^2 + \frac{\lambda}{16} v^4 = -\frac{\mu^2}{4}\frac{2\mu^2}{\lambda} + \frac{\lambda}{16}\left(\frac{2\mu^2}{\lambda}\right)^2 = -\frac{\mu^4}{2\lambda} + \frac{\mu^4}{4\lambda} = -\frac{\mu^4}{4\lambda} \neq 0.$$

(2) Linear term

$$-\frac{\mu^2}{2}\eta v + \frac{\lambda}{16} 4\eta v^3 = \eta v\left(-\frac{\mu^2}{2} + \frac{\lambda}{4} v^2\right) = \eta v\left(-\frac{\mu^2}{2} + \frac{2\mu^2}{4}\right) = 0.$$

as expected from the minimization condition.

(3) The full potential becomes:

$$V = \tfrac{1}{2}\mu^2\eta^2 - \tfrac{\mu^4}{4\lambda} + V_{\text{int}}, \quad m_\eta^2 = \mu^2 = \tfrac{1}{2}(\lambda v)^2 \tag{4.10}$$

$$\boxed{V_{\text{int}} = \tfrac{\lambda}{16}\left[\left(X_1^2 + X_2^2 + \eta^2 + I^2\right)^2 + 4\eta\left(X_1^2 + X_2^2 + \eta^2 + I^2\right)\sqrt{\tfrac{2\mu^2}{\lambda}}\right].} \tag{4.11}$$

The above results can be obtained by expanding the potential in powers, e.g.

$$m_\eta^2 = \left.\frac{\partial^2 V}{\partial \eta^2}\right|_{X_1=X_2=\eta=I=0},$$

$$\frac{\partial V}{\partial \eta} = -\frac{\mu^2}{2}(\eta + v) + \frac{\lambda}{16}4\left(X_1^2 + X_2^2 + (\eta+v)^2 + I^2\right)(\eta + v),$$

$$\frac{\partial^2 V}{\partial \eta^2} = -\frac{\mu^2}{2} + \frac{\lambda}{4}\left(X_1^2 + X_2^2 + (\eta+v)^2 + I^2 + 2(\eta+v)^2\right),$$

$$\left.\frac{\partial^2 V}{\partial \eta^2}\right|_{X_1=X_2=\eta=I=0} = -\frac{\mu^2}{2} + \frac{\lambda}{4}\left[v^2 + 2v^2\right] = -\frac{\mu^2}{2} + \frac{3\lambda}{4}v^2$$

$$= -\frac{\mu^2}{2} + \frac{3}{4}2\mu^2 = -\frac{\mu^2}{2} + \frac{3}{2}\mu^2 = \mu^2 \Rightarrow m_\eta = \mu.$$

$$\frac{\partial V}{\partial z} = -\frac{\mu^2}{2}I + \frac{\lambda}{4}\left[X_1^2 + X_2^2 + (R+v)^2 + I^2\right]z, \quad z = X_1, X_2, I,$$

$$\frac{\partial^2 V}{\partial z^2} = -\frac{\mu^2}{2} + \frac{\lambda}{4}\left(X_1^2 + X_2^2 + (R+v)^2 + I^2 + 2I^2\right)_{X_1=X_2=R=I=0}$$

$$= -\frac{\mu^2}{2} + \frac{\lambda}{4}v^2 = -\frac{\mu^2}{2} + \frac{\mu^2}{2} = 0, \quad z = z = X_1, X_2, I.$$

The off diagonal elements also vanish, i.e. all terms other than the first are zero. Thus we have one massive field and three Goldstone bosons. The doublet becomes:

$$\phi = \begin{pmatrix} \phi^+ \\ \frac{\eta + v + iI}{\sqrt{2}} \end{pmatrix}.$$

Had we considered:

$$\beta = \frac{\pi}{2}, \text{ i.e. the minimum at } R = 0,\ I = v,$$

we would have:

$$\phi = \begin{pmatrix} \phi^+ \\ \frac{R + iv + iI}{\sqrt{2}} \end{pmatrix}, \quad \langle \phi \rangle = \begin{pmatrix} 0 \\ \frac{iv}{\sqrt{2}} \end{pmatrix}.$$

Now the imaginary part becomes massive.

4.1.4 *Symmetry restoration*

Let us now consider

$$V(\phi) = -\frac{\mu^2}{2}\left(\phi^\dagger\phi\right) + \frac{\lambda}{4}\left(\phi^\dagger\phi\right)^2 + \lambda'\left(\phi^\dagger\phi\right)\chi^2$$

in other words the Higgs potential considered above plus an interaction term with another singlet scalar field χ. It is interesting to consider the case in which the later field is in equilibrium in the early universe with temperature T. Then $\chi^2 \to T^2$ and the above potential takes the form:

$$V(\phi) = -\frac{\mu^2}{2}\left(\phi^\dagger\phi\right)\left(1 - \left(\frac{T}{T_c}\right)^2\right) + \frac{\lambda}{4}\left(\phi^\dagger\phi\right)^2, T_c^2 = \mu^2\lambda', \tag{4.12}$$

where $T_c = \sqrt{\mu^2/(2\lambda')}$. Then we see for $T > T_c$ the symmetry is not broken. As soon as, however, the temperature drops below T_c we have a spontaneous breakdown of the symmetry as discussed above. On the contrary in a situation in which the temperature is rising the symmetry is restored as soon as the temperature exceeds T_c (see Fig. 4.3).

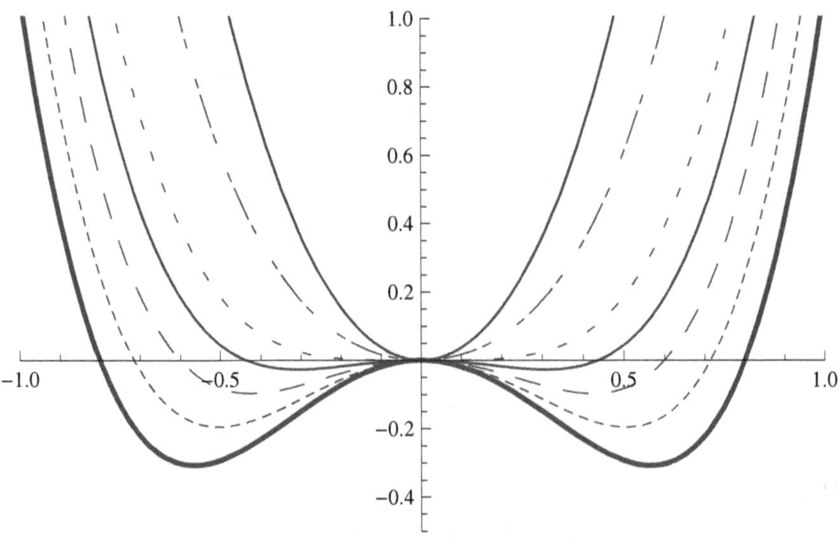

Fig. 4.3: As the temperature increases, from bottom to top, above some critical temperature the Higgs isodoublet no longer acquires a vacuum expectation value and the symmetry is restored.

4.2 The Higgs mechanism in gauge theories

We have seen thar in the standard model we encounter a scalar field transforming like a doublet under $SU_I(2)$

$$\phi = \begin{pmatrix} \phi^+ \\ \phi_0 \end{pmatrix}, I = \frac{1}{2}, Y = 1, \quad \phi^* = \begin{pmatrix} \phi^- \\ \phi^{0*} \end{pmatrix},$$

$$\tilde{\phi} = i\sigma_2\phi^* = \begin{pmatrix} \phi^{0*} \\ -\phi^- \end{pmatrix}, I = \frac{1}{2}, Y = -1,$$

where Y is the hypercharge quantum number associated with $U(1)$ connected with the usual charge via

$$Q = I_3 + \frac{Y}{2}$$

4.2.1 *The spontaneous symmetry braking (SSB); the unitary gauge*

Now we can show that the most general field ϕ can be transformed into

$$\begin{pmatrix} 0 \\ \frac{v+\eta}{\sqrt{2}} \end{pmatrix} \tag{4.13}$$

by gauge transformation. Indeed consider

$$e^{i\tau\cdot\xi/v} \begin{pmatrix} 0 \\ \frac{v+\eta}{\sqrt{2}} \end{pmatrix},$$

with

$$\tau\cdot\xi = \tau_+\xi_- + \tau_-\xi_+ + \tau_3\xi_3 = \begin{pmatrix} \xi_3 & \xi_1 - i\xi_2 \\ \xi_1 + i\xi_2 & -\xi_3 \end{pmatrix}.$$

We recall that

$$e^{i\tau\cdot\xi/v} \simeq 1 + i\frac{\tau\cdot\xi}{v}.$$

Thus we get

$$\begin{aligned} e^{i\tau\cdot\xi/v} \begin{pmatrix} 0 \\ \frac{v+\eta}{\sqrt{2}} \end{pmatrix} &\simeq \begin{pmatrix} 0 \\ \frac{v+\eta}{\sqrt{2}} \end{pmatrix} + \begin{pmatrix} \frac{i\xi_3}{v} & \frac{i\xi_1+\xi_2}{v} \\ \frac{i\xi_1-\xi_2}{v} & -\frac{i\xi_3}{v} \end{pmatrix} \begin{pmatrix} 0 \\ \frac{v}{\sqrt{2}} \end{pmatrix} \\ &= \begin{pmatrix} 0 \\ \frac{v+\eta}{\sqrt{2}} \end{pmatrix} + \begin{pmatrix} \frac{i\xi_1+\xi_2}{\sqrt{2}} \\ -\frac{i\xi_3}{\sqrt{2}} \end{pmatrix} \\ &= \begin{pmatrix} \frac{i\xi_1+\xi_2}{\sqrt{2}} \\ \frac{v+\eta-i\xi_3}{\sqrt{2}} \end{pmatrix}. \end{aligned}$$

This completes the proof by observing that:

$$\begin{aligned}\phi^+ &= \tfrac{1}{\sqrt{2}}\left(X_1 + iX_2\right) \\ \phi_0 &= \tfrac{1}{\sqrt{2}}\left(R + iI\right)\end{aligned} \quad \Rightarrow$$

$$X_1 = \xi_2,\ X_2 = \xi_1,\ R = v + \eta,\ I = -\xi_3.$$

In other words the above fields are equivalent

$$\begin{pmatrix} 0 \\ \frac{v+\eta}{\sqrt{2}} \end{pmatrix} \Leftrightarrow \begin{pmatrix} \phi^+ \\ \phi^0 \end{pmatrix} \tag{4.14}$$

i.e. they are related by a unitary gauge transformation. So, if the theory is invariant under such a transformation, it is adequate to consider the field given by Eq. (4.13). This is known as the unitary gauge.

The mass of this physical Higgs field is, of course, what we have found above when we considered the local transformation.

A very interesting pattern of Higgs mechanism is found in superymmetric extensions of the Standard Model (see section 8.3).

4.2.2 *How the gauge bosons acquire a mass*

Consider now the expression $(D_\mu \phi)^\dagger D^\mu \phi$ with

$$iD_\mu = i\partial_\mu + \left(\frac{g}{2}\tau \cdot \mathbf{A}_\mu + \frac{g'}{2}B_\mu\right) = i\partial_\mu + Q_\mu,$$

$$Q_\mu = \frac{g}{2}\tau \cdot \mathbf{A}_\mu + \frac{g'}{2}B_\mu = \frac{1}{2}\left(g(\tau_1 A^1_\mu + \tau_2 A^2_\mu + \tau_3 A^3_\mu) + g' B_\mu\right) \tag{4.15}$$

Define now

$$W^\pm = \frac{1}{\sqrt{2}}(A^1_\mu \pm iA^2_\mu), \Leftrightarrow A^1_\mu = \frac{1}{\sqrt{2}}(W^+_\mu + W^-\mu),\ A^2_\mu = \frac{1}{i\sqrt{2}}(W^+_\mu - W^-_\mu) \tag{4.16}$$

Then

$$\tau_1 A^1_\mu + \tau_2 A^2_\mu = \frac{1}{\sqrt{2}}\left((\tau_1 - i\tau_2)W^+_\mu + (\tau_1 + i\tau_2)W^-_\mu\right) = \sqrt{2}(\tau_+ W^-_\mu + \tau_- W^+_\mu) \tag{4.17}$$

with

$$\tau_+ = \frac{\tau_1 + i\tau_2}{2} = \begin{pmatrix} 0\,1 \\ 0\,0 \end{pmatrix},\ \tau_- = \frac{\tau_1 - i\tau_2}{2} = \begin{pmatrix} 0\,0 \\ 1\,0 \end{pmatrix} \tag{4.18}$$

that is

$$Q_\mu = \frac{1}{2}g\tau \cdot \mathbf{A}_\mu + \frac{1}{2}g' B_\mu = g\frac{\tau_+ W^-_\mu + \tau_- W^+_\mu}{\sqrt{2}} + g\frac{\tau_3}{2}A^3_\mu + \frac{1}{2}g' B_\mu.$$

now

$$Q_\mu \begin{pmatrix} 0 \\ \frac{1}{\sqrt{2}}(\eta + \upsilon) \end{pmatrix} = \frac{1}{\sqrt{2}}(\eta + \upsilon) Q_\mu \begin{pmatrix} 0 \\ 1 \end{pmatrix}$$

and

$$Q_\mu \begin{pmatrix} 0 \\ 1 \end{pmatrix} = \frac{g}{\sqrt{2}} \left(W_\mu^- \tau_+ \begin{pmatrix} 0 \\ 1 \end{pmatrix} + W_\mu^+ \tau_- \begin{pmatrix} 0 \\ 1 \end{pmatrix} \right) + \frac{g}{2} A_\mu^3 \tau_3 \begin{pmatrix} 0 \\ 1 \end{pmatrix} + \frac{g'}{2} B_\mu \begin{pmatrix} 0 \\ 1 \end{pmatrix} =$$

$$\frac{g}{\sqrt{2}} \left(W_\mu^- \begin{pmatrix} 1 \\ 0 \end{pmatrix} + 0 \right) + \frac{g}{2} A_\mu^3 \begin{pmatrix} 0 \\ -1 \end{pmatrix} + \frac{g'}{2} B_\mu \begin{pmatrix} 0 \\ 1 \end{pmatrix} = \begin{pmatrix} \frac{g}{\sqrt{2}} W_\mu^- \\ \frac{1}{2}(-gA_\mu^3 + g' B_\mu) \end{pmatrix}.$$

Thus

$$(iD^\mu \phi)^+ iD_\mu \phi = \frac{1}{2} \partial^\mu \eta \partial_\mu \eta + \frac{1}{2}(\eta + \upsilon)^2 \left(\frac{g}{\sqrt{2}} W^{+\mu}, \frac{1}{2}(-gA^{3\mu} + g' B^\mu) \right) \begin{pmatrix} \frac{g}{\sqrt{2}} W_\mu^- \\ \frac{1}{2}(-gA_\mu^3 + g' B_\mu) \end{pmatrix} \tag{4.19}$$

(the linear term in the derivative vanishes since the Higgs field is real). As a result

$$(iD^\mu \phi)^+ iD_\mu \phi = \frac{1}{2} \partial^\mu \eta \partial_\mu \eta$$

$$+ \frac{1}{4}(\eta + \upsilon)^2 \left(g^2 (W^{+\mu} W_\mu^- + W^{-\mu} W_\mu^+) + (-gA^{3\mu} + g' B^\mu)(-gA_\mu^3 + g' B_\mu) \right)$$

or

$$(iD^\mu \phi)^+ iD_\mu \phi = \frac{1}{2} \partial^\mu \eta \partial_\mu \eta +$$

$$\frac{1}{4}(\eta^2 + 2\eta\upsilon) \left(g^2 (W^{+\mu} W_\mu^- + W^{-\mu} W_\mu^+) + (-gA^{3\mu} + g' B^\mu)(-gA_\mu^3 + g' B_\mu) \right)$$

$$+ \frac{1}{4}\upsilon^2 \left(g^2 (W^{+\mu} W_\mu^- + W^{-\mu} W_\mu^+) + (-gA^{3\mu} + g' B^\mu)(-gA_\mu^3 + g' B_\mu) \right) \tag{4.20}$$

The first term of the previous equation is the kinetic energy term of the physical Higgs η. The second term is the interaction term between the Higgs field and the vector bosons. The last term is the mass term. The first part of the mass term can be cast in the form:

$$\frac{1}{2} \left(\frac{g\upsilon}{2} \right)^2 (W_\mu^+ W^{\mu -} + W_\mu^- W^{\mu +}).$$

It corresponds to the charged matrix:

$$\frac{v^2}{4} \quad \begin{array}{c|cc} & W_\mu^- & W_\mu^+ \\ \hline W^{+\mu} & g^2 & 0 \\ W^{-\mu} & 0 & g^2 \end{array} \quad \text{which is diagonal} \rightarrow \quad \boxed{M_W^2 = \left(\frac{g}{2}\right)^2 v^2}$$

Thus

$$\frac{1}{2} M_W^2 \left(W_\mu^+ W^{\mu -} + W_\mu^- W^{\mu +} \right),$$

i.e. the charged bosons have a mass $m_W = \frac{gv}{2}$ Similarly the second part of the mass term leads to the non-diagonal neutral boson matrix:

$$\frac{v^2}{2}\frac{1}{2} \quad \begin{array}{c|cc} & A_\mu^3 & B_\mu \\ \hline A^{3\mu} & g^2 & -gg' \\ B^\mu & -gg' & g'^2 \end{array} \quad \rightarrow$$

This matrix has two eigenvalues 0 and $g^2 + (g')^2$.

$$\text{for eigenvalue } 0 \;\Leftrightarrow A_\mu \Leftrightarrow \begin{bmatrix} g^2 & -gg' \\ -gg' & g'^2 \end{bmatrix} \begin{pmatrix} x \\ y \end{pmatrix} = 0 \Rightarrow \begin{array}{r} g^2 x = gg' y \\ x = \frac{g'}{g} y \end{array}$$

$$\text{normalized eigenvector: } \begin{pmatrix} \frac{g'}{\sqrt{g^2+g'^2}} \\ \frac{g}{\sqrt{g^2+g'^2}} \end{pmatrix} \Rightarrow A_\mu = \frac{g' A_\mu^3 + g B_\mu}{\sqrt{g^2 + g'^2}}.$$

Eigenvalue $g^2 + (g')^2 \Leftrightarrow Z_\mu$

$$\text{normalized eigenvector: } \begin{pmatrix} \frac{g}{\sqrt{g^2+g'^2}} \\ -\frac{g'}{\sqrt{g^2+g'^2}} \end{pmatrix} \Rightarrow Z_\mu = \frac{g A_\mu^3 - g' B_\mu}{\sqrt{g^2 + g'^2}}.$$

The Z-boson mass can be written:

$$\frac{v^2}{2}\frac{1}{4}\left(g^2 + g'^2\right) = \frac{1}{2} M_Z^2,$$

or

$$\boxed{M_Z^2 = \frac{g^2 v^2}{4}\left[1 + \left(\frac{g'}{g}\right)^2\right]},$$

that is

$$M_Z = M_W \sqrt{1 + \left(\frac{g'}{g}\right)^2} \quad \text{or} \quad \boxed{M_Z = \frac{M_W}{\cos\theta_W}}$$

where

$$\frac{g'}{g} = \tan\theta_W \qquad \text{(Weinberg angle)} \tag{4.21}$$

which is often called weak angle.

Thus the field A_μ, which is massless, can be identified with the photon. Noting that in Eq. (4.20) the other two neutral fields appear in the form of Z_μ we can write:

$$(iD^\mu\phi)^+ iD_\mu\phi = \frac{1}{2}\left(\partial_\mu\eta\partial^\mu\eta\right) \text{ (kinetic term for the Higgs) } +$$

$$\left.\begin{aligned}&\frac{1}{2}M_W^2\left(W_\mu^+W^{\mu-}+W_\mu^-W^{\mu+}\right)\\&+\frac{1}{2}M_Z^2\left(Z_\mu Z^\mu\right)\end{aligned}\right\}\text{mass term}$$

$$+\frac{\eta^2+2v\eta}{v^2}\left[\frac{1}{2}M_W^2\left(W_\mu^+W^{\mu-}+W_\mu^-W^{\mu+}\right)+\frac{1}{2}M_Z^2\left(Z_\mu Z^\mu\right)\right], \tag{4.22}$$

4.2.3 *The electroweak Lagrangian after the SSB*

The Higgs potential is given by Eqs (4.10) and (4.11). Thus using Eq. (4.22) we find

$$\mathcal{L}_\phi = \frac{1}{2}\left(\partial_\mu\eta\partial^\mu\eta\right)+\frac{1}{2}M_W^2\left(W_\mu^+W^{\mu-}+W_\mu^-W^{\mu+}\right)+$$
$$\frac{1}{2}\mu^2\eta^2+\frac{1}{2}M_Z^2\left(Z_\mu Z^\mu\right)-\left(V_{\text{int}}-\frac{\mu^4}{4\lambda}\right), \tag{4.23}$$

with

$$V_{\text{int}} = \frac{1}{2}\frac{\eta^2+2v\eta}{v^2}\left[M_W^2\left(W_\mu^+W^{\mu-}+W_\mu^-W^{\mu+}\right)+M_Z^2\left(Z_\mu Z^\mu\right)\right]$$
$$+\frac{\lambda}{16}\eta^3\left(\eta+4\sqrt{\frac{2\mu^2}{\lambda}}\right). \tag{4.24}$$

Diagrammatically these interactions are exhibited in Figs 4.4, 4.5 and 4.6. We note that:

i) By defining

$$Q = I_3 + \frac{Y}{2} \qquad \text{charge operator}$$

we find

$$Q = \begin{pmatrix} 1/2 & 0 \\ 0 & -1/2 \end{pmatrix} + \begin{pmatrix} 1/2 & 0 \\ 0 & 1/2 \end{pmatrix} = \begin{pmatrix} 1 & 0 \\ 0 & 0 \end{pmatrix}.$$

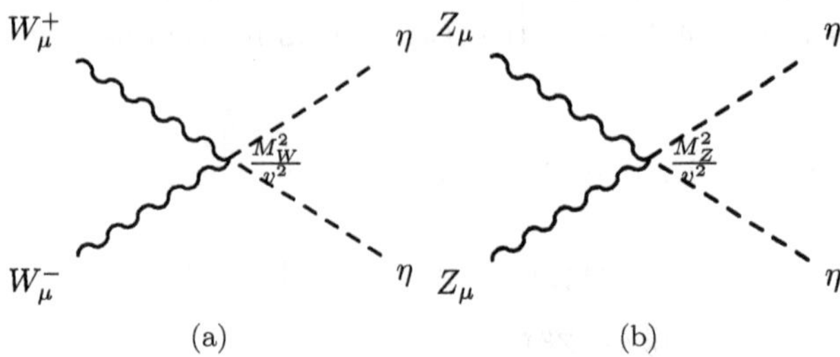

Fig. 4.4: The quartic charged gauge boson (a) and neural gauge boson (b) couplings to Higgs.

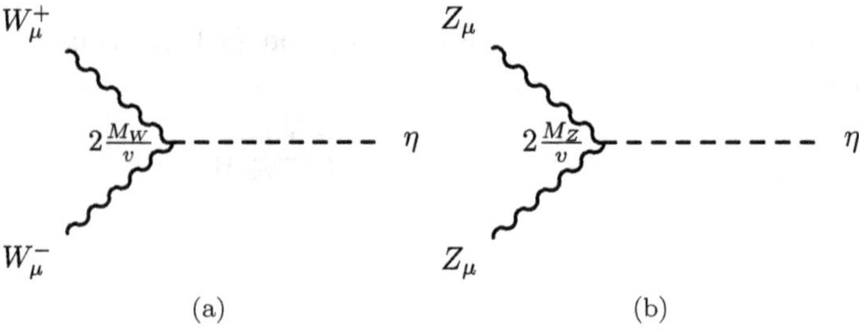

Fig. 4.5: The cubic charged gauge boson (a) and neural gauge boson (b) couplings to Higgs.

Thus

$$Q\begin{pmatrix} 0 \\ v/\sqrt{2} \end{pmatrix} = 0 \ .$$

In other words the charge operator remains unbroken.

ii) A symmetry $U(1)$ is preserved. This is electromagnetism $U(1)_{EM}$. We write:

$$\underbrace{SU_I\,(2) \otimes U_Y\,(1)}_{\text{broken spontaneously}} \otimes SU_C\,(3) \to U_{EM} \otimes SU_C\,(3)$$

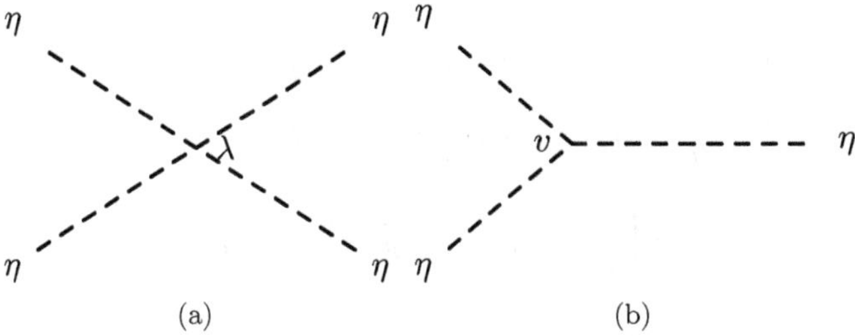

Fig. 4.6: The quartic (a) and cubic (b) Higgs self couplings.

4.2.4 *Summary*

Summarizing we have:

(1) One physical Higgs η with mass $m_\eta = \mu = v\sqrt{\frac{\lambda}{2}}$. No Goldstone bosons.
(2) One massless gauge boson to be identified with the photon.
(3) Two charged gauge bosons W^+, W^- with mass $M_W^2 = \frac{1}{4}(gv)^2$.
(4) One massive gauge boson Z_μ with mass $M_Z^2 = \frac{1}{4}(gv)^2 / \cos^2\theta_W$ or

$$\boxed{M_Z = \frac{M_W}{\cos\theta_W}.}$$

(5) 8 massless gauge bosons associated wih with the group $SU_C(3)$ which remain unbroken. These are called gluons (see chapter 6).
(6) The Goldstone bosons disappeared. Their degrees of freedom were eaten by the massive gauge bosons (see next section). Here is an accounting

Before:

$1 \times 4 + 4\times\ 2 = 4 + 8 = 12$	
$\downarrow$	$\downarrow$ polarization for massless
scalar	vector boson

After:

$1 \times 1 + 1 \times 2$	$+3\times$	$3 = 1 + 2 + 9 = 12$ $\checkmark$
$\downarrow$		$\downarrow$
polarization for massless boson		polarization for massive spin 1

The Goldstone bosons were eaten up by the gauge bosons! (see next section)

4.3 What happened to the Goldstone bosons?

To see this consider the Lagrangian:

$$\mathcal{L} = (D_\mu \phi)^\dagger D^\mu \phi - \frac{1}{4} F_{\mu\nu} F^{\mu\nu} - V(\phi)$$
$$D_\mu = \partial_\mu - i \left[\frac{g}{\sqrt{2}} \left(A_\mu^+ \tau_- + A_\mu^- \tau_+ \right) + \frac{g}{2} A_\mu^3 \tau_3 + \frac{g'}{2} B_\mu \right].$$

Consider first the charged part of ϕ in an arbitrary gauge: $\begin{pmatrix} \phi^+ \\ v/\sqrt{2} \end{pmatrix}$. Then

$$D_\mu \phi \sim \left[\partial_\mu - i \frac{g}{\sqrt{2}} \left(A_\mu^+ \tau_- + A_\mu^- \tau_+ \right) \right] \begin{pmatrix} \phi^+ \\ v/\sqrt{2} \end{pmatrix},$$
$$\tau_- \begin{pmatrix} \phi^+ \\ v/\sqrt{2} \end{pmatrix} = \begin{pmatrix} 0 \\ \phi^+ \end{pmatrix},$$
$$\tau_+ \begin{pmatrix} \phi^+ \\ v/\sqrt{2} \end{pmatrix} = \begin{pmatrix} v/\sqrt{2} \\ 0 \end{pmatrix},$$
$$D_\mu \phi \sim \begin{pmatrix} \partial_\mu \phi^+ \\ 0 \end{pmatrix} - i \frac{g}{\sqrt{2}} A_\mu^+ \begin{pmatrix} 0 \\ \phi^+ \end{pmatrix} - i \frac{g}{\sqrt{2}} \frac{v}{\sqrt{2}} A_\mu^- \begin{pmatrix} 1 \\ 0 \end{pmatrix}$$
$$= \begin{pmatrix} \partial_\mu \phi^+ - i \frac{gv}{2} A_\mu^- \\ -i \frac{g}{\sqrt{2}} A_\mu^+ \phi^+ \end{pmatrix},$$

with A_μ^+ the gauge boson as before. Then the η independent term becomes:

$$(D_\mu \phi)^+ (D^\mu \phi) \sim \left(\partial_\mu \phi^- + i \frac{gv}{2} A_\mu^- \right) \left(\partial^\mu \phi^+ - i \frac{gv}{2} A^{\mu +} \right) + \frac{1}{2} g^2 A_\mu^+ A^{-\mu} \phi^- \phi^+,$$

where the second term yields a boson-scalar interaction, which will be absorbed in a complete treatment by the inclusion of η, which here was ignored. To see the meaning of the first term, we define:

$$W_\mu^+ = A_\mu^+ + i \frac{\partial_\mu \phi^+}{gv/2} \qquad \text{(eats up the positive Higgs)},$$
$$W_\mu^- = A_\mu^- - i \frac{\partial_\mu \phi^-}{gv/2} \qquad \text{(eats up the negative Higgs)}.$$

Then we get :

$$(D_\mu \phi)^+ (D^\mu \phi) \sim m_W^2 W_\mu^+ W^{\mu -} = \frac{1}{2} m_W^2 \left(W_\mu^+ W^{\mu -} + W_\mu^- W^{\mu +} \right)$$

we thus recover the familiar mass term for the charged boson with $m_W = (1/2)gv$. For the neutral components we proceed similarly.

$$D_\mu\phi = \left[\partial_\mu - i\left(\frac{g}{2}A^3_\mu\tau_3 + \frac{g'}{2}B_\mu\right)\right]\begin{pmatrix} 0 \\ \frac{v+ix}{\sqrt{2}} \end{pmatrix}$$

$$= \begin{pmatrix} 0 \\ i\frac{\partial_\mu x}{\sqrt{2}} \end{pmatrix} + \begin{pmatrix} 0 \\ -i\left(-\frac{g}{2}A^3_\mu + \frac{g'}{2}B_\mu\right)\frac{v+ix}{\sqrt{2}} \end{pmatrix}$$

$$= \frac{i}{\sqrt{2}}\begin{pmatrix} 0 \\ \partial_\mu x + \left(\frac{g}{2}A^3_\mu - \frac{g'}{2}B_\mu\right)v \end{pmatrix},$$

$$\left(D_\mu\phi\right)^\dagger\left(D^\mu\phi\right) \sim \frac{1}{2}\left[\partial_\mu x + \left(\frac{g}{2}A^3_\mu - \frac{g'}{2}B_\mu\right)v\right]\left[\partial^\mu x + \left(\frac{g}{2}A^{\mu 3} - \frac{g'}{2}B^\mu\right)v\right].$$

Now define

$$Z_\mu = \underbrace{\cos\theta_W A^3_\mu - \sin\theta_W B_\mu}_{\tilde{Z}_\mu} + \frac{\partial_\mu x}{gv/2\cos\theta_W}.$$

where $\tilde{Z}_\mu$ the neutral massive boson in the unitary gauge. Then

$$\left(D_\mu\phi\right)^\dagger\left(D^\mu\phi\right) = \frac{1}{2}\left[\partial_\mu x + \frac{gv}{2\cos\theta_W}\left(\overbrace{A^3_\mu\cos\theta - B_\mu\sin\theta_W}^{\tilde{Z}_\mu}\right)\right]\times$$

$$\left[\partial^\mu x + \frac{gv}{2\cos\theta_W}\left(\overbrace{A^{\mu 3}\cos\theta - B^\mu\sin\theta_W}^{\tilde{Z}^\mu}\right)\right]$$

$$= \frac{1}{2}\frac{g^2v^2}{4\cos^2\theta_W}\left[\overbrace{A^3_\mu\cos\theta - B_\mu\sin\theta_W}^{\tilde{Z}_\mu} + \frac{\partial_\mu x}{gv/2\cos\theta_W}\right]\times$$

$$\left[\overbrace{A^{\mu 3}\cos\theta - B^\mu\sin\theta_W}^{\tilde{Z}^\mu} + \frac{\partial^\mu x}{gv/2\cos\theta_W}\right]$$

or

$$\left(D_\mu\phi\right)^\dagger\left(D^\mu\phi\right) = \left(\frac{gv}{2\cos\theta_W}\right)^2\left(Z_\mu Z^\mu\right) = \frac{1}{2}M_Z^2 Z_\mu Z^\mu.$$

Considering the terms containing η we recover the interaction terms contained in Eq. (4.22).

Note that

$$\begin{aligned}
\partial_\mu W_\nu^- - \partial_\nu W_\mu^- &= \partial_\mu A_\nu^- - i\partial_\mu \frac{\partial_\nu \phi}{gv/2} - \partial_\nu A_\mu^- + i\partial_\nu \frac{\partial_\mu \phi}{gv/2} \\
&= \partial_\mu A_\nu^- - \partial_\nu A_\mu^- \qquad \text{(unchanged)}, \\
\partial_\mu W_\nu^+ - \partial_\nu W_\mu^+ &= \partial_\mu A_\nu^+ - \partial_\nu A_\mu^+ \\
\partial_\mu Z_\nu - \partial_\nu Z_\mu &= \partial_\mu \tilde{Z}_\nu - \partial_\nu \tilde{Z}_\mu \qquad \text{(unchanged)}.
\end{aligned}$$

What happened to the Higgs ϕ^+, ϕ^-, X? They were eaten up by the gauge bosons to get the extra degree of freedom, i.e. the longitudinal component of the polarization associated with the massive bosons. The accounting has been given in the previous section.

4.4 The ρ parameter

One usually defines the parameter:

$$\rho = \frac{m_W^2}{\cos^2\theta_w m_Z^2}.$$

In the standard model $\rho = 1$. One can show that extending the standard model by introducing more than one isodoublets preserves this relation. This, however, is not true in general. Thus, e.g., it is violated by the introduction of a Higgs isotriplet. To see this we note that there are no terms of the type

$$(D_\mu \phi_1)^\dagger D_\mu \phi_2, \quad \phi_1 = \text{isodoublet}, \quad \phi_2 = \text{isotriplet}.$$

Thus the introduction of the isotriplet $(\Sigma^+, \Sigma^0, \Sigma^-)$, acquiring a vacuum expectation value in the neutral direction, only affects the mass of the charged bosons and leaves the mass of the neutral gauge bosons unaffected. So it only changes the ratio of the charge to neutral gauge bosons.

In fact one can show that:

$$\left[\frac{g}{\sqrt{2}}\left(I_+ W_\mu^- + I_- W_\mu^+\right) + \frac{g}{2} I_3 A_\mu^3\right] \begin{pmatrix} 0 \\ 1 \\ 0 \end{pmatrix} \frac{v_2}{\sqrt{2}}$$

$$= \frac{gv_2}{\sqrt{2}} \left[\begin{pmatrix} 1 \\ 0 \\ 0 \end{pmatrix} W_\mu^- + \begin{pmatrix} 0 \\ 0 \\ 1 \end{pmatrix} W_\mu^+ + 0\right] = \frac{gv_2}{\sqrt{2}} \begin{pmatrix} W_\mu^- \\ 0 \\ W_\mu^+ \end{pmatrix},$$

$$\frac{gv_2}{\sqrt{2}} \left(W_\mu^-, 0, W_\mu^+\right) \frac{gv_2}{\sqrt{2}} \begin{pmatrix} W_\mu^+ \\ 0 \\ W_\mu^- \end{pmatrix} = \frac{1}{2} g^2 v_2^2 \left(W_\mu^- W_\mu^+ + W_\mu^+ W_\mu^-\right)$$

$$\left(M_W^2\right)_2 = \left(gv_2\right)^2.$$

One, of course, has in addition the usual isodoublet contribution to the charged vector boson mass. Thus

$$M_W^2 = \frac{1}{4}\left(gv_1\right)^2 + \left(gv_2\right)^2 = \frac{1}{4}g^2v^2,$$
$$v^2 = v_1^2 + 4v_2^2, \quad \rho = 1 + \frac{v_2^2}{v_1^2}.$$

So in this case one has initially $4+3=7$ degrees of freedom for the scalars, but ends up with 2 massive Higgs, 3 degrees of freedom eaten up by the gauge bosons and two surviving Goldstone bosons.

The situation is, in general, described by the following theorem:

4.5 Fundamental theorem of Higgs mechanism

i) Let a single algebra L with g generators $T^{(a)}$ describing a group G, which has g gauge bosons A_μ^a. Let there be N Higgs scalars transforming as an irreducible representation of G such that the scalar potential $V(\phi)$ has a minimum of $\langle\phi\rangle = (v_1, v_2, \cdots, v_N) = v$.

ii) Let n generators of G such that

$$T^{(a)}v = 0, \qquad a = 1, \cdots, n.$$

Then

1) The generators $T^{(a)}$, $\quad a = 1, \cdots, n$ form a subalgebra L' of L.
2) n components of ϕ acquire a mass.
3) n gauge bosons remain massless.
4) $d - n$ gauge bosons become massive $d = \min(N, g)$.
5) Furthermore we distinguish the cases:

   (1) $g > N \to$ additional $g - N$ gauge bosons remain massless.
   (2) $N > g \to$ there exist $N - g$ Goldstone bosons.
   (3) $N = g \to n$ gauge bosons with mass zero.

Consider $g = 4$, $N = 4$, $n = 1$, $d = 4$

1 massive ϕ	1 massless gauge
3 ϕ eaten up	3 massive gauge
1 Higgs	4 gauge bosons

Sketch of the proof
Let T^a $a = 1, 2 \cdots, g$ and $v^{(l)}$ $l = 1, \cdots, N$ such that

$$T^a \left| v^{(l)} \right\rangle = 0, \qquad l = 1, \cdots, n \qquad a = 1, \cdots, n,$$

then

$$T^a \left| v^{(l)} \right\rangle \neq 0, \qquad l = n+1, \cdots, N \qquad a = n+1, \cdots, g.$$

Let $v^{(l)} = (v_1, v_2, \cdots v_N)$ then

$$\frac{\partial V}{\partial \phi_i} \delta\phi_i = \frac{\partial V}{\partial \phi_i} \epsilon^a T^a_{ij} \phi_j, \quad \text{all } a$$

from this we get:

$$\frac{\partial}{\partial \phi_k} \left(\frac{\partial V}{\partial \phi_i} \delta\phi_i \right) \delta\phi_k = \epsilon^a \frac{\partial V}{\partial \phi_k} T^a_{kj} \phi_j \epsilon^b T^b_{k\ell} \phi_\ell + \epsilon^a \epsilon^b \frac{\partial^2 V}{\partial \phi_i \partial \phi_k} T^a_{ij} \phi_j T^b_{k\ell} \phi_k$$

At the minimum the above expression vanishes and so does the first term. Recalling now that

$$m_{ik} = \frac{1}{2} = \left. \frac{\partial^2 V}{\partial \phi_i \partial \phi_k} \right|_{\phi_i = v^{(i)}, \phi_k = v^{(k)}}$$

we get

$$m_{ik} \epsilon^a \epsilon^b \left(T^a v^{(i)} \right) \left(T^b v^{(k)} \right) = 0$$

Now by assumption

$$T^a v^{(i)} = 0, \quad T^b v^{(k)} = 0, \quad a, b = 1, 2 \cdots n \Rightarrow m_{ik} \neq 0$$

$$T^a v^{(i)} \neq 0, \quad T^b v^{(k)} \neq 0, \quad a, b = n+1, n+2, \cdots g \Rightarrow m_{ik} = 0$$

4.6 Problems

(1) They have been attempts to use the group $O(3)$ as a gauge group, with the Higgs scalars in the $I = 1$ representation of the group.
- Write down an expression giving the covariant derivative and the transformation law for the gauge fields.
- In this case one has three real scalar fields. Assuming that the Higgs potential is an even function, go through the Higgs mechanism and find the physical Higgs and the number of the would be Golstone bosons.
- Obtain the mass spectrum of the gauge bosons.
- Discuss why this model is not physically acceptable.

(2) Consider an extended Higgs sector.

- Is the relationship between the W and the Z boson mass effected, if there exist more than one isodoublet of non interacting complex Higgs scalars, acquiring a vacuum expectation value?
- Is it affected, if, in addition to the standard complex isodoublet, one introduces a real isotriplet?

Chapter 5

Fermion Masses and Currents

5.1 Fermion masses

Naively one would be tempted to write:

$$m_f = m\bar{f}f$$

where f represents the Fermion. This is a Lorentz scalar, but it is not acceptable in the SM, because a chiral basis has already been incorporated into the SM.

This, however, can be overcome. We notice that for any Fermion f:

$$\bar{f}_L f_L = f^+ \frac{1-\gamma_5}{2}\gamma_0 \frac{1-\gamma_5}{2} f.$$

But γ_0 anticommutes with γ_5 and $\gamma_5^2 = 1$. Thus:

$$\bar{f}_L f_L = f^+ \gamma_0 \frac{1+\gamma_5}{2}\frac{1-\gamma_5}{2} f = \bar{f}\frac{1+\gamma_5}{2}\frac{1-\gamma_5}{2} f = 0$$

Similarly:

$$\bar{f}_R f_R = 0$$

Furthermore

$$\bar{f}_L f_R = f^+ \tfrac{1-\gamma_5}{2}\gamma_0 \tfrac{1+\gamma_5}{2} f = f^+ \gamma_0 \tfrac{1+\gamma_5}{2}\tfrac{1+\gamma_5}{2} f = \bar{f}\tfrac{1+\gamma_5}{2}\tfrac{1+\gamma_5}{2} f = \bar{f}\tfrac{1+\gamma_5}{2} f$$

$$\bar{f}_R f_L = \bar{f}\tfrac{1-\gamma_5}{2} f$$

Thus

$$\bar{f}_L f_R + \bar{f}_R f_L = \bar{f}\frac{1+\gamma_5}{2} f + \bar{f}\frac{1-\gamma_5}{2} f = \bar{f}f$$

In other words

$$m_f \bar{f}f = m_f(\bar{f}_L f_R + \bar{f}_R f_L)$$

The expression on the tight hand side, however, is not an $SU(2)$ scalar but an SU(2) doublet (recall that a pin zero times a spin 1/2 yields a spin 1/2)

The Higgs field, however, comes to our rescue!

$$\phi = \begin{pmatrix} \phi^+ \\ \phi^0 \end{pmatrix}, \, Y = 1, \, Q = I_3 + \frac{Y}{2},$$

Instead of its conjugate one uses the field:

$$\tilde{\phi} = i\tau_2 \phi^+ = \begin{pmatrix} 0 \; 1 \\ -1 \; 0 \end{pmatrix} \begin{pmatrix} \phi^- \\ \phi_0^* \end{pmatrix} = \begin{pmatrix} \phi_0^* \\ -\phi^- \end{pmatrix}, \, Y = -1$$

(the higher charge state is on top!) This allows the following Yukawa interactions:

(1) Consider the case of lepton (assume, for simplicity, one generation):

$$\bar{e}_L e_R \to \mathcal{L}_Y^e = f_{ee} \left(\bar{\nu}_L^0 \; \bar{e}_L^0 \right) \begin{pmatrix} \phi^+ \\ \phi^0 \end{pmatrix} e_R^0 = f_{ee} (\bar{\nu}_L^0 \phi^+ e_R^0 + \bar{e}_L^0 \phi^0 e_R^0)$$

and

$$\bar{e}_R e_L \to f_{ee}^* \bar{e}_R^0 \left(\phi^- \; \phi_0^* \right) \begin{pmatrix} \nu_L^0 \\ e_L^0 \end{pmatrix} = f_{ee}^* (\bar{e}_R^0 \phi^- \nu_L^0 + \bar{e}_R^0 \phi_0^* e_L^0).$$

(The Hermitian conjugate of the previous one).

Notice that we cannot construct an analogous term for the neutrinos, since ν_R is absent in the SM.

(2) Quarks:
we encounter two possibilities:

- Down right-handed quarks:

$$\mathcal{L}_Y^d = f_{dd} \left(\bar{u}_L^0 \; \bar{d}_L^0 \right) \begin{pmatrix} \phi^+ \\ \phi^0 \end{pmatrix} d_R^0 = f_{dd} (\bar{u}_L^0 \phi^+ d_R^0 + \bar{d}_L^0 \phi^0 d_R^0)$$

and its Hermitian conjugate $L \Longleftrightarrow R$

$$f_{dd}^* \bar{d}_R^0 \left(\phi^- \; \phi_0^* \right) \begin{pmatrix} u_L^0 \\ d_L^0 \end{pmatrix} = f_{dd}^* (\bar{d}_R^0 \phi^- u_L^0 + \bar{d}_R^0 \phi_0^* d_L^0)$$

- Up right-handed quarks:

$$\mathcal{L}_Y^u = f_{uu} \left(\bar{u}_L^0 \; \bar{d}_L^0 \right) \begin{pmatrix} \phi_0^* \\ -\phi^- \end{pmatrix} u_R^0 = f_{uu} (\bar{u}_L^0 \phi_0^* u_R^0 - \bar{d}_L^0 \phi^- u_R^0)$$

and its Hermitian conjugate:

$$f_{uu}^* \bar{u}_R^0 \left(\phi^- \; \phi_0^* \right) \begin{pmatrix} u_L^0 \\ d_L^0 \end{pmatrix} = f_{uu}^* (\bar{u}_R^0 \phi_0 u_L^0 + \bar{u}_R^0 \phi^+ d_L^0).$$

Suppose that the Higgs acquires a vacuum expectation value. Then in the unitary gauge:

$$\phi = \begin{pmatrix} 0 \\ \frac{v+\eta}{\sqrt{2}} \end{pmatrix}, \tilde{\phi} = \begin{pmatrix} \frac{v+\eta}{\sqrt{2}} \\ 0 \end{pmatrix}.$$

Now

$$\mathcal{L}_Y^e = f_{ee} \left(\bar{\nu}_L^0 \; \bar{e}_L^0 \right) \begin{pmatrix} 0 \\ \frac{v+\eta}{\sqrt{2}} \end{pmatrix} e_R^0 = f_{ee} \frac{v}{\sqrt{2}} \bar{e}_L^0 e_R^0 + f_{ee} \frac{\eta}{\sqrt{2}} \bar{e}_L^0 e_R^0$$

Let

$$m_e = f_{ee} \frac{v}{\sqrt{2}} \Longrightarrow \frac{f_{ee}}{\sqrt{2}} = \frac{m_e}{v},$$

then

$$\mathcal{L}_Y^e = (m_e \bar{e}_L^0 e_R^0 + H.C.) + \left(m_e \frac{\eta}{v} \bar{e}_L^0 e_R^0 + H.C. \right).$$

Similarly for down-quark:

$$\mathcal{L}_Y^d = m_d \bar{d}_L^0 d_R^0 + m_d \frac{\eta}{v} \bar{d}_L^0 d_R^0 + H.C.$$

and for up-quark:

$$\mathcal{L}_Y^u = m_u \bar{u}_L^0 u_R^0 + m_u \frac{\eta}{v} \bar{u}_L^0 u_R^0 + H.C. \quad .$$

The above Yukawa interactions can be generalized to many generations.

$$\mathcal{L}_Y^e = f_{ee} \left(\bar{\nu}_L^0 \; \bar{e}_L^0 \right) \begin{pmatrix} \phi^+ \\ \phi^0 \end{pmatrix} e_R^0 + HC \Rightarrow \mathcal{L}_Y^e = \sum_{i,j=e,\mu,\tau} f_{ij} \left(\bar{\nu}_{Li}^0 \; \bar{l}_{Li}^0 \right) \begin{pmatrix} \phi^+ \\ \phi^0 \end{pmatrix} e_{Rj}^0 + HC.$$

These after the Higgs acquires a vacuum expectation become:

$$\mathcal{L}_Y^e = \sum_{i,j=e,\mu,\tau} m_{ij}^e \left(\left(\bar{e}_{iL}^0 e_{jR}^0 + \bar{e}_{iL}^0 e_{jR}^0 \frac{\eta}{v} \right) \right) + HC$$

($m_{ij}^e = f_{ij}^e \frac{v}{\sqrt{2}}$). The first term yields the mass matrix, in general non diagonal, while the second gives the charged lepton Higgs interaction.

Similarly for the quarks:

$$\mathcal{L}_Y^d = \sum_{i,j} m_{ij}^d \left(\left(\bar{\kappa}_{iL}^0 \kappa_{jR}^0 + \bar{\kappa}_{iL}^0 \kappa_{jR}^0 \frac{\eta}{v} \right) \right) + HC, \quad \kappa_i = (d, s, b)$$

($m_{ij}^d = f_{ij}^d \frac{v}{\sqrt{2}}$),

$$\mathcal{L}_Y^u = \sum_{i,j} m_{ij}^u \left(\left(\bar{\alpha}_{iL}^0 \alpha_{jR}^0 + \bar{\alpha}_{iL}^0 \alpha_{jR}^0 \frac{\eta}{v} \right) \right) + HC, \quad \alpha_i = (u, c, t)$$

($m_{ij}^u = f_{ij}^u \frac{v}{\sqrt{2}}$).

No mass for the neutrino. The right handed neutrino does not exist!

The Fermion mass matrices are shown in Table 5.1. The physical masses are obtained by diagonalizing the above matrices.

These, however, are symmetric, but need not be Hermitian. We will first examine the case of them being real, i.e. Hermitian, and next the more complicated case when they are complex (non Hermitian).

Table 5.1: The Fermion mass matrices

	u_R^0	c_R^0	t_R^0
$\bar{u}_L^0$	$m_{11}^{(u)}$	$m_{12}^{(u)}$	$m_{13}^{(u)}$
$\bar{c}_L^0$	$m_{21}^{(u)}$	$m_{22}^{(u)}$	$m_{23}^{(u)}$
$\bar{t}_L^0$	$m_{31}^{(u)}$	$m_{32}^{(u)}$	$m_{33}^{(u)}$

;

	d_R^0	s_R^0	b_R^0
$\bar{d}_L^0$	$m_{11}^{(d)}$	$m_{12}^{(d)}$	$m_{13}^{(d)}$
$\bar{s}_L^0$	$m_{21}^{(d)}$	$m_{22}^{(d)}$	$m_{23}^{(d)}$
$\bar{b}_L^0$	$m_{31}^{(d)}$	$m_{32}^{(d)}$	$m_{33}^{(d)}$

;

	e_R^0	μ_R^0	τ_R^0
$\bar{e}_L^0$	$m_{11}^{(e)}$	$m_{12}^{(e)}$	$m_{13}^{(e)}$
$\bar{\mu}_L^0$	$m_{21}^{(e)}$	$m_{22}^{(e)}$	$m_{23}^{(e)}$
$\bar{\tau}_L^0$	$m_{31}^{(e)}$	$m_{32}^{(e)}$	$m_{33}^{(e)}$

5.1.1 *Hermitian mass matrices*

Consider first the case leptons.

$$\mathcal{L}_Y^e = \sum_{i,j=1,2,3} m_{ij}^e \left(\left(\bar{\ell}_{iL}^0 \ell_{jR}^0 + \bar{\ell}_{iL}^0 \ell_{jR}^0 \frac{\eta}{v} \right) \right) + HC,\ \ell_1 = e,\ \ell_2 = \mu,\ \ell_3 = \tau \tag{5.1}$$

The mass matrix m^e can be diagonalized by a similarity transformation:

$$S^{e+} m^e S^e = \text{diagonal}(m^e, m^\mu, m^\tau). \tag{5.2}$$

More explicitly the transformation matrix is

$$\begin{pmatrix} e_L^0 \\ \mu_L^0 \\ \tau_L^0 \end{pmatrix} = S^{(e)} \begin{pmatrix} e_L \\ \mu_L \\ \tau_L \end{pmatrix},\ \begin{pmatrix} e_R^0 \\ \mu_R^0 \\ \tau_R^0 \end{pmatrix} = (S^{(e)})^* \begin{pmatrix} e_R \\ \mu_R \\ \tau_R \end{pmatrix}. \tag{5.3}$$

where the states without subscripts refer to the eigenvectors.

One further step is needed to bring the above expressions to a more convenient form. Since the mass matrix is Hermitian after diagonalization we get, e.g. for the leptons:

$$L_{mass}^e = m_e \left(\bar{e}_L e_R + \bar{e}_R e_L \right) + m_\mu \left(\bar{\mu}_L \mu_R + \bar{\mu}_R \mu_L \right) + m_\tau \left(\bar{\tau}_L \tau_R + \bar{\tau}_R \tau_L \right)$$

It is like in quantum mechanics, with the exception that the bra and the ket are different, they differ in chirality. Note, however, that :

$$m_e(\bar{e}_L e_R + \bar{e}_R e_L) = m_e \left((e)^+ \left(\frac{1-\gamma_5}{2} \gamma_0 \frac{1+\gamma_5}{2} e \right) + (e)^+ \left(\frac{1+\gamma_5}{2} \gamma_0 \frac{1-\gamma_5}{2} e \right) \right)$$

$$= m_e \left((e)^+ \gamma_0 \gamma_0 \left(\frac{1-\gamma_5}{2} \gamma_0 \frac{1+\gamma_5}{2} e \right) + (e)^+ \gamma_0 \gamma_0 \left(\frac{1+\gamma_5}{2} \gamma_0 \frac{1-\gamma_5}{2} e \right) \right)$$

$$= m_e \left(\bar{e}\gamma_0 \left(\frac{1-\gamma_5}{2} \gamma_0 \frac{1+\gamma_5}{2} e \right) + \bar{e}\gamma_0 \left(\frac{1+\gamma_5}{2} \gamma_0 \frac{1-\gamma_5}{2} e \right) \right)$$

Since γ_5 anticommutes with γ_0 and $\gamma_5^2 = 1$ we get

$$m_e = \left(\bar{e} \left(\frac{1+\gamma_5}{2} \frac{1+\gamma_5}{2} e \right) + \bar{e} \left(\frac{1-\gamma_5}{2} \frac{1-\gamma_5}{2} e \right) \right) =$$

$$m_e \left(\bar{e} \left(\frac{1+\gamma_5}{2} e \right) + \bar{e} \left(\frac{1-\gamma_5}{2} e \right) \right)$$

That is

$$m_e(\bar{e}_L e_R + \bar{e}_R e_L) = m_e \bar{e} e.$$

$$\mathcal{L}^{\text{leptons}}_{\text{mass}} \Leftrightarrow \mathcal{L}^{e}_{\text{mass}} = m_e \bar{e} e + m_\mu \bar{\mu} \mu + m_\tau \bar{\tau} \tau. \tag{5.4}$$

The analysis for the quarks proceeds similarly:

$$\mathcal{L}^d_Y = \sum_{i,j=1,2,3} m^d_{ij}((\bar{\kappa}^0_{iL} \kappa^0_{jR} + \bar{\kappa}^0_{iL} \kappa^0_{jR} \frac{\eta}{v})) + HC,\ \kappa_1 = d,\ \kappa_2 = s,\ \kappa_3 = d, \tag{5.5}$$

$$S^{d+} m^d S^d = \text{diagonal}(m^d, m^s, m^b). \tag{5.6}$$

More explicitly the transformation matrix is

$$\begin{pmatrix} d^0_L \\ s^0_L \\ b^0_L \end{pmatrix} = S^{(d)} \begin{pmatrix} d_L \\ s_L \\ b_L \end{pmatrix}, \begin{pmatrix} d^0_R \\ s^0_R \\ b^0_R \end{pmatrix} = (S^{(d)})^* \begin{pmatrix} d_R \\ s_R \\ d_R \end{pmatrix}. \tag{5.7}$$

$$\mathcal{L}^d_{\text{mass}} = m_d \bar{d} d + m_s \bar{s} s + m_b \bar{b} b. \tag{5.8}$$

Similarly for the up type quarks

$$\mathcal{L}^u_Y = \sum_{i,j=1,2,3} m^u_{ij}((\bar{\alpha}^0_{iL} \alpha^0_{jR} + \bar{\alpha}^0_{iL} \alpha^0_{jR} \frac{\eta}{v})) + HC,\ \alpha_1 = u,\ \alpha_2 = c,\ \alpha_3 = t, \tag{5.9}$$

$$S^{u+} m^u S^u = \text{diagonal}(m^u, m^c, m^t). \tag{5.10}$$

More explicitly the transformation matrix is

$$\begin{pmatrix} u^0_L \\ c^0_L \\ t^0_L \end{pmatrix} = S^{(d)} \begin{pmatrix} u_L \\ c_L \\ t_L \end{pmatrix}, \begin{pmatrix} u^0_R \\ c^0_R \\ t^0_R \end{pmatrix} = (S^{(d)})^* \begin{pmatrix} u_R \\ c_R \\ t_R \end{pmatrix}. \tag{5.11}$$

$$\mathcal{L}^u_{\text{mass}} = m_u \bar{u} u + m_c \bar{c} c + m_t \bar{t} t. \tag{5.12}$$

The Higgs Fermion interaction in the physical basis is analogous and it will be given below.

5.1.2 *Non Hermitian mass matrices; left-right similarity transformations*

In general a complex symmetric matrix, when the nature of the left and right states is different, can be diagonalized by separate left and right 3×3 unitary matrices[1]. Thus, e.g., for the up quark matrix we write:

$$S_L^+ m^u S_R = \text{diagonal}(m^u, m^c, m^t).$$

How do we find S_L and S_R? We consider the matrices

$$m_u^+ m_u \text{ and } m_u m_u^+$$

which are clearly Hermitian.

$$S_R^+ m_u^+ m_u S_R = (m_u^2, m_c^2, m_t^2).$$

$$S_L^+ m_u m_u^+ S_L = (m_u^2, m_c^2, m_t^2).$$

Having obtained S_L and S_R, we observe that:

$$S_R^+ m_u^+ m_u S_R = (S_R^+ m_u^+ S_L)(S_L^+ m_u S_R) = \left|S_L^+ m_u S_R\right|^2 = (m_u^2, m_c^2, m_t^2).$$

So the eigenvalues are the square root of those of the matrix $m_u^+ m_u$. We write $S^{(u)} = S_L$, then $S_R = (S^{(u)})^*$. Thus we relate the eigenvectors u, c, t with the original (weak) basis u^0, c^0, t^0 as follows:

$$\begin{pmatrix} u_L^0 \\ c_L^0 \\ t_L^0 \end{pmatrix} = S^{(u)} \begin{pmatrix} u_L \\ c_L \\ t_L \end{pmatrix}, \quad \begin{pmatrix} u_R^0 \\ c_R^0 \\ t_R^0 \end{pmatrix} = (S^{(u)})^* \begin{pmatrix} u_R \\ c_R \\ t_R \end{pmatrix}.$$

[1] In general the full mass matrix can be written as a 6×6 Hermitian matrix of the form

$$\mathcal{M}_f = \left(\begin{array}{c|cc} & f_L & f_R \\ \hline \bar{f}_L & 0 & m_f \\ \bar{f}_R & m_f^* & 0 \end{array}\right) \tag{5.13}$$

where m_f is a symmetric 3×3 Fermion matrix of the type we discussed above for the Fermion f. Clearly $\mathcal{M}_f$ is a Hermitian matrix and it can be diagonalized by a similarity transformation. The eigenvalues are real and occur in pairs with the same absolute value. Suppose we order the eigenvalues so that, e.g., the positive come first. Then the mixing 6×6 matrix can be cast in the form:

$$S = \frac{1}{\sqrt{2}} \left(\begin{array}{c|c} S_L^1 & S_L^2 \\ \hline S_R^1 & S_R^2 \end{array}\right) \tag{5.14}$$

Then the 3×3 matrices we are after are:

$$S_L = S_L^1 \text{ or } S_L^2,\ S_R = S_R^1 \text{ or } S_R^2 \tag{5.15}$$

Similarly for the down quark matrix

$$\begin{pmatrix} d_L^0 \\ s_L^0 \\ b_L^0 \end{pmatrix} = S^{(d)} \begin{pmatrix} d_L \\ s_L \\ b_L \end{pmatrix}, \begin{pmatrix} d_R^0 \\ s_R^0 \\ b_R^0 \end{pmatrix} = (S^{(d)})^* \begin{pmatrix} d_R \\ s_R \\ b_R \end{pmatrix}.$$

and the electron matrix

$$\begin{pmatrix} e_L^0 \\ \mu_L^0 \\ \tau_L^0 \end{pmatrix} = S^{(e)} \begin{pmatrix} e_L \\ \mu_L \\ \tau_L \end{pmatrix}, \begin{pmatrix} e_R^0 \\ \mu_R^0 \\ \tau_R^0 \end{pmatrix} = (S^{(e)})^* \begin{pmatrix} e_R \\ \mu_R \\ \tau_R \end{pmatrix}.$$

One further step is needed to bring the above expressions to a more convenient form. For the leptons, e.g., we notice:

$$L^e_{mass} = m_e \bar{e}_L e_R + m_e^* \bar{e}_R e_L + m_\mu \bar{\mu}_L \mu_R + m_\mu^* \bar{\mu}_R \mu_L + m_\tau \bar{\tau}_L \tau_R + m_\tau^* \bar{\tau}_R \tau_L$$

Now define

$$m_e = |m_e|\, e^{i\varphi_e},\ e = e_L + e^{i\varphi_e} e_R,$$

$$m_\mu = |m_\mu|\, e^{i\varphi_\mu},\ \mu = \mu_L + e^{i\varphi_\mu} \mu_R,$$

$$m_\tau = |m_\tau|\, e^{i\varphi T},\ \tau = \tau_L + e_R^{i\varphi_e} \tau_R.$$

Then

$$\begin{aligned} |m_e|\,\bar{e}e &= |m_e|\,(\bar{e}_L + \bar{e}_R e^{-i\varphi_e})(e_L + e^{i\varphi_e} e_R) \\ &= |m_e|\, e^{i\varphi_e} \bar{e}_L e_R + |m_e|\, e^{-i\varphi_e} \bar{e}_R e_L \\ &= m_e \bar{e}_L e_R + m_e^* \bar{e}_R e_L. \end{aligned}$$

Thus we obtain the familiar form:

$$L^e_{mass} = |m_e|\,\bar{e}e + |m_\mu|\,\bar{\mu}\mu + |m_\tau|\,\bar{\tau}\tau.$$

Note, however, that we the above redefinition the left handed component of the fields remains unchanged, but the right handed component is multiplied by a phase:

$$e_L \to e_L,\ \mu_L \to \mu_L,\ \tau_L \to \tau_L,\ e_R \to e^{-i\phi_e} e_R,\ \mu_R \to e^{-i\phi_\mu} \mu_R,\ \tau_R \to e^{-i\phi_\tau} \tau_R.$$

but this change is of no consequence.

For the quarks we have:

$$\mathcal{L}^d_{mass} = m_d \bar{d}_L d_R + m_d^* \bar{d}_R d_L + m_s \bar{s}_L s_R + m_s^* \bar{s}_R s_L + m_b \bar{b}_L b_R + m_b^* \bar{b}_R b_L,$$

$$\mathcal{L}^u_{mass} = m_u \bar{u}_L u_R + m_u^* \bar{u}_R u_L + m_c \bar{c}_L c_R + m_c^* \bar{c}_R c_L + m_t \bar{t}_L t_R + m_t^* \bar{t}_R t_L.$$

Proceeding as in the case of leptons above, we find:

$$\mathcal{L}^u_{mass} = |m_u|\,\bar{u}u + |m_c|\,\bar{c}c + |m_t|\,\bar{t}t,$$

$$\mathcal{L}^d_{mass} = |m_d|\,\bar{d}d + |m_s|\,\bar{s}s + |m_b|\,\bar{b}b.$$

In an analogous fashion we obtain the interactions of Fermions with the physical Higgs:

$$\begin{aligned}(\mathcal{L}_Y)_{int} &= \frac{\eta}{v}\,|m_e|\,\bar{e}e + \frac{\eta}{v}\,|m_\mu|\,\bar{\mu}\mu + \frac{\eta}{v}\,|m_\tau|\,\bar{\tau}\tau \\ &+ \frac{\eta}{v}\,|m_u|\,\bar{u}u + \frac{\eta}{v}\,|m_c|\,\bar{c}c + \frac{\eta}{v}\,|m_t|\,\bar{t}t \\ &+ \frac{\eta}{v}\,|m_d|\,\bar{d}d + \frac{\eta}{v}\,|m_s|\,\bar{s}s + \frac{\eta}{v}\,|m_b|\,\bar{b}b. \end{aligned} \tag{5.16}$$

These are known as **Yukawa interactions**. We note that in the SM the Yukawa interactions are proportional to the Fermion masses. We will discuss the Higgs decay into Fermion anti-Fermion pairs after we fix the vacuum expectation value from beta decays.

5.2 The currents

We have seen in section 3.7.1, in connection with the Standard Model, that:

$$D^L_\mu f_L = \left(\partial_\mu - i\frac{g}{2}\vec{\tau}\cdot\vec{A}_\mu - i\frac{g^{'}}{2}YB_\mu\right) f_L,$$

$$D^R_\mu f_R = \left(\partial_\mu - i\frac{g^{'}}{2}YB_\mu\right) f_R,$$

We will examine each case separately.

5.2.1 *The left handed currents*

$$\mathcal{L}_L = \sum_f \bar{f}_L i\gamma^\mu D^L_\mu f_L.$$

In the above expressions, Fermions are expressed in the weak basis, i.e. the superscript 0 is understood.

The electroweak part is:

$$D^L_\mu = \partial_\mu - i[\frac{g}{\sqrt{2}}(\tau_+ W^-_\mu + \tau_- W^+_\mu) + \frac{g}{2}A^{(3)}_\mu \tau_3 + \frac{g^{'}}{2}(2Q - \tau_3)B_\mu].$$

After the spontaneous breakdown of symmetry the charged vector bosons do not get admixed. The neutral ones are admixed with the mixing given by the Weinberg angle:

$$g^{'} = g\tan\theta_W,$$

i.e.:

$$A_\mu^{(3)} = \sin\theta_W A_\mu + \cos\theta_W Z_\mu,$$

$$B_\mu = \cos\theta_W A_\mu - \sin\theta_W Z_\mu$$

$$\begin{pmatrix} A_\mu^{(3)} \\ B_\mu \end{pmatrix} = \begin{pmatrix} \sin\theta_W & \cos\theta_W \\ \cos\theta_W & -\sin\theta_W \end{pmatrix} \begin{pmatrix} A_\mu \\ Z_\mu \end{pmatrix},$$

Thus

$$\frac{g}{2}A_\mu^{(3)}\tau_3 + \frac{g^{'}}{2}(2Q-\tau_3)B_\mu =$$

$$\frac{g}{2}\tau_3(\sin\theta_W A_\mu + \cos\theta_W Z_\mu) + \frac{g\tan\theta_W}{2}(2Q-\tau_3)(\cos\theta_W A_\mu - \sin\theta_W Z_\mu)$$

$$= \frac{\tau_3}{2}(g\sin\theta_W A_\mu - g\sin\theta_W A_\mu) + \frac{g\tan\theta_W}{2}2Q\cos\theta_W A_\mu$$

$$+\left(\left(\frac{g}{2}\cos\theta_W + \frac{g}{2}\tan\theta_W\sin\theta_W\right)\tau_3 Z_\mu + \frac{g}{2}\tan\theta_W 2Q(-\sin\theta_W)\right)Z_\mu =$$

$$g\sin\theta_W Q A_\mu + \frac{g}{2\cos\theta_W}\tau_3 Z_\mu + \frac{g}{2\cos\theta_W}\left(-2Q\sin^2\theta_W\right)Z_\mu =$$

$$eQA_\mu + \frac{g}{\cos\theta_W}\left(\frac{\tau_3}{2} - Q\sin^2\theta_W\right)Z_\mu.$$

or

$$\frac{g}{2}A_\mu^{(3)}\tau_3 + \frac{g^{'}}{2}(2Q-\tau_3)B_\mu = eQA_\mu + \frac{g}{\cos\theta_W}\left(\tau_3 - Q\sin^2\theta_W\right)Z_\mu,$$

where $e = g\sin\theta_W$ is the unit of the electric charge. Thus

$$\mathcal{L}_L = \sum_f \Bigg(\bar{f}_L i\gamma^\mu\partial_\mu f_L + eQA_\mu\bar{f}_L\gamma^\mu f_L + \frac{g}{\cos\theta_W}Z_\mu\bar{f}_L\gamma^\mu(\tau_3 - Q\sin\theta_W)f_L$$

$$+\frac{g}{\sqrt{2}}(\bar{f}_L\gamma^\mu\tau_- f_L)W_\mu^+ + \frac{g}{\sqrt{2}}(\bar{f}_L\gamma^\mu\tau_+ f_L)W_\mu^- \Bigg)$$

5.2.2 The right handed currents

$$\mathcal{L}_R = \sum_f \bar{f}_R D_\mu^R f_R,$$

$$D_\mu^R = \partial_\mu - i\frac{g'}{2} Y B_\mu =$$

$$\partial_\mu - i\frac{g}{2}\tan\theta_W 2Q\cos\theta_W\left(\cos\theta_W A_\mu - \sin\theta_W Z_\mu\right) =$$

$$\partial_\mu - i\left(eQA_\mu - \frac{gQ}{\cos\theta_W}\sin^2\theta_W Z_\mu\right).$$

Thus

$$\mathcal{L}_R = \sum_f \left[(\bar{f}_R i\gamma^\mu \partial_\mu f_R) + \bar{f}_R\gamma^\mu(eQA_\mu - \frac{g}{\cos\theta_W}Q\sin^2\theta_W Z_\mu)f_R\right].$$

5.2.3 Both left and right handed currents combined

$$L_L + L_R = \sum_f \{\bar{f}i\gamma^\mu\partial_\mu f + eQ\bar{f}\gamma^\mu A_\mu f$$

$$+\frac{g}{\sqrt{2}}(\bar{f}_L\gamma^\mu\tau_- f_L)W_\mu^+ + \frac{g}{\sqrt{2}}(\bar{f}_L\gamma^\mu\tau_+ f_L)W_\mu^-$$

$$+\frac{g}{\cos\theta_W}\left[\bar{f}_L\gamma^\mu\left(\tau_3 - Q\sin^2\theta_W\right)f_L + \bar{f}_R(-Q\sin^2\theta_W)\gamma^\mu f_R\right]Z_\mu\Big\}.$$

Now after defining:

$$j_1 = \frac{g}{\sqrt{2}}(\bar{f}_L\gamma^\mu\tau_- f_L)W_\mu^+ = \frac{g}{2\sqrt{2}}\bar{f}\gamma^\mu\tau_-(1-\gamma^5)fW_\mu^+, \tag{5.17}$$

$$j_2 = \frac{g}{\sqrt{2}}(\bar{f}_L\gamma^\mu\tau_+ f_L)W_\mu^- = \frac{g}{2\sqrt{2}}\bar{f}\gamma^\mu\tau_+(1-\gamma^5)fW_\mu^-, \tag{5.18}$$

i.e.

$$j_1 + j_2 = \frac{g}{2\sqrt{2}}\left(J_\mu W_\mu^+ + (J_\mu)^+ W_\mu^-\right), \quad J^\mu = \bar{f}\gamma^\mu\tau_-(1-\gamma^5)f \tag{5.19}$$

and

$$J_Z^\mu = \bar{f}\left(\gamma^\mu(1-\gamma^5)\tau_3 - 4Q\sin^2\theta_W\gamma^\mu\right)f, \tag{5.20}$$

the latter via the relation:

$$\frac{g}{\cos\theta_W}\left\{\bar{f}_L\left(\gamma^\mu(\tau_3 - Q\sin^2\theta_W\right)f_L + \bar{f}_R(-Q\sin^2\theta_W)f_R\right\}Z_\mu =$$

$$\frac{g}{\cos\theta_W}\left\{\bar{f}\left(\frac{1}{2}\gamma^\mu(1-\gamma^5)\frac{\tau_3}{2} - Q\sin^2\theta_W\gamma^\mu\right)f\right\}Z_\mu =$$

$$\frac{g}{2\cos\theta_W}\left\{\bar{f}\left(\gamma^\mu(1-\gamma^5)\frac{\tau_3}{2} - 2Q\sin^2\theta_W\gamma^\mu\right)f\right\}Z_\mu =$$

$$\frac{g}{4\cos\theta_W}J_Z^\mu Z_\mu,$$

we are in a position to specialize the above currents as follows.

5.2.4 *The EM current*

$$
\begin{aligned}
J^{\mu}_{EM} &= \sum Q\bar{f}_L\gamma^{\mu}f_L + \sum Q\bar{f}_R\gamma^{\mu}f_R \\
&= \frac{1}{2}[\sum Q\bar{f}\gamma^{\mu}(1-\gamma_5)f + \sum Q\bar{f}\gamma^{\mu}(1+\gamma_5)f] \\
&= \sum Q\bar{f}\gamma^{\mu}f \\
&= \frac{2}{3}(\bar{u}^0\gamma^{\mu}u^0 + \bar{c}^0\gamma^{\mu}c^0 + \bar{t}^0\gamma^{\mu}t^0) - \frac{1}{3}(\bar{d}^0\gamma^{\mu}d^0 + \bar{s}^0\gamma_{\mu}s^0 + \bar{b}^0\gamma_{\mu}b^0) \\
&- (\bar{e}^0\gamma^{\mu}e^0 + \bar{\mu}^0\gamma^{\mu}\mu^0 + \bar{\tau}^0\gamma_{\mu}\tau^0).
\end{aligned}
$$

5.2.5 *The charged currents; the KM matrix*

Let us now discuss about charged current. It is given by the $SU(2)$ symmetry in the weak basis. In the case of the quarks we have

$$
\begin{aligned}
J^{\lambda} &= 2\sum_{i,j}(\bar{d}^0_{iL}\gamma^{\lambda}u^0_{jL}) = 2(\bar{d}^0_L, \bar{s}^0_L, \bar{b}^0_L)\gamma^{\lambda}\begin{pmatrix} u^0_L \\ c^0_L \\ t^0_L \end{pmatrix} \\
&= 2(\bar{d}_L, \bar{s}_L, \bar{b}_L)(S^d)^{+}\gamma^{\lambda}S^u\begin{pmatrix} u_L \\ c_L \\ t_L \end{pmatrix}. \qquad (5.21)
\end{aligned}
$$

Define $(S^d)^+S^u = U^{(KM)}$ (Kobayashi-Maskawa matrix), then

$$
J^{\lambda} = 2\sum_{\kappa,\alpha} U^{(KM)}_{\kappa,\alpha}(\bar{\kappa}_L\gamma^{\lambda}\alpha_L),\ \kappa = d,\ s,\ b,\ \alpha = u, c, t,
$$

i.e. the KM matrix expresses the current in the physical basis.

Historically two generations were considered, in the early Cabbibo theory, that is the quark matrix was taken to be

$$
U = \begin{pmatrix} \cos\theta_c & \sin\theta_c \\ -\sin\theta_c & \cos\theta_c \end{pmatrix}, \sin\theta_c \backsimeq 0.23,\ \theta_c = \text{Cabbibo angle},
$$

which is a good approximation for low energies.

$$
\begin{aligned}
J^{\lambda} &= \cos\theta_c\left((\bar{d}\gamma^{\lambda}(1-\gamma_5)u) + (\bar{s}\gamma^{\lambda}(1-\gamma_5)c)\right) \\
&+ \sin\theta_c\left((\bar{d}\gamma^{\lambda}(1-\gamma_5)c) - (\bar{s}\gamma^{\lambda}(1-\gamma_5)u)\right). \qquad (5.22)
\end{aligned}
$$

So, $u \leftrightarrow d$ strong, (the coupling of u and d is large)
$c \leftrightarrow s$ strong, (the coupling of c and s is large)
the other two couplings $u \leftrightarrow s$ and $c \leftrightarrow d$ are suppressed.

Experimentally[2]:

$$|U_{ud}| = 0.9434 \pm 0.0025,$$

$$|U_{us}| = 0.279 \pm 0.003,$$

$$|U_{ud}| + |U_{us}| = 0.996 \pm 0.004,$$

$$\Longrightarrow |U_{ub}| = 1 - (0.996 \pm 0.004) = 0.004 \pm 0.004.$$

Since the neutrinos are massless, any linear combination of them is as good as any other. So define:

$$\nu_L = (S^e_L)^+ \nu^0_L,\ \nu^0_L = (\nu^0_e, \nu^0_\mu, \nu^0_\tau)_L,\ \nu_L = (\nu_e, \nu_\mu, \nu_\tau)_L.$$

Thus

$$J^\lambda = 2\bar{\nu}^0_L \gamma^\lambda S^e_L e^0_L = 2(S^e_L)^+ \bar{\nu}^0_L \gamma^\lambda e^0_L = 2\bar{\nu}_L \gamma^\lambda e_L,\ e_L = (e, \mu, \tau)_L.$$

So the leptonic charged current remains diagonal, more specifically:

$$J^\lambda = \bar{\nu}_e \gamma^\lambda (1-\gamma_5) e + \bar{\nu}_\mu \gamma^\lambda (1-\gamma_5)\mu + \bar{\nu}_\tau \gamma^\lambda (1-\gamma_5)\tau.$$

5.2.6 *The neutral currents*

We have seen that the interaction of Fermions with the Z-boson is given by:

$$\mathcal{L}_f = \frac{g}{4\cos\theta_W} J^\lambda Z_\lambda,$$

$$J^\lambda = \bar{f}^0[\gamma^\lambda(1-\gamma_5)\tau_3 - 4Q\sin^2\theta_W] f^0.$$

After the diagonalization of the Fermion fields, $f^0 = (S_f)f$, we find:

$$\bar{f} S^+_f \left(\gamma^\lambda(1-\gamma_5)\tau_3 - 4Q\sin^2\theta_W\right) S_f f = \bar{f}\left(\gamma^\lambda(1-\gamma_5)\tau_3 - 4Q\sin^2\theta_W\right) f,$$

i.e. it remains diagonal since $S^+_f S_f = 1$.

We specialize it for the various Fermions:

- $f = \nu$ (neutrino):

$$J^\lambda = (\bar{\nu}_e, \bar{\nu}_\mu, \bar{\nu}_\tau)\,\gamma^\lambda(1-\gamma_5)\begin{pmatrix}\nu_e\\ \nu_\mu\\ \nu_\tau\end{pmatrix}. \tag{5.23}$$

[2]One, of course, has to transform the currents from he quark to the hadronic level

$$(1-\gamma_5) \to (g_V - g_A\gamma_5),\quad g_V = 1,\ g_A \approx 1.24,$$

the well known V-A theory.

- $f = e$ electron:

$$J_Z^\lambda(\text{electron}) = (\bar{e}, \bar{\mu}, \bar{\tau})\,[-\gamma^\lambda(1-\gamma_5) + 4\sin^2\theta_w \gamma^\lambda] \begin{pmatrix} e \\ \mu \\ \tau \end{pmatrix}$$

$$= (\bar{e}, \bar{\mu}, \bar{\tau})\,\gamma^\lambda(g_V - g_A\gamma_5) \begin{pmatrix} e \\ \mu \\ \tau \end{pmatrix}, \tag{5.24}$$

$$g_V = -1 + 4\sin^2\theta_W, \quad g_A = -1\,.$$

- f =up quark:

$$J_Z^\lambda(\text{up quark}) = (\bar{u}, \bar{c}, \bar{t})\,[\gamma^\lambda(1-\gamma_5) - \frac{8}{3}\sin^2\theta_w \gamma^\lambda] \begin{pmatrix} u \\ c \\ t \end{pmatrix}$$

$$= (\bar{u}, \bar{c}, \bar{t})\,\gamma^\lambda(g_V - g_A\gamma_5) \begin{pmatrix} u \\ c \\ t \end{pmatrix}, \tag{5.25}$$

$$g_V = 1 - \frac{8}{3}\sin^2\theta_W, \quad g_A = 1$$

- f =down quark.

$$J_Z^\lambda(\text{down quark}) = (\bar{d}, \bar{s}, \bar{b})\,[-\gamma^\lambda(1-\gamma_5) + \frac{4}{3}\sin^2\theta_W \gamma^\lambda] \begin{pmatrix} d \\ s \\ b \end{pmatrix}$$

$$= (\bar{d}, \bar{s}, \bar{b})\,\gamma^\lambda(g_V - g_A\gamma_5) \begin{pmatrix} d \\ s \\ b \end{pmatrix} \tag{5.26}$$

$$g_V = -1 + \frac{4}{3}\sin^2\theta_W, \quad g_A = -1\,.$$

For the proton: *uud* we have:

$$(1-\gamma_5) - \left(\frac{8}{3} + \frac{8}{3} - \frac{4}{3}\right)\sin^2\theta_W = (1-\gamma_5) - 4\sin^2\theta_W.$$

Thus

$$J_Z^\lambda(\text{proton}) = \bar{p}\gamma^\lambda(g_V - g_A\gamma_5)p, \tag{5.27}$$
$$g_V = 1 - 4\sin^2\theta_W \approx 0, \quad g_A = 1\,.$$

For the neutron:

$$-(1-\gamma_5) - \left(\frac{8}{3} - \frac{4}{3} - \frac{4}{3}\right)\sin^2\theta_W = -(1-\gamma_5),$$
$$J_Z^\lambda(\text{neutron}) = \bar{n}\gamma^\lambda(g_V - g_A\gamma_5)n \tag{5.28}$$
$$g_V = -1, \quad g_A = -1\,.$$

5.3 The contact interaction

We have seen that the Fermions interact with the gauge bosons. Thus they can interact with other Fermions via exchanging an intermediate (virtual) gauge boson (see, e.g., Fig. 5.1).

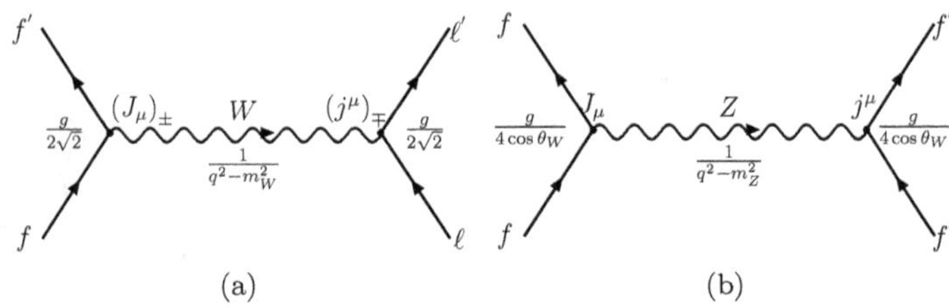

Fig. 5.1: A vector boson exchange between two currents, when at least one of them is leptonic. A W boson is exvanged between two currents (a). The most interesting case is when one of the leptons is a neutrino, i.e. $\ell = \nu$ or $\ell^{'} = \nu$ and the other is the corresponding charged lepton. We also show the exchange of a Z boson between two currents (b). The most interesting case is when one of the currents involves neutrinos, e.g. $f^{'} = \nu$.

At low energies q^2 is much smaller than the mass of the gauge boson squared, $m_W = 80$ GeV. Then the propagator can be approximated by $1/m_W^2$ and the interaction becomes:

- Charged gauge boson:

$$\mathcal{L} = \left(\frac{g}{2\sqrt{2}}\right)^2 \frac{1}{m_W^2} J^c_\mu j^{c\mu} = \frac{1}{\sqrt{2}} G_F J^c_\mu j^{c\mu}, \tag{5.29}$$

with

$$\frac{g^2}{m_W^2} = 4\sqrt{2} G_F. \tag{5.30}$$

In the special case of neutron decay the currents are:

$$J^c_\mu = \bar{\Psi}_p \gamma_\mu \left(g_V - g_A \gamma_5\right) \Psi_n, \quad j^{c\mu} = \bar{\Psi}_e \gamma_\mu \left(1 - \gamma_5\right) \Psi_\nu. \tag{5.31}$$

This contact interaction, an interaction with almost zero range, is of particular historical interest, known from the Fermi theory of beta decay. From a number of beta decay experiments the value of G_F has been determined, $G_F \approx 1.05 \times 10^{-5} m_p^{-2}$ (see Section 13.1 for the units).

- Gauge boson Z_μ

$$\mathcal{L} = \left(\frac{g}{4\cos\theta_W}\right)^2 \frac{1}{m_Z^2} J^Z_\mu j^{Z\mu} = \frac{1}{2}\frac{1}{\sqrt{2}} G_F J^Z_\mu j^{Z\mu} \tag{5.32}$$

(note the factor of 1/2). Specializing it in the case of the neutral particles, the neutrinos and the neutron, which do not feel the electromagnetic interaction, we get

$$j^Z_\mu = \bar{\nu}_\alpha \gamma^\mu (1-\gamma_5)\nu_\alpha \text{ or, } j^Z_\mu = \bar{\tilde{\nu}}_\alpha \gamma^\mu (1+\gamma_5)\tilde{\nu}_\alpha,\ \alpha = e, \mu, \tau,$$

$$J^Z_\mu = \bar{n}\gamma_\mu (g_V - g_A\gamma_5) n.$$

This was the neutral current interaction predicted by the SM. It was discovered at CERN by the historic Gargamelle experiment in 1973 via ν_μ nucleon elastic scattering:

$$\nu_\mu(\bar{\nu}_\mu) + N \to \nu_\mu(\bar{\nu}_\mu) + \text{hadrons}$$

This cannot proceed via the exchange W bosons, since the interaction of neutrinos via the charged current cannot be elastic. However the energies must approach the Z mass to see a significant number of events through the neutral current. This had, of course, been achieved at CERN at the time of their discovery.

This contact interaction as proposed by Fermi was fine at low energies, but it caused problems at high energies, which necessitated the introduction of the gauge bosons as mediators and eventually to the SM.

5.4 Determination of the Standard Model parameters

The contact interaction allows one to determine experimentally the value of the vacuum expectation value υ. Indeed

$$m_W = \frac{1}{2} g\upsilon,\ \frac{g^2}{m_W^2} = 4\sqrt{2}G_F \Rightarrow \upsilon = 2^{-1/4} G_F^{-1/2}$$

Thus from the experimentally known value $G_F \approx 1.05\times 10^{-5} m_p^{-2} = 1.166\times 10^{-5}\text{GeV}^{-2}$ we find

$$\upsilon = 246 \text{ GeV}$$

$$m_\eta = 245\frac{1}{2}\sqrt{2\lambda} \text{ GeV}$$

Furthermore from the relations

$$e = g\sin\theta_W \Leftrightarrow \sqrt{4\pi\alpha} = g\sin\theta_W$$

$$\Rightarrow m_W = \frac{1}{2}\frac{\sqrt{4\pi\alpha}}{\sin\theta_W}2^{-1/4}G_F^{-1/2} = \frac{1}{\sin\theta_W}\left(\frac{\pi\alpha}{\sqrt{2}G_F}\right)^{-1/2} .$$

Thus

$$m_W = \frac{37.3}{\sin\theta_W}, \quad m_Z = \frac{m_W}{\cos\theta_w} = \frac{74.6}{\sin 2\theta_W}.$$

The Weinberg angle was first estimated using the muon neutrino and anti-neutrino scattering data from the Gargamelle cloud chamber (see previous section). The estimation is based on the Paschos-Wolfenstein formula for an isoscalar target:

$$\frac{\sigma^{\nu}_{NC} - \sigma^{\tilde{\nu}}_{NC}}{\sigma^{\nu}_{CC} - \sigma^{\tilde{\nu}}_{CC}} = \frac{1}{2} - \sin^2\theta_W \tag{5.33}$$

Where σ_{NC} is the the total neutral current cross section mediated by the Z boson (for example $\nu_\mu + p \to \nu_\mu + p + \pi + +\pi^-$ etc.), and, σ_{CC} is the charged current processes (Mediated by the W bosons) total cross section (for example $\nu_\mu + n \to \mu^- +$ Hadrons).

Thus from the experimentally determined value[3] of the Weinberg angle $\sin^2_{\theta_W} \approx 0.23$ we find:

$$m_W = 79 \text{ GeV}, \quad m_Z = 89 \text{ GeV}, \quad g = 0.63 .$$

Higher order loop corrections lead to slightly different values

$$m_W = 80 \text{ GeV}, \quad m_Z = 91 \text{ GeV},$$

which must be compared to the experimentally determined values:

$$m_W = (80.938 \pm 0.025) \text{ GeV}, \quad m_Z = (91.1867 \pm 0.0021) \text{ GeV}.$$

In the summer of 2012 a particle was found at LHC at CERN with a mass

$$m_\eta = 126.2 \pm -0.6(\text{stat.}) \pm 0.2(\text{syst.}) \text{ GeV},$$

which almost certainly is the expected Higgs scalar. Anyway this was the opinion of the Nobel committee, which awarded the 2013 physics prize to be shared by the theorists Peter Higgs, from the UK, and Francois Englert from Belgium. The of observed channels were $h \to \gamma\gamma$ by Atlas and the channel $h \to$ two charged leptons and their anti-leptons by CMS. The first does not proceed at the tree level (there is no direct $h\gamma\gamma$ coupling), but it can occur at the loop level. The second can proceed via the allowed hZZ coupling, which, as we have seen, proceeds directly with a coupling proportional to

[3]Actually the Weinberg angle is a running quantity, i.e. it changes with energy. The value adopted here as the scale of m_Z.

m_Z. No experimentalist was included in the Nobel prize, partly because the number of the leading people involved in these huge experiments was large but also, perhaps, because some more experimental work needs be done to establish that its spin is not two or higher. Furthermore the branching ratios to the Fermionic channels must be explored to confirm that they proceed with an amplitude proportional to the Fermion mass as expected for a Higgs scalar.

Anyway the detection of the Higgs leads to

$$\lambda = 2\left(\frac{m_\eta}{245\text{GeV}}\right)^2 \approx \frac{1}{2}.$$

This allows a perturbative treatment of the scalar interaction (see Eq. (4.24)).

5.5 The vector boson self-couplings

Now that the mass of the vector bosons is determined and their possible decay channels are understood, it is time to obtain their self-couplings analogous to those we have found for gluons (see Eq. (3.68)). Proceeding in an analogous way we find for the $SU(2)$ bosons

$$\begin{aligned}\mathcal{L}_{VB} = &-\frac{1}{4}\left(\partial_\mu A_\nu^\alpha - \partial_\nu A_\mu^\alpha\right)\left(\partial^\mu A^{\nu\alpha} - \partial^\nu A^{\mu\alpha}\right) \\ &- g\epsilon_{\beta\gamma\alpha}\partial_\mu A_\nu^\alpha A^{\mu\beta} A^{\nu\gamma} - \frac{1}{4} g^2 f_{\beta\gamma\alpha}\epsilon_{\rho\sigma\alpha} A_\mu^\beta A_\nu^\gamma A^{\mu\rho} A^{\nu\sigma} \qquad (5.34)\end{aligned}$$

We must now make the transformation into the physical basis:

$$A_\mu^1 = \frac{1}{\sqrt{2}}\left(W_\mu^+ + W_\mu^-\right),\ A^2 = -i\frac{1}{\sqrt{2}}\left(W_\mu^+ - W_\mu^-\right),\ A^3 = \sin\theta_W A_\mu + \cos\theta_W Z_\mu$$

In doing this we first note the antisymmetric structure functions and charge conservation imply:

- there must be at least two W bosons.
- there must be at most two neutral bosons

No self couplings without the charged gauge boson!

Thus the only combinations allowed are:

$$WWA, A^3 \rightarrow (WWZ, WWA) \text{ for the cubic terms },$$

$$(WWWW, WWA^2, A^4) \rightarrow (WWWW, WWZZ, WWAZ, WWAA) \text{ quartic terms}$$

The calculation is straightforward but a bit tedious. Thus for the cubic terms we get:

$$\mathcal{L}_{VB}^c = ig$$

$$\partial\mu W_\nu^{(+))}\left(W^{(-)\mu}\left(\sin\theta_W A^\nu + \cos\theta_W Z^\nu\right) - \left(\sin\theta_W A^\mu + \cos\theta_W Z^\mu\right) W^{(-)\nu}\right) +$$

$$-g\ i\partial\mu W_\nu^{(-))}$$

$$\left(W^{(+)\mu}\left(\sin\theta_W A^\nu + \cos\theta_W Z^\nu\right) - \left(\sin\theta_W A^\mu + \cos\theta_W Z^\mu\right) W^{(+)\nu}\right) +$$

$$-g\ i\partial\mu\left(\sin\theta_W A_\nu + \cos\theta_W Z_\nu\right)\left(W^{(+)\mu}W^{(-)\nu} - W^{(-)\mu}W^{(+)\nu}\right)$$

The corresponding diagrams are shown in Fig. 5.2. The expression for the quartic coupling is obtained analogously. We find

$$\mathcal{L}_{VB}^q = \frac{1}{4}g^2\left(W_\mu^{(+)}W_\nu^{(-)} - W_\mu^{(-)}W_\nu^{(+)}\right)\left(W^{(+)\mu}W^{(-)\nu} - W^{(-)\mu}W^{(+)\nu}\right) +$$

$$-\frac{1}{2}g^2\left(\sin\theta_W A_\mu + \cos\theta_W Z_\mu\right)\left(\sin\theta_W A^\nu + \cos\theta_W Z^\nu\right)$$

$$\left(W^{(+)\mu}W^{(-)\nu} + W^{(-)\mu}W^{(+)\nu}\right) + \frac{1}{2}g^2\left(\sin\theta_W A_\mu + \cos\theta_W Z_\mu\right)$$

$$\left(\sin\theta_W A^\mu + \cos\theta_W Z^\mu\right)\left(W_{(+)\nu}W^{(-)\nu} + W_{(-)\nu}W^{(+)\nu}\right)$$

The corresponding diagrams are shown in Fig. 5.3. There is a difference

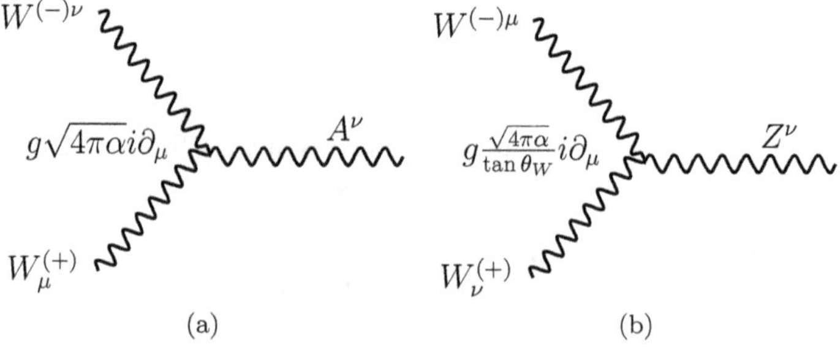

Fig. 5.2: The cubic self-couplings of the physical electroweak gauge bosons. Here $g\sin\theta_W = e = \sqrt{4\pi\alpha}$, the unit of electric charge. All other Lorentz contractions are understood.

in comparison to the gluons of QCD. The electroweak gauge bosons, except for the photon, are quite massive. So their existence can be exploited only at very high energies.

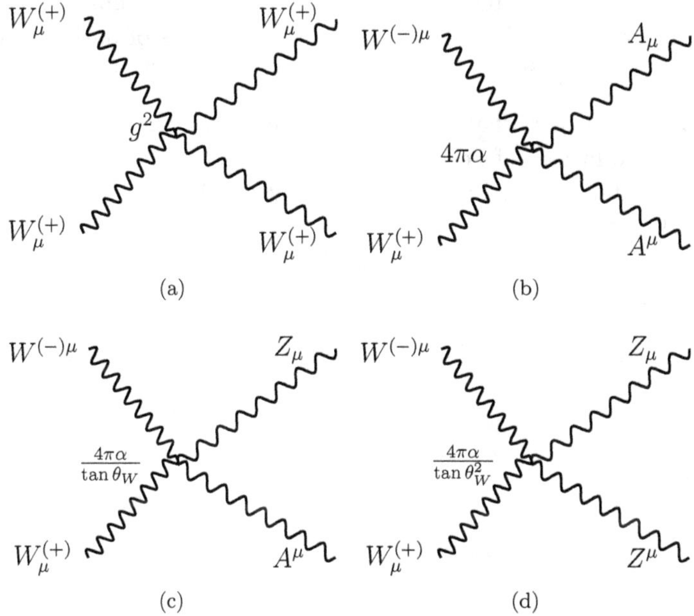

Fig. 5.3: The quartic self-couplings of the physical electroweak gauge bosons. Here $g\sin\theta_W = e = \sqrt{4\pi\alpha}$, the unit of electric charge. All other Lorentz contractions are understood.

5.6 Summary

The Standard model, a model with a gauge symmetry $G_S = SU_I(2) \otimes U_Y(1) \otimes SU_c(3)$:

(1) Predicted the existence of neutral currents.
(2) It predicted the masses of the gauge bosons M_W and M_Z and the coupling g in terms of for Weinberg angle and the Fermi constant G_F. The Weinberg angle can only be specified by experiment since the symmetry G_S is not simple.
(3) It unified the weak and electromagnetic interactions.
(4) All the particles it predicted have been found, including the long sought Higgs, which almost certainly has been found at CERN in 2013. Its mass could only be determined experimentally.
(5) It is in agreement with all experimental data except i) The neutrinos, which against its predictions, are massive (see chapter 11).

ii) It cannot explain the baryon anti-baryon asymmetry observed in nature. iii) It cannot explain the existence of dark matter and dark energy in the Universe (see section 10.11).

It suffers by the fact that it has too many parameters, it cannot, e.g., determine the masses of Fermions etc. It has not unified the electroweak and strong interactions. Finally it leaves out gravity.

Chapter 6

The SM SU(3) Group; Quantum Chromodynamics

6.1 Quantum chromodynamics QCD

We have seen that the strong interaction is governed by the group $SU_c(3)$ of the standard model with a set of traceless generator generators, the Gell-Mann matrices λ^a, such that $tr(\lambda^a)^2 = 2$, $a = 1 \cdots 8$. We have also seen that associated with this symmetry we have a set of 8 gauge bosons[1] G^a_λ, with λ a Lorentz index and $a = 1 \cdots 8$, the **gluons**. The interactions mediated by the gluons is given in Fig. 6.1.

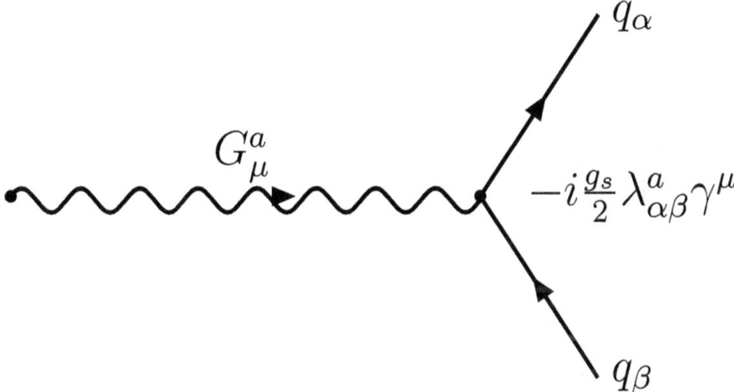

Fig. 6.1: A Feynman diagram indicating the gluon mediated interaction.

A comparison of the gluon exchange and the W exchange interaction is given in Fig. 6.2

[1]Sometimes the Lorentz index is understood and omitted and the gluons are written in the form of a 3×3 matrix G^μ_λ with μ and λ color degrees of freedom taking values, e.g. r, g, b. The two diagonal matrices have the form diag(1,-1,0) and $(1/\sqrt{3})$diag(1,1,-2).

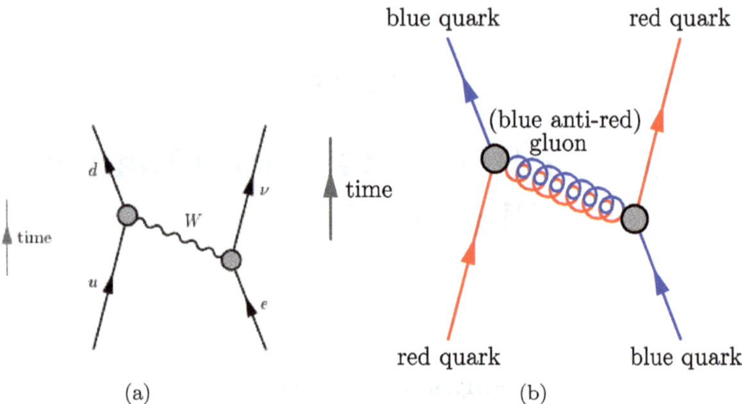

Fig. 6.2: A comparison of a W exchange (weak interaction) (a) and a gluon exchange (strong interaction) (b). Note the propagation of color via the gluons.

The gluons remain massless even after the spontaneous symmetry breaking. Thus this symmetry remains unbroken. Since, however, it is not Abelian, it does not lead to any long range force. The emerging theory is called **Quantum Chromodynamics**. This theory is described by the Lagrangian:

$$\mathcal{L}_c = -\frac{1}{4} F^j_{\mu\nu} F^{j\mu\nu} + \sum_r \bar{q}_{r\alpha} \gamma^\mu \left(D_\mu\right)^\alpha_\beta q^\beta_r, \tag{6.1}$$

where r is a quark flavor index, α and β are color indices and D^μ is the covariant derivative

$$\left(D_\mu\right)^\alpha_\beta = \partial_\mu \delta^\alpha_\beta - i g_s G^j_\mu \frac{1}{2} \left(\lambda^j\right)^\alpha_\beta \tag{6.2}$$

and

$$F^j_{\mu\nu} = \partial_\mu G^j_\nu - \partial_\nu G^j_\mu - g_s f_{jk\ell} G^k_\mu G^\ell_\nu, \tag{6.3}$$

where g_s is the croup gauge coupling constant[2] and f_{ijk} are the structure constants of $SU(3)$ given by:

$$\left[\lambda^j, \lambda^k\right] = 2i f_{jk\ell} \lambda^\ell. \tag{6.4}$$

Note that, in contrast to the other unbroken Abelian $U_{EM}(1)$ symmetry, the presence of the SU(3) F^2 term leads to three and four point gluon self couplings, which is due to the couplings $f_{jk\ell}$ (see Eq. (3.68) and 3.2). This effect results in some technical difficulties in QCD as compared to the electrodynamics.

[2] We use in this chapter g_s instead of g_3 used earlier.

- The gluons tend to polarize the medium. In other words it becomes energetically cheaper to create quark anti-quark pairs out of the vacuum (see Fig. 6.3). So we have the production of hadron jets (see Fig. 6.4).
- The polarization of the vacuum is not like the dipole creation familiar from dielectrics, which tends to decrease the interaction between two charges. Instead it behaves more like quadrupoles in dielectrics, hard to make in macroscopic scale, which tend to increase the interaction between the colored quarks (see Fig. 6.5). In other words at high energies, or small distances, the quarks behave almost like being free, i.e. we have **asymptotic freedom**. On the other hand at low energies (large distances of the order of fm) the interaction becomes very strong. Thus we encounter **confinement**, perpetual quark slavery.
- Since the interaction becomes strong at low energies, one cannot invoke perturbation theory. So multi-gluon exchange or pair creation out of the vacuum diagrams become important (see Fig. 6.3).

The strength of the interaction depends on the energy and the number n_f of active quark pairs:

$$\alpha_s(E) = \frac{4\pi}{\left(-33 + 2n_f \ln \frac{E^2}{\mu^2}\right)}, \tag{6.5}$$

where $\left\{ \begin{array}{l} n_f = \text{the number of quarks active pair production (up to 6)} \\ \mu = \text{experimentally determined scale} \approx 0.2\text{GeV} \end{array} \right. .$

Another common parametrization is given by

$$\alpha_s(q^2) = \frac{\alpha_s(\mu^2)}{1 + \beta\alpha_s(\mu^2)\left(\ln \frac{q^2}{\mu^2}\right)}, \beta = \frac{33 - 2n_f}{12\pi}. \tag{6.6}$$

These formulas yield $\alpha_s = 0.12$ at $q^2 = (100\text{MeV})^2$.

6.2 The color structure of the one gluon exchange potential involving quarks

Let us suppose that the two quarks are $q_\alpha(1)q_\beta(2)$, where α and β are color indices, taking values r, g, b, and (1) (2) label the particles. For simplicity of notation we will drop the particle index, with the understanding that particle one will be first and particle second, by writing $q_\alpha(1)q_\beta(2) \Leftrightarrow \alpha\beta = rr, rg,$ etc. The Feynman diagrams leading to the interaction between two quarks are exhibited in Fig. 6.6.

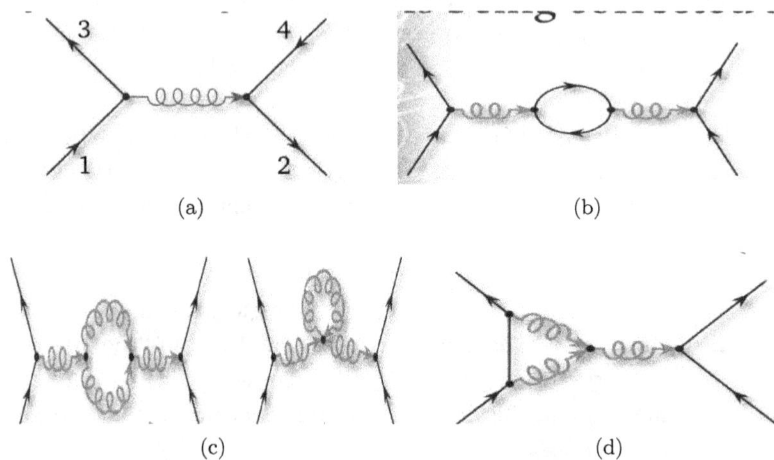

Fig. 6.3: The one gluon exchange (a). At some point it becomes cheaper to create quark-antiquark pairs out of the vacuum (b). As a result the interaction between the quarks is screened and decreased. This is analogous to QED and leads to a decrease in the interaction between to charged particles in dielectrics. In QCD, however, we also have virtual gluon creation out of the vacuum. For some cubic and quartic couplings, which are present in a non Abelian theory, this is exhibited in (c). Another diagram involving a cubic coupling is shown in (d). Such virtual gluons lead to anti-screening, i.e. the interaction increases, "color dielectric". This has dramatic effects at low energies (large distances). As a result we have confinement of colored particles.

Ignoring the overall strength (color charge, $g_r = g_s = 4\pi\alpha_s$) and omitting the gluon propagator, the interaction between similar colors is

$$\langle rr|V|rr\rangle = \langle gg|V|gg\rangle = \left(\frac{1}{6} + \frac{1}{2}\right) = \frac{2}{3}$$

The first contribution comes from the exchange of the gluon G^8 and the second one comes from G^3 (see Fig. 6.6b). Similarly from the exchange of the gluon G^8 (see Fig. 6.6c) we get

$$\langle bb|V|bb\rangle = \frac{1}{6}2^2 = \frac{2}{3}$$

In a similar fashion we get

$$\langle gr|V|gr\rangle = \langle rg|V|rg\rangle = \left(\frac{1}{6} - \frac{1}{2}\right) = -\frac{1}{3}(\text{see Fig. 6.6b})$$

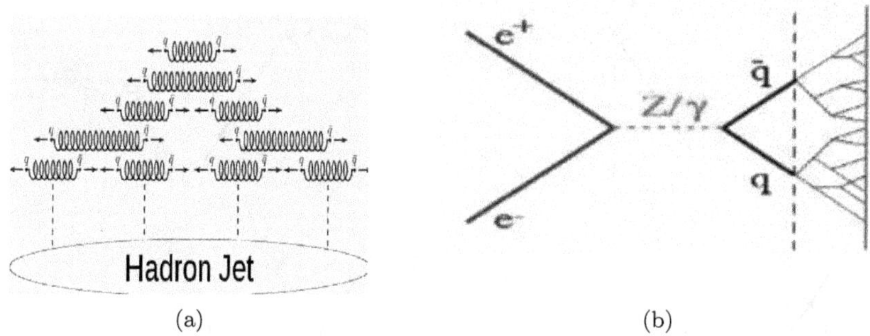

(a) (b)

Fig. 6.4: Hadron jets produced via the strong interaction (a). Hadron jets produced in an e^+, e^- (electron-positron) collider (b).

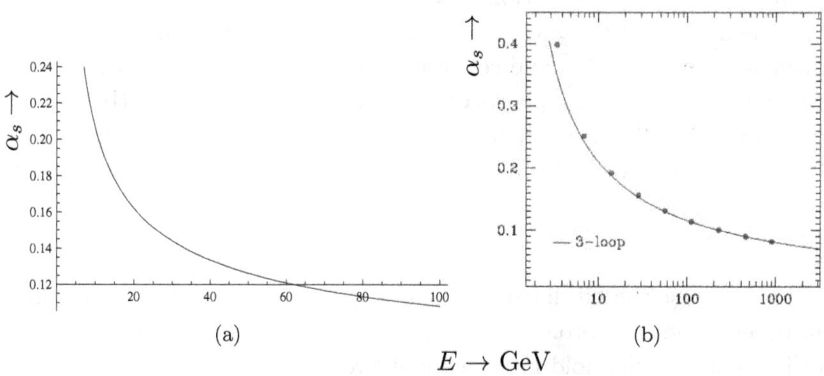

(a) (b)

$E \rightarrow$ GeV

Fig. 6.5: The strong coupling strength deceases as a function of the energy scale. At large scale it becomes very weak (asymptotic freedom), while at small scales it becomes very strong (imposed slavery). Shown is the prediction of the simple formula (Eq. 6.5) (a) and the experimental data of the ALPHA collaboration (b).

$$\langle rb|V|rb\rangle = \langle br|V|br\rangle = \frac{1}{6}(-2) = -\frac{1}{3}(\text{see Fig. 6.6c})$$

Furthermore

$$\langle \alpha\beta|V|\beta\alpha\rangle = 1,\, \alpha \neq \beta(\text{see Fig. 6.6a})$$

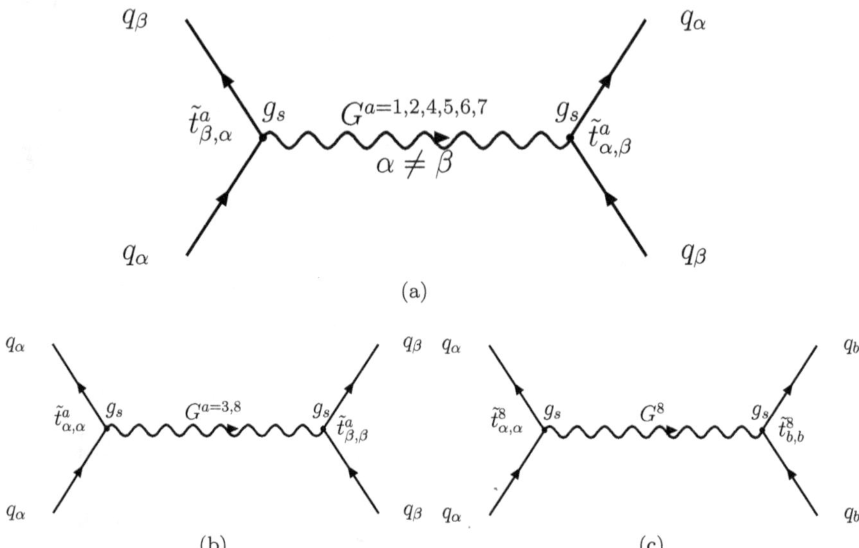

Fig. 6.6: The strong interaction mediated by gluons, with the Lorentz index suppressed. Shown are the color changing gluon interaction ($\tilde{t}^a_{\alpha\beta}$, $\alpha \neq \beta$, which destroys a quark β and converts to a quark α) (a), the color conserving interaction involving quarks other than b ($\tilde{t}^{3,8}_{\alpha\beta}, \alpha, \beta = r, g$) (b) and the color conserving interaction in which at least one of the quarks is blue ($\tilde{t}^{3,8}_{\alpha\alpha}$, $\alpha = r, g, b$) (c). Note that all operators are normalized to unity $tr(\tilde{t}^a \tilde{t}^b) = \delta_{a,b}$.

In this last case the off diagonal gluon is exchanged, $G^{1,2,4,5,6,7}$. All other matrix elements are zero.

The same results hold in the case of two antiquarks.

The above product of two quarks constitutes a basis in color space, $\underline{3} \otimes \underline{3}$, which yields a reducible nine dimensional representation of $SU(3)$. Diagonalizing this interaction in the nine dimensional space one finds thee states with eigenvalue $-\frac{4}{3}$ and six states with eigenvalue $\frac{2}{3}$ as follows:

$$-\frac{4}{3} \Leftrightarrow \tilde{b} = \frac{1}{2}(rg - gr), \tilde{g} = \frac{1}{2}(rb - br), \tilde{r} = \frac{1}{2}(gb - bg)$$

$$\frac{2}{3} \Leftrightarrow \frac{1}{2}(rg + gr), \frac{1}{2}(rb + br), \frac{1}{2}(gb + bg), rr, gg, bb.$$

The first is an irreducible three-dimensional indexcolor anti-triplet representation (anti-triplet) and the second is a six dimensional representation (sextet), which is symmetric in the color indices.

One uses the anti-triplet and the sextet as a basis and thus the color interaction takes the form

$$\boldsymbol{\lambda}_1.\boldsymbol{\lambda}_2 = \frac{1}{2}\left(C(\lambda,\mu) - 2C(1,0)\right)$$

where λ and μ specify the irreducible representation of $SU(3)$, which are non negative integers. $C(\lambda,\mu)$ is the value of the Casimir operator of $SU(3)$, given by:

$$C(\lambda,\mu) = \frac{2}{3}\left(\lambda^2 + \mu^2 + \lambda\mu + 3(\lambda+\mu)\right)$$

Thus:

$$\text{triplet:}(\lambda,\mu) = (1,0) \rightarrow C(\lambda,\mu) = \frac{8}{3}$$

$$\text{anti-triplet:}(\lambda,\mu) = (0,1) \rightarrow C(\lambda,\mu) = \frac{8}{3} \rightarrow \boldsymbol{\lambda}_1.\boldsymbol{\lambda}_2 = -\frac{4}{3}$$

$$\text{sextet:}(\lambda,\mu) = (2,0) \rightarrow C(\lambda,\mu) = \frac{20}{3} \rightarrow \boldsymbol{\lambda}_1.\boldsymbol{\lambda}_2 = \frac{2}{3}$$

Let us now examine the interaction between quarks and antiquarks. The antiquark viewed as the antisymmetric combination of two quarks given above. Than we can evaluate the matrix element of the gluon between the relevant three quark states, noting that one of the quarks of the antisymmetric combination is a spectator (not interacting). Thus we find the non vanishing independent ME involving different colors are:

$$\langle r\bar{b}|V|r\bar{b}\rangle = \frac{1}{3},\ \langle r\bar{g}|V|r\bar{g}\rangle = \frac{1}{3},\ \langle g\bar{b}|V|g\bar{b}\rangle = \frac{1}{3}$$

For the similar color combination we get:

$$\langle r\bar{r}|V|r\bar{r}\rangle = \langle g\bar{g}|V|g\bar{g}\rangle = \langle b\bar{b}|V|b\bar{b}\rangle = -\frac{2}{3},$$

$$\langle r\bar{r}|V|g\bar{g}\rangle = \langle r\bar{r}|V|b\bar{b}\rangle = \langle g\bar{g}|V|b\bar{b}\rangle - 1$$

Note the change in sign compared to the $q-q$ interaction. This usually is attributed to the "opposite color charge of anti-quarks", but note the notational difference in the last equation. We have seen, however, that it comes as a transformation from the picture of anti-quarks as an antisymmetric combination of quarks.

The product of a quark and anti-quark transforms under $SU(3)$ like $\underline{3} \otimes \underline{3}^*$, which is reducible. One can easily see that the above above six combinations $q_\alpha \bar{q}_\beta$, $\alpha \neq \beta$ transform as the six members of the octet (see also problem 14, Chapter 2). Diagonalizing the above matrix in the case of similar color indices we find the eigensolutions:

$$\epsilon_0 = -\frac{8}{3} \leftrightarrow r_0 = \frac{1}{\sqrt{3}}(|r\bar{r}\rangle + |g\bar{g}\rangle + b\bar{b}\rangle),$$

$$\epsilon_1 = \frac{1}{3} \leftrightarrow r_8^1 = \frac{1}{\sqrt{3}}(|r\bar{r}\rangle - |g\bar{g}\rangle),$$

$$\epsilon_2 = \frac{1}{3} \leftrightarrow r_8^2 = \frac{1}{\sqrt{6}}(|r\bar{r}\rangle + |g\bar{g}\rangle - 2b\bar{b}\rangle)$$

The first transforms like the singlet, while the other two are the additional members of the octet. The similarity with the expressions of the $SU(3)$ should not come as a surprise.

In summary we have seen that:

$$\underline{3} \otimes \underline{3}^* = \underline{1} + \underline{8}$$

As a result we can wite the quark antiquark interaction as :

$$\boldsymbol{\lambda}_1.\boldsymbol{\lambda}_2 = \frac{1}{2}\left(C(\lambda,\mu) - C(1,0) - C(0,1)\right)$$

$$\text{singlet:}(\lambda,\mu) = (0,0) \rightarrow C(\lambda,\mu) = 1 \rightarrow \boldsymbol{\lambda}_1.\boldsymbol{\lambda}_2 = -\frac{8}{3}$$

$$\text{octet:}(\lambda,\mu) = (1,1) \rightarrow C(\lambda,\mu) = \frac{18}{3} \rightarrow \boldsymbol{\lambda}_1.\boldsymbol{\lambda}_2 = \frac{1}{3}$$

Before concluding this section we should mention that in the context of the one gluon exchange potential between the above allowed color combinations is proportional to $\boldsymbol{\lambda}_1.\boldsymbol{\lambda}_2\alpha_s$

6.3 Approximations at low energies; interaction potentials between quarks

We know that the quarks do not appear free and all the observed hadrons are colorless. So all experimental information regarding quarks is necessarily indirect and complicated manifestations of chromodynamics. It looks as though our only access to electrodynamics came from Van der Waals forces between molecules.

We have already seen that due to anti-screening the interaction between quarks become very strong at low energies so perturbative techniques are not going to be effective. So some approximations have to be made. It common to assume that the quarks have a mass, which is cannot be directly deternined since they are never free. They are obtained from fits of the appropriate spectra of hadrons viewed as bound states of quarks. This is achieved by assuming a confining potential, which is attractive, with a strength that increases with the distance between the interacting quarks. The most popular are the linear, $V(r) \propto r$, logarithmic $V(r) \propto \ln(r/a)$ and quadratic $V(r) \propto r^2$ The latter is going to be discussed below (see next section). On top of this one supposes an interaction between the quarks.

In this section[3] we are going to derive the effective potential between quarks in the one gluon approximation in the non relativistic limit up to including terms of second order in the quark momenta, i.e. of order of p^2/m_q^2. To this end we express the 4-spinor forms into matrix elements involving two component wave functions. Clearly this approximation will be applicable to heavy quarks or assuming a constituent masses, about a third of the nucleon mass, for light quarks.

6.3.1 *One gluon exchange potential in a process involving only baryons*

The non relativistic reduction of the one gluon exchange amplitude [Henley *et al.* (1986)] (see Fig. 6.2) leads to the effective 2-body operator:

$$\begin{aligned} \tilde{V} = &-\frac{1}{(2\pi)^3}\boldsymbol{\lambda}_1.\boldsymbol{\lambda}_2\delta(\tilde{\mathbf{q}}_1+\tilde{\mathbf{q}}_2)\frac{4\pi\alpha_s}{\tilde{\mathbf{q}}_1^2} \\ &\left[1-\frac{1}{(2m_q^2}\left(\tilde{\mathbf{Q}}_1.\tilde{\mathbf{Q}}_1+i\boldsymbol{\sigma}_1.(\tilde{\mathbf{q}}_1\times\tilde{\mathbf{Q}}_2)+i\boldsymbol{\sigma}_2.(\tilde{\mathbf{q}}_2\times\tilde{\mathbf{Q}}_1)\right.\right. \\ &- \left.\left.(\boldsymbol{\sigma}_1\times\tilde{\mathbf{q}}_1).(\boldsymbol{\sigma}_2\times\tilde{\mathbf{q}}_2)\right)\right], \end{aligned} \tag{6.7}$$

where $\boldsymbol{\lambda}_1.\boldsymbol{\lambda}_2$ the SU(3) invariant, $\tilde{\mathbf{q}}_i = \mathbf{p}_i^{'} - \mathbf{p}_i$ and $\tilde{\mathbf{Q}}_i = \mathbf{p}_i^{'} + \mathbf{p}_i$, $i = 1, 2$. It is traditional to perform a Fourier transform and go to the coordinate space. Thus we get

$$V = -\alpha_s\boldsymbol{\lambda}_1.\boldsymbol{\lambda}_2\left[\frac{1}{r}+\frac{1}{(2m_q)^2}\left(\frac{2}{3}V_S+\frac{1}{3}V_T+\frac{1}{r}V_{QQ}+\frac{1}{r^3}V_{qQ}\right)\right], \tag{6.8}$$

with α_s treated as a parameter to be fitted to the spectra. It is assumed to be of order 1, i.e. about five times larger than a typical value used in high energy physics (see Fig. 6.5). Furthermore

$$V_S = 4\pi\delta(\mathbf{r})\boldsymbol{\sigma}_1.\boldsymbol{\sigma}_2,$$

[3]The material of this section is taken from the work: D. Strottman and J.D. Vergados, The six quark clusters in QCD (to be published)

$$V_T = 3\boldsymbol{\sigma}_1.\hat{\mathbf{r}}.\boldsymbol{\sigma}_2.\hat{\mathbf{r}} - \boldsymbol{\sigma}_1.\boldsymbol{\sigma}_2,$$

$$V_{QQ} = \nabla_{r_1}(\leftarrow).\nabla_{r_2}(\leftarrow) + \nabla_{r_1}(\rightarrow).\nabla_{r_2}(\rightarrow) - \nabla_{r_1}(\leftarrow).\nabla_{r_2}(\rightarrow)$$
$$-\nabla_{r_1}(\rightarrow).\nabla_{r_2}(\leftarrow).$$
$$V_{qQ} = (\sigma_1 + \sigma_2).\left(\ell(\rightarrow) - \ell(\leftarrow)\right)$$
$$-\frac{1}{2}\left((\boldsymbol{\sigma}_2 - \boldsymbol{\sigma}_1).(i\, r \times \nabla_R(\leftarrow) - (\boldsymbol{\sigma}_2 - \boldsymbol{\sigma}_1).(i\, r \times \nabla_R(\rightarrow)\right).$$

In the above expressions the arrows in parenthesis indicate the direction (bra or ket) on which the non local operator acts. As usual $\ell = \ell_2 - \ell_1$ is the relative orbital angular momentum, $\mathbf{r} = \mathbf{r_2} - \mathbf{r_1}$ and $\mathbf{R} = \frac{1}{2}(\mathbf{r_2} + \mathbf{r_1})$. The relative orbital angular momentum gives no contribution since the relevant matrix element is diagonal (this part of the operator does not depend on the CM coordinates).

6.3.2 *One gluon exchange potential in processes involving the creation of a $q\bar{q}$ pair*

In this case in the place of the diagrams of Fig. 6.6 one obtains a new diagram by replacing a a $q-q$ line by a $q-\bar{q}$ pair as in Fig. 6.7. This diagram, combined with a number of "spectator" (non interacting) quarks The non

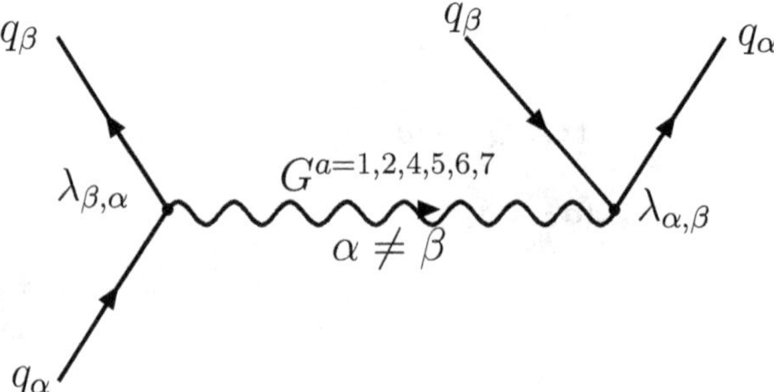

Fig. 6.7: A typical diagram involving the creation of a $q-\bar{q}$ pair by a gluon (for notation see Fig. 6.6).

relativistic reduction of the one gluon exchange amplitude [Henley *et al.* (1986)], resulting from this diagram, leads to the effective 2-body operator:

$$\tilde{V} = -\frac{1}{(2\pi)^3}\boldsymbol{\lambda}_1.\boldsymbol{\lambda}_2\delta(\tilde{\mathbf{q}}_1 - \tilde{\mathbf{Q}}_2)\frac{4\pi\alpha_s}{\tilde{\mathbf{q}}_1^2}\left[\frac{\boldsymbol{\sigma}_2.\mathbf{q}_1}{2m_q} + \frac{\boldsymbol{\sigma}_2.\mathbf{Q}_1}{2m_q} + i\frac{1}{2m_q}\mathbf{q}_1.(\boldsymbol{\sigma}_2 \times \boldsymbol{\sigma}_1)\right]. \tag{6.9}$$

Furthermore this diagram, combined with a number of "spectator" (non interacting) interacting quarks, can lead to processes involving only color singlet states (see Fig. 6.8). Another approach is to consider the $^3P_0$, which is derived from the quark string model[4], yielding a $q-\bar{q}$ pair, "created out of the vacuum".

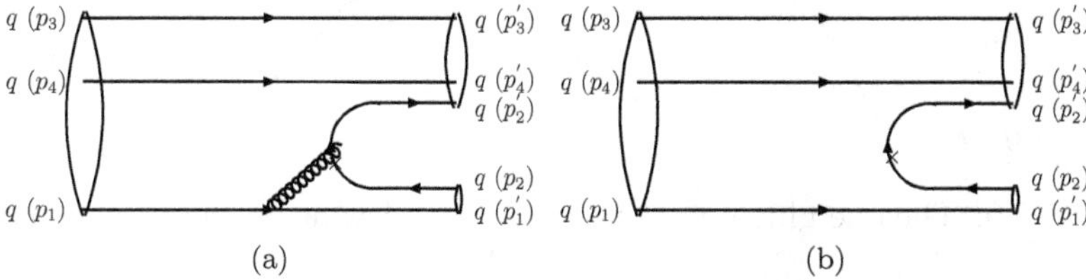

Fig. 6.8: The 1 gluon exchange potential acting between the quarks labeled $q(p_1)$, $q(p_1^{'})$, $q(p_2)$ and $q(p_2^{'})$, while the other quarks do not participate in the interaction (they are spectators) (a). The same effect can be accomplished by creating a $q\bar{q}$ pair out of the vacuum, marked by $\times$, (b), in which case the pair is normally in an $S=1, \ell=1, J=0$ state ($^3P_0$ model). The process exhibited describes the baryon meson coupling. If the initial baryon is heavier than the final, it provides the decay amplitude $B_1 \rightarrow B_2M$ (the lens like curves indicate a baryon or meson bound state). The same diagram is applicable in any process in which the interaction causes an $1q \rightarrow 2q - 1\bar{q}$ transition. For example, if the middle horizontal line is missing and the top arrow is reversed, it describes the decay of one meson into two mesons.

6.3.3 *One gluon exchange potential in processes involving meson spectra*

The non relativistic reduction of the one gluon exchange amplitude (see Fig. 6.9) in this case leads to the effective potential:

$$\begin{aligned}\tilde{V} &= -\frac{1}{(2\pi)^3}\boldsymbol{\lambda}_1.\boldsymbol{\lambda}_2\delta(\tilde{\mathbf{Q}}_1-\tilde{\mathbf{Q}}_2)\frac{4\pi\alpha_s}{\tilde{\mathbf{Q}}_1^2}\\ &\left\{\boldsymbol{\sigma}_1.\boldsymbol{\sigma}_2+\frac{1}{2}\frac{1}{(2m_q)^2}\left[\boldsymbol{\sigma}_2.(\tilde{\mathbf{q}}_1\times\tilde{\mathbf{Q}}_1)+\boldsymbol{\sigma}_1.(\tilde{\mathbf{q}}_2\times\tilde{\mathbf{Q}}_1)\right]\right.\\ &\left.+\frac{1}{4}\frac{1}{(2m_q)^2}\left[2\tilde{Q}_1^2-\tilde{q}_1^2-\tilde{q}_2^2+(\tilde{\mathbf{q}}_1-\tilde{\mathbf{q}}_2)\times\tilde{\mathbf{Q}}_1).(\boldsymbol{\sigma}_1\times\boldsymbol{\sigma}_2)\right.\right\}. \quad (6.10)\end{aligned}$$

[4]See, e.g., D. H. Perkins, Introduction to High Energy physics, 3nd Edition, sec. 8.10, Addison Wesley.

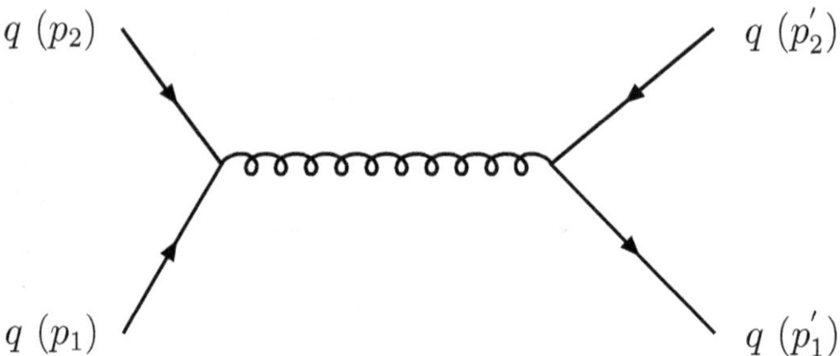

Fig. 6.9: The one gluon effective potential relevant for meson spectra.

6.4 Low energy formalism

One would like to make suitable approximations to microscopically study the usual baryons, e.g., the proton, neutron, Δ resonances etc , and mesons, pions, ρ-mesons,J/Ψ, Y and their excitations (Ψ', Y' etc), etc called quarkonia.

Many phenomenologically realistic models exist, see e.g. the pioneer early works [Chodos *et al.* (1974)]-[Eichten *et al.* (1980)]. For purposes of illustration we will consider the much simpler approach using a harmonic oscillator basis.

The kinetic energy part is easy to evaluate in momentum space, especially, if a convenient harmonic oscillator basis is adopted, since in this case one can separate the relative from the center of mass motion. The spin isospin quantum numbers can be treated in the usual way for the three body (baryon) and the two body system (meson). So we will concentrate on the orbital parts.

6.4.1 *The orbital part at the quark level*

Orbital wave functions in momentum space are expressed in terms of Jacobi coordinates:

$$\psi_{\mathbf{P}_M} = \sqrt{2E_\pi}\left(2\sqrt{2}\right)^{1/2}(2\pi)^{3/2}\,\delta\left(\sqrt{2}\mathbf{Q}_M - \mathbf{P}_M\right)\phi_M(\boldsymbol{\rho}), \tag{6.11}$$

$$\psi_{\mathbf{P}_B} = \left(3\sqrt{3}\right)^{1/2}(2\pi)^{3/2}\,\delta\left(\sqrt{3}\mathbf{Q} - \mathbf{P}_B\right)\phi_B(\boldsymbol{\xi},\boldsymbol{\eta}), \tag{6.12}$$

where $\mathbf{P}_M$, and $\mathbf{P}_B$ are the momenta of the meson and the baryon respectively and

$$\boldsymbol{\xi} = \frac{1}{\sqrt{2}}(\mathbf{p}_1 - \mathbf{p}_2)\,,\quad \boldsymbol{\eta} = \frac{1}{\sqrt{6}}(\mathbf{p}_1 + \mathbf{p}_2 - 2\mathbf{p}_3)\,,\quad Q = \frac{1}{\sqrt{3}}(\mathbf{p}_1 + \mathbf{p}_2 + \mathbf{p}_3), \tag{6.13}$$

$$\boldsymbol{\rho} = \frac{1}{\sqrt{2}}(\mathbf{p}_1 - \mathbf{p}_2)\ ,\ Q_M = \frac{1}{\sqrt{2}}(\mathbf{p}_1 + \mathbf{p}_2), \tag{6.14}$$

where $\mathbf{p_i}$, $\mathbf{i} = \mathbf{1} \cdots \mathbf{3}$ are the momenta of the three quarks of the baryon, or $\mathbf{p_i}$, $\mathbf{i} = \mathbf{1}, \mathbf{2}$ are the momenta of the quark and anti-quark in the meson.

The above wave functions were normalized in the usual way:

$$\langle \psi_{\mathbf{P}_B} | \psi_{\mathbf{P}'_B} \rangle = (2\pi)^3 \delta(\mathbf{P}_B - \mathbf{P}'_B)\ ,\ \langle \psi_{\mathbf{P}_M} | \psi_{\mathbf{P}'_M} \rangle = 2E_M (2\pi)^3 \delta(\mathbf{P}_M - \mathbf{P}'_M). \tag{6.15}$$

The harmonic oscillator wave functions describing the relative coordinates are well known. Thus, e.g. for $0s$ astes take the form:

$$\phi_B(\ \boldsymbol{\xi},\ \boldsymbol{\eta}) = \phi(\xi)\phi(\eta), \tag{6.16}$$

$$\phi(\xi) = \phi(0) e^{-(b_B^2 \xi^2)/2},\ \phi(0) = \sqrt{\frac{b_B^3}{\pi\sqrt{\pi}}}\ \text{etc}\ ,$$

$$\phi_M(\rho) = \phi_M(0) e^{-(b_M^2 \rho^2)/2},\ \phi_M(0) = \sqrt{\frac{b_M^3}{\pi\sqrt{\pi}}}. \tag{6.17}$$

6.4.2 *The kinetic energy part*

The mass of the quarks is assumed to be the constituent quark mass, about a third of the mass of the nucleon for the light quarks u, d and s. For the heavy quarks one extracts their masses from the mass of the corresponding mesons J/Ψ and Y.

6.4.3 *The confining potential*

We will assume a confining potential of the form:

$$V_c = -\boldsymbol{\lambda}_1.\boldsymbol{\lambda}_2 k (\mathbf{r_1} - \mathbf{r_2})^{\mathbf{2}}, \tag{6.18}$$

where $\boldsymbol{\lambda}_1.\boldsymbol{\lambda}_2$ is the two-body part of the SU)3) Casimir operator, quadratic $SU(3)$ invariant $g(\lambda, \mu)$, given by

$$\boldsymbol{\lambda}_1.\boldsymbol{\lambda}_2 = C_{(\lambda,\mu)},$$

$$C_{(\lambda,\mu)} = \frac{1}{2}\left(g(\lambda,\mu) - 2g(1,0)\right),\ g(\lambda,\mu) = \frac{2}{3}\left(\lambda^2 + \mu^2 + \lambda\mu + 3(\lambda+\mu)\right). \tag{6.19}$$

The integers λ and μ characterize the $SU(3)$ representation, e.g.,

$$(\lambda, \mu) = (1,0),\ (0,1),\ (2,0),\ (1,1)$$

for the fundamental $\underline{3}$ (triplet), $\underline{3}^*$ (anti-triplet), $\underline{6}$ (sextet) and the $\underline{8}$ (octet or adjoined) representations respectively. $k = \omega^2 m_q$ is a constant to be fitted by yielding the correct value of the mass of the nucleon for a given size parameter a (see below).

6.4.4 *Fitting the strength of the confining potential*

The strength k of the confining potential can be determined by considering the nucleon as a three quark system. One must first compute the mass of the baryons. To this end we consider:

- The kinetic energy:

$$\epsilon_q = \frac{1}{2m_q}\left(\mathbf{p}_1^2 + \mathbf{p}_2^2 + \mathbf{p}_3^2\right), \quad m_q = \frac{m_p}{3}. \tag{6.20}$$

  For harmonic oscillator wave functions in the internal variables we find in the rest frame of the baryon:

$$\begin{aligned}\langle\psi_N|\epsilon_q|\psi_N\rangle &= \langle\psi_N|\frac{1}{2m_q}\left(\frac{1}{2}(\mathbf{p}_1-\mathbf{p}_2)^2 + \frac{1}{6}(\mathbf{p}_1+\mathbf{p}_2-2\mathbf{p}_3)^2\right)|\psi_N\rangle \\ &= \frac{1}{2m_q}\left(\frac{3}{2}\frac{1}{b^2} + \frac{3}{2}\frac{1}{b^2}\right) = \frac{9}{2m_p}\frac{1}{b^2}.\end{aligned} \tag{6.21}$$

- The average confining potential.
  In this case we find $ME = 4/3\langle V_c\rangle$, with $\langle V_c\rangle$ the expectation of the radial part of the potential. The negative value is due to the fact that only the color isotriplet pairs contribute ($C_{(0,1)} = -4/3$). It is straightforward to show that

$$\langle V_c\rangle = 3\frac{1}{2}kb^2, \quad b = \text{the nucleon size} \Rightarrow ME = 2kb^2. \tag{6.22}$$

We thus arrive at the equation:

$$m_p = \frac{9}{2m_p b^2} + 2kb^2 \tag{6.23}$$

or

$$ka^2 = \frac{m_p}{2}\left(-\frac{9}{2m_p^2 b^2} + 1\right). \tag{6.24}$$

For $a = 1$ fm we find:

$$kb^2 = 0.05 m_p = 50\ \text{MeV} \rightarrow k = 50\frac{\text{MeV}}{\text{fm}^2}.$$

6.5 Matrix elements involving two quarks

We need not worry about the 1-body terms arising from the kinetic terms, since it is trivial to compute them for any hadron.

6.5.1 *The confining potential*

The confining potential is spin independent and does not change color. So the two quark matrix elements are independent of isospin. Thus:

$$\langle (n_1 l_1, n_2 l_2)L; [f_{cs}](\lambda,\mu)S; JI|V_c|(n_1' l_1', n_2' l_2')L'; [f_{cs}'](\lambda',\mu')S'; JI\rangle =$$

$$\delta_{L,L'}\delta_{S,S'}\delta_{\lambda,\lambda'}\delta_{\mu,\mu'}\delta_{[f_{cs}],[f_{cs}']}(-)C_{\lambda,\mu}\langle (n_1 l_1, n_2 l_2)L|V_c|(n_1' l_1', n_2' l_2')L\rangle,$$

where $\langle (n_1 l_1, n_2 l_2)L|V_c|(n_1' l_1', n_2' l_2')L\rangle$ is the radial integral involving the potential and

$$C_{(\lambda,\mu)} = -\frac{4}{3} \text{ for } (\lambda,\mu) = (0,1) \text{ (anti-triplet)}$$

$$C_{(\lambda,\mu)} = \frac{2}{3} \text{ for } (\lambda,\mu) = (2,0) \text{ (sextet).}$$

6.5.2 *The one gluon exchange potential*

Even though the above operator in the coordinate space is non local, one can still compute the needed matrix elements by taking gradients of the state vectors and employing the standard Racah algebra techniques. We find it more convenient, however, to work in momentum space. Furthermore most of the elementary interactions are derived in momentum space. So in many other applications one would like to have developed a formalism permitting exploitation of the advantages of momentum space.

We first introduce the dimensionless variables $\mathbf{Q}_i = \frac{1}{\sqrt{2}}\tilde{\mathbf{Q}}_i b$, $\mathbf{q}_i = \frac{1}{\sqrt{2}}\tilde{\mathbf{q}}_i b$, $i = 1, 2$, with b being the harmonic oscillator size parameters. Then the above operator associated with baryons only (see Eq. (6.7)) can be brought into a more convenient form:

$$\begin{aligned}\tilde{V} = &-\frac{1}{(2\pi)^3}\boldsymbol{\lambda}_1.\boldsymbol{\lambda}_2\frac{1}{2\sqrt{2}}\delta(\mathbf{q}_1+\mathbf{q}_2)\frac{4\pi\alpha_s}{2b}\frac{1}{\mathbf{q}_1^2}\\ &[1 - 2\kappa(\mathbf{Q}_1.\mathbf{Q}_1 + i\sigma_1.(\mathbf{q}_1\times\mathbf{Q}_2) + i\sigma_2.(\mathbf{q}_2\times\mathbf{Q}_1)\\ &-(\sigma_1\times\mathbf{q}_1).(\sigma_2\times\mathbf{q}_2))],\end{aligned} \tag{6.25}$$

with $\alpha_s = \frac{g_r^2}{4\pi}$ and $\kappa = \frac{1}{(2m_q b)^2}$. In terms of tensor operators it is conveniently rewritten as follows:

$$\begin{aligned}\tilde{V} = &-\frac{1}{(2\pi)^3}\boldsymbol{\lambda}_1.\boldsymbol{\lambda}_2\frac{1}{2\sqrt{2}}\delta(\mathbf{q}_1+\mathbf{q}_2)\frac{4\pi\alpha_s}{2b}\frac{1}{\mathbf{q}_1^2}\\ &\left\{1 - 2\kappa\left(\mathbf{Q}_1.\mathbf{Q}_1 - \sqrt{2}\sigma_1.[(\mathbf{Q}_2\times\mathbf{q}_1)]^1 - \sqrt{2}\sigma_2.[(\mathbf{Q}_1\times\mathbf{q}_2)]^1 -\right.\right.\\ &\left.\left.\left(\frac{2}{3}\sigma_1.\sigma_2 + \frac{1}{3}T(\sigma_1,\sigma_2,\hat{q}_1)\right)q_1^2\right)\right\},\end{aligned} \tag{6.26}$$

with T the tensor operator defined by

$$T(\sigma_1, \sigma_2, \hat{q}) = 3\sigma_1.\hat{q}\sigma_2.\hat{q} - \sigma_1.\sigma_2 = \sqrt{\frac{6}{5}}\sqrt{4\pi}Y^2(\hat{q}).[\sigma_1 \times \sigma_2]^2.$$

In other words the operator in momentum space indeed takes a very simple form.

The same procedure can be applied in the case of the operators of Eqs. (6.9) and (6.10).

Chapter 7

Rates and Cross Sections in Electroweak Theory

In this chapter, using the tools previously developed, we will evaluate some quantities of phenomenological interest. Those not familiar with introductory particle physics may want to look at the appendix at the end of the book before proceeding further.

7.1 The Feynman diagrams and rules

The Feynman diagram is an ingenious way of summarizing pictorially the results of Field Theory (see, e.g., [Majore (2015)] and the classic books by S. Weinberg, [Weinberg (1995.)],[Weinberg (1996.)]), an essential tool in understanding the various particle physics processes. It is an economic way to obtain via a perturbative expansion the relative amplitudes, which lead to the determination of of the experimental observables.

We are not going to derive these rules and diagrams, only list them and give some heuristic arguments. The Feynman diagrams contain lines, which may or may not intersect. The point of intersection is called o knot and will be indicated by a full circular disc. The diagrams contain both external lines, with one open end, and internal lines. Out of the many possible diagrams we only need consider the topologically non-equivalent, i.e. those that cannot by brought to each other by a continuous deformation, which leaves the external lines unaffected. We should not forget that the orientation of these lines is important.

Before proceeding further the following correspondence in notation is useful:

- Incoming fermion with momentum $\mathbf{p}$ and spin $s \Leftrightarrow u(\mathbf{p}, s)$
- Outgoing fermion with momentum $\mathbf{p}$ and spin $s \Leftrightarrow \bar{u}(\mathbf{p}, s)$
- Incoming anti-fermion with momentum $\mathbf{p}$ and spin $s \Leftrightarrow \bar{v}(\mathbf{p}, s)$

- Outgoing anti-fermion with momentum $\mathbf{p}$ and spin $s \Leftrightarrow \upsilon(\mathbf{p}, s)$
- Incoming photon with Lorentz index μ, momentum $\mathbf{k}$ and helicity $\lambda \Leftrightarrow \epsilon_\mu(\mathbf{k}, \lambda)$
- Outgoing photon with Lorentz index μ, momentum $\mathbf{k}$ and helicity $\lambda \Leftrightarrow (\epsilon_\mu(\mathbf{k}, \lambda))^*$

7.1.1 *Propagators*

We will begin with lines indicating the propagation of particles, **propagators**, for fermions, bosons with spin one and scalars (see Fig. 7.1). We should mention that the arrow of the fermion line does not indicate the direction of motion, but the flow of charge. It coincides with the direction of motion of particles, but it is opposite to that of antiparticles.

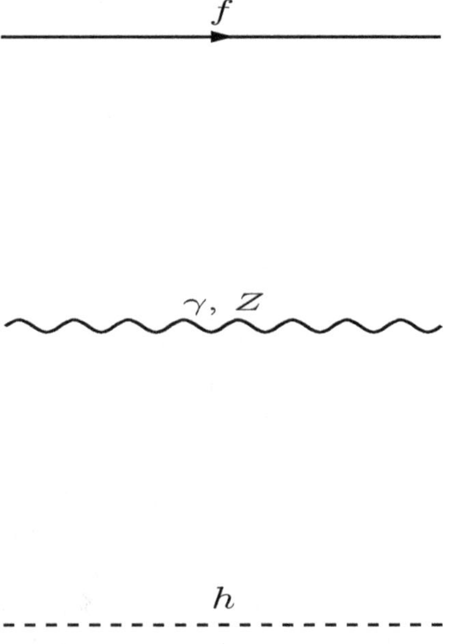

Fig. 7.1: i) The propagator for a fermion (top) $\Leftrightarrow \frac{i}{\not{p}-m+i\epsilon}$, ii) The photon and vector boson propagator (middle), $\Leftrightarrow \frac{-ig_{\lambda,\mu}}{k^2+i\epsilon}$ for the photon and $\frac{-ig_{\lambda,\mu}}{k^2-m_Z^2+i\epsilon}$ for the Z-boson, both of which maybe different in other gauges and iii) the scalar propagator (bottom) $\Leftrightarrow \frac{i}{p^2-m_h^2+i\epsilon}$.

7.1.2 *The simplest diagrams in quantum electrodynamics; the photon-charge interaction*

The interaction of the photon with a charge is shown by the diagrams of Fig. 7.2. The relevant amplitudes are:

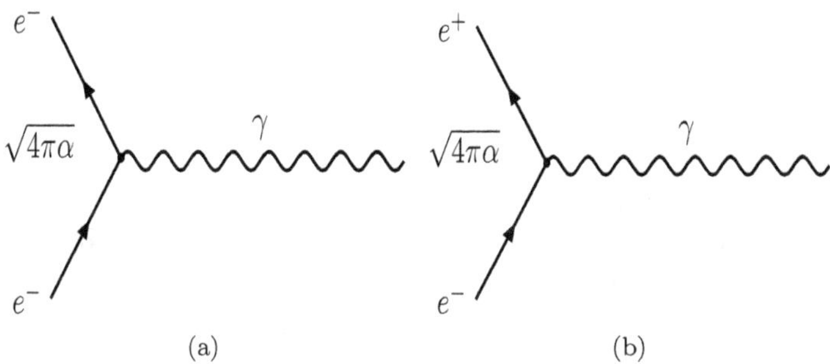

Fig. 7.2: (a) An oncoming electron interacts with a photon and exits (outgoing). (b) An electron and a positron collide to produce a photon. In the vertex we indicate the strength of the interaction $e = \sqrt{4\pi\alpha}$. Note the direction of the arrow. All lines are "external".

$$\mathcal{M}_a = \bar{u}(\mathbf{p}_1')\gamma^\mu u(\mathbf{p}_1)\epsilon_\mu(\mathbf{k},\lambda),\ \mathcal{M}_b = \bar{v}(\mathbf{p}_1')\gamma^\mu u(\mathbf{p}_1)\epsilon_\mu(\mathbf{k},\lambda) \tag{7.1}$$

corresponding to Figs. 7.2a and 7.2b respectively. This process by itself, as it stands, is not allowed by conservation of energy and momentum. More than one photon must be produced

7.1.3 *Quantum electrodynamics; the interaction between two charges*

For the simple case of electron scattering the situation is shown in Fig. 7.3. The situation is similar when the two oncoming particles are not the same, e.g. a proton and an electron, but then the diagram (b) is not possible. The mathematical structure of the amplitudes are:

$$\mathcal{M}_a = 4\pi\alpha\bar{u}(\mathbf{p}_1')\gamma_\lambda u(\mathbf{p}_1)\frac{i}{(p_1'-p_1)^2+i\epsilon}\bar{u}(\mathbf{p}_2')\gamma^\lambda u(\mathbf{p}_2) \tag{7.2}$$

$$\mathcal{M}_b = 4\pi\alpha\bar{u}(\mathbf{p}_2')\gamma_\lambda u(\mathbf{p}_1)\frac{i}{(p_2'-p_1)^2+i\epsilon}\bar{u}(\mathbf{p}_1')\gamma^\lambda u(\mathbf{p}_2) \tag{7.3}$$

associated with diagrams 7.3a and 7.3b respectively.

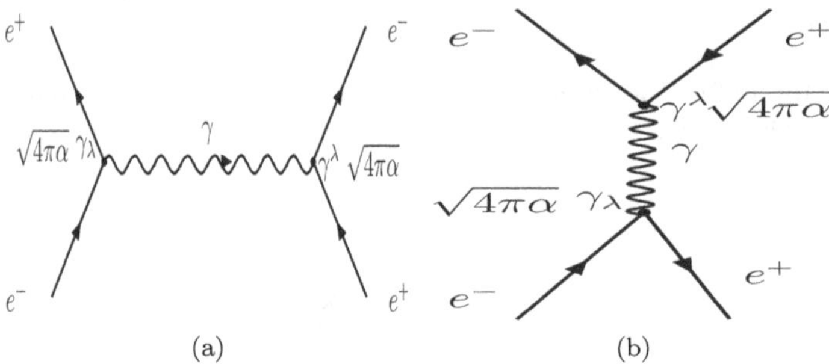

Fig. 7.3: The interaction between two electrons (a) and (b) to second order in the interaction. The sign of the diagram (b) is opposite to that of (a). The fermion lines are external, the photon line internal.

For the electron positron collision: $e^- + e^+ \to e^- + e^+$ the Feynman diagrams are shown in Fig. 7.4. Given sufficient energy in an electron positron accelerator the produced outgoing particles can also be μ^-, μ^+ or τ^-, τ^+ .

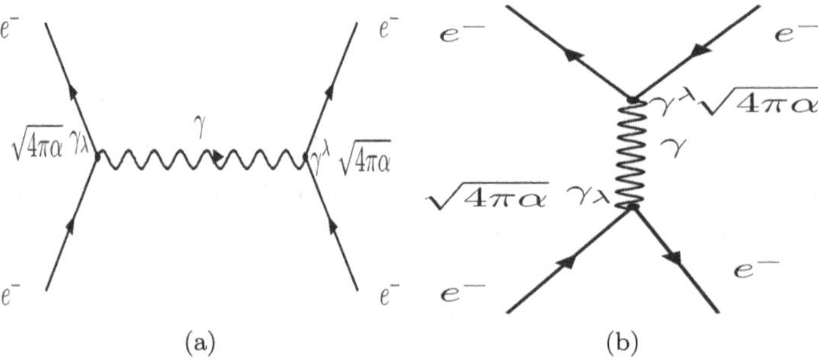

Fig. 7.4: The diagrams involved in electron positron collisions to second order in the interaction. The sign of the diagram (b) is opposite to that of (a). The fermion lines are external, the photon line internal.

The structure of the relevant mathematical amplitudes is:

$$\mathcal{M}_a = 4\pi\alpha\bar{v}(\mathbf{p}_1')\gamma_\lambda u(\mathbf{p}_1)\frac{i}{(p_1' + p_1)^2 + i\epsilon}\bar{v}(\mathbf{p}_2')\gamma^\lambda u(\mathbf{p}_2) \tag{7.4}$$

$$\mathcal{M}_b = 4\pi\alpha\bar{v}(\mathbf{p'}_2)\gamma_\lambda u(\mathbf{p}_1)\frac{i}{(p_2 - p_1)^2 + i\epsilon}\bar{v}(\mathbf{p}_1')\gamma^\lambda u(\mathbf{p}_2) \tag{7.5}$$

associated with diagrams 7.4a and 7.4b respectively.

One can draw similar diagrams, beyond those of QED. The procedure is similar, but the relevant interactions must be incorporated in the appropriate vertices as dictated by the standard model or its extensions.

7.1.4 *The Feynman rules*

In calculating the invariant amplitude one can apply the Feynman rules based on the Feynman diagrams discussed above. We list them here:

(1) **Vertex coupling.** In most cases this is the coupling constant which has been defined previously in connection with the standard model.

(2) **Vertex function.** In each vertex write an a four momentum conservation δ function

$$(2\pi)^4\delta(k_1 + k_2 + +k_n)$$

with the chosen sign of momentum for each particle appearing in the vertex.

(3) **Propagator**. For each internal line, usually representing a scalar or vector boson carrying momentum q, write a factor:

$$\frac{1}{q^2 - m^2}$$

(4) **Internal line**. For each such line write down a factor:

$$\frac{1}{(2\pi)^4}\int d^4q$$

and integrate over all internal momenta.

(5) **Energy momentum conservation**. The obtained result should include an overall δ function:

$$(2\pi)^4\delta(p_1 + p_2 + +p_n)$$

involving all external particles.

(6) **Sum over all diagrams**. If there exist many diagrams with the same external lines you add them up.

What remains after you eliminate the overall δ function is the invariant amplitude $\mathcal{M}$. Note that:

- Usually 2. and 4. above cancel and 5. may not appear. Just be careful.
- If they are loops in the diagram, consider each loop as a separate diagram and apply to it the same rules as above. Then insert the obtained result in the expression for the full diagram.

7.2 Further elementary considerations

We will consider tow possibilities:

- The decay rate of a particle A into a number of particles, in practice two or three:

$$A \to B + C,\ A \to B + C + D$$

- The cross section of two particles A and B leading to two or more particles:

$$A + B \to C + D$$

We want to know the number of particles $d\mathcal{N}$ scattered in a given direction per unit time. This is proportional to the number of scattering centers N and the flax Φ of the oncoming particles, i.e. the number of particles per unit area per unit time, in a given direction:

$$d\mathcal{N} = N\Phi d\sigma. \tag{7.6}$$

The proportionality constant is called **cross section**. It is one of the most important quantities in physics and has dimensions of length^2. The flux of the incident particles is usually written as $\Phi = n_i \upsilon_i$, where n_i is the particle density, number of particles per unit volume, and υ_i is the particle velocity. In most cases the particle density is taken to be one per unit volume, $1/V$. In other words the cross section per scatterer is given by:

$$d\sigma = \frac{1}{\Phi} d\mathcal{N} = \frac{1}{\upsilon_i} V d\mathcal{N} \tag{7.7}$$

If the target particles are not stationary, $\upsilon_i \to |\vec{\upsilon}_1 - \vec{\upsilon}_2|$

Furthermore if a process is allowed, its inverse is also allowed

$$A + B \to C + D \Rightarrow C + D \to A + B \text{ (principle of detailed balance)}$$

Charge conjugation symmetry allows the existence of particle $\bar{B}$, if B exists, with opposite additive quantum numbers. Then from a given dynamically allowed process, other dynamically allowed processes follow by removing a particle from one side and adding its conjugate on the other side, e.g.

$$A + B \to C + D \Rightarrow A \to \bar{B} + C + D \Rightarrow A + \bar{B} \to \bar{C} + D \text{ etc} \tag{7.8}$$

In addition, of course, energy (including rest energy) and momentum conservation must be satisfied. If does not, the process is not observed and we say that it is not **kinematically allowed**. Sometimes it is allowed above some energy, which is called threshold energy. Thus the observation of neutron decay $n \to p + e^- + \bar{\nu}_e$ implies the observation of $\nu_e + n \to p + e^-$ (inverse beta decay). On the other hand the process $p \to n + e^+ + \nu_e$ is not kinematically allowed, since the proton is lighter than the sum of the mass of the products implied by its decay.

Thus charge conservation guarantees the absolute stability of the free electron, since it is the lightest charged particle. Similarly baryon number conservation guarantees matter stability, since the proton is the lightest baryon. Baryon number, however, is due to a global symmetry, not a gauge symmetry, and, therefore, it is not absolutely sacred. It may be broken at some level.

7.3 Decay widths and cross sections

Here we will start with some expressions without proof. The reader who is interested in such issues is referred to the literature, see e.g. the standard textbook by Bjorken and Drell [Bjorken and Drell (1964)].

7.3.1 *Some useful expressions*

The rate for an initial meson of mass M and 4-momentum p to decay into n fermions with masses m_i and 4-momenta k_i, $i = 1, 2, \ldots, n$ is given by the expression:

$$dw = \frac{1}{|\boldsymbol{v}_1 - \boldsymbol{v}_2|} \frac{m_1}{E_1} \frac{m_2}{E_2} \cdots \frac{m_n}{E_n} \frac{d^3k_1}{(2\pi)^3} \frac{d^3k_2}{(2\pi)^3} \cdots \frac{d^3k_n}{(2\pi)^3} (2\pi)^4 \delta^4 \left(p - \sum_i k_i \right), \tag{7.9}$$

where the δ function formally expresses the energy momentum conservation.

The cross section for two particles, with masses m_1' and m_2' and four momenta p_1' and p_2', colliding to produce n particles is also given by:

$$d\sigma = \frac{1}{|\boldsymbol{v}_1 - \boldsymbol{v}_2|} |\mathcal{M}|^2 \frac{m_1'}{E_1'} \frac{m_2'}{E_2'} \frac{m_1}{E_1} \frac{m_2}{E_2} \cdots \frac{m_n}{E_n}$$
$$\frac{d^3k_1}{(2\pi)^3} \frac{d^3k_2}{(2\pi)^3} \cdots \frac{d^3k_n}{(2\pi)^3} (2\pi)^4 \delta^4 \left(p_1' + p_2' - \sum_i k_i \right). \tag{7.10}$$

$\mathcal{M}$ is the invariant amplitude, which depends on the assumed theory and the mechanism under consideration, which will obtained in our applications by the techniques previously developed. In the computation $|\mathcal{M}|^2$ we average average over initial spins, sum over final spins and polarizations.

In the above expression $|\boldsymbol{v}_1 - \boldsymbol{v}_2|$ is the relative velocity of the colliding particles. For relativistic particles we make the substitution:

$$\frac{1}{|\boldsymbol{v}_1 - \boldsymbol{v}_2|} \Longrightarrow \frac{m_1 m_2}{\sqrt{(p_1 \cdot p_2)^2 - m_1^2 m_2^2}}.$$

Indeed

$$\frac{m_1 m_2}{\sqrt{(p_1 \cdot p_2)^2 - m_1^2 m_2^2}} = \frac{1}{c\sqrt{\gamma_1^2\gamma_2^2\left(1 - \boldsymbol{\beta}_1.\boldsymbol{\beta}_2\right)^2 - 1}}, \; \gamma_i = \frac{1}{\sqrt{1-\beta_i^2}}, \; \boldsymbol{\beta}_i = \frac{1}{c}\boldsymbol{v}_i$$

Now

$$\gamma_1^2\gamma_2^2\left(1 - \hat{\beta}_1.\hat{\beta}_2\right)^2 - 1 = \gamma_1^2\gamma_2^2 - 1 + \gamma_1^2\gamma_2^2\left(-2\boldsymbol{\beta}_1.\boldsymbol{\beta}_2 + (\boldsymbol{\beta}_1.\boldsymbol{\beta}_2)^2\right) \approx |\boldsymbol{\beta}_1 - \boldsymbol{\beta}_2|^2,$$

since for sufficiently low velocities[1]:

$$\gamma_1^2\gamma_2^2 - 1 \approx \beta_1^2\beta_2^2, \; \gamma_1^2\gamma_2^2\left(-2\boldsymbol{\beta}_1.\boldsymbol{\beta}_2 + (\boldsymbol{\beta}_1.\boldsymbol{\beta}_2)^2\right) \approx -2\boldsymbol{\beta}_1.\boldsymbol{\beta}_2.$$

We see from these formulas that the calculation involves two essential steps:

- The evaluation of the invariant amplitude.
  This depends on the specific process in question as we have seen in the discussion of the standard model.
- The evaluation of the phase space integrals.
  These depend pretty much only on the number of particles. Their evaluation in the case of two particles in the final state is straight forward, especially if the invariant amplitude is independent of the momenta.

7.3.2 *The phase space integrals*

We will try to evaluate the phase space integrals if there exist only two particles in the final state. We will consider two possibilities:

[1] Sometimes the next order in velocities is employed. In other words

$$\gamma_1^2\gamma_2^2\left(1 - \hat{\beta}_1.\hat{\beta}_2\right)^2 - 1 = \gamma_1^2\gamma_2^2 - 1 + \gamma_1^2\gamma_2^2\left(-2\boldsymbol{\beta}_1.\boldsymbol{\beta}_2 + (\boldsymbol{\beta}_1.\boldsymbol{\beta}_2)^2\right)$$

$$\approx \beta_1^2 + \beta_2^2 + \beta_1^2\beta_2^2 - \frac{1}{2}\left(\beta_1^4 + \beta_2^4\right) + (1+\beta_1^2)(1+\beta_2^2)(-2\boldsymbol{\beta}_1.\boldsymbol{\beta}_2) + (\boldsymbol{\beta}_1.\boldsymbol{\beta}_2)^2 \approx$$

$$|\boldsymbol{\beta}_1 - \boldsymbol{\beta}_2|^2 - |\boldsymbol{\beta}_1 \times \boldsymbol{\beta}_2|^2 - (\beta_1 - \beta_2)^2\beta_1\beta_2 - 2(\beta_1^2 + \beta_2)^2\left(\boldsymbol{\beta}_1.\boldsymbol{\beta}_2 - \frac{1}{2}\beta_1\beta_2\right) - \frac{1}{2}\left(\beta_1^4 + \beta_2^4\right)$$

Under certain conditions the last three quartic terms can be neglected. Thus:

$$c\sqrt{\gamma_1^2\gamma_2^2\left(1 - \hat{\beta}_1.\hat{\beta}_2\right)^2 - 1} \approx \sqrt{|\boldsymbol{v}_1 - \boldsymbol{v}_2|^2 - \frac{1}{c^2}|\boldsymbol{v}_1 \times \boldsymbol{v}_2|^2} = v_{\text{Moel}}$$

This velocity is known as Moeller velocity.

i) The decay of a particle of mass M:
It simpler if we work in the initial particle's rest frame. The quantity of interest is:
$$dD^{(2)} = \frac{1}{(2\pi)^2}\frac{m_1}{E_1}\frac{m_2}{E_2}d^3k_1 d^3k 2\delta(M-E_1-E_2)\delta(\mathbf{k}_1+\mathbf{k}_2) \quad (7.11)$$
The integration of one of the momenta, e.g. $\mathbf{k}_2$, is trivial due to the three dimensional δ function. We thus find:
$$dD^{(2)} = \frac{1}{(2\pi)^2}\frac{m_1}{E_1}\frac{m_2}{E_2}d^3k_1\delta(M-E_1-E_2)), \quad E_2(\mathbf{k}_2)\to E_2(-\mathbf{k}_1) \quad (7.12)$$
The energy conserving δ function can be written:
$$\delta(M-\sqrt{k_1^2+m_1^2}-\sqrt{k_1^2+m_2^2}) = \frac{\delta(k_1-\xi)}{\left|\frac{k_1}{\sqrt{k_1^2+m_1^2}}+\frac{k_1}{\sqrt{k_1^2+m_2^2}}\right|_{k_1\to\xi}}$$
That is
$$\frac{m_1}{E_1}\frac{m_2}{E_2}k_1^2\delta(M-\sqrt{k_1^2+m_1^2}-\sqrt{k_1^2+m_2^2}) = \left.\frac{m_1m_2}{M}k_1\right|_{k_1\to\xi} \quad (7.13)$$
where we have used the well known expression:
$$\delta(f(x)) = \sum_i \frac{\delta(x-\xi_i)}{|f'(\xi_i)|}, \text{ with } \xi_i \text{ the roots of } f(x)=0$$
In our example only the positive root of the equation contributes which is
$$\xi = \frac{\sqrt{(-m_1-m_2+M)(m_1-m_2+M)(-m_1+m_2+M)(m_1+m_2+M)}}{2M}$$
Note that the root is valid for $m_1+m_2<M$ as expected from energy conservation. The above expression can equivalently be written as:
$$\xi = \frac{\sqrt{M^4-2(m_1^2+m_2^2)M^2+(m_1^2-m_2)^2}}{2M}$$
We thus find:
$$dD^{(2)} = \frac{1}{8\pi^2}\frac{m_1m_2}{M^2}\sqrt{M^4-2(m_1^2+m_2^2)M^2+(m_1^2-m_2)^2 d\Omega}$$
with $d\Omega$ the solid angle. This formula gets very simplified in some simple cases. The first is:
$$\frac{1}{8\pi^2}m^2\sqrt{1-\frac{4m^2}{M^2}}d\Omega, \ m_1=m_2=m$$
The second occurs, if one of the final particles is much lighter than the other, e.g. $m_1=m$, $m_2=\mu<<m$. Then
$$\frac{1}{8\pi^2}m\,\mu\sqrt{1-\frac{m^2}{M^2}}d\Omega, \ m_1=m, m_2=\mu<<m$$

ii) Two particles interacting to yield two particles.
In this case it is simpler to perform the calculation in the center of momentum. In this case the 4-momenta of the four particles are:
$$p_1^{'} = (E_1^{'}, \mathbf{p}^{'}),\ p_2^{'} = (E_2^{'}, -\mathbf{p}^{'}),\ k_1 = (E_1, \mathbf{k}),\ k_2 = (E_2, -\mathbf{k})$$
We also find:
$$E_1^{'} + E_2^{'} = \sqrt{s^{'}},\ E_1 + E_2 = \sqrt{s}$$
where s and $s^{'}$ are the Maldestam invariants (see section 13.6). Then we proceed as above and notice that energy conservation implies $\sqrt{s^{'}} = \sqrt{s}$. Furthermore after integration with respect to k_2 with the end of the delta function we obtain $k_1^2 dk_1 d\Omega = k^2 dk d\Omega$, but we have the relation $\sqrt{k^2 + m_1^2} + \sqrt{k^2 + m_2^2} = \sqrt{s}$. This relation is the same as the one we found above for the decay with M replaced by $\sqrt{s}$. Integration then over the energy, $\sqrt{s} \rightarrow \sqrt{s^{'}}$, yields:
$$k = \frac{1}{2\sqrt{s^{'}}}\sqrt{(s^{\prime})^2 - 2\left(m_1^2 + m_2^2\right) s^{\prime} + (m_1^2 - m_2)^2}$$
with $\sqrt{s^{'}} = \sqrt{(p^{'})^2 + \left(m_1^{'}\right)^2} + \sqrt{(p^{'})^2 + \left(m_2^{'}\right)^2}$ referring to the original particles, i.e. this quantity is the center of mass energy of the initial system. Thus the analogous equation for the phase space integrals in the case of scattering becomes:
$$dS^{(2)} = \frac{1}{8\pi^2}\frac{m_1 m_2}{s^{'}}\sqrt{(s^{\prime})^2 - 2\left(m_1^2 + m_2^2\right) s^{\prime} + (m_1^2 - m_2)^2 d\Omega}$$
If all final particles have the same mass equal to m we get:
$$dS^{(2)} = \frac{1}{8\pi^2} m^2 \sqrt{1 - \frac{4m^2}{s^{'}}} d\Omega.$$
If the masses of the interacting particles can be neglected $s^{'} = 2E$, with E the energy of each of the two colliding beams.
We will now proceed in evaluating the invariant amplitude in some cases of special interest.

7.4 Rutherford scattering

In this case a spinless particle of mass m and charge z and momentum p is scattered elastically with a particle with mass M and charge Ze. The invariant amplitude is
$$\mathcal{M} = \frac{4\pi\alpha z Z}{q^2}$$

since the scattering is elastic we have $\mathbf{p}_f = \mathbf{p}_i$ and

$$q^2 = q_0^2 - \mathbf{q}^2 \approx -\mathbf{q}^2 = -\left(p^2 + p^2 - 2p^2 cos\theta\right) = -2p^2\cos\theta = -\left(2p\sin\frac{\theta}{2}\right)^2,$$

where θ is scattering angle, i.e. the angle between the oncoming and outgoing light particle. Thus the differential cross section becomes:

$$d\sigma = \frac{m}{p}\frac{(4\pi\alpha zZ)^2}{q^4}\frac{d^3p'}{(2\pi)^3}\frac{d^3P}{(2\pi)^3}(2\pi)^4\delta\left(\mathbf{p} - \mathbf{p}' - \mathbf{P}\right)\delta(E - E').$$

In deriving this formula we assumed that the outgoing heavy particle does not carry a significant amount of energy. The integration is trivial with the use of the momentum conserving delta function. Thus

$$d\sigma = \frac{(4\pi\alpha zZ)^2}{q^4}\frac{1}{(2\pi)^2}(p')^2dp'd\Omega\delta(E - E') = \frac{(4\pi\alpha zZ)^2}{q^4}\frac{1}{(2\pi)^2}d\Omega m^2$$

This is often written as[2]:

$$\frac{d\sigma}{d\Omega} = f\left(\mathbf{q}^2\right),\ f\left(\mathbf{q}^2\right) = (2\alpha zZ)^2\frac{m^2}{\mathbf{q}^4} \tag{7.14}$$

or

$$\frac{d\sigma}{d\Omega} = \frac{(\alpha zZ)^2}{16}\left(\frac{\hbar c}{E}\right)^2\frac{1}{\sin^4\frac{\theta}{2}}$$

There is a problem with this formula, since it leads to an infinite total cross section:

$$\sigma = \int\frac{d\sigma}{d\Omega}d\Omega = \frac{(\alpha zZ)^2}{16}\left(\frac{\hbar c}{E}\right)^2 2\pi\int_0^\pi\frac{\sin\theta}{\sin^4\frac{\theta}{2}}d\theta = \infty.$$

The problem arises by the fact that we treated the particles involved as point particles with an interaction

$$V = (4\pi\alpha zZ)\frac{\hbar c}{r}$$

we know however that the presence of the electrons screens this interaction leading to

$$V_{\text{eff}} = (4\pi\alpha zZ)\frac{\hbar c}{r}e^{-r/a}$$

where a a length of the order of the Bohr orbit. Its Fourier transform is

$$\tilde{V}_{\text{eff}} = (4\pi\alpha zZ)\hbar c\frac{1}{(2\pi)^3}\int e^{\mathbf{k}.\mathbf{r}}\frac{1}{r}e^{-r/a}d^3\mathbf{r} = (4\pi\alpha zZ)\frac{\hbar c}{2\pi^2}\frac{1}{k^2 + 1/a^2}$$

[2]For a different derivation using the Born approximation see [Frauenfelder and Henley (1974)], section 8.3.

$$\tilde{V}_{\text{eff}} = (4\pi\alpha zZ)\frac{\hbar c}{2\pi^2}\frac{\hbar^2}{\mathbf{q}^2 + \hbar^2/a^2}$$

Thus the function $f\left(\mathbf{q}^2\right)$ becomes:

$$f\left(\mathbf{q}^2\right) = (2\alpha zZ)^2\frac{m^2}{\left(\mathbf{q}^2 + \hbar^2/a^2\right)^2} \tag{7.15}$$

Thus

$$\sigma = \frac{(\alpha zZ)^2}{16}\left(\frac{\hbar c}{E}\right)^2 2\pi\int_0^\pi \frac{\sin\theta}{\left(\sin^2\frac{\theta}{2} + \hbar/(2pa)^2\right)^2}d\theta =$$

$$\frac{(\alpha zZ)^2}{16}\left(\frac{\hbar c}{E}\right)^2 2\pi\left(\frac{2pa}{\hbar}\right)^2$$

or

$$\sigma = (\alpha zZ)^2\frac{mc^2}{2}E\pi a^2$$

These results remain essentially unchanged, if the charge of the nucleus is distributed uniformly within a sphere of radius $R << a$.

The number of α particles scattered per unit time is:

$$\frac{dN}{dt} = Id\sigma,$$

where I is the alpha particle flux, the number of particles falling on the target per unit time. The solid angle is $d\Omega = dS/r^2$ where r is the distance between the target and the detector and dS the surface spanned by $d\Omega$ at distance r (see Fig. 7.5a for a sketch of the apparatus). The number of the scatterers in the foil of the detector is nd, where n is the particle density, i.e. the number of particles per unit length, and d the thickness of the foil (the probability for one particle shading another in an non-crystal-material is negligible). Thus the number of particles scattered per unit length is given by:

$$\frac{dN}{dtdS} = Id\sigma\frac{(\alpha zZ)^2}{16r^2}\left(\frac{\hbar c}{mc^2}\right)^2\left(\frac{1}{\upsilon/c}\right)^3\frac{1}{\sin^4\frac{\theta}{2}} \tag{7.16}$$

where m is the mass and υ the velocity of the α particles.

This expression is similar to the celebrated Rutherford-Geiger - Marsden formula obtained classically. The essential feature is that an appreciable number of particles are scattered at angles $\theta \neq 0$, even backwards, contrary to the uniform positive charge fluid inside the atom, in which the electrons swim, predicted by the then prevailing hybrid Thomson model. The relative differential rate is shown in Fig. 7.5b.

Rutherford's idea has been applied in subsequent experiments to determine the possible substructure of other subatomic particles, like the proton etc.

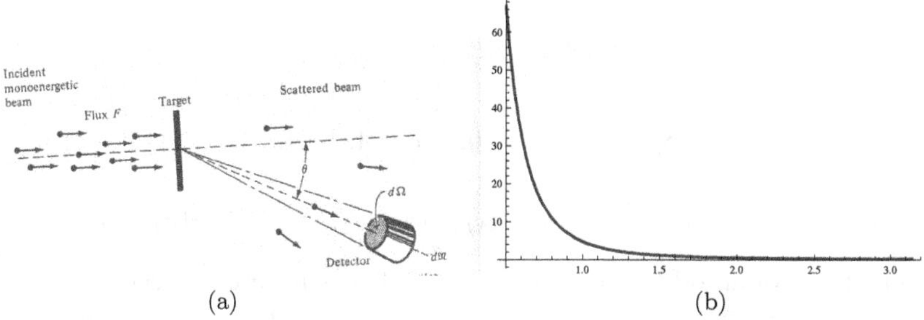

(a) (b)

Fig. 7.5: A diagram of the apparatus employed in the Rutherford scattering (a) and the differential cross-section as a function of the angle θ divided by that at $\theta = \pi/2$ (b).

7.5 Brief review of the trace technique

We begin by reviewing some useful formulae (for more details and conventions see the standard textbook [Bjorken and Drell (1964)]).

$$u = \sqrt{\frac{E+m}{2m}} \begin{pmatrix} \phi \\ \frac{\vec{\sigma}\cdot\vec{p}}{E+m}\phi \end{pmatrix}, \bar{u}u = 1$$

$$v = \sqrt{\frac{E+m}{2m}} \begin{pmatrix} \frac{\vec{\sigma}\cdot\vec{p}}{E+m}\chi \\ \chi \end{pmatrix}, \bar{v}v = -1,$$

$$\sum_s u(ps)\bar{u}(ps) = \frac{\not{p}+m}{2m}, \quad \not{p} = p^\mu \gamma_\mu,$$

$$\sum_s v(ps)\bar{v}(ps) = \frac{\not{p}-m}{2m},$$

$$\sum_s u_\alpha(ps)\bar{u}_\beta(ps) = \left(\frac{\not{p}+m}{2m}\right)_{\alpha\beta},$$

$$\sum_s v_\alpha(ps)\bar{v}_\beta(ps) = \left(\frac{\not{p}-m}{2m}\right)_{\alpha\beta},$$

$$\not{a}\not{b} = a_\mu b_\nu \gamma^\mu \gamma^\nu,$$

$$\gamma^\lambda \gamma_\lambda = 4,$$

$$\gamma^\lambda\gamma^\mu\gamma_\lambda = -2\gamma^\mu,$$

$$\gamma^\lambda\gamma^\mu\gamma^\nu\gamma_\lambda = 4g^{\mu\nu},$$

$$tr\gamma_\mu = 0,\ tr\gamma_5 = 0,\ trI = 4,\ tr(\not{a}\not{b}) = 4ab,$$

$$tr\left(\not{a}_1\not{a}_2\not{a}_3\not{a}_4\right) = 4\left((a_1\cdot a_2)(a_3\cdot a_4) + (a_1\cdot a_4)(a_2\cdot a_3) - (a_1\cdot a_3)(a_2\cdot a_4)\right).$$

The trace of an odd number of Dirac matrices is zero. Indeed:

$$\gamma_5\gamma_1,\gamma_2\cdots\gamma_{2k+1} = -\gamma_1\gamma_5\cdots\gamma_{2k+1} = -\gamma_1,\gamma_2\cdots\gamma_{2k+1}\gamma_5.$$

But at the same time:

$$tr(\gamma_5\gamma_1,\gamma_2\cdots\gamma_{2k+1}) = tr(\gamma_1,\gamma_2\cdots\gamma_{2k+1}\gamma_5),$$

that is

$$tr(\gamma_5\gamma_1,\gamma_2\cdots\gamma_{2k+1}) = 0.$$

Also

$$tr(\gamma_1,\gamma_2\cdots\gamma_{2k+1}) = tr(\gamma_5^2\gamma_1,\gamma_2\cdots\gamma_{2k+1}) = -tr(\gamma_5\gamma_1,\gamma_2\cdots\gamma_{2k+1}\gamma_5) =$$

$$-tr(\gamma_1,\gamma_2\cdots\gamma_{2k+1}\gamma_5^2) = -tr(\gamma_1,\gamma_2\cdots\gamma_{2k+1}) \Rightarrow tr(\gamma_1,\gamma_2\cdots\gamma_{2k+1}) = 0.$$

Furthermore

$$tr(\gamma_5\not{a}\not{b}) = 0,$$

$$tr(\gamma_5\not{a}\not{b}\not{c}\not{d}) = 4i\epsilon_{\alpha\beta\gamma\delta}a^\alpha b^\beta c^\gamma d^\delta = 4i\epsilon^{\alpha\beta\gamma\delta}a_\alpha b_\beta c_\gamma d_\delta,$$

$$\epsilon^{0123} = 1, \quad \epsilon_{0123} = -1\ .$$

7.6 The decay of a vector boson

7.6.1 *The decay widths of vector bosons*

We have seen that the amplitude for

$$W^- \to \ell^- + \tilde{\nu}_\ell,\ W^+ \to \ell^+ + \nu_\ell,\ \ell = e, \mu, \tau$$

is given by:

$$\frac{g}{2\sqrt{2}}\bar{u}(\nu_e, p_\nu)\gamma_\lambda(1-\gamma_5)u(e, p_e)\epsilon^\lambda(k,\lambda)$$

and thus:

$$|\mathcal{M}|^2 = \frac{g^2}{8}\frac{1}{3}$$

$$\sum_{\text{pol},E_e>0,E_\nu<0} (\bar{v}(p_\nu)\gamma_\nu(1-\gamma_5)u(p_e))\,\epsilon^{*\nu}(k,\lambda)\,(\bar{v}(p_\nu)\gamma_\mu(1-\gamma_5)u(p_e))^*\,\epsilon^\mu(k,\lambda). \tag{7.17}$$

The factor 1/3 comes from the average over initial spins. Note that

$$[\bar{v}(p_\nu)\gamma_\nu(1-\gamma_5)u(p_e)][\bar{v}(p_\nu)\gamma_\mu(1-\gamma_5)u(p_e)]^* =$$

$$[\bar{v}(p_\nu)\gamma_\nu(1-\gamma_5)\bar{u}(p_e)u(p_e)\overline{\gamma_\mu(1-\gamma_5)}v(p_\nu),$$

with

$$\overline{\gamma_\mu(1-\gamma_5)} = \gamma^0[\gamma_\mu(1-\gamma^5)]^+\gamma^0. \tag{7.18}$$

Now

$$\gamma^0[\gamma_\mu(1-\gamma^5)]^+\gamma^0 = \gamma^0(1-\gamma^5)\gamma_\mu^+\gamma^0 = \gamma^0\gamma_\mu^+\gamma^0(1-\gamma^5),$$

but

$$\gamma_\mu^+ = \begin{cases} \gamma_0, & \mu=0 \\ -\gamma_\mu, & \mu\neq 0, \end{cases}$$

that is

$$\gamma^0\gamma_\mu^+ = \begin{cases} \gamma^0\gamma^0, & \mu=0 \\ -\gamma_\mu\gamma^0 = \gamma^0\gamma_\mu, & \mu\neq 0 \end{cases}$$

Now we can perform the sum:

$$\sum_{\text{pol},E_e>0,E_\nu<0} \bar{u}(p_e)\gamma_\nu(1-\gamma_5)v(p_\nu)\bar{v}(p_\nu)\gamma_\mu(1-\gamma_5)v(p_e)\epsilon^\nu(k,\lambda)[\epsilon^\mu(k,\lambda)]^*.$$

The sum over the helicities can easily performed:

$$\sum_\lambda \epsilon_\mu(k,\lambda)\epsilon_\nu^*(k,\lambda) = \begin{cases} -g_{\mu\nu} \text{ (photon)} \\ -g_{\mu\nu} + \frac{k_\mu k_\nu}{m^2}, \text{ (massive boson)} \end{cases}.$$

So for boson in the rest frame we get $-g_{\mu\nu}$.

Now before summing over the fermion spins we must a way to deal with to deal with the constraint over the signs of the energy. Recall that the spinors obey the Dirac equation

$$(\not{p}-m)u=0, \quad (\not{p}+m)v=0$$

and the projection operators:

$$\frac{\not p + m}{2m}, \quad \frac{\not p - m}{2m},$$

$$\frac{\not p_e + m_e}{2m_e} u = \begin{cases} u(p_e), E > 0 \\ 0, E < 0 \end{cases},$$

$$\frac{\not p_\nu - m_\nu}{2m_\nu} v = \begin{cases} 0, E > 0 \\ v, E < 0 \end{cases}.$$

Thus the previous sum over the spin polarizations is simplified to read:

$$\begin{aligned}
S &= \sum_{spin} \left[\bar{u}(p_e)\gamma_\nu(1-\gamma_5)\frac{\not p_\nu - m_\nu}{2m_\nu} u(p_\nu)\bar{u}(p_\nu)\gamma_\mu(1-\gamma_5)\frac{\not p_e + m_e}{2m_e} v(p_e) \right] \\
&= tr \left[\gamma_\nu(1-\gamma_5)\frac{\not p_\nu - m_\nu}{2m_\nu}\gamma_\mu(1-\gamma_5)\frac{\not p_e + m_e}{2m_e} \right] \\
&= tr \left[\gamma_\nu \frac{\not p_\nu - m_\nu}{2m_\nu}\gamma_\mu(1-\gamma_5)\frac{\not p_e + m_e}{2m_e}(1+\gamma_5) \right] \\
&= 2tr \left[\gamma_\nu \frac{\not p_\nu + m_\nu}{2m_\nu}\frac{\not p_e + m_e}{2m_e}(1+\gamma_5) \right] \\
&= 2tr \left[\gamma_\nu \frac{\not p_\nu}{m_\nu}\gamma_\mu \frac{\not p_e}{m_e}(1+\gamma_5) \right].
\end{aligned}$$

$$S = 2\frac{p_\nu^\alpha p_e^\beta}{m_e m_\nu} tr\left[\gamma_\nu \gamma_\alpha \gamma_\mu \gamma_\beta (1+\gamma_5)\right], \tag{7.19}$$

where we ignored m_ν in front of $\not p_\nu$ and noting the fact that the trace of on odd number of Dirac matrices is zero.

Now we use the properties of the Dirac matrices to get some trace relations: Thus we have:

$$\begin{aligned}
S &= 2p_\nu^\alpha p_e^\beta \left[(g_{\nu\alpha}g_{\mu\beta} + g_{\mu\alpha}g_{\nu\beta} - g_{\mu\nu}g_{\alpha\beta}) + i\epsilon_{\nu\alpha\mu\beta} \right] \\
&= \frac{2}{m_e m_\nu} \left[(p_e)_\nu (p_\nu)_\mu + (p_e)_\mu (p_\nu)_\nu - g_{\lambda\nu}(p_e \cdot p_\nu) + i\epsilon_{\nu\alpha\mu\beta}(p_e)^\alpha (p_\nu)^\beta \right] \\
&= \frac{g^2}{8}\frac{2(-1)g^{\mu\nu}}{m_e m_\nu} \left[(p_e)_\nu (p_\nu)_\mu + (p_e)_\mu (p_\nu)_\nu - g_{\lambda\nu}(p_e \cdot p_\nu) + i\epsilon_{\nu\alpha\mu\beta}(p_e)^\alpha (p_\nu)^\beta \right] \\
&= \frac{g^2}{8}\frac{2}{m_e m_\nu} \left[4(p_e \cdot p_\nu) - 2(p_e \cdot p_\nu) \right] \\
&= \frac{g^2}{8}\frac{2}{m_e m_\nu} 2(p_e \cdot p_\nu),
\end{aligned}$$

$$
\begin{aligned}
|\mathcal{M}|^2 = & \frac{1}{2M_W}\frac{1}{3}\frac{g^2}{8}\frac{m_e}{E_e}\frac{m_\nu}{E_\nu}\frac{4}{m_e m_\nu}(p_e \cdot p_\nu) \\
& \frac{d^3p_e}{(2\pi)^3}\frac{d^3p_\nu}{(2\pi)^3}(2\pi)^3\delta^3(\vec{p}_e+\vec{p}_\nu)(2\pi)\delta(M_W - E_e - E_\nu).
\end{aligned}
$$

The phase space integral has been evaluated above. It is, however, instructive to be repeated here. The integration over one of the momenta is trivial due to the δ function. This delta function implies that

$$p_e p_\nu = E_\nu E_e - \vec{p}_e\vec{p}_\nu = E_\nu E_e + (\vec{p})^2, \vec{p} = \vec{p}_\nu.$$

Thus we get

$$
\begin{aligned}
dw &= \frac{1}{2M_W}(\frac{g^2}{8})4\frac{d^3p}{(2\pi)^2}\frac{(E_\nu E_e + (\vec{p})^2)}{E_e E_\nu}\delta(M_W - \sqrt{p^2+m_e^2} - p) \\
&\approx \frac{g^2}{8}\frac{1}{2M_W}\frac{4}{3}\frac{1}{2\pi}\frac{2p^2}{p^2}p^2 dp\delta(M_W - 2p),
\end{aligned}
$$

where we have neglected the fermion masses, i.e. we set $E_e E_\nu = \sqrt{p^2+m_e^2}p \approx p^2$. The remaining integration is trivial again due to the energy conserving delta function. Thus finally:

$$
\begin{aligned}
w = \Gamma &= \frac{g^2}{8}\frac{1}{2M_W}\frac{4}{3}\frac{1}{2\pi}2\left(\frac{M_W}{3}\right)^2 \\
&= \frac{g^2}{8}\frac{1}{2M_W}\frac{2}{3}\frac{1}{2\pi}M_W^2 \\
&= \frac{g^2}{16}\frac{1}{3\pi}M_W = \frac{1}{16}\frac{1}{3\pi}4\sqrt{2}M_W^3 G_F = \frac{1}{6\sqrt{2}\pi}G_F M_W^3.
\end{aligned}
$$

That is:

$$\Gamma = 211\text{MeV}.$$

If the W is not at rest we get the modification

$$
\begin{aligned}
& (p_e p_\nu)(2\pi)^3\delta^3(\vec{p} - \vec{p}_e - \vec{p}_\nu)\delta(E_W - E_e - E_\nu), \\
& [E_e E_\nu - \vec{p}_e(\vec{p} - \vec{p}_e)]\delta(E_W - E_e - \sqrt{(p-p_e)^2 + m_e^2}) \\
\backsimeq\; & \frac{E_e(E_W - E_e) + E_e^2 - \vec{p}_e \cdot \vec{p}}{E_e(E_W - E_e)}\delta(E_W - 2E_e) \\
=\; & \frac{E_W^2}{2}.
\end{aligned}
$$

Thus the rate is as before, expect that

$$M_W \longrightarrow \frac{E_W^2}{M_W}.$$

In other words

$$\Gamma = \frac{1}{16}\frac{1}{3\pi}4\sqrt{2}\frac{E_W^2}{M_W}M_W^2 G_F = \frac{1}{6\sqrt{2}\pi}G_F E_W^2 M_W.$$

The above result holds for only one channel. Note, however, that in the decay of W^-:

- There exist three lepton channels:

$$\mu^-\bar{\nu}_\mu, \tau^-\bar{\nu}_\tau, e^-\bar{\nu}_e.$$

- The hadronic channels:

$$d\bar{u}, s\bar{c}, b\bar{t}.$$

  The $b\bar{t}$ channel, however, is not open (it is energetically forbidden). Each of the allowed hadronic channels must be multiplied by 3, since there exist tree colors

The total width, therefore, is:

$$\Gamma_{tot} = (3 + 2 \times 3)\,\Gamma = 9\Gamma = 1.90 \text{ GeV}.$$

The calculation of the Z-decay width proceeds similarly via the neutral current, with the modifications:

- The overall rate is a factor of 1/2 smaller, since:

$$\left(\frac{g}{4\cos\theta_W}m_Z\right)^2 : \left(\frac{g}{2\sqrt{2}}m_W\right)^2 = \frac{1}{2}.$$

- In Eq. (7.19):

$$(1 + \gamma_5) \to \frac{1}{2}\left(g_V^2 + g_A^2 + 2g_V g_A \gamma_5\right).$$

Thus we get:

(1) The decay into neutrinos: $Z \longrightarrow \nu\bar{\nu}$ one flavor. The width is

$$\Gamma = \frac{G_F M_W^3}{12\sqrt{2}\pi}.$$

(2) The decay into charged leptons, e.g.,: $Z \longrightarrow e^+e^-$

$$\Gamma = \frac{G_F M_W^3}{12\sqrt{2}\pi}\frac{1}{2}\left((-1 + 4\sin^2\theta_W)^2 + 1\right).$$

(3) Decay into hadrons:
$Z \longrightarrow q\bar{q}$ with a decay width

$$\Gamma = \frac{G_F M_W^3}{12\sqrt{2}\pi}\frac{1}{2}\begin{cases}\left((1-\frac{8}{3}\sin^2\theta_W)^2+1\right)u\bar{u}, c\bar{c}\\ \left((1-\frac{4}{3}\sin^2\theta_W)^2+1\right)d\bar{d}, s\bar{s}, b\bar{b}\end{cases}$$

($t\bar{t}$ is not allowed)

$$\begin{aligned}\Gamma_{(Z\longrightarrow all)} &= \frac{G_F M_W^3}{12\sqrt{2}\pi}\\ &\left\{3+\frac{1}{2}\left[3\left(((1-4\sin\theta_W)^2)^2+1\right)+2\times 3\left(\left((1-\frac{8}{3}\sin\theta_W)^2\right)^2+1\right)\right.\right.\\ &\left.\left.+3\times 3\left(\left((1-\frac{4}{3}\sin\theta_W)^2\right)^2+1\right)\right]\right\}\\ &= \frac{G_F M_W^3}{12\sqrt{2}\pi}\left(\frac{39}{2}-28\sin^2\theta_W+\frac{88}{3}\sin^4\theta_W\right)=\\ &= 14.6\frac{G_F M_W^3}{12\sqrt{2}\pi}=3.00\text{GeV}.\end{aligned}$$

Experimentally:

$$\Gamma_Z = 2.490 \pm 0.007GeV,$$

which is not bad, since higher order terms were neglected.

7.6.2 *The discovery of vector bosons*

It is perhaps clear that the best mode for the vector bosons production are:

$$q_1\bar{q}_2 = (u\bar{d},\, d\bar{u} \text{ and} \frac{1}{\sqrt{2}}\left(u\bar{u}+d\bar{d}\right) \text{ for } W^+,\, W^- \text{ and } Z \text{ respectively}$$

This quite easy to achieve in $p\bar{p}$ colliders provided that they reach an energy $\sqrt{s} \approx 3m_Z = 270$ GeV (the factor of three comes because the production occurs at the quark level so 1/3 of the available energy can be exploited). Thus the CERN collider with an energy of $2 \times 270 = 540$ GeV was the first to be able to do this.

The best decay channels are

$$W^+ \to \ell^+\nu_\ell,\; W^- \to \ell^-\bar{\nu}_\ell,\; Z \to \ell^+\ell^-$$

even though the branching ratio for these is expected to be small.

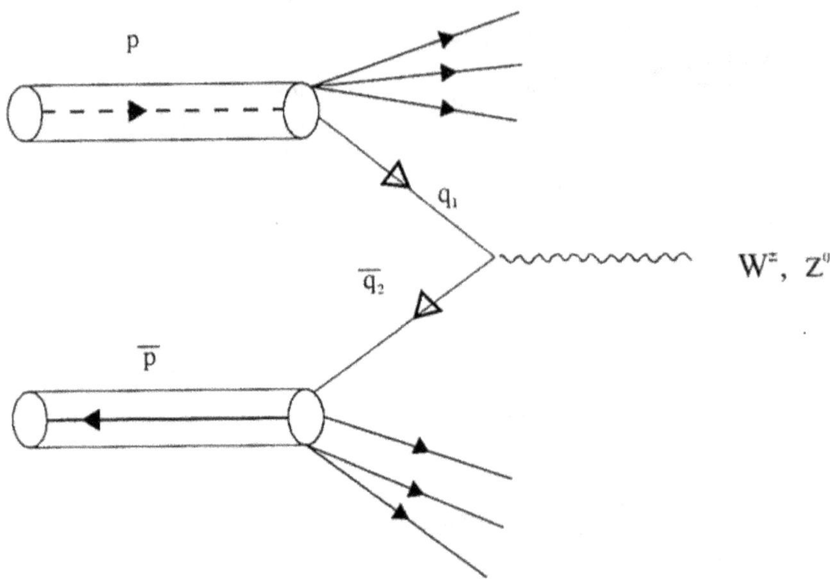

Fig. 7.6: The most likely mode if production of gauge bosons.

The Z boson was first observed in the channel e^-e^+ in the experiments UA1 and UA2 of CERN (see Fig. 7.7). These experiments concluded that

$$M_Z = (91.187 \pm 0.007)\text{GeV/c}^2,$$

with a width $\Gamma_Z = (2.490 \pm 0.007)$ GeV

The leptonic width of the Z boson is much more interesting. Since the neutrinos escape, the total neutrino width cannot be directly measured, but it can be inferred from the calculated total lepton width assuming N_ν neutrino flavors. Indeed measuring all charged decay channels $Z^0 \to \sum_\ell \ell^+\ell^-$, where $\ell = e, \mu, \tau$ one finds (0.0838 ± 0.0003) GeV, which is much smaller than the total leptonic width expected theoretically. In other words this leaves room for $\Gamma\left(Z^0 \to \sum_i \nu_i \tilde{\nu}_c\right) = 0.166$ GeV for neutrinos with energies below $M_Z/2$. We have seen that the contribution of each neutrino is almost double of that of the charged lepton, lrpton i.e. 2(0.084/3)=0.48.From this one deduces, $N_\nu = 0.166/0.048 \approx 3$. Thus:

$$N_\nu = 3.09 \pm 0.013 \text{ for } m_\nu \leq M_Z/2$$

.

The detection of the charged gauge bosons is a bit harder, since the neutrino escapes observation. The detection of the charged bosons involves

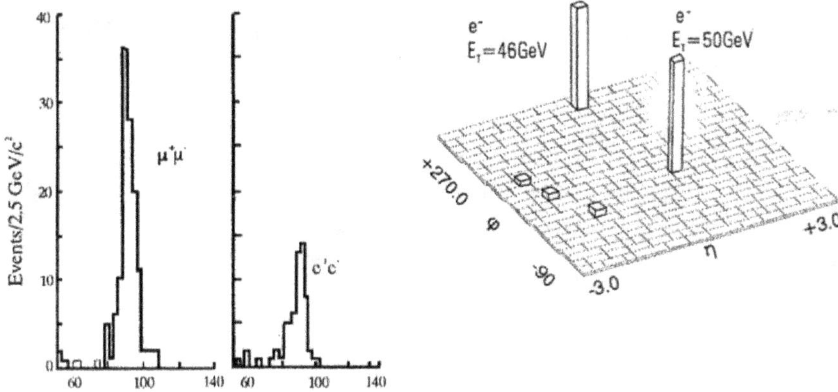

Fig. 7.7: The observation of the Z boson as a resonance in the channels $\mu^+\mu^-$ and e^+e^- as seen in FERMILAB after their initial discovery at CERN.

a very difficult analysis and it will not be discussed here. The detection focuses on the observation in the transverse plane, where leptons are produced preferentially and the background events are fewer. In other words

(1) The charged lepton has large transverse momentum in relation to the beam direction.
(2) There is a large deficiency in the transverse momentum (a large fraction is taken away by the escaping neutrino).

As we mentioned the exact analysis is complicated, but the result of the UA1 experiment as well as those following it is interesting:

$$m_W = 80.41 \pm 0.10 \text{ GeV/c}^2, \quad \Gamma_W = (2.06 \pm 0.06) \text{ GeV} \qquad (7.20)$$

This corresponds to a life time 3.2×10^{-26}s. This time is not too small compared to the large mass of this particle.

7.7 Muon-antimuon production in electron-positron colliders

The amplitude for $e^- + e^+ \to \mu^- + \mu^+$ (see Fig. 7.8a) is proportional to:

$$\mathcal{M} = [\bar{u}(p_3)\gamma_\lambda v(p_4)][\bar{v}(p_2)\gamma^\lambda u(p_1)]\frac{4\pi\alpha}{q^2},$$

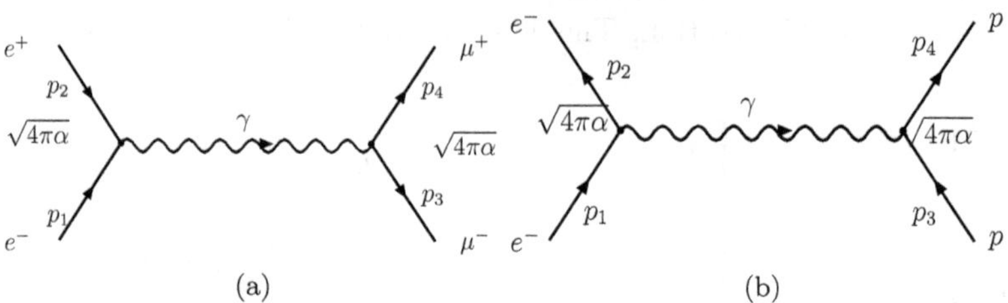

Fig. 7.8: A diagram describing the μ^-, μ^+ production in electron-positron colliders (a) and electron-proton scattering (b).

which yields:

$$\mathcal{M}^2 = \frac{(4\pi\alpha)^2}{q^4}\frac{1}{2}\frac{1}{2}$$
$$\sum_{spin}[\bar{u}(p_3)\gamma_\lambda v(p_4)][\bar{v}(p_2)\gamma^\lambda u(p_1)][\bar{u}(p_3)\gamma_\mu v(p_4)]^*[\bar{v}(p_2)\gamma^\mu u(p_1)]^*$$

In the above sum the constraints on the energy are

$$E_1 > 0, E_3 > 0, E_2 < 0, E_4 < 0.$$

Let

$$\begin{aligned}
\Lambda_{\lambda\mu}(p_3, p_4) &= \sum_{spin}[\bar{u}(p_3)\gamma_\lambda v(p_4)][\bar{u}(p_3)\gamma_\mu v(p_4)]^* \\
&= \sum_{spin} \bar{u}(p_3)\gamma_\lambda v(p_4) v^+(p_4)(\gamma_\mu)^+\gamma_0 u(p_3) \\
&= \sum_{spin} \bar{u}(p_3)\gamma_\lambda v(p_4)\bar{v}(p_4)\gamma_0(\gamma_\mu)^+\gamma_0 u(p_3).
\end{aligned}$$

But

$$\gamma_0(\gamma_\mu)^+\gamma_0 = \gamma_\mu,$$

that is

$$\begin{aligned}
\Lambda_{\lambda\mu}(p_3, p_4) &= \sum_{spin} \bar{u}(p_3)\gamma_\lambda v(p_4)\bar{v}(p_4)\gamma_\mu u(p_3) \\
&= \sum_{spin} \bar{u}(p_3)\gamma_\lambda \frac{\not{p}_\mu - m_\mu}{2m_\mu} v(p_4)\bar{v}(p_4)\gamma_\mu \frac{\not{p}_\mu + m_\mu}{2m_\mu} u(p_3)
\end{aligned}$$

$$= \frac{1}{2m_\mu}\frac{1}{2m_\mu} tr[\gamma_\lambda(\not p_4 - m_\mu)\gamma_\mu(\not p_3 + m_\mu)]$$
$$= \frac{1}{m_\mu^2}\left[tr(\gamma_\lambda \not p_4 \gamma_\mu \not p_3) - m_\mu^2 tr(\gamma_\lambda\gamma_\mu)\right]$$
$$= \frac{1}{m_\mu^2}\left[(p_4)_\lambda(p_3)_\mu + (p_3)_\lambda(p_4)_\mu - g_{\lambda\mu}(p_3\cdot p_4) - m_\mu^2 g_{\lambda\mu}\right]$$
$$= \frac{(p_4)_\lambda(p_3)_\mu + (p_3)_\lambda(p_4)_\mu - g_{\lambda\mu}(p_3\cdot p_4)}{m_\mu^2} - g_{\lambda\mu},$$

Similarly we obtain the companion expression $\Lambda^{\lambda\mu}(p_1, p_2)$:

$$\Lambda^{\lambda\mu}(p_1, p_2) = \frac{(p_1)^\lambda(p_2)^\mu + (p_2)^\lambda(p_1)^\mu - g^{\lambda\mu}(p_1\cdot p_2)}{m_e^2} - g^{\lambda\mu}.$$

Thus

$$\mathcal{M}^2 = \frac{(4\pi\alpha)^2}{q^4}\frac{1}{2}\frac{1}{2}\Lambda_{\lambda\mu}\Lambda^{\lambda\mu}.$$

It is a bit tedious, but straightforward, to proceed further. One can show that:

$$\begin{gathered}(p_1p_4)(p_2p_3) + (p_2p_4)(p_1p_3) - (p_3p_4)(p_1p_2)\\ +(p_2p_4)(p_1p_3) + (p_1p_4)(p_2p_3) - (p_3p_4)(p_1p_2)\\ -(p_3p_4)(p_1p_2) - (p_3p_4)(p_1p_2) + 4(p_3p_4)(p_1p_2)\end{gathered}$$

$$= \frac{2(p_1p_4)(p_2p_3) + 2(p_2p_4)(p_1p_3)}{m_e^2 m_\mu^2} - \frac{-2p_3p_4 + 4p_3p_4}{m_\mu^2} + \frac{-2p_1p_2 + 4p_1p_2}{m_e^2} + 4$$
$$= \frac{2(p_1p_4)(p_2p_3) + 2(p_2p_4)(p_1p_3)}{m_e^2 m_\mu^2} + \frac{2p_3p_4}{m_\mu^2} + \frac{2p_1p_2}{m_e^2} + 4.$$

Thus

$$\mathcal{M}^2 = \frac{(4\pi\alpha)^2}{4}\frac{2}{q^4}\left[\frac{(p_1p_4)(p_2p_3) + (p_2p_4)(p_1p_3)}{m_e^2 m_\mu^2} + \frac{p_3p_4}{m_\mu^2} + \frac{p_1p_2}{m_e^2} + 2\right],$$

$$q^2 = (p_1 + p_2)^2 = (E_1 + E_2)_{cm}^2 = s,$$

$$2p_1p_2 + p_1^2 + p_2^2 = s,$$

$$2p_1p_2 = s - 2m_e^2,$$
$$2p_3p_4 = s - 2m_\mu^2$$

$$\mathcal{M}^2 = \frac{(4\pi\alpha)^2}{4}\frac{1}{s^2}$$
$$\left[2\frac{(p_1p_4)(p_2p_3)+(p_2p_4)(p_1p_3)}{m_e^2m_\mu^2}+\frac{s-2m_\mu^2}{m_\mu^2}+\frac{s-2m_e^2}{m_e^2}+4\right]=$$
$$\frac{(4\pi\alpha)^2}{4}\frac{1}{s^2}\left[2\frac{(p_1p_4)(p_2p_3)+(p_2p_4)(p_1p_3)}{m_e^2m_\mu^2}+\frac{2s}{m_\mu^2}+\frac{2s}{m_e^2}\right]=$$
$$\frac{(4\pi\alpha)^2}{4}\frac{2}{s^2}\left[\frac{(p_1p_4)(p_2p_3)+(p_2p_4)(p_1p_3)+s(m_e^2+m_\mu^2)}{m_e^2m_\mu^2}\right]$$

The flux factor is

$$f = \frac{m_1m_2}{\sqrt{(p_1p_2)^2-m_1^2m_2^2}}.$$

We have seen above that:

$$p_1p_2 = \frac{s}{2}-m_e^2.$$

Thus

$$f = \frac{m_e^2}{\sqrt{(\frac{s}{2}-m_e^2)^2-m_e^4}} = \frac{m_e^2}{\sqrt{\frac{s^2}{4}-sm_e^2}} = \frac{2m_e^2}{\sqrt{s^2-4sm_e^2}} \backsimeq \frac{2m_e^2}{s}.$$

Furthermore

$$E_4 = \sqrt{p_4^2+m_\mu^2} = \sqrt{p_3^2+m_\mu^2} = E_3.$$

Thus neglecting the particle masses and noting that $\mathbf{p}_2 = -\mathbf{p}_1$, $\mathbf{p}_3 = -\mathbf{p}_4$ we get

$$(p_1p_4)(p_2p_3) = (E_1E_4-\mathbf{p}_1\cdot\mathbf{p}_4)(E_2E_3-\mathbf{p}_2\cdot\mathbf{p}_3) \approx E^4\left(1+\hat{p}_1\cdot\hat{p}_3\right)^2 = \frac{s^2}{16}(1+\xi)^2$$

$$(p_1p_3)(p_2p_4) = (E_1E_3-\mathbf{p}_1\cdot\mathbf{p}_3)(E_2E_4-\mathbf{p}_2\cdot\mathbf{p}_4) \approx E^4\left(1-\hat{p}_1\cdot\hat{p}_3\right)^2 = \frac{s^2}{16}(1-\xi)^2.$$

This leads to:

$$\mathcal{M}^2 = \frac{(4\pi\alpha)^2}{16}(1+\xi^2)\frac{1}{m_e^2m_\mu^2}$$

Then the differential cross section is given by:

$$d\sigma = f\frac{(4\pi\alpha)^2}{16}(1+\xi^2)\frac{1}{m_e^2 m_\mu^2}\int \frac{m_e}{E_1}\frac{m_e}{E_2}\frac{m_\mu}{E_3}\frac{m_\mu}{E_4}\frac{d^3p_3}{(2\pi)^3}\frac{d^3p_4}{(2\pi)^3}$$
$$(2\pi)^4\delta^3(\vec{p}_3+\vec{p}_4)\delta(2E-E_3-E_4)$$
$$= f\frac{(4\pi\alpha)^2}{16}\frac{1+\xi^2}{(2\pi)^2}\int d^3p_3\frac{1}{E_1^2}\frac{1}{E_3^2}\delta(2E-2E_3)$$
$$= f\frac{(4\pi\alpha)^2}{16}\frac{1+\xi^2}{(2\pi)^2}\int p_3^2dp_3d\Omega\frac{1}{E_1^2}\frac{1}{E_3^2}\frac{1}{2}\delta(E-E_3)$$
$$= f\frac{(4\pi\alpha)^2}{16}\frac{1+\xi^2}{(2\pi)^2}\frac{1}{2}\int p_3E_3dE_3d\Omega\frac{1}{E_1^2}\frac{1}{E_3^2}\delta(E-E_3)$$
$$= f\frac{(4\pi\alpha)^2}{16}\frac{1+\xi^2}{(2\pi)^2}\frac{1}{2}(2\pi)d\xi E\sqrt{E-m_\mu^2}\frac{1}{E_1^2}\frac{1}{E_3^2}$$
$$= f\frac{(4\pi\alpha)^2}{16}\frac{1+\xi^2}{(4\pi)}d\xi\frac{1}{E^3}\sqrt{E-m_\mu^2}$$

or

$$d\sigma = f\frac{(4\pi\alpha)^2}{16}\frac{1}{4\pi}\frac{1}{E^3}\sqrt{E-m_\mu^2}(1+\xi^2)d\xi.$$

Noting that $E^2 = \frac{1}{4}s$ we find:

$$\sigma = f\frac{(4\pi\alpha)^2}{16}\frac{1}{2\pi}\frac{4}{s}\sqrt{1-\frac{4m_\mu^2}{s}}\frac{4}{3}$$
$$= f\frac{(4\pi\alpha)^2}{16}\frac{1}{2\pi}\frac{4}{s}\frac{4}{3}\sqrt{1-\frac{4m_\mu^2}{s}}$$
$$= f\frac{8\alpha^2}{3\pi}\frac{1}{s}\sqrt{1-\frac{4m_\mu^2}{s}}.$$

Neglecting the masses we obtain:

$$\sigma_0 = \frac{2m_e^2}{s}\frac{8}{3\pi}\frac{\alpha^2}{s}.$$

Thus we can write:

$$\frac{\sigma}{\sigma_0} = \frac{1+\frac{b}{3}+8\frac{m_\mu^2+m_e^2}{s}}{1+\frac{1}{3}}\sqrt{1-\frac{4m_\mu^2}{s}},$$

with

$$b = \left(1-\frac{4m_\mu^2}{s}\right)\left(1-\frac{4m_\mu^2}{s}\right).$$

But

$$b \simeq 1-\frac{4m_\mu^2}{s},$$

that is

$$\frac{\sigma}{\sigma_0} = \frac{1 + \frac{1}{3}(1 - \frac{4m_\mu^2}{s}) + \frac{8m_\mu^2}{s}}{\frac{4}{3}} \sqrt{1 - \frac{4m_\mu^2}{s}}$$

$$= \left[\frac{3}{4} + \frac{1}{4}(1 - \frac{4m_\mu^2}{s}) + \frac{6m_\mu^2}{s}\right] \sqrt{1 - \frac{4m_\mu^2}{s}}$$

$$= \left(1 - \frac{1}{4}\frac{4m_\mu^2}{s} + \frac{3}{2}\frac{4m_\mu^2}{s}\right) \sqrt{1 - \frac{4m_\mu^2}{s}}$$

or

$$\frac{\sigma}{\sigma_0} = \left(1 + \frac{5}{4x}\right) \sqrt{1 - \frac{1}{x}}.$$

This function is exhibited in Fig. 7.9. It has a maximum at $x = \frac{s}{4m_\mu} = \frac{5}{2}$, with value 1.16.

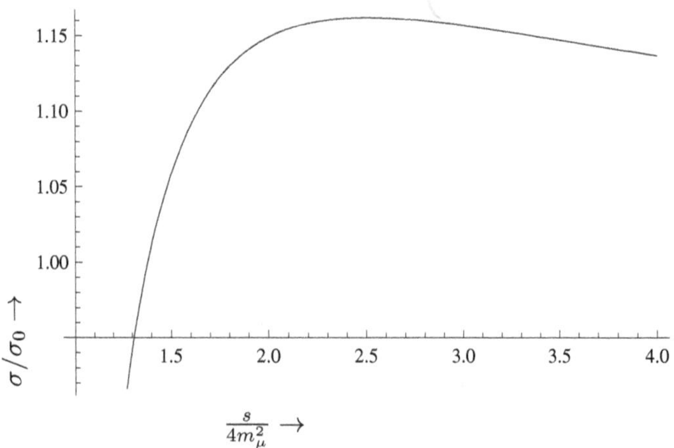

Fig. 7.9: The function $\frac{\sigma}{\sigma_0}$ as a function of $\frac{s}{4m_\mu^2}$.

7.8 Electron proton scattering

The relevant Feynman diagram is exhibited in Fig. 7.8b. We will begin by assuming that the proton is an elementary particle. Then we will examine the case that the proton has a structure described by a form factor and we will finish this section by considering electron proton inelastic scattering.

7.8.1 *The proton as elementary particle*

Then we proceed as in the previous section except that now

$$\mathcal{M}^2 = \frac{(4\pi\alpha)^2}{q^4}\frac{1}{2}\frac{1}{2}$$
$$\sum_{spin}[\bar{u}(p_3)\gamma_\lambda u(p_4)][\bar{u}(p_2)\gamma^\lambda u(p_1)][\bar{u}(p_3)\gamma_\mu u(p_4)]^*[\bar{u}(p_2)\gamma^\mu u(p_1)]^*$$

In the above sum the constraints on the energy are

$$E_1 > 0, E_3 > 0, E_2 > 0, E_4 > 0.$$

Thus

$$\Lambda_{\lambda\mu} = \frac{1}{2m_p}\frac{1}{2m_p}tr[\gamma_\lambda(\not{p}_4 + m_p)\gamma_\mu(\not{p}_3 + m_p)]$$
$$= \frac{(p_4)_\lambda(p_3)_\mu + (p_3)_\lambda(p_4)_\mu - g_{\lambda\mu}(p_3 \cdot p_4)}{m_p^2} + g_{\lambda\mu}, \quad (7.21)$$

$$\Lambda^{\lambda\mu} = \frac{(p_1)^\lambda(p_2)^\mu + (p_2)^\lambda(p_1)^\mu - g^{\lambda\mu}(p_1 \cdot p_2)}{m_e^2} + g^{\lambda\mu}. \quad (7.22)$$

Thus

$$\mathcal{M}^2 = \frac{(4\pi\alpha)^2}{4}\frac{2}{q^4}\left[\frac{(p_1p_4)(p_2p_3) + (p_2p_4)(p_1p_3)}{m_e^2m_p^2} - \frac{p_3p_4}{m_p^2} - \frac{p_1p_2}{m_e^2} + 2\right],$$

with

$$q^2 = (p_2 - p_1)^2 = (p_4 - p_3)^2.$$

To proceed further we will make the following assumptions:

- The initial proton is at rest, i.e. $p_3 = (m_p, 0, 0, 0)$
- The electron is relativistic $p_1 = E(1, \hat{p_1})$, $p_2 = E^{'})(1, \hat{p_2})$. Thus

$$q^2 = (E^{'} - E)^2 - \left(E^{'}\hat{p}_2 - E\hat{p_1}\right)^2 = -2EE^{'}(1 - \cos\theta) = -4EE^{'}\sin^2\frac{\theta}{2}$$

It is convenient to choose q and $p = p_3$ as convenient kinematical variables. Thus

$$(p_1p_4)(p_2p_3) = (p_1q)(p_2p_3) + (p_1p)(p_2p_3),$$

$$(p_1p_3)(p_2p_4) = (p_2q)(p_1p_3) + (p_1p_3)(p_2p_3)$$

$$(p_1q)(p_2p_3) = \left[E(E^{'} - E) - E\hat{p_1} \cdot (E^{'}\hat{p}_2 - E\hat{p_1})\right] m_pE^{'}$$

$$= \left[EE' - EE'\cos\theta\right] m_p E' = 2m_p E(E')^2 \sin^2\frac{\theta}{2}$$

$$(p_1 q)(p_2 p_3) = \left[E(E' - E) - E\hat{p}_1 \cdot (E'\hat{p}_2 - E\hat{p}_1)\right] m_p E = 2m_p E^2 E' \cos^2\frac{\theta}{2}$$

$$(p_1 p_3)(p_2 p_3) = m_p^2 EE',\ p_3 p_4 = p_3 q + p_3^2 = m_p(E' - E) + m_p^2,\ p_1 p_2$$

$$= 2EE'\sin^2\frac{\theta}{2}$$

We thus get:

$$\mathcal{M}^2 = \frac{(4\pi\alpha)^2}{4}\frac{2}{\left(4EE'\sin^2\frac{\theta}{2}\right)^2}\frac{2EE'}{m_e^2}\left[\left(\frac{E'}{m_p} - 1\right)\sin^2\frac{\theta}{2} + \frac{E}{m_p}\cos^2\frac{\theta}{2}\right],$$

where we have neglected in the square bracket a small term proportional to m_e^2. We also find $f = m_e/E$. Thus the differential cross section in a given solid angle $d\Omega$, after performing the phase space integrals, becomes:

$$d\sigma = f\frac{(4\pi\alpha)^2}{4}\frac{2}{\left(4EE'\sin^2\frac{\theta}{2}\right)^2}\frac{2EE'}{m_e^2}\left[\left(\frac{E'}{m_p} - 1\right)\sin^2\frac{\theta}{2} + \frac{E}{m_p}\cos^2\frac{\theta}{2}\right]$$

$$\frac{1}{(2\pi)^2}\frac{m_e^2 m_p E'\, d\Omega}{E^2(\cos\theta - E'/E)}$$

The energy E' is not independent of the other variables, but it is determined from energy conservation as the solution of the equation:

$$m_p + E - E' - \sqrt{m_p^2 + E^2 + (E')^2 - 2EE'\cos\theta} = 0$$

which yield

$$E' = \frac{E}{1 + 2(E/m_p)\sin^2\frac{\theta}{2}}. \tag{7.23}$$

Thus after setting the flux factor $f = 1/\sqrt{(E/m_e)^2 - 1} \approx \frac{m_e}{E}$ we obtain:

$$\frac{d\sigma}{d\Omega} = \frac{(4\pi\alpha)^2}{16}\frac{1}{(2\pi)^2}\frac{m_e}{E}\left[\left(\frac{E'}{m_p} - 1\right)\sin^2\frac{\theta}{2} + \frac{E}{m_p}\cos^2\frac{\theta}{2}\right]$$
$$\frac{m_p}{E^3}\frac{1}{\sin^2\frac{\theta}{2}}\frac{2\left(1 - (E/mp)\cos\theta\right)}{1 + 2(E/m_p)\sin^2\frac{\theta}{2}} \tag{7.24}$$

Using Eq. (7.23) the above equation can be cast in the form:

$$\frac{d\sigma}{d\Omega} = \frac{(4\pi\alpha)^2}{16}\frac{1}{(2\pi)^2}\frac{m_e}{E}\left[\left(\frac{\frac{E}{m_p}\cos(\theta)-1}{\frac{E}{m_p}(-\cos(\theta))+\frac{E}{m_p}+1}\right)\sin^2\frac{\theta}{2}+\frac{E}{m_p}\cos^2\frac{\theta}{2}\right]$$

$$\frac{m_p}{E^3}\frac{1}{\sin^2\frac{\theta}{2}}\frac{2\left(1-(E/m_p)\cos\theta\right)}{1+2(E/m_p)\sin^2\frac{\theta}{2}} \tag{7.25}$$

The needed phase space integrals are of the form:

$$I = \int \frac{m_e}{E}\frac{m_e}{E_2}\frac{m_p}{m_p}\frac{m_p}{E_4}\frac{d^3p_2}{(2\pi)^3}\frac{d^3p_4}{(2\pi)^3}(2\pi)^4\delta^3(\mathbf{p}_1-\mathbf{p}_2-\mathbf{p}_4)\delta(m_p+E-E_2-E_4)$$

$$= \frac{1}{(2\pi)^2}\frac{m_e}{E}\frac{m_e}{E_2}\frac{m_p}{m_p+E-E_2}E_2p_2dE_2d\Omega$$

$$\delta\left(m_p+E-E_2-\sqrt{m_p^2+(\mathbf{p}_1-\mathbf{p}_2)^2}\right)$$

Using now the relation $m_p + E - E_2 - \sqrt{m_p^2+(\mathbf{p}_1-\mathbf{p}_2)^2} \approx m_p^2 + \sqrt{E^2+E_2^2-2EE_2\cos\theta}$ together with familiar property of the δ function

$$\delta(g(x)) = \frac{\delta(x-x_0)}{g^{'}(x_0)}$$

with x_0 being the root of the equation $g(x)=0$ we obtain:

$$I = \frac{1}{(2\pi)^2}\frac{m_e}{E}\frac{m_e}{E_2}\frac{m_p}{m_p+E-E_2}E_2\sqrt{E_2^2-m_e^2}dE_2d\Omega\delta(E_2-E^{'})\frac{m_p+E-E^{'}}{E^{'}-E\cos\theta}$$

$$= \frac{1}{(2\pi)^2}\frac{m_e}{E}\frac{m_e}{E^{'}}E^{'}\sqrt{(E^{'})^2-m_e^2}d\Omega\frac{m_p}{E^{'}-E\cos\theta}$$

$$\approx \frac{1}{(2\pi)^2}\frac{m_e}{E}\frac{m_e}{E}d\Omega\frac{m_pE^{'}}{(\cos\theta-E^{'}/E)}$$

with the value $E^{'}$ found above. Since these replacements have already been incorporated in the invariant amplitude, we obtain the result given by Eq. (7.24). The obtained results are exhibited in Fig. 7.10.

We should stress that this equation was obtained with the assumption that the proton is an elementary particle. Any deviation from this result will indicate that the proton is a composite particle. Then the scattering can be either elastic or inelastic. In the last case the proton can break up into various fragments.

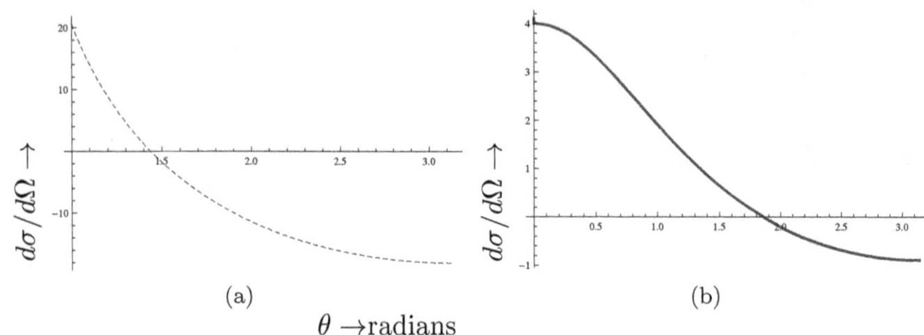

Fig. 7.10: The differential cross section $\frac{d\sigma}{d\Omega}$ for electron-proton collisions. in arbitrary units, as a function of the scattering angle θ assuming a point like proton. The results corresponding to a fixed proton target at electron energies of 0.5 GeV (a) and 1.0 GeV (b).

7.8.2 *The proton form factors*

The experimental data suggested that the proton is not an elementary particle, but it has a structure. This can be described in terms of two form factors, two functions of the momentum transfer. In other words the invariant amplitude takes the form:

$$\mathcal{M}^2 = \frac{(4\pi\alpha)^2}{4} \frac{2}{\left(4EE' \sin^2 \frac{\theta}{2}\right)^2} F_p^2(q), \quad F_p^2(q) = \left[2K_1(q^2) \sin^2 \frac{\theta}{2} + K_2(q) cos^2 \frac{\theta}{2}\right] \tag{7.26}$$

Using this form one can compute the differential cross section for elastic scattering as above. By suitably choosing the scattering angle one can separate these two functions. It turns out that these functions are adequately described by a dipole shape form factor,

$$F_p^2(q) = \left(\frac{1}{1 - 1.25(q^2/m_p^2)}\right)^2, \ q^2 < 0 \tag{7.27}$$

The result is exhibited in Fig. 7.11.

The differential cross section becomes:

$$\frac{d\sigma}{d\Omega} = \frac{(4\pi\alpha)^2}{16} \frac{1}{(2\pi)^2} \frac{m_e}{E} \frac{m_p}{E^3} \frac{1}{\sin^4 \frac{\theta}{2}} \left[2K_1 \sin^2 \frac{\theta}{2} + K_2 cos^2 \frac{\theta}{2}\right] \tag{7.28}$$

7.8.3 *Inelastic electron-proton scattering*

Since the proton has a momentum dependent form factor, it has internal structure. Therefore during high energy collisions it can break up or it

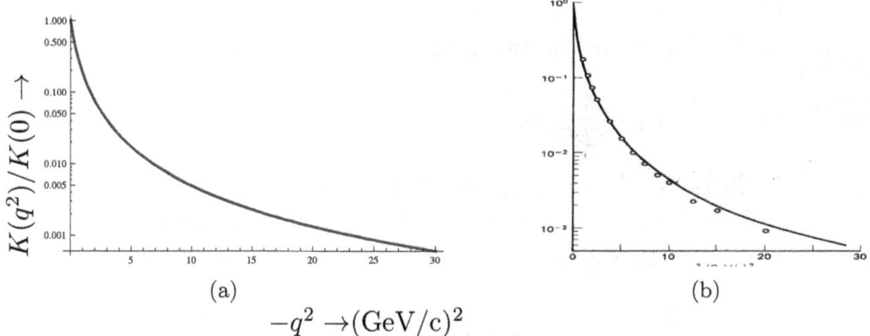

Fig. 7.11: The dipole shape form factor $K(q^2)/K(0)$ of the proton (a) and a fit to the experimental data, based on data by of P. N. A Kirk *et al.* Phys. Rev. **D8** (1973)60, (b).

can be transformed to excited baryon states. Thus one can have inelastic scattering, e.g.

$$e^- + p \rightarrow e^- + X$$

where X stands for a number of particles with momenta $p_4, p_5, \cdots, p_n$. This is exhibited in Fig. 7.12b.

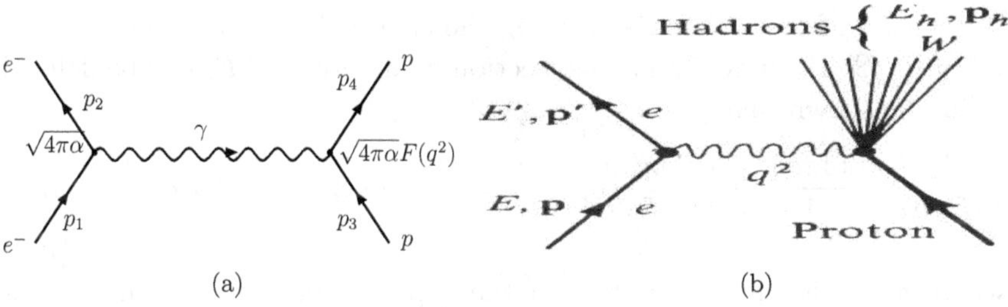

Fig. 7.12: The elastic scattering of a proton with a substructure described by dipole shape form factor $F(q^2)/$ (a) and the inelastic electron-proton scattering (b).

We will not discuss this possibility here. The interested reader is referred to standard textbooks, e.g. [Griffiths (1987)], sec. 8.4. We will briefly summarize the ingredients here.

The function $\Lambda_{\lambda\mu}$ is no longer given by Eq. (7.22). We do not know it, but we anticipate its form, i.e. we know that it is a symmetric second rank tensor constructed out of the two momenta $p = p_3$ and $q = p_4 - p_3$:

$$\Lambda_{\lambda\mu} = W_1 g_{\lambda\mu} + \frac{W_2}{M^2} p_\lambda p_\mu + \frac{W_4}{M^2} q_\lambda q_\mu + \frac{W_5}{M^2}\left(p_\lambda q_\mu + q_\lambda p_\mu\right),$$

where the functions W_i are functions of the two invariant quantities p^2 and pq and have dimensions of mass squared and M coincides with the proton mass. We demand that $q^\mu \Lambda_{\lambda\mu} = 0$. This leads to the conditions:

$$\left(W_1 + \frac{W_4}{M^2} q^2 + \frac{W_5}{M^2} pq\right) q_\lambda + \left(\frac{W_2}{M^2} pq + \frac{W_5}{M^2} q^2\right) p_\lambda = 0,$$

which must hold for Lorentz indices. We thus get the relations:

$$W_1 + \frac{W_4}{M^2} q^2 + \frac{W_5}{M^2} pq = 0, \; \frac{W_2}{M^2} pq + \frac{W_5}{M^2} q^2 = 0$$

In other words:

$$W_5 = -W_2 \frac{pq}{q^2}, \; W_4 = -W1 \frac{M^2}{q^2} - W_5 \frac{pq}{q^2} = -W1 \frac{M^2}{q^2} + W_2 \left(\frac{pq}{q^2}\right)^2$$

and

$$\Lambda_{\lambda\mu} = W_1 \left(g_{\lambda\mu} - \frac{q_\lambda q_\mu}{M^2}\right) + W_2 \left[\left(p_\lambda - \frac{pq}{q^2} q_\lambda\right)\left(p_\mu - \frac{pq}{q^2} q\mu\right)\right]$$

In the case of the inelastic scattering the energy E_2 is no longer specified, $E_2 \neq E'$. So the differential cross section in an interval dE_2 can be defined. It can be shown that

$$\frac{d\sigma}{dE_2 d\Omega} = \frac{(4\pi\alpha)^2}{16} \frac{1}{(2\pi)^2} \frac{m_e}{E} \frac{m_p}{E^3} \frac{1}{\sin^4 \frac{\theta}{2}} \frac{1}{m_p^2} \left[2W_1 \sin^2 \frac{\theta}{2} + W_2 cos^2 \frac{\theta}{2}\right] g(E_2), \tag{7.29}$$

where $g(E_2)$ is the distribution of the expected energy E_2. In the case $g(E_2) = \delta(E_2 - E')$ the last equation is reduced to Eq. (7.28), taking into account the different normalization of the structure functions.

The functions W_1 and W_2 are functions of q^2 and pq. From an experimental point of view instead of pq the variable x, defined by

$$x = \frac{-q^2}{2qp} \tag{7.30}$$

is preferred.

7.9 Neutrino electron scattering

We will discuss the electron neutrino scattering at low energies, in which case the current-current formulation is appropriate.

In the case of electron neutrinos this process can also proceed via the charged current (see Fig. 7.13) leading to the interaction:

$$\mathcal{L}^c = \frac{G_F}{\sqrt{2}}\bar{u}(p'_\nu)\gamma_\lambda(1-\gamma_5)u(p_e)\bar{u}(p'_e)\gamma_\lambda(1-\gamma_5)u(p_\nu), \tag{7.31}$$

while in the case of any neutrino flavor we have the neutral current interaction via the Z-exchange:

$$\mathcal{L}_Z = \frac{G_F}{2\sqrt{2}}\bar{u}(p'_\nu)\gamma_\lambda(1-\gamma_5)u(p_\nu)\bar{u}(p'_e)\gamma_\lambda\left[-(1-\gamma_5)+4\sin^2\theta_W\right]u(p_e) \tag{7.32}$$

(See Fig. 7.13). We can combine Eqs. (7.31) and (7.32) using the the Fierz transformation:

$$\bar{u}_L(p'_\nu)\gamma_\lambda u_L(p_e)\bar{u}_L(p'_e)\gamma_\lambda u_L(p_\nu) = \bar{u}_L(p'_\nu)\gamma_\lambda u_L(p_\nu)\bar{u}_L(p'_e)\gamma_\lambda u_L(p_e), \tag{7.33}$$

which in the case of the neutrino electron leads to the total:

$$\mathcal{L} = \frac{G_F}{2\sqrt{2}}\bar{u}(p'_\nu)\gamma_\lambda(1-\gamma_5)u(p_\nu)\bar{u}(p'_e)\gamma_\lambda\left[g_V - g_A\gamma_5\right]u(p_e), \tag{7.34}$$

with

$$g_V = 1+4\sin^2\theta_W, \quad g_A = 1. \tag{7.35}$$

We can now proceed in a fashion analogous to that of the previous section in computing the differential cross section $d\sigma/dT$, where T is the final electron kinetic energy, in the case of experimental interest, i.e. if the initial electron is at rest and the initial neutrino has a given energy E_ν, sufficiently low. We will ignore the binding of the electron in the atom and the coulomb interaction. Proceeding as in the previous sections we need to compute:

$$\Lambda_{\lambda\mu}(p_e,p'_e,m_e) = \sum_{spins}\bar{u}(p'_e)\gamma_\lambda(g_V-g_A\gamma_5)u(p_e)\left(\bar{u}(p'_e)\gamma_\mu(g_V-g_A\gamma_5)u(p_e)\right)^*$$

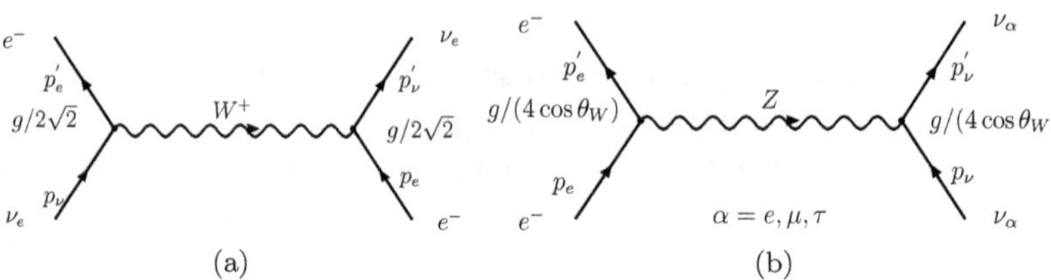

Fig. 7.13: Neutrino- electron scattering. In the case of electron neutrino we have contribution from both W-exchange (a) and Z exchange (b). For the other two flavors only the Z-exchange (b) contributes.

$$= \sum_{spins} \bar{u}(p_e')\gamma_\lambda(g_V - g_A\gamma_5)u(p_e)\bar{u}(p_e)\gamma_\mu(g_V - g_A\gamma_5)u(p_e')$$

$$= \bar{u}(p_e')\gamma_\lambda(g_V - g_A\gamma_5)u(p_e)\frac{\not{p}_e + m_e}{2m_e}\bar{u}(p_e)\gamma_\mu(g_V - g_A\gamma_5)\frac{(\not{p}_e)^{'} + m_e}{2m_e}u(p_e')$$

$$= \bar{u}(p_e')(g_V + g_A\gamma_5)\gamma_\lambda u(p_e)\frac{\not{p}_e + m_e}{2m_e}\bar{u}(p_e)\gamma_\mu\frac{(\not{p}_e)^{'}}{2m_e}(g_V + g_A\gamma_5)u(p_e')$$

$$+ \bar{u}(p_e')(g_V + g_A\gamma_5)\gamma_\lambda u(p_e)\frac{\not{p}_e + m_e}{2m_e}\bar{u}(p_e)\gamma_\mu\frac{1}{2}(g_V - g_A\gamma_5)u(p_e')$$

$$= \bar{u}(p_e')\gamma_\lambda u(p_e)\frac{\not{p}_e + m_e}{2m_e}\bar{u}(p_e)\gamma_\mu\frac{(\not{p}_e)^{'}}{2m_e}(g_V^2 + G_A^2 + 2g_Vg_A\gamma_5)u(p_e')$$

$$+ \bar{u}(p_e')\gamma_\lambda u(p_e)\frac{\not{p}_e + m_e}{2m_e}\bar{u}(p_e)\gamma_\mu\frac{1}{2}(g_V^2 - g_A^2)u(p_e').$$

Applying now the trace techniques and noting that the trace of an odd number of Dirac matrices is zero we find:

$$\Lambda_{\lambda\mu}(p_e, p_e', m_e) = \frac{1}{m_e^2}\left[\left((p_e)_\lambda(p_e^{'})_\mu + (p_e)_\mu(p_e^{'})_\lambda - g_{\lambda\mu}p_ep_e^{'}\right)(g_V^2 + g_A^2)\right.$$
$$\left. + g_{\lambda\mu}m_e^2((g_V^2 - g_A^2)) + 2ig_Vg_A\epsilon_{\lambda\alpha\mu\beta}(p_e)^\alpha(p_e^{'})^\beta\right].$$

Proceeding similarly with the neutrino component and noting in this case $g_V = 1$, $g_A = 1$ we get:

$$\Lambda^{\lambda\mu}(p_\nu, p_\nu', m_\nu) = \frac{1}{m_\nu^2}\left[2\left((p^\nu)^\lambda(p_\nu^{'})^\mu + (p^\nu)^\mu(p_\nu^{'})_\lambda - g^{\lambda\mu}p_\nu p_\nu^{'}\right)\right.$$
$$\left. + 2i\epsilon^{\lambda\alpha\mu\beta}(p_e)_\alpha(p_e^{'})_\beta\right].$$

It is straightforward, but a bit tedious to do the contractions. Obviously the symmetric can contract with the symmetric and the antisymmetric with the antisymmetric. For the latter one can show the identity:

$$\epsilon_{\lambda\alpha\mu\beta}\epsilon^{\lambda\rho\mu\sigma}(p_e)^{\alpha}(p_e^{'})^{\beta}(p_\nu)_\rho(p_\nu^{'})_\sigma = -\left((p_e p_\nu)(p_e^{'} p_\nu^{'}) - (p_e p_\nu^{'})(p_e^{'} p_\nu)\right).$$

Thus

$$\Lambda_{\lambda\mu}(p_e, p_e^{'}, m_e)\Lambda^{\lambda\mu}(p_\nu, p_\nu^{'}, m_\nu) = \frac{4}{m_e^2 m_\nu^2}\Lambda^2,$$

$$\Lambda^2 = \left((g_V + g_A)^2(p_\nu p_e)(p_e^{'} p_\nu^{'}) + (g_V - g_A)^2(p_\nu p_e^{'}) + m_e^2(g_V^2 - g_A^2)(p_\nu p_\nu^{'})\right). \tag{7.36}$$

The cross section now takes the form:

$$d\sigma = \int \frac{d^3\mathbf{p_e}^{'}}{(2\pi)^3}\frac{d^3\mathbf{p}_\nu^{'}}{(2\pi)^3}\frac{m_\nu}{E_\nu}\frac{m_\nu}{E_\nu^{'}}\frac{m_e}{E_e}\frac{m_e}{E_e^{'}}\frac{1}{4}\frac{1}{2}\frac{G_F^2}{2}\frac{4}{m_e^2 m_\nu^2}\Lambda^2(2\pi)^4\delta^2(p_e + p_\nu - p_e^{'} - p_\nu^{'}).$$

The factor $1/4$ is a statistical factor coming from the average over the initial spins.

We will find it convenient to work in he laboratory frame in which the initial electron is at rest. Then

$$\begin{aligned}\Lambda^2 &= (g_V + g_A)^2 m_e E_\nu\left((E_e^{'} - m_e)m_e + E_e^{'} E_\nu - \mathbf{p}_e^{'}.\mathbf{p}_\nu\right)\\ &+ (g_V^2 - g_A^2)m_e^2\left(m_e E_\nu - (p_\nu p_e^{'})\right) + (g_V - g_A)^2\left(E_\nu + m_e - E_e^{'}\right)\\ &\left(E_e^{'} E_\nu - \mathbf{p}_e^{'}.\mathbf{p}_\nu\right).\end{aligned} \tag{7.37}$$

The integration over the outgoing neutrino can be done trivially, due to the δ function, to yield:

$$d\sigma = \frac{1}{4(2\pi)^2}\frac{G_F^2}{m_e E_e^{'} E_\nu(E_\nu + m_e - E_e^{'})}$$
$$\int d^3\mathbf{p}_e^{'}\Lambda^2\delta\left(E_\nu + me - E_E^{'} - \sqrt{E_\nu^2 - 2E_\nu p_e^{'}\xi + (p_e^{'})^2}\right),$$

where $\xi = \hat{p}_\nu.\hat{p}_e^{'}$. The integration over the angles can be done via the δ function yielding:

$$d\sigma = \frac{1}{8\pi}G_F^2\Lambda^2\frac{1}{m_e E_\nu^2}dT, \quad T = E_e^{'} - m_e.$$

Substituting: $\mathbf{p}_e^{'}.\mathbf{p}_\nu = T(m_e + E_\nu)$ in the above expression for Λ^2 we find:

$$\Lambda^2 = \left((g_V + g_A)^2 E_\nu^2 + (g_V - g_A)^2(E_\nu^2 - T^2) + m_e(g_V^2 - g_A^2)T\right)m_e^2.$$

Thus our final expression for (ν_e, e) scattering becomes:

$$\frac{d\sigma}{dT} = \frac{1}{8\pi} G_F^2 m_e \left((g_V + g_A)^2 + (g_V - g_A)^2 \left(1 - \left(\frac{T}{E_\nu} \right)^2 \right) \right.$$
$$\left. + \, (g_V^2 - g_A^2) \frac{m_e T}{E_\nu^2} \right) . \tag{7.38}$$

Note some people use factors g_V and g_A which are a factor of two smaller than ours. Fuhtheromore:

- For all other flavors, i.e (ν_α, e) scattering, $\nu_\alpha \neq \nu_e$, we use the above expressions except that $g_A = -1 + 4\sin^2\theta_W \approx 0$.
- for the proton: $g_V = 1 - 4\sin^2\theta_W \approx 0$, $g_A = -1$
- for the neutron: $g_V = -1$, $g_A = 1$.
- for anti-neutrinos: $g_A \rightarrow -g_A$.

With the obvious modifications it can also be applied in reactions involving the charged current interaction.

7.10 Compton scattering

Compton scattering, i.e. electrons produced when γ rays are scattered from electrons in a material, has played a crucial role in the history of physics, since it demonstrated that the energy of photons is quantized. In other words no matter how big the intensity of light is, it cannot eject electrons, if the frequency of radiation is less than some critical frequency which depends on the work function, the energy it takes to ionize the material. This phenomenon was explained by Einstein by proposing that light is composed of photons with energy proportional to the frequency,

$$E = h\nu, h = \text{ Planck's constant} \tag{7.39}$$

This formula had already been proposed by Planck, but he aaumed this was a property of the radiating system, not of light itself.

In our days the process is described by Fig. 7.14. The cross section involved Compton scattering [Compton (1923)] is evaluated by considering the Feynman diagram Fig. 7.14b and has been evaluated long time ago. It is given by the famous Klein-Nishima [Klein and Nishima (1929)] formula[3]:

$$\frac{d\sigma_{K-N}}{d\Omega} = \frac{1}{2} r_0^2 Z^2 \frac{\epsilon_1}{\epsilon} \left(\frac{\epsilon_1}{\epsilon} + \frac{\epsilon}{\epsilon_1} - \sin^2\theta \right) \tag{7.40}$$

[3]With the tools available the reader can calculate the relevant cross section. It can, however, be found in standard textbooks, see, e.g., [Seiden (2005)], section 3.6.2.

with $r_0 \approx \alpha(\hbar/m_e c) = 2.82 \times 10^{-13}$cm the classical electron radius, Z the charge of the atom, ϵ and ϵ_1 the energies of the initial and scattered γ ray respectively and θ the scattering angle (see Fig. 7.14a). These variables are not independent but related by energy conservation

$$\epsilon_1 = \frac{\epsilon}{1 + (\epsilon/(m_e c^2))\,(1 - \cos\theta)} \Leftrightarrow \Delta\lambda = \lambda_0(1 - \cos\theta), \lambda = \frac{h}{\nu}, \lambda_0 = \frac{h}{m_e c}. \tag{7.41}$$

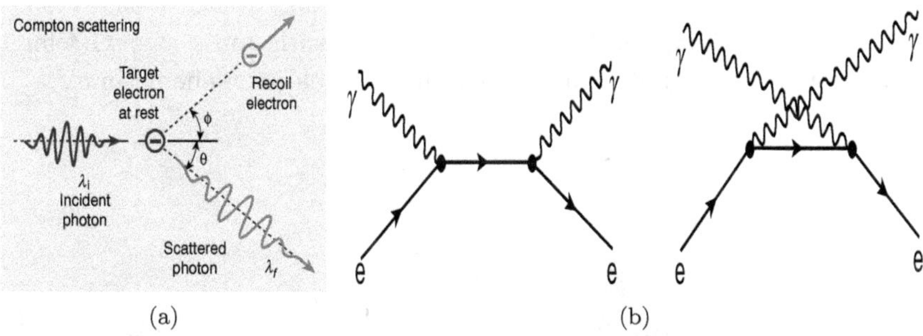

(a) (b)

Fig. 7.14: The Kinematics involved in Compton scattering (a) and the relevant Feynman Diagrams (b). Note the presence of a direct and an exchange diagram.

For our purposes it will be more convenient to work with the electron kinetic energy $T = \epsilon - \epsilon_1$ and the initial γ energy ϵ expressed in units of the electron mass, i.e. $z = T/(m_e c^2)$ and $y = \epsilon/(m_e c^2)$. Then

$$\frac{d\sigma_{K-N}}{dz} = \frac{\pi r_0^2 Z^2}{y(y-z)} \left(\left(-\frac{1}{y-z} + \frac{1}{y} + 1 \right)^2 + \frac{y-z}{y} + \frac{y}{y-z} - 1 \right). \tag{7.42}$$

Note that this cross section is singular both when the energy of the oncoming photon is zero or when the electron takes all the available energy (then the outgoing photon has zero energy) (see Fig. 7.15).

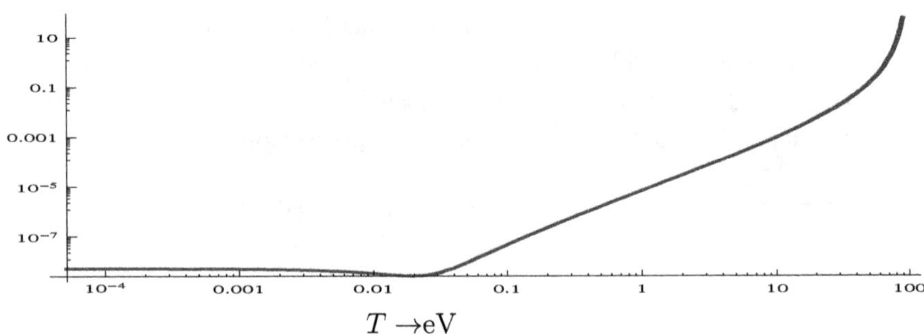

Fig. 7.15: The Compton differential cross srction in units of $\frac{1}{2}r_0^2Z^2$ for a photon of energy of 100 eV as a function of the electron energy in eV.

Chapter 8

Supersymmetry for Pedestrians

8.1 Introduction

As it is well known we can classify the particles in general and the elementary particles in particular into two main groups. Those with integral spin (in units of $\hbar$) called bosons and those with half integral spin, called fermions. In the standard model we cannot relate bosons and fermions, we cannot group together fermions and bosons. Supersymmetry (SUY) attempts to do this, i.e. relate the fermions and bosons via a symmetry. Thus to every member there corresponds a superpartner, i.e. to every particle with spin s one associates a superpartner with spin $|s-1/2|$ and vice versa. Thus for every gauge boson of spin 1 there corresponds a **gaugino** with spin 1/2, to every scalar with spin 0 (Higgs), there corresponds a fermion with spin 1/2 the **Higgsino**. Finally to every fermion with spin 1/2 there corresponds an **s-fermion** usually indicated as $\tilde{f}$ with spin 0, (q, quark, $\Leftrightarrow$ **s-quark**, $\tilde{q}$, lepton $\Leftrightarrow$ **s-lepton**, e (electron) $\Leftrightarrow$ **s-electron** $\tilde{e}$ etc). It should be mentioned that the s-particles are replicas of the particles, i.e. they carry the same internal quantum numbers, they only differ in their space time structure.

Some people find annoying the doubling of the number of particles. We should reply that this is precisely what was done with Dirac's antiparticle theory. Not only it is no longer annoying, but its aesthetic beauty, connected with the new symmetry, was accompanied by a real triumph, when the anti-particles of the then existing particles were discovered. So in our days in almost all particle tables we do not list the antiparticles of the known particles.

We will see that supersymmetry possesses aesthetic beauty, but clearly such a symmetry does not appear to exist in ordinary physics, no superparners have been found. So supersymmetry, if it exists at all, it should be broken at low energies. For the time being, however, we are satisfied with the fact that it provides elegant solutions [Haber and Kane (1985)] to the following:

- It leads to a unification of the three coupling constants of the S-M. Without it the three coupling constants do not intersect into one point.
- It provides a neat solution to the two scale problem. Without it normal higher order loop corrections will drive the low mass to a high value. Supersymmetry achieves this since in every bosonic loop there exists a fermionic loop with opposite sign. It thus allows the sustenance of a low mass Higgs in the presence of a high scale.
- It provides a good dark matter candidate (the universe is dominated by dark matter, i.e. exotic matter which cannot radiate) [Ross (1982)].

Admittedly, in spite of its aesthetic beauty and the above pretty phenomenological ingredients, its adoption is conditional on finding some superpartners at high energies, hopefully at LHC.

8.2 The particle content of MSSM

We will consider the minimal supersymmetric extension of the standard model(MSSM). So the gauge symmetry will still be $G_S = SU_c(3) \times SU_I(2) \times U_Y(1)$.

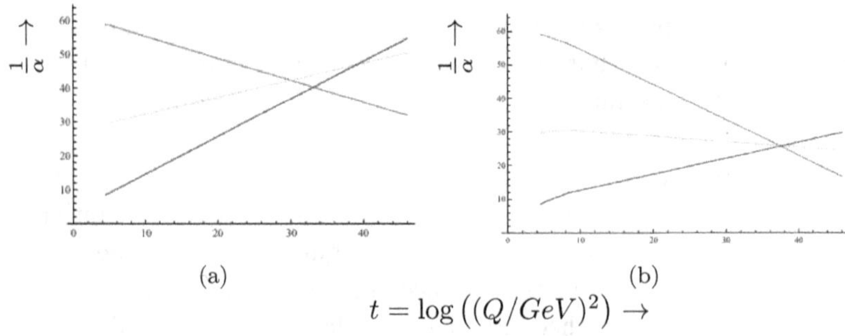

Fig. 8.1: The inverse of the three coupling constants $1/\alpha$ as functions of the energy scale in the case of standard GUTs (a) and SUSY GUTs (b) at the one loop level. In the case of SUSY all three couplings intersect at the scale of $Q \approx 10^{16}$ GeV. At the scale $Q = M_Z$ we have $\alpha_3^{-1} < \alpha_2^{-1} < \alpha_1^{-1}$.

8.2.1 *Gauge particles*

In addition to the gauge bosons we have their supersymmetic partners the gauginos:

$(G_\mu)^\beta_\alpha$ (gluons), W^+_μ, W^-_μ W^3_μ, B_μ (of spin 1) , g (graviton of spin 2) .

particles with spin 1/2 (gauginos):

(gauginos): $\tilde{g}^\beta_\alpha$ (gluinos), $\tilde{W}^+$, $\tilde{W}^-$, $\tilde{W}^3$ (winos), $\tilde{B}$(bino),

a spin 3/2 particle:

$\tilde{g}$ (gravitino).

8.2.2 *Fermions and s-fermions*

Two possibilities:

- Hadrons in addition to the quarks with spin 1/2 we have the s-quarks with spin 0:

  $$\begin{pmatrix} u_{\alpha L} \\ d_{\alpha L} \end{pmatrix}, \begin{pmatrix} c_{\alpha L} \\ s_{\alpha L} \end{pmatrix}, \begin{pmatrix} t_{\alpha L} \\ b_{\alpha L} \end{pmatrix} \Leftrightarrow \begin{pmatrix} \tilde{u}_{\alpha L} \\ \tilde{d}_{\alpha L} \end{pmatrix}, \begin{pmatrix} \tilde{c}_{\alpha L} \\ \tilde{s}_{\alpha L} \end{pmatrix}, \begin{pmatrix} \tilde{t}_{\alpha L} \\ \tilde{b}_{\alpha L} \end{pmatrix},$$

  $u^c_{\alpha L}, d^c_{\alpha L}, c^c_{\alpha L}, s^c_{\alpha L}, t^c_{\alpha L}, b^c_{\alpha L} \Leftrightarrow \tilde{u}^{*c}_{\alpha L}, \tilde{d}^* c_{\alpha L}, \tilde{c}^* c_{\alpha L}, \tilde{s}^{*c}_{\alpha L}, \tilde{t}^{*c}_{\alpha L}, \tilde{b}^{*c}_{\alpha L},$
  where $\alpha = r,\ g,\ b$ is a color index. For convenience in notation we have retained the helicity label for the super-partners even though, as spinless, they have no helicity.
- Leptons. In addition to the standard leptons we have their super-parners (s-leptons) with spin 0.

  $$\begin{pmatrix} \nu_{eL} \\ e^-_L \end{pmatrix}, \begin{pmatrix} \nu_{\mu L} \\ \mu^-_L \end{pmatrix}, \begin{pmatrix} \nu_{\tau L} \\ \tau^-_L \end{pmatrix} \Leftrightarrow \begin{pmatrix} \tilde{\nu}_{eL} \\ \tilde{e}^-_L \end{pmatrix}, \begin{pmatrix} \tilde{\nu}_{\mu L} \\ \tilde{\mu}^-_L \end{pmatrix}, \begin{pmatrix} \tilde{\nu}_{\tau L} \\ \tilde{\tau}^-_L \end{pmatrix},$$

  $$e^{c+}_L,\ \mu^{c+}_L,\ \tau^{c+}_L, \Leftrightarrow \tilde{e}^{*+}_L,\ \tilde{\mu}^{*+}_L,\ \tilde{\tau}^{*+}_L.$$

8.2.3 *The Higgs content*

In supersymmetry we need to introduce two doublets [Castano *et al.* (1994)]:

$$H_u = \begin{pmatrix} H^0_u \\ H^-_u \end{pmatrix}, \quad \text{with } Y = -1 \text{ and } H_d = \begin{pmatrix} H^+_d \\ H^0_d \end{pmatrix} \text{ with } Y = 1$$

(their adjoints, of course, also appear). The reason is the superpotential must be an analytic function of the super-fields. So it should not contain the conjugate fields[1]. To the Goldstone bosons there correspond **Goldstinos**. An artist's of of the particles in the MSSM are exhibited in Fig. 8.2.

[1]Recall that for a function $g(x,y) = g((Z+Z^*)/2,(Z-Z^*)/2i)$ to be an analytic function of Z, the variable Z^* should not appear.

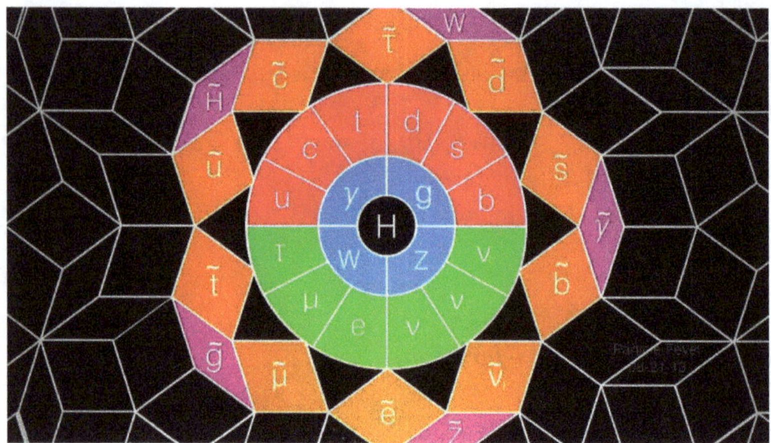

Fig. 8.2: The particle content of the MSSM (Minimal Supersymmetic Standard Model). In this panoramic view amidst the beauty of geometrical symmetry the artist could not find room for the second Higgs and Higgsino.

8.3 The Higgs mechanism

The Higgs potential ignoring loop corrections takes the form:

$$
\begin{aligned}
V &= m_1^2(H_d)^+H_d + m_2^2(H_u)^+H_u + B\mu\,(H_u).(H_d) + HC) \\
&+ \frac{g^2+g'^2}{8}\left((H_u)^+H_u - (H_d)^+H_d\right)^2 + \frac{g^2}{2}\left(\left((H_u)^+H_d\right)\left((H_d)^+H_u\right)\right),
\end{aligned}
\tag{8.1}
$$

with [Shirman (2009)]

$$(H_u).(H_d) = H_u^- H_d^+ - H_u^0 H_d^0,\; (H_u)^+H_d = H_u^{0*}H_d^+ + H_u^+ H_d^0,$$

$$(H_d)^+H_u = H_d^- H_u^0 + H_u^- H_d^{0*}, \quad m_1^2 = m_{H_d}^2 + |\mu|^2,\; m_2^2 = m_{H_u}^2 + |\mu|^2.$$

Note that in Supersymmetry the quartic couplings are the $SU_I(2)$ and $U_Y(1)$ gauge couplings [Castano *et al.* (1994)]. The charged members of the Higgs doublet cannot, of course, acquire a vacuum expectation value and, like in the standard model, we will choose that only the real parts of the neutral components can attain a vacuum expectation value, i.e.:

$$\langle H_u\rangle = \frac{1}{\sqrt{2}}\begin{pmatrix} \upsilon_u \\ 0 \end{pmatrix}, \quad \langle H_d\rangle = \frac{1}{\sqrt{2}}\begin{pmatrix} 0 \\ \upsilon_d \end{pmatrix}.$$

In such a case, for the minimization procedure, it is adequate to consider the potential:

$$V = m_1^2|H_d^0|^2 + m_2^2|H_u^0|^2 - B\mu\left(H_u^0 H_d^0 + HC\right) + \frac{g^2+g'^2}{8}\left(|H_u^0|^2 - |H_d^0|^2\right)^2. \tag{8.2}$$

The extrema of the potential are given by the conditions:

$$\left.\frac{\partial V}{\partial Re(H_u^0)}\right|_{Re(H_u^0)=\upsilon_u} = 0, \qquad \left.\frac{\partial V}{\partial Re(H_d^0)}\right|_{Re(H_d^0)=\upsilon_d} = 0,$$

These correspond to maxima subject to the constraints:

$$\frac{1}{4}\Big\{16\mu^4 + 8\left(-2B^2 + 2m_2^2 + \left(\upsilon_d^2+\upsilon_u^2\right)\left(g^2+(g')^2\right)\right)\mu^2 -$$

$$16B\upsilon_d\upsilon_u\left(g^2+(g')^2\right)\mu - 4m_1^2\left(-4\mu^2 - 4m_2^2 + \left(\upsilon_d^2 - 3\upsilon_u^2\right)\left(g^2+(g')^2\right)\right) -$$

$$\left(g^2+(g')^2\right)\left(4\left(\upsilon_u^2 - 3\upsilon_d^2\right)m_2^2 + 3\left(\upsilon_d^2-\upsilon_u^2\right)^2\left(g^2+(g')^2\right)\right)\Big\} \geq 0,$$

$$m_{H_d}^2 + m_{H_u}^2 + 2\mu^2 \geq |B\mu|. \tag{8.3}$$

In fact at the minimum we find the relations:

$$m_{H_d}^2 = \frac{-\frac{1}{2}(g^2+g'^2)\upsilon_d^3 - 2\mu^2\upsilon_d + \frac{1}{2}(g^2+g'^2)\upsilon_u^2\upsilon_d + 2B\mu\upsilon_u}{2\upsilon_d}, \tag{8.4}$$

$$m_{H_u}^2 = \frac{-\frac{1}{2}(g^2+g'^2)\upsilon_u^3 - 2\mu^2\upsilon_u + \frac{1}{2}(g^2+g'^2)\upsilon_d^2\upsilon_u + 2B\mu\upsilon_d}{2\upsilon_u}. \tag{8.5}$$

Instead of υ_u and υ_d, we use two new variables υ and β such that

$$\upsilon_u = \upsilon\sin\beta,\ \upsilon_d = \upsilon\cos\beta,\ \ tan\beta = \frac{\upsilon_u}{\upsilon_d} \tag{8.6}$$

with υ to be identified with that of the standard model, i.e.:

$$\upsilon = \frac{\sqrt{2}m_z}{\sqrt{g^2+g'^2}}. \tag{8.7}$$

Then Eqs. (8.4) and (8.5) become:

$$m_{H_d}^2 = -\frac{1}{2}m_Z^2\cos(2\beta) - \mu^2 + B\mu\tan\beta,\ \ m_{H_u}^2 = \frac{1}{2}m_Z^2\cos(2\beta) - \mu^2 + B\mu\cot\beta. \tag{8.8}$$

The first of the above constraints for the minimum becomes:

$$B\mu\cos(2\beta)\cot(2\beta) \geq 0\ .$$

Sometimes in the literature the following relations are given [Castano *et al.* (1994)]:

$$\frac{1}{2}m_Z^2 = \frac{m_1^2 - m_2^2 \tan^2\beta}{\tan^2\beta - 1}, \quad B\mu = \frac{1}{2}(m_1^2 + m_2)\sin 2\beta \tag{8.9}$$

$$-2B\mu = \left(m_{H_d}^2 - m_{H_u}^2\right)\tan(2\beta) + m_Z^2 \sin(2\beta), \tag{8.10}$$

$$|\mu|^2 = \frac{m_{H_d}^2 - m_{H_u}^2 \tan^2\beta}{\tan^2\beta - 1} - \frac{1}{2}M_Z^2. \tag{8.11}$$

As we have seen in the discussion of the standard model, once the potential is known and the minimum of the potential has been located, i.e. the non zero vacuum expectation values are known, the mass matrix for the Higgs is given by

$$\mathcal{M}_{H_u,H_d} = \left.\left(\frac{\partial^2 V}{\partial x_i \partial x_j}\right)\right|_{x_1=\upsilon_1,\cdots,x_8=\upsilon_8}$$

$$x_i, x_j = \mathrm{Re}H_d^0, \mathrm{Re}H_u^0, \mathrm{Im}H_d^0, \mathrm{Im}H_u^0, H_u^+, H_d^-, H_u^-, H_d^+. \tag{8.12}$$

In general this is an 8×8 matrix, but the charged particles do not mix with the neutrals. In the particular case of the above potential the 4×4 matrix involving the neutral particles splits into two 2×2 matrices.

(1) The mass matrix of the imaginary parts of the neutrals takes the form:

$$\mathcal{M}_I = \left(\begin{array}{c|cc} & \Im(H_d) & \Im(H_u) \\ \hline \Im(H_d) & \frac{g^2+g'^2}{4}(\upsilon_d^2 - \upsilon_u^2) + m_d^2 + \mu^2) & B\mu \\ \Im(H_u) & B\mu & \frac{g^2+g'^2}{4}(\upsilon_d^2 - \upsilon_u^2) + m_u^2 + \mu^2 \end{array}\right) \tag{8.13}$$

Then using Eqs. (8.6) and (8.7) we obtain

$$\mathcal{M}_I = \begin{pmatrix} B\upsilon_u\mu/\upsilon_v & B\mu \\ B\mu & B\upsilon_d\mu/\upsilon_u \end{pmatrix} = \begin{pmatrix} B\mu\tan\beta & B\mu \\ B\mu & B\mu\cot\beta \end{pmatrix}. \tag{8.14}$$

One of the eigenvalues is zero and corresponds to a would be Goldstone boson G^0 and the other is a physical pseudoscalar particle A with mass $m_A^2 = B\mu/(\sin\beta\cos\beta)$. They are expressed as:

$$\frac{G^0}{\sqrt{2}} = -\cos\beta\,\mathrm{Im}(H_d^0) + sin\beta\mathrm{Im}(H_u^0), \quad \frac{A}{\sqrt{2}} = \sin\beta\mathrm{Im}(H_d^0) + cos\beta\mathrm{Im}(H_u^0). \tag{8.15}$$

(2) The matrix of the real parts of the neutral components is:

$$\mathcal{M}_R = \left(\begin{array}{c|cc} & \Re(H_d) & \Re(H_u) \\ \hline \Re(H_d) & \frac{g^2+g'^2}{4}(3v_d^2 - v_u^2) + m_{H_d}^2 + \mu^2) & \frac{1}{4}(g^2+g'^2)v_d v_u - B\mu \\ \Re(H_u) & \frac{1}{4}(g^2+g'^2)v_d v_u - B\mu & \frac{g^2+g'^2}{(}3v_u^2 - v_d^2) + m_{H_u}^2 + \mu^2 \end{array}\right) \tag{8.16}$$

Then using Eqs. (8.6), (8.7) and (8.8) we obtain

$$\mathcal{M}_R = \begin{pmatrix} m_Z{}^2 \cos^2\beta + B\mu \tan\beta & -\cos\beta \sin\beta m_Z{}^2 - B\mu \\ -\cos\beta \sin\beta m_Z{}^2 - B\mu & sin^2\beta m_Z{}^2 + B\mu \cot\beta \end{pmatrix}, \tag{8.17}$$

with eigenvalues:

$$m_h^2 = \frac{1}{2} m_Z{}^2 + B\mu \csc(2\beta)$$

$$-\sqrt{m_Z{}^4/4 + B\mu \csc(2\beta)\left(B\mu \csc(2\beta) - m_Z{}^2/2 \cos(4\beta)\right)}$$

$$= \frac{1}{2}\left(M_A^2 + m_Z^2\right) - \sqrt{(m_A^2 + m_Z^2)^2 - 4m_A^2 m_Z^2 \cos^2 2\beta},$$

$$m_H^2 = \frac{1}{2} m_Z{}^2 + B\mu \csc(2\beta)$$

$$+\sqrt{m_Z{}^4/4 + B\mu \csc(2\beta)\left(B\mu \csc(2\beta) - m_Z{}^2/2 \cos(4\beta)\right)} =$$

$$\frac{1}{2}\left(M_A^2 + m_Z^2\right) + \sqrt{(m_A^2 + m_Z^2)^2 - 4m_A^2 m_Z^2 \cos^2 2\beta}.$$

The mixing angle α is given by

$$\tan 2\alpha = \frac{(\sin(2\beta) m_Z^2 + 2B\mu)\tan(2\beta)}{\sin(2\beta) m_Z^2 - 2B\mu} = \frac{m_A^2 + m_Z^2}{m_A^2 - m_Z^2} \tan 2\beta, \tag{8.18}$$

while the eigenvectors are:

$$\begin{aligned} \frac{H}{\sqrt{2}} &= \left(\mathrm{Re}(H_d^0) - \frac{v_d}{\sqrt{2}}\right)\cos\alpha + \left(\mathrm{Re}(H_u^0) - \frac{v_u}{\sqrt{2}}\right)\sin\alpha, \\ \frac{h}{\sqrt{2}} &= -\left(\mathrm{Re}(H_d^0) - \frac{v_d}{\sqrt{2}}\right)\sin\alpha + \left(\mathrm{Re}(H_u^0) - \frac{v_u}{\sqrt{2}}\right)\cos\alpha. \end{aligned} \tag{8.19}$$

(3) The matrix of the charged Higgs
We come, finally, to the charged components of the Higgs, noticing that now the $B\mu$ as as well as the last term of Eq. (8.1) make a contribution. We find that the mass matrix due to the interaction term now takes the form:

$$\mathcal{M}_c = \frac{g^2}{2} \left(\begin{array}{c|cc} & H_d^- & H_u^- \\ \hline H_d^+ & H_u^{0*} H_u^0 & H_u^{0*} H_d^{0*} \\ H_u^+ & H_u^u H_d^0 & H_d^{0*} H_d^0 \end{array} \right) \tag{8.20}$$

Replacing the fields by their vacuum expectation values and including the mass terms, the analogous to Eq. (8.8) expression becomes:

$$\begin{aligned} m_{H_d}^2 + \mu^2 &= \frac{1}{2}(gv)^2 \sin^2\beta + B\mu \tan\beta, \\ m_{H_u}^2 + \mu^2 &= \frac{1}{2}(gv)^2 \cos^2\beta + B\mu \cot\beta. \end{aligned} \tag{8.21}$$

Thus the full mass matrix including the $B\mu$ term for the mixing terms we get:

$$\mathcal{M}_c = \begin{pmatrix} \frac{1}{2}g^2v^2\sin^2\beta + B\mu\tan\beta & \frac{1}{2}g^2\cos\beta\sin\beta v^2 + B\mu \\ \frac{1}{2}g^2\cos\beta\sin\beta v^2 + B\mu & \frac{1}{2}g^2v^2\cos^2\beta + B\mu\cot\beta \end{pmatrix}, \tag{8.22}$$

with eigenvalues

$$\left(0, \frac{g^2 v}{2} + B\mu \csc\beta \sec\beta \right) = (0, \frac{g^2}{g^2 + (g')^2} m_Z^2 + M_A^2$$

The first is a would be Goldstone boson $G^\pm$ and the second is a charged physical particle $H^\pm$. One finds:

$$G^\pm = -\cos\beta H_d^\pm + \sin\beta H_u^\pm, \quad H^\pm = \sin\beta H_d^\pm + \cos\beta H_u^\pm. \tag{8.23}$$

8.4 The Fermion masses

Proceeding as in the standard model we get

$$\mathcal{L}_Y = f^{(u)} \left(\bar{u}_L, \bar{d}_L\right) \begin{pmatrix} H_u^0 \\ H_u^- \end{pmatrix} u_R + HC = f^{(u)} \left(\bar{u}_L u_R H_u^0 + \bar{d}_L u_R H_u^-\right) + HC.$$

When the Higgs acquires a vacuum expectation we get:

$$\mathcal{L}_Y \to m^{(u)} \bar{u}u, \quad m^{(u)} = f^{(u)} \frac{v}{\sqrt{2}} \sin\beta = f^{(u)} \frac{m_W}{g} \sqrt{2} \sin\beta.$$

Similarly for the down quarks:

$$\mathcal{L}_Y = f^{(d)} \left(\bar{u}_L, \bar{d}_L\right) \begin{pmatrix} H_d^+ \\ H_d^0 \end{pmatrix} d_R + HC = f^{(d)} \left(\bar{d}_L d_R H_d^0 + \bar{u}_L d_R H_d^+\right) + HC,$$

$$\mathcal{L}_Y \to m^{(d)}\bar{d}d, \quad m^{(d)} = f^{(d)}\frac{\upsilon}{\sqrt{2}}\cos\beta = f^{(d)}\frac{m_W}{g}\sqrt{2}\cos\beta.$$

In the presence of generation mixing the relevant mass matrices become:

$$m^d_{ij} = f^d_{i,j}\cos\beta\sqrt{2}\frac{m_W}{g}\bar{d}_{iL}d_{jR} + HC, \quad i,j = \text{generation indices}, \quad (8.24)$$

where we used $\upsilon_d = \upsilon\cos\beta$ and proceeded as in the SM. Similarly for the upper quarks

$$m^u_{ij} = f^u_{i,j}\sin\beta\sqrt{2}\frac{m_W}{g}\bar{u}_{iL}u_{jR} + HC, \quad i,j = \text{generation indices} \quad (8.25)$$

($\upsilon_u = \upsilon\sin\beta$). Thus, if $f^d_{i,j} \approx f^u_{i,j}$, values of $\tan\beta$ larger than unity are required to make the upper quarks more massive. Proceeding in an analogous fashion we get the Yukawa interactions involving the imaginary neutral Higgs.

$$\mathcal{L}_Y = f^{(u)}\bar{u}_L u_R\left(Re(H^0_u) + iIm(H^0_u)\right) + HC = f^{(u)}Re(H^0_u)\bar{u}u - iIm(H^0_u)\bar{u}\gamma_5 u,$$
$$\mathcal{L}_Y = f^{(d)}\bar{d}_L d_R\left(Re(H^0_d) + iIm(H^0_d)\right) + HC = f^{(d)}Re(H^0_d)\bar{u}u - iIm(d^0_u)\bar{d}\gamma_5 d.$$

We have seen in the previous section that fields $Re(H^0_u)$ and $Re(H^0_d)$ are not eigenstates. They are given in terms of the eigenstates H and h as follows:

$$Re(H^0_d) = \frac{1}{\sqrt{2}}\left(\cos\alpha\, H - \sin\alpha\, h\right), \quad Re(H^0_u) = \frac{1}{\sqrt{2}}\left(\sin\alpha\, H + \cos\alpha\, h\right).$$

Expressing the couplings $f^{(d)}$ and $f^{(u)}$ in terms of the corresponding quark masses we get:

$$\mathcal{L}^{\text{even}}_Y = \frac{g\, m^{(u)}}{2m_W\sin\beta}\left(H\sin\alpha + h\cos\alpha\right)\bar{u}u + \frac{g\, m^{(d)}}{2m_W\cos\beta}\left(H\cos\alpha - h\sin\alpha\right)\bar{d}d.$$

The fields $Im(H^0_u)$ and $Im(H^0_u)$ can similarly be expressed in terms of eigenstates:

$$Im(H^0_d) = \frac{1}{\sqrt{2}}\left(-\cos\beta\, G_0 + \sin\beta\, A\right) \quad Im(H^0_u) = \frac{1}{\sqrt{2}}\left(\sin\beta\, G_0 + \cos\alpha\, A\right),$$

where G_0 is the would be Goldstone boson and A is the CP odd Higgs Field. Then proceeding as above we get:

$$\mathcal{L}^{\text{odd}}_Y = i\left\{\frac{g\, m^{(u)}}{2m_W}\cot\beta\bar{u}\gamma_5\, u + \frac{g\, m^{(d)}}{2m_W}\tan\beta\bar{d}\gamma_5\, d\right\}A.$$

Returning to the charged Higgs sector we get:

$$\mathcal{L}_Y = f^{(u)}\bar{u}_R d_l H^+_u + f^{(d)}\bar{u}_L d_R H^+_d + HC,$$

These can be expressed in terms of the following two eigenstates:

$$G^{\pm} = -\cos\beta\, H^{\pm}_d + \sin\beta\, H^{\pm}_u \quad H\pm = \sin\beta\, H^{\pm}_d + \cos\beta\, H^{\pm}_u,$$

where the $G^{\pm}$ are the would be Goldstone bosons and $H^{\pm}$ the physical Higgs. So the above Yukawa interaction can be expressed in terms of the physical Higgs:

$$\mathcal{L}_Y = \left\{\frac{gm_u}{2\sqrt{2}m_W}\cot\beta\,\bar{u}(1-\gamma_5)d + \frac{gm_d}{2\sqrt{2}m_W}\tan\beta\,\bar{u}(1+\gamma_5)d\right\}H^+ + HC.$$

8.5 Supersymmetry breaking

Supersymmetry cannot be an exact symmetry. It should be broken at low energies since the superpartners have not been found. The breaking of supersymmetry can occur via soft supersymmetry breaking [Georgi and Dimpopoulos (1981)]. We know that gauge symmetry does not allow to put explicit fermion masses. This is not the case, however, for the scalar superpartners. So we give them a separate mass, but preserve their couplings to be the same as those of their ordinary partners. One could have terms like:

$$\begin{aligned} V_{\text{soft}} &= m^2_{H_d}(H_d)^+ H_d + m^2_{H_u}(H_u)^+ H_u + B\mu\,(H_u).(H_d) + HC) \\ &+ \sum_i \left(m^2_{Q_i}\tilde{Q}^+_i \tilde{Q}_i + m^2_{\tilde{L}_i}\tilde{L}^+_i \tilde{L}_i + m^2_{\tilde{u}_i}\tilde{u}^*_i \tilde{u}_i + m^2_{\tilde{d}_i}\tilde{d}^*_i \tilde{d}_i + m^2_{\tilde{e}_i}\tilde{e}^*_i \tilde{e}_i \right) \\ &+ \sum_i \left(A^{ij}_{u_i}\tilde{u}^*_i H_u \tilde{Q}_j + A^{ij}_{d_i}\tilde{d}^*_i H_d \tilde{Q}_j + A^{ij}_{e_i}\tilde{e}^*_i H_d \tilde{L}_j + HC \right). \end{aligned} \tag{8.26}$$

The upper case symbols indicate an isodoublet while the lower an isosinglet. In addition a term like:

$$V_{\text{gaugino}} = \frac{1}{2}\sum_i M_i \lambda_i \lambda_i + HC, \tag{8.27}$$

where λ_i are the gauginos with the G_S quantum numbers understood.

When couplings between ordinary particles and their superpartners are allowed, they should preserve a discreet symmetry, **R-parity**, defined by:

$$R = (-1)^{2s+3B+L} = \begin{cases} +1 \text{ (for ordinary particles)}, \\ -1 \text{ (for SUSY particles)} \end{cases}, \tag{8.28}$$

where s=spin, B=baryon number, L=lepton number. R-parity conservation guarantees that the lightest SUSY particle is absolutely stable.

8.6 Some remarks about the particle spectrum

As in the case of the SM the MSSM cannot predict the particle spectrum. This done by phenomenology guided by some general ideas. One assumes a certain set of parameters at high energies, which evolve down according to renormalization group equations (RGM). Their values at the weak scale are then set by experimental constraints. At the GUT scale one assumes some kind of universality, i.e. four independent parameters:

- A common mass $m_{1/2}$ for the fermions,

- a common mass m_0 for the scalars,
- a common trilinear coupling A_0,
- a value for $\tan\beta$, see Eq. (8.6), and
- the sign of μ.

These values are constraint by experiment according to their implications at low energies. This defines the so-called **allowed SUSY parameter space**.

Some simple predictions are:

$$m^2_{\tilde{Q}} = m_0^2 + 6m^2_{1/2} \text{ (s-quark)}, \quad m^2_{\tilde{L}} = m_0^2 + \frac{1}{2}m^2_{1/2} \text{ (s-lepton), for isodoublets,}$$

$$m^2_{\tilde{e}} = m_0^2 + 0.15m^2_{1/2} \text{ (isosiglet s-lepton)}$$

and

$$M_2 = \frac{4}{5}m_{1/2} \text{ (wino)}, M_3 = \frac{\alpha_3}{\alpha_2}M_2 \text{ (gluino)}, \quad M_1 = \frac{\alpha_1}{\alpha_2}M_2 \text{ (bino)}.$$

No such simple predictions apply in the case of the Higgs, as we have seen above. No such simple formula exists for the mass of the superparner of the top quark (the s-top). The value of this mass is crucial in the resolution of the two scale problem. The lightest fermion typically is initially the bino. For dark matter searches the most interesting dark matter is the **neutralino**, a linear combination of the neutral gauginons and Higgsinos. So we will briefly discuss the mixing of gauginos and Higgsinos.

(1) The mixing of charged gauginos and Higsinos.

As we know from our experience with the standard model the fermion masses involve mixing between the left chirality and the right chirality via the Higgs field. Thus we are lead to the isoscalar couplings

$$\mathcal{L} = g_2\bar{\tilde{W}}^-_L \cdot \left[H_u \times \tilde{H}_{uR}\right]^1 \to g_2 \upsilon_u \bar{\tilde{W}}^-_L \tilde{H}_{uR}, \tag{8.29}$$

(after Higgs acquires a v.e.v).

$$\mathcal{L} = g_2\bar{\tilde{H}}^-_L . H_d \tilde{W}^+_{uR} \text{ (after Higgs acquires a v.e.v)} \to -g_2\upsilon_d\bar{\tilde{H}}^-_L\tilde{W}^+_R. \tag{8.30}$$

We also have the mixing between the Higgsinos via the μ term in a fashion analogous to the Higgs case. We thus arrive to the mass matrix:

$$\mathcal{M}_{charged} = \left(\begin{array}{c|cc} & \tilde{W}^+R & \tilde{H}^+_{uR} \\ \hline \tilde{W}^-_L & M_2 & g_2\upsilon\sin\beta \\ \tilde{H}^-_{dL} & -g_2\upsilon\cos\beta & \mu \end{array}\right). \tag{8.31}$$

This matrix can be diagonalized by separate right and left unitary transformations resulting, from the diagonalization of the Hermitian matrices

$(\mathcal{M}_{charged})^+ . \mathcal{M}_{charged}$, and $\mathcal{M}_{charged} . (\mathcal{M}_{charged})^+$ respectively.

One finds the mixing matrices for the right and left eigenvectors are:

$$\frac{1}{2}\tan 2\theta_R = \frac{\sqrt{2}\left(\mu\cos(\beta) - \sin(\beta)M_2\right)M_W}{-\mu^2 + M_2^2 + 2\cos(2\beta)M_W^2},$$

$$\frac{1}{2}\tan 2\theta_L = \frac{\sqrt{2}\left(\mu\sin(\beta) - \cos(\beta)M_2\right)M_W}{\mu^2 - M_2^2 + 2\cos(2\beta)M_W^2}.$$

Thus we obtain the two Dirac fermions:

$$\begin{aligned}\chi_1^+ &= -\sin\theta_R\tilde{W}_R^+ + \cos\theta_R\tilde{H}_{uR}^+ - \sin\theta_L\tilde{W}_L^+ + \cos\theta_L\tilde{H}_{dL}^+,\\ \chi_2^+ &= \cos\theta_R\tilde{W}_R^+ + \sin\theta_R\tilde{H}_{uR}^+ + \cos\theta_L\tilde{W}_L^+ + \sin\theta_L\tilde{H}_{dL}^+\end{aligned} \tag{8.32}$$

and their conjugates with eigenvalues:

$$M_{c_1}^2 = \frac{1}{2}\Bigg(\mu^2 + M_2^2 + 2M_W^2 - \sqrt{4\cos^2(2\beta)M_W^4 + 4\left(\mu^2 - 2\sin(2\beta)M_2\mu + M_2^2\right)M_W^2 + \left(\mu^2 - M_2^2\right)^2}\Bigg), \tag{8.33}$$

$$M_{c_2}^2 = \frac{1}{2}\Bigg(\mu^2 + M_2^2 + 2M_W^2 + \sqrt{4\cos^2(2\beta)M_W^4 + 4\left(\mu^2 - 2\sin(2\beta)M_2\mu + M_2^2\right)M_W^2 + \left(\mu^2 - M_2^2\right)^2}\Bigg). \tag{8.34}$$

(2) The neutral fermions (neutralinos)

Proceeding as above we obtain the matrix:

$$\mathcal{M}_{\text{neutral}} = \left(\begin{array}{c|cccc} & \tilde{B}_R & \tilde{W}_R^0 & \tilde{H}_u^0 & \tilde{H}_d^0 \\ \hline \tilde{\bar{B}}_L & M_1 & 0 & -m_z c_\beta s_w & m_z s_\beta s_w \\ \tilde{\bar{W}}_L^0 & 0 & M_2 & m_z c_\beta c_w & -m_z s_\beta c_w \\ \tilde{\bar{H}}_{uL}^0 & -m_z c_\beta s_w & m_z c_\beta c_w & 0 & -\mu \\ \tilde{\bar{H}}_{dL}^0 & m_z s_\beta s_w & -m_z s_\beta c_w & -\mu & 0 \end{array}\right) \tag{8.35}$$

In the above expressions $c_W = \cos\theta_W$, $s_W = \sin\theta_W$, $c_\beta = \cos\beta$, $s_\beta = \sin\beta$, where $\tan\beta = \langle\upsilon_2\rangle/\langle\upsilon_1\rangle = \langle\upsilon_u\rangle/\langle\upsilon_d\rangle$ is the ratio of the vacuum expectation values of the Higgs scalars H_u^0 and H_d^0. μ is a dimensionful coupling constant which is not specified by the theory (not even its sign).

In the absence of supersymmetry breaking ($M_1 = M_2 = M$ and $\mu = 0$) the photino is one of the eigenstates with mass M. One of the remaining eigenstates has a zero eigenvalue and is a linear combination of $\tilde{H}_d^0$ and $\tilde{H}_u^0$ with mixing angle β. In this case the photino is given a a linear combination of the bino and the wino as

$$\tilde{\gamma} = \cos\theta_W \tilde{B} + \sin\theta_W \tilde{W}_3 \tag{8.36}$$

and the massless Higgsino as

$$\tilde{H} = -\sin\beta \tilde{H}_3 + \cos\beta \tilde{H}_2 \tag{8.37}$$

In the presence of SUSY breaking terms the $\tilde{B}, \tilde{W}_3$ basis is superior since the lowest eigenstate χ_1 or LSP is expected to be primarily $\tilde{B}$. From our point of view the most important parameters are the mass m_x of LSP and the mixing $C_{j1}, j = 1, 2, 3, 4$ which yield the χ_1 content of the initial basis states.

By diagonalizing the above matrix we obtain a set of eigenvalues m_j and the diagonalizing matrix C_{ij} as follows

$$\begin{pmatrix} \tilde{B}_R \\ \tilde{W}_{3R} \\ \tilde{H}_{1R} \\ \tilde{H}_{2R} \end{pmatrix} = (C_{ij}^R) \begin{pmatrix} \chi_{1R} \\ \chi_{2R} \\ \chi_{3R} \\ \chi_{4R} \end{pmatrix} \qquad \begin{pmatrix} \tilde{B}_L \\ \tilde{W}_{2L} \\ \tilde{H}_{1L} \\ \tilde{H}_{2L} \end{pmatrix} = (C_{ij}) \begin{pmatrix} \chi_{1L} \\ \chi_{2L} \\ \chi_{3L} \\ \chi_{4L} \end{pmatrix}, \tag{8.38}$$

with $C_{ij}^R = C_{ij}^* e^{i\lambda_j}$ The phases are $\lambda_i = 0, \pi$ depending on the sign of the eigenmass. Anyway, if one wants, one can express the above results in photino-zino basis $\tilde{\gamma}, \tilde{Z}$ via

$$\tilde{W}_3 = sin\theta_W \tilde{\gamma} - cos\theta_W \tilde{Z}, \quad \tilde{B} = cos\theta_W \tilde{\gamma} + sin\theta_W \tilde{Z}. \tag{8.39}$$

The resulting eigenvectors are Majorana fermions

$$\chi_j = \chi_{jL} + e^{i\lambda_j} \chi_{jR}^c, \quad j = 1, \cdots, 4 \,. \tag{8.40}$$

The eigenstate with the lowest mass is called LSP (lightest supersymmetric particle) and is a viable dark matter candidate. If μ is large, it is primarily a bino with interactions governed by the coupling g_1. There exists, however, a region of the parameter space with small μ. In this case the dark matter candidate is primarily a Higgsino. We will not elaborate further on the dark matter phenomenology [Vergados (2007)].

Chapter 9

Grand Unification; the SU(5) Example

9.1 Mathematical Introduction

We have discussed the main theoretical deficiencies of the standard model (SM), namely:

- The symmetry group $G_S = SU_c(3) \otimes SU_I(2) \otimes U_Y(1)$ is not simple. So it contains three coupling constants, which can only be determined experimentally.
- It does not lead to a unification of all possible interactions.
- It does not explain charge quantization.
- It does not provide a framework for the observed baryon asymmetry in the Universe.

For these reasons there have been efforts to go beyond the SM. The most attractive path was to proceed by extending the symmetry G_s to a simple group. This way one hopes to achieve unification of all interactions. For this reason the emerging framework is called Grand Unified Theory or simply GUT.

In proceeding towards GUTs one notices that the group $G_S = SU_c(3) \otimes SU_I(2) \otimes U_Y(1)$ has rank $\ell = 4$, which is equal to the sum of the ranks of its individual parts, i.e. $\ell = 2 + 1 + 1 = 4$. Thus a suitable simple group must have at least rank equal to 4. Furthermore it is desirable that the group admits complex representations, i.e. the conjugate T^* of such a representation T should not to be equivalent to it, i.e. the two should not be related via a similarity transformation [Slansky (1981)]. Under these conditions the lowest rank groups are the $SU(5)$, $SO(10)$ and E_6 with ranks 4, 5 and 6 respectively. $SU(5)$ is the simplest of these and, for this reason, it will be selected for exhibiting the main ideas of Grand Unification, see, e.g. [Georgi and Georgi (1974)], [Langacker (1981)], [Vergados (1986)], [Ellis (1980)], even though $SO(10)$ [Fritsch and Minkowski (1975)] and E_6 [Slansky (1981)] may possess some advantages.

9.2 The structure of the GUT $SU(5)$

The group $SU(5)$ is characterized by $n^2 - 1 = 5^2 - 1 = 24$ parameters, i.e. by 24 generators represented by 24×24 traceless hermitian matrices. A convenient basis can be obtained by extending our experience with the simpler groups $SU(3)$ and $SU(2)$. We need four traceless diagonal elements. These can be chosen to be the diagonal elements of the algebra:

$$H_1 = (1, -1, 0, 0, 0),\ H_2 = \frac{1}{\sqrt{3}}(1, 1, -2, 0, 0),$$

$$H_3 = (0, 0, 0, 1, -1),\ H_4 = \frac{1}{\sqrt{15}}(2, 2, 2, -3, -3).$$

In this fashion of writing them, we recognize the first two as the generators of $SU(3)$, the third as the generator of $SU(2)$. The last one commutes with all the generators of $SU(3)$ and $SU(2)$ of G_S (it is proportional to the identity in the relevant subspaces). It can be identified with the generator of $U(1)$. In fact we will see that it is proportional to the hypercharge. All these are normalized so that $tr\left((H_i)^2\right) = 2$.

For our purposes we can take the hypercharge operator to be[1]

$$Y = (2/3, 2/3, 2/3, -1, -1), \tag{9.3}$$

which is an obvious from the particle content.

For the off diagonal elements we proceed in an analogous way:

[1]The hypercharge operator must be diagonal. So it can be written as a linear combination of of the diagonal matrices $h_i = e_{i,i} - e_{i+1,i+1},\ i, 1 \cdots 4$ with $e_{i,i}$ is a diagonal matrix with 1 in position i and zero everywhere else. It must commute with all the elements of the $SU_c(3)$ and $SU_I(2)$. It can easily be shown that it can take the form:

$$Y = k(2h_1 + 4h_2 + 6h_3 + 3h_4). \tag{9.1}$$

The proportionality constant can be chosen $k = -1/3$. A state of an $SU(5)$ representation can be described in by the weight $|m\rangle = m_{\alpha_1}, m_{\alpha_2}, m_{\alpha 3}, m_{\alpha_4}$, which are the eigenvalues af the operators h_i, i.e.

$$h_i|m\rangle = m_{\alpha_i}|m\rangle.$$

Thus we have:

$$Y|m\rangle = -\frac{2}{3}m_{\alpha_1} - \frac{4}{3}m_{\alpha_2} - 2m_{\alpha_3} - m_{\alpha_4}. \tag{9.2}$$

For the weights $(0, 1, -1, 0), (1, -1, 0, 0), (-1, 0, 0, 0), (0, 0, 0, 1), (0, 0, 1, -1)$ of $\underline{5}^*$ (see below) we obtain the eigenvalues (2/3,2/3,2/3,-1,-1) respectively.

- Consider
a) the operators $T_{ij}^{(1)}$, $i < j$, $j \le 5$ with zeros everywhere except the number one in row i and column j and one in row j and column i (the analog extension of the Pauli matrix σ_1) and
b) the operators $T_{ij}^{(2)}$, $i < j$, $j \le 5$ with $-i$ in row i and column j and i in row j and column i(the analog of the Pauli matrix σ_2).
These operators are normalized so that $tr\left((T_{ij})^2\right) = 2$.
- Alternatively the operators:
a) The step up operators E_{ij}^+, $i < j$, $j \le 5$ with zeros everywhere except in row i and column j, where there exists one (the analog of the matrix σ^+) and
b) The step down (lowering) operators E_{ij}^-, which are the Hermitian conjugates of the previous ones, i.e. $E_{ij}^- = \left(E_{ij}^+\right)^+$.
These operators are normalized to unity

$$\left(tr\left(\left(E_{ij}^-\right)^2\right) = tr\left(\left(E_{ij}^+\right)^2\right) = 1\right).$$

9.3 The particle content

Like in the case of the SM the particle content, except for the gauge bosons, is not specified by the symmetry. Before proceeding in discussing the particle content of the theory, a few facts about the structure of the group representations are needed. The lowest dimensional representation of the $SU(5)$ is the fundamental or vector representation, which is five dimensional[2], $\underline{5} \equiv [1]$ and its adjoined $\underline{5^*} \equiv [1,1,1,1] \equiv [1^4]$.

Taking the product of two vector representation one gets a 25 dimensional representation, which splits into an antisymmetric 10-dimensional and a symmetric 15 dimensional one:

$$\underline{5} \otimes \underline{5} \to \underline{10} \oplus \underline{15} \text{ or } [1] \otimes [1] \to [2] \oplus [1,1], \tag{9.4}$$

$$\underline{5^*} \otimes \underline{5} \to \underline{24} \oplus \underline{1} \text{ or } [1^4] \otimes [1] \to [2,1^3] \oplus [0]. \tag{9.5}$$

Note that the $\underline{24}$ is the regular representations encountered in the case of the generators. Note also that by combining the $\underline{10}$ and $\underline{10}^*$ one can also get the $\underline{24}$.

[2]It is customary in particle physics to indicate a representation by its dimension. This is not mathematically elegant and often ambiguous. It suffices, however, if only the lowest dimension representations are used. Strictly speaking one should use the Young Tableaux specified by a set of positive integers $[n_1, n_2, \cdots n_\ell]$ or the weights (Dynkin labels [Vergados (2016)] like $|m\rangle = m_{\alpha_1}, m_{\alpha_2}, \cdots m_{\alpha_\ell}$, with m_{α_i} integers. ℓ is the rank of the group.

9.3.1 *The Fermions*

No new fermions were introduced. Thus the fifteen per generation fermions of the SM model were put into the two simplest multiplets, $\underline{5}^*$ and $\underline{10}$ as follows[3]:

$$\underline{5^*} = \left(d_L^{\alpha c}, (\nu_L, e_L^-)\right)^T, \quad \alpha = r, g, b \text{ color index.} \tag{9.6}$$

The rest were put in the $\underline{10}$

$$\underline{10} = (u_L^{\alpha c}, (u_L^\alpha, d_L^\alpha), e_L^c), \quad \alpha = r, g, b \text{ color index.} \tag{9.7}$$

The particles of the $\underline{10}$ are not arranged in the form of a vector, but are placed in the appropriate positions of an antisymmetric matrix with 10 elements as follows:

$$\underline{10} = \frac{1}{\sqrt{2}} \begin{pmatrix} 0 & u_1^c & -u_2^c & u_1 & d_1 \\ -u_1^c & 0 & -u_3^c & u_1 & d_1 \\ u_2^c & -u_3^c & 0 & u_3 & d_3 \\ -u_1 & -u_2 & -u_3 & 0 & e^+ \\ -d_1 & -d_2 & -d_3 & -e^+ & 0 \end{pmatrix}. \tag{9.8}$$

Note the absence of left handed anti-neutrino. Note also that the bilinear $\bar{\psi}_L^a \gamma_\lambda \psi_L^a$, with a indicating that it is a member of $\underline{5}^*$ or of the $\underline{10}$ can have the quantum numbers of the adjoint representation of $SU(5)$. So such a current can couple to the gauge bosons. Note also that since quarks and leptons co-exist in the same multiplet the gauge bosons can cause both lepton and baryon number violation.

The G_S quantum numbers (color, $SU(2)$ and hypercharge) of the fermions are obvious.

9.3.2 *The gauge bosons*

The gauge bosons transform like the adjoint representation. So they can be decomposed in terms of the generators of the group. We will do the ordering and classification according to the subgroup $G_S = SU_c(3) \otimes SU_I(2) \otimes U_Y(1)$

(1) Gauge bosons with subgroup $SU(3)$ quantum numbers.
The gauge bosons associated with the generators H_1, H_2, E_{12}^+, E_{13}^+, E_{23}^+ and their adjoint transform like the gluons (octets under $SU_c(3)$) and will be indicated as $(G_\mu)_\alpha^\beta$, where α and β are

[3] It is customary to include only left handed fermion and anti-fermions. Thus instead of the right handed fermions we include the left handed anti-fermions.

color indices. They are placed in the upper left 3×3 segment of the 5×5 matrix. These gauge bosons are singlets under $SU(2)$ and have hypercharge $Y = 0$.

(2) Gauge bosons with subgroup $SU(2)$ quantum numbers.
The gauge boson associated with H_3, E^+_{45} and E^-_{45} are identified with A^3_μ, W^+_μ and W^-_μ respectively and they occupy the lowest right hand 2×2 segment of the matrix. The G_s quantum numbers of these gauge bosons are obvious (singlets under $SU(3)$, triplets under $SU_I(2)$ and have $Y = -1$)

(3) Gauge bosons with both $SU(3)$ and $SU(2)$ quantum numbers.
These are associated with the generators $E^+_{\alpha r}$, $\alpha = 1, 2, 3$; $r = 4, 5$. These will be indicated as $(X_\mu)^\alpha$ and $(Y_\mu)^\alpha$ associated with $r = 4$, 5 respectively. They occupy the positions indicated by α and r. The gauge bosons associated with $E^-_{\alpha r}$ are the adjoined of the above and will be indicated as $(X_\mu)_\alpha$ and $(Y_\mu)_\alpha$ or equivalently as $\left(\bar{X}_\mu\right)^\alpha$ and $\left(\bar{Y}_\mu\right)^\alpha$. The gauge bosons X and Y are triplets under $SU_c(3)$ and doublets under $SU_I(2)$ and have hypercharge $Y = 2/3 + 1 = 5/3$. Thus the X and Y have a charge $Q = (1/2) + 5/6 = 4/3$ and $Q = (-1/2) + 5/6 = 1/3$ respectively.

(4) A gauge boson associated with the generator H_4 sometimes indicated as T_{24}. This is indicated as

$$B_\mu = \frac{1}{\sqrt{30}}(2, 2, 2, -3, -3)B. \tag{9.9}$$

In a matrix form we have the form of the Table 9.1:

At this point it instructive to mention that the above colored gauge bosons must couple to the fermion currents, which are bilinear in terms of the Fermion fields. These are classified in terms of their quantum numbers as follows:

(1) $\bar{u}^c_{\alpha L}\gamma_\lambda q_{\beta L}$, ($\underline{3}^*$ or $\underline{6}$), $\underline{2}$, 5/3, $q_L = \left(u_{\beta L}, d_{\beta L}\right)^T$,
in the order dimension of color representation, of isospin representation and hypercharge Y. Explicitely the color triplet is of the form:

$$\text{color triplet} = \epsilon^{\alpha\beta\rho}\bar{u}^c_{\alpha L}\gamma_\lambda q_{\beta L}.$$

(2)

$$\bar{d}^c_{\alpha L}\gamma_\lambda q_{\beta L}, \quad (\underline{3}^* \text{ or } \underline{6}),\ \underline{2},\ -1/3$$

with the color anti-triplet again of the form: $\epsilon^{\alpha\beta\rho}\bar{d}^c_{\alpha L}\gamma_\lambda q_{\beta L}$.

Table 9.1: The SU(5) Gauge bosons arranged in the form of a matrix. Note that B has elements in all places along the diagonal.

$$\underline{24} = \begin{pmatrix} \frac{(G_1)^1_1}{\sqrt{2}} + \frac{(G_2)^1_1}{\sqrt{6}} + \frac{2B}{\sqrt{30}} & (G)^2_1 & (G)^3_1 & X^1 & Y^1 \\ (G)^1_2 & -\frac{(G_1)^2_2}{\sqrt{2}} + \frac{(G_2)^2_2}{\sqrt{6}} + \frac{2B}{\sqrt{30}} & (G)^3_2 & X^2 & Y^2 \\ (G)^1_3 & (G)^2_3 & \frac{2}{\sqrt{6}}(G_2)^3_3 + \frac{2B}{\sqrt{30}} & X^3 & Y^3 \\ \bar{X}^1 = X_1 & \bar{X}^2 = X_2 & \bar{X}^3 = X_3 & \frac{A^3}{\sqrt{2}} - \frac{3B}{\sqrt{30}} & W^+ \\ \bar{Y}^1 = Y_1 & \bar{Y}^2 = Y_2 & \bar{Y}^3 = Y_3 & W^- & -\frac{A^3}{\sqrt{2}} - \frac{3B}{\sqrt{30}} \end{pmatrix}.$$

(3) $\bar{q}_{\alpha L}\gamma_\lambda e^c_L$, $\underline{3}^*$, $\underline{2}$, $5/3$, $\ell_L = (\nu_L, e^-_L)^T$.
(4) $\bar{\ell}_{\alpha L}\gamma_\lambda d^c_{\beta L}$, $\underline{3}^*$, $\underline{2}$, $5/3$.
(5) $\bar{\ell}_{\alpha L}\gamma_\lambda u^c_{\beta L}$, $\underline{3}^*$, $\underline{2}$, $-1/3$.

We will see below that only the color anti-triplets are relevant for proton decay since one of the outgoing particles must be a lepton.

9.3.3 *The Higgs content*

The $SU(5)$ gauge symmetry will be broken spontaneously. In the first step from $SU(5)$ to G_S and in the second step from the breaking of G_S as discussed before in chapter 4. This scenario can be implemented in a number of ways. The most economic is to consider a Higgs scalar H belonging to the regular (adjoined) 24-dimensional representation of $SU(5)$. The second by a Higgs Φ belonging into a quinteplet of $SU5$. The Higgs H takes the form:

$$H = \begin{pmatrix} & & & H_{X_1} & H_{Y_1} \\ & H^\beta_\alpha + 2\upsilon\delta^\beta_\alpha & & H_{X_2} & H_{Y_2} \\ & & & H_{X_3} & H_{Y_3} \\ H_{X_1} & H_{X_2} & H_{X_3} & & \\ H_{Y_1} & H_{Y_2} & H_{Y_3} & & (H_3)^r_s - 3\upsilon\delta^r_s \end{pmatrix}, \tag{9.10}$$

where the upper left and the lower right segments are understood to be 3×3 and 2×2 in the $SU_c(3)$ and $SU_I(2)$ spaces. The parameter υ will be specified later. The complex scalar field takes the form

$$\Phi = \left(\phi_\alpha, (\phi^+, \phi^0)\right)^T, \quad \alpha = \text{color index}. \tag{9.11}$$

9.4 The Higgs mechanism

What we want to do is: Given a Higgs potential whether it is possible for the potential to attain a minimum for some non zero values of the Higgs scalars.

The Higgs potential $V(H, \Phi)$ is supposed to have a discreet symmetry such that:

$$V(H, \Phi) = V(H, -\Phi) = V(-H, \Phi)$$

Thus it can be cast in the form:

$$V(H, \Phi) = V(H) + V(\Phi) + V_{\text{int}}(H, \Phi), \tag{9.12}$$

where

$$V(H) = -m_1^2 tr\left((H)^2\right) + \lambda_1 tr\left(\left((H)^2\right)^2\right) + \lambda_2\left(tr(H)^4\right), \tag{9.13}$$

with the parameters λ_1 and λ_2 satisfying a suitable constraint so that the potential is bounded from below. The other two terms are given by:

$$V(\Phi) = -m_2^2(\Phi^+.\Phi) + \lambda_3\left((\Phi^+.\Phi)\right)^2,$$
$$V_{\rm int}(H,\Phi) = \lambda_4 tr\left((H)^2\right)(\Phi^+.\Phi) + \lambda_5\Phi^+\left(tr\left((H)^2\right)\Phi\right). \quad (9.14)$$

We recall that since the Higgs have been put into the regular representation of SU(5), they transform like the generstors of $SU(5)$. The possible invariants are the Casimir operators, which, see section 1.4, take the form:

$$tr(A^2) = \sum_{i,j} A_{ij}Aji, \quad tr(A^4) = \sum_{i,j,k,\ell} A_{ij}AjkA_{k\ell}A_{\ell i}. \quad (9.15)$$

Before proceeding further with the minimization procedure, we will save some time, if we observe that the off diagonal elements of H possess some kind of charge (electric charge, color etc) so that they cannot attain a non zero vacuum expectation value. It is safe, therefore, to consider H to be diagonal. In such a case it can be written as a linear combination of the diagonal generators of $SU(5)$. Thus

$$H = a_1H_1 + a_2H_2 + a_3H_3 + a_4H_4. \quad (9.16)$$

Thus

$$tr(H^2) = \left(-a_3 - \sqrt{\frac{3}{5}}a_4\right)^2 + \left(a_3 - \sqrt{\frac{3}{5}}a_4\right)^2 +$$

$$\left(a_1 + \frac{\alpha_2}{\sqrt{3}} + \frac{2a_4}{\sqrt{15}}\right)^2 + \left(-a_1 + \frac{\alpha_2}{\sqrt{3}} + \frac{2a_4}{\sqrt{15}}\right)^2 + \left(-\frac{2a_2}{\sqrt{3}} + \frac{2a_4}{\sqrt{15}}\right)^2,$$

$$(tr(H^2))^2 = \left(\left(-a_3 - \sqrt{\frac{3}{5}}a_4\right)^2 + \left(a_3 - \sqrt{\frac{3}{5}}a_4\right)^2 +\right.$$

$$\left.\left(a_1 + \frac{\alpha_2}{\sqrt{3}} + \frac{2a_4}{\sqrt{15}}\right)^2 + \left(-a_1 + \frac{\alpha_2}{\sqrt{3}} + \frac{2a_4}{\sqrt{15}}\right)^2 + \left(-\frac{2a_2}{\sqrt{3}} + \frac{2a_4}{\sqrt{15}}\right)^2\right)^2,$$

$$tr(H^4) = \left(-a_3 - \sqrt{\frac{3}{5}}a_4\right)^4 + \left(a_3 - \sqrt{\frac{3}{5}}a_4\right)^4 +$$

$$\left(a_1 + \frac{\alpha_2}{\sqrt{3}} + \frac{2a_4}{\sqrt{15}}\right)^4 + \left(-a_1 + \frac{\alpha_2}{\sqrt{3}} + \frac{2a_4}{\sqrt{15}}\right)^4 + \left(-\frac{2a_2}{\sqrt{3}} + \frac{2a_4}{\sqrt{15}}\right)^4.$$

Now:

(1) As a first step we minimize the potential $V(H)+V_{\rm int}(H,\Phi)$ subject to the above conditions.

$$\frac{\partial V(H)}{\partial a_i}=0,\ \det\left(\frac{\partial^2 V(H)}{\partial a_i \partial a_j}\right)>0;\ i,j=1,\,2,\,3,\,4. \tag{9.17}$$

The above set of 4 equations admits 80 non zero solutions $a_i = \upsilon_i \neq 0$. One can choose any one of them, since the rest can be obtained from it by a gauge transformation. We have chosen:

$$a_1=a_2=a_3=0,\ a_4=\upsilon_4;\quad \upsilon_4=\frac{\sqrt{15}}{\sqrt{2}}\frac{1}{\sqrt{(30\lambda_1+7\lambda_2)}}m_1, \tag{9.18}$$

provided that

$$\lambda_2>0,\quad 30\lambda_1+7\lambda_2>0\ . \tag{9.19}$$

This leads to the vacuum expectation value:

$$\langle H\rangle=\frac{\upsilon_4}{\sqrt{15}}\begin{pmatrix} 2 & 0 & 0 & 0 & 0\\ 0 & 2 & 0 & 0 & 0\\ 0 & 0 & 2 & 0 & 0\\ 0 & 0 & 0 & -3 & 0\\ 0 & 0 & 0 & 0 & -3\end{pmatrix}. \tag{9.20}$$

The mass matrix:

$$m_H^2 \Leftrightarrow M_{ij}^2=\left.\frac{\partial^2 V(H)}{\partial a_i \partial a_j}\right|_{a_1=a_2=a_3=0,\ a_4=\upsilon_4} \tag{9.21}$$

turns out to be diagonal. One finds:

$$m_H^2=\left\{8m_1^2,\frac{20m_1^2\lambda_2}{30\lambda_1+7\lambda_2},\frac{20m_1^2\lambda_2}{30\lambda_1+7\lambda_2},\frac{80m_1^2\lambda_2}{30\lambda_1+7\lambda_2}\right\}. \tag{9.22}$$

Choosing m_1 to be very large the Higss fields become very heavy. We see that the the diagonal elements of the Higgs field H is just the G_s invariant $\langle H\rangle$.

(2) As a next step we consider the the term of the potential:

$$V_1(\Phi,H)=-m_2^2(\Phi^+.\Phi)+\lambda_4 tr\left((H)^2\right)(\Phi^+.\Phi)+\lambda_5\Phi^+\left(tr\left((H)^2\right)\Phi\right). \tag{9.23}$$

Under the substitution $H->\langle H\rangle>$ it takes the form:

$$V(\phi_\alpha,\phi_d)=\frac{1}{15}\lambda_5\left(9|\phi_d|^2+4|\phi_\alpha|^2\right)\upsilon_4^2+\left(2\lambda_4\upsilon_4^2-m_2^2\right)\left(|\phi_d+^2+|\phi_\alpha|^2\right), \tag{9.24}$$

with $|\phi_\alpha|^2$ and $|\phi_d|^2$ the modulus first three and the last two components of the quinteplet, .i.e. $\Phi^+.\Phi = |\phi_\alpha|^2 + |\phi_d|^2$. From this we construct the mass matrix

$$(\mathcal{M})_{i,j} = \frac{\partial^2 V_1(\phi_\alpha, \phi_d)}{\partial x_i \partial x_j}, \quad x_i, x_j = \phi_\alpha, \phi_d. \tag{9.25}$$

By choosing $\lambda_5 = -(10/3)\lambda_4$ we find the component ϕ_α becomes super heavy while ϕ_d remains light with negative mass, namely:

$$m_a^2 = -m_2^2 + \frac{10}{9}\lambda_4 v_4^2 \approx \frac{10}{9}\lambda_4 v_4^2, \quad m_d^2 = -m_2^2. \tag{9.26}$$

(3) In the second step we proceed to minimize the potential, using the above results.

We only have to minimize the potential:

$$V_0 = -m_2^2(\Phi^+.\Phi) + \lambda_3(\Phi^+.\Phi)^2, \tag{9.27}$$

as we have done in the standard model. It is adequate to consider the case $\phi_d = (0, h)$ or $\Phi = (0, 0, 0, 0, h)$, since the other components cannot acquire a vacuum expectation value. This way we find that the minimum of the potential occurs at $h = v$, with v and the mass of the physical Higgs are given by:

$$v = \frac{m_2}{\sqrt{2\lambda_3}}, \ (\lambda_3 > 0), \quad m_h^2 = 2m_2^2. \tag{9.28}$$

9.5 The gauge boson masses

With the above judicious choice of $\langle H \rangle$ in the first step the symmetry G_S is not broken. Thus the gauge bosons associated with this symmetry remain massless. The X and Y bosons, however, acquire a mass. We find that

$$M_{X_i}^2 = \frac{g}{2}\left(\frac{v_4}{15}\right)^2 (2,2,2,-3,-3)\left(\left(E_{i4}\right)^+ E_{i4}\right)(2,2,2,-3,-3)^T. \tag{9.29}$$

Where E_{i4} is a matrix with 1 in the location of X_i in table 9.1 and zeros everywhere else. Thus $\left(E_{i4}\right)^+ E_{i4}$ is a diagonal matrix with 1 in the ith row and column and zeros everywhere else. Thus get

$$M_{X_i}^2 = \frac{g}{2}\left(\frac{1}{15}\right)^2 2v_4^2 = \left(\frac{1}{15}\right)^2 g v_4^2. \tag{9.30}$$

Similarly

$$M_{Y_i}^2 = \frac{g}{2}\left(\frac{v_4}{15}\right)^2 (2,2,2,-3,-3)\left(\left(E_{i5}\right)^+ E_{i5}\right)(2,2,2,-3,-3)^T = \left(\frac{1}{15}\right)^2 g v_4^2. \tag{9.31}$$

These masses will be fitted phenomenologically.

The masses of the other gauge bosons are obtained after the second symmetry breaking $\langle \Phi \rangle$, exactly like in the standard model.

9.6 Gauge couplings

The possible gauge couplings of $SU(5)$ are:

$$\mathcal{L} = \frac{g}{\sqrt{2}} \left\{ \bar{\psi}^a_L \gamma^\lambda \left(A_\lambda\right)^b_a \psi_{bL} - \bar{\psi}_{abL} \gamma^\lambda \left(\left(A_\lambda\right)^a_d \psi^{db}_L + \left(A_\lambda\right)^b_d \psi^{ad}_L \right) \right\}, \quad (9.32)$$

$a, b, d = 1, 2, ..., 5$. The gauge couplings of interest for baryon violation are those involving the gauge bosons X and Y, namely:

$$\mathcal{L} = \frac{g}{\sqrt{2}} \left\{ X^\rho_\lambda \left(\epsilon_{\alpha\beta\rho} u^{\bar{\alpha}c}_L \gamma_\lambda u^\beta_L + \bar{d}_{\rho L} \gamma_\lambda e^c_L - \bar{e}_L \gamma_\lambda d^c_{\rho L} \right) \right.$$
$$\left. Y^\rho_\lambda \left(\epsilon_{\alpha\beta\rho} d^{\bar{\alpha}c}_L \gamma_\lambda u^\beta_L - \bar{u}_{\rho L} \gamma_\lambda e^c_L - \bar{\nu}_L \gamma_\lambda d^c_{\rho L} \right) \right\} \quad + HC. \quad (9.33)$$

The antisymmetric tensors $\epsilon_{\alpha\beta\rho}$ etc guarantee that the two quarks involved are couplet to the antisymmetric color representation $\underline{3}^*$, which subsequently is couplet to the gauge boson to yield a color singlet. These couplings lead to the diagrams of Fig. 9.1. Combining them we get a contact like interaction leading to the proton decay diagram of 9.2.

The expected life time of the proton is given by

$$\tau = \frac{1}{\alpha^2_{GUT}} k_{\text{eff}} \left(\frac{m_X}{m_p} \right)^4 m_p^{-1} \approx 10^{31}\text{y} \left(\frac{m_X}{4 \times 10^{14}\text{GeV}} \right)^4, \quad (9.34)$$

where k_{eff} is a number of order unity which depends on the hadronic ME involved [Vergados (1986)]. At this point we should mention that the search for proton instability had started a long time ago [Reines *et al.* (1954)], [Bionta *et al.* (1983)], [Battistony *et al.* (1983)] and the discussion is still on [Hakaya (2005)], [Hayato *et al.* (2005)]. No proton decay events have been found. A lower limit on the proton life time has been set $\tau \geq 5 \times 10^{33}$y. This seems to exclude the minimal $SU(5)$ model outlined above.

9.7 Baryon asymmetry

Dirac's theory and the SM imply a symmetry between matter and antimatter, which is not what is observed. In the Universe we have only matter. Antimatter is produced only in the laboratory. It has been known for a long time that this can be explained under three conditions proposed by Sakharov [Sakharov (1991)]:

- The presence of baryon violating interactions.
- The existence of CP violating interactions.

  CP violation has in fact been observed a long time ago in the $K^0, \bar{K}^0$ system and recently in the B_d and B_s systems, see chapter

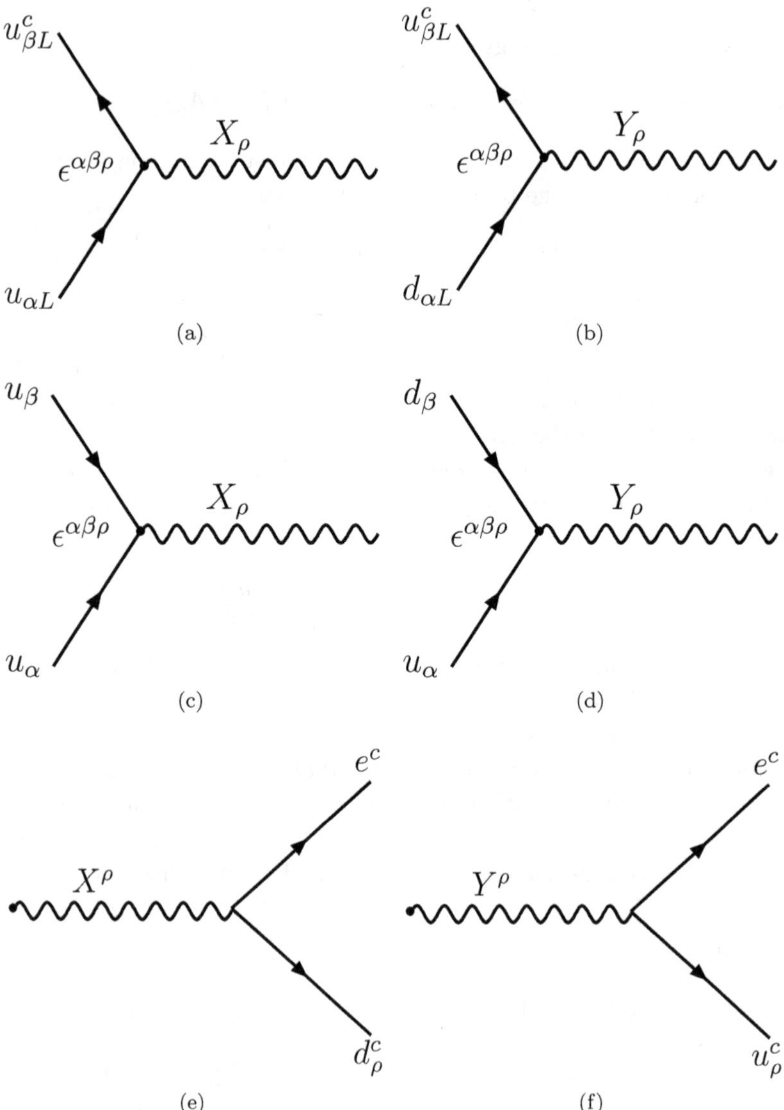

Fig. 9.1: Diagrams leading to baryon violation via the X (a) and Y bosons. (a) (b) (c) (d) vertices involve the interacting quarks of the baryon, while vertices (e) and (f) and involve the quark lepton pair.

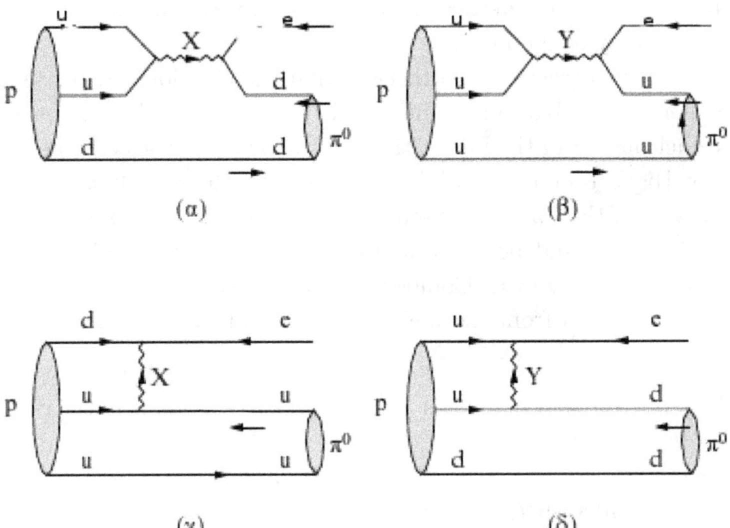

Fig. 9.2: Typical diagrams which may lead to proton decay $p \to e^+ + \pi^0$. In all cases there appears a spectator quark, i.e. a quark, which does not participate in the interaction. The ellipsoid curve merely indicates the the relevant hadrons are bound into a colorless state. The lepton is an outgoing positron.

12. It can proceed even in the standard model via fermion loops, with the only source of CP violation being the known CKM phase. Anyway all CP violation observed to date can be explained by the Kobayashi-Maskawa phase in the quark mixing matrix (see chapter 1, section 1.2.2). Unfortunately, it is known that this source of CP violation, acting in the conditions of the early universe, produces a net baryon number that is too small by about 10 orders of magnitude to explain the observed density of matter in the universe. So there must be other CP violating interactions in nature. It is possible that this new source of CP violation is associated with the neutrino masses. This requires CP violation in the neutrino sector. Long-baseline neutrino experiments are now being designed to search for that effect. However, it is not known how to relate the

phase that might be observed in these experiments to the baryon asymmetry of the universe.

A more attractive hypothesis is that the CP violating phase that generates the baryon asymmetry is present in the Higgs boson interactions, or in the interactions of the new particles that generate the Higgs boson potential. In this case, the full dynamics that generates the baryon asymmetry would occur at the TeV energy scale and would be accessible to the LHC and the planned ILC (International Linear Collider) experiments.

- The deviation from thermal equilibrium in the Universe, at some crucial stage.

We will begin by the first condition and attempt to explain it in the context of GUTs. At time $t = 10^{-35}$ s after the Big Bang the temperature of the Universe was $T = 3 \times 10^{27}\ {}^0$K and the energy $E = 3 \times 10^{17}$ MeV= 3×10^{14} GeV. Under such conditions the then existing X and Y bosons participate in the reactions:

$$X \underset{r}{\rightarrow} u + u \quad \left(\Delta B = \frac{2}{3}\right), \quad X \underset{r'}{\rightarrow} e^+ + \bar{d} \quad \left(\Delta B = -\frac{1}{3}\right), \tag{9.35}$$

$$Y \underset{r}{\rightarrow} u + d \quad \left(\Delta B = \frac{2}{3}\right), \quad Y \underset{r'}{\rightarrow} e^+ + \bar{u} \quad \left(\Delta B = -\frac{1}{3}\right), \tag{9.36}$$

where r and $r' = 1 - r$ are the probabilities to observe these reactions. ΔB is the change in baryon number, Assuming that the proton carries baryon number 1. At the same time the reactions

$$u + u \rightarrow X, \quad e^+ + \bar{d} \rightarrow X, \tag{9.37}$$

$$u + d \rightarrow Y, \quad e^+ + \bar{u} \rightarrow Y, \tag{9.38}$$

which used to take place previously are now energetically forbidden.

In the presence of antimatter we can have:

$$\bar{X} \underset{\bar{r}}{\rightarrow} \bar{u} + \bar{u} \quad \left(\Delta B = -\frac{2}{3}\right), \quad \bar{X} \underset{\bar{r}'}{\rightarrow} e^- + d \quad \left(\Delta B = \frac{1}{3}\right), \tag{9.39}$$

$$\bar{Y} \underset{\bar{r}}{\rightarrow} \bar{u} + \bar{d} \quad \left(\Delta B = -\frac{2}{3}\right), \quad \bar{Y} \underset{\bar{r}'}{\rightarrow} e^- + u \quad \left(\Delta B = \frac{1}{3}\right). \tag{9.40}$$

The decay of X and $\bar{X}$ leads to a baryon violation

$$\Delta B_X = \frac{2}{3} r + \left(-\frac{1}{3}(1 - r)\right) - \frac{2}{3}\bar{r} + \frac{1}{3}(1 - \bar{r}) = r - \bar{r}. \tag{9.41}$$

For the Y and $\bar{Y}$ we get $\Delta B_Y = r^{'} - \bar{r}^{'}$. Thus finally:

$$\Delta B = r - \bar{r} + r^{'} - \bar{r}^{'} . \tag{9.42}$$

In other words, if $r = \bar{r}$ and $r^{'} = \bar{r}^{'}$, we have no baryon number violation. Thus observation forces us to assume that, depending on the model, they are different.

The baryon number ΔN_B is given by:

$$\Delta N_B = \Delta B n_i, \quad i = X, Y, \bar{X}, \bar{Y}, \quad n_X = n_Y = n_{\bar{X}} = n_{\bar{Y}}, \tag{9.43}$$

where n is the particle density, which of the same order with the densisity of the then existing photons, but today it is much smaller. The above results can be modified if, in addition, we consider the reactions;

$$u + u \rightarrow e^+ + \bar{d}, \quad \Delta B = -1, \quad X \text{ exchange} , \tag{9.44}$$

$$u + d \rightarrow e^+ + \bar{u}, \quad \Delta B = -1, \quad Y \text{ exchange} , \tag{9.45}$$

(see Fig. 9.3). We expect, however, that they are negligible compared to

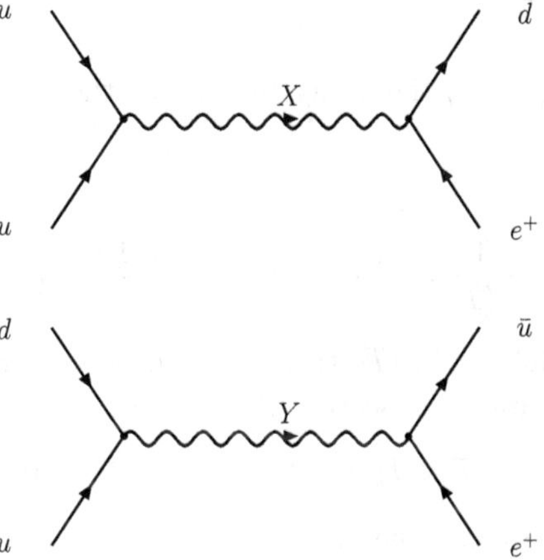

Fig. 9.3: Two diagrams, which cause baryon number violation.

the gauge boson decays. In fact the decays of the X $\bar{X}$, Y and $\bar{Y}$ bosons at temperature T proceed with a rate:

$$\Gamma_i^{(1)} = \frac{1}{12}\alpha_{\text{GUT}}\frac{T^2 + M_i^2}{M_i}, \; i = X, Y, \bar{X}, \bar{Y} \tag{9.46}$$

The relevant rate can be computed in a fashion analogous to the decay of the W-boson discussed in section 7.6, noting, of course, that now the particle is not at rest.

For $T < M_i$, the exchange process proceeds with a rate:

$$\Gamma_i^{(2)} = \alpha_{\text{GUT}}^2 T \left(\frac{T}{M_i} \right)^4 . \tag{9.47}$$

The above estimates depend on the model dependent parameter α_{GUT}, which, however, is expected to lie in the range $10^{-6} - 10^{-4}$. In any case for the period we are interested in we have $\Gamma_i^{(1)} \gg \Gamma_i^{(2)}$. In other words such exchange reactions do not affect our estimate. Similarly the annihilation of X with $\bar{X}$ and Y with $\bar{Y}$ is not very likely during this period.

As we have already mentioned another condition to achieve baryon violation is to be away from thermal equilibrium. Otherwise we have time reversal invariance, which, due to CPT invariance, leads to CP invariance. In this case the matter and the antimatter rates should be the same and, as a result, no net baryon violation results.

Deviation from thermal Equilibrium exists if the expansion rate of the Universe is much larger than the rate of the decay of the X and Y bosons, that is:

$$T_i > t, \;\; T_i = \frac{1}{\Gamma_i}, \quad T_i \text{ the decay time and } t \text{ the age of the Universe.} \tag{9.48}$$

In other words:

$$\Gamma_i t < 1 \Rightarrow \alpha_{\text{GUT}} \frac{1}{n(T)} 10^{-1} \left(\frac{T_P}{T} \right)^2 < 1 \Rightarrow \frac{10^{-8}}{n(T)} \left(\frac{T_P}{T} \right)^3 < 1, \tag{9.49}$$

where $T_P = 10^{19}$GeV and $n(T)$ the density of states. For $SU(5)$ one finds $n(T) = 150$ (see chapter 10, section 10.7). Thus

$$10^{-5} T_P < T < T_i \text{ or } 10^{-16}\text{GeV} < T < 10^{-14}\text{GeV}, \tag{9.50}$$

which can, in principle, be satisfied. The above procedure leads to

$$\frac{\Delta N_B}{B_\gamma} = (r - \bar{r}) \frac{N_X}{N_{\text{tot}}} = \frac{\Gamma_X - \Gamma_{\bar{X}}}{\Gamma_X} \frac{N_X}{N_{\text{tot}}}, \tag{9.51}$$

where N_γ is the number of photons, N_X the number of the heavy bosons and N_{tot} the number of all elementary particles, which existed and could produce photons. A reasonable estimate is

$$\frac{N_X}{N_{\text{tot}}} = 10^{-1}.$$

Furthermore reasonable models yield:

$$\frac{\Gamma_X - \Gamma_{\bar{X}}}{\Gamma_X} \approx 10^{-9} \rightarrow \frac{\Delta N_B}{N_\gamma} \approx 10^{-10}, \tag{9.52}$$

in satisfactory agreement with experiment:

$$\frac{\Delta N_B}{N_\gamma} \approx 10^{-10} = 6.1^{+0.3}_{-0.2} \times 10^{-10}.$$

All these, of course, happen at $t = 10^{-35}$s. Eventually the above reactions freeze. The net process is exhibited in Fig. 9.4.

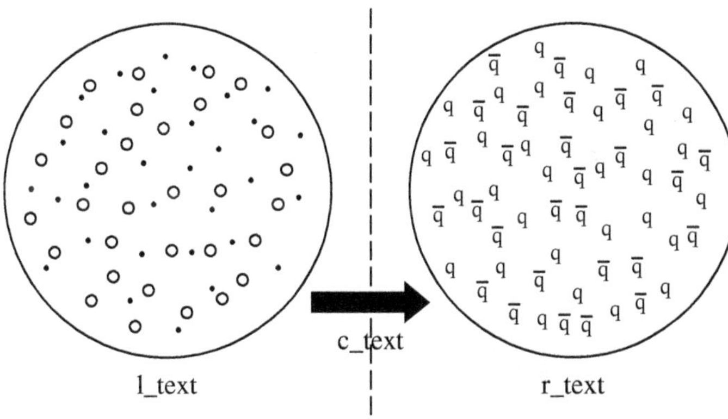

Fig. 9.4: At about 10^{-35}s after the Big Bang there exists an about equal number of X and Y bosons indicated by $(\cdot)$ and their antiparticles $\bar{X}$ and $\bar{Y}$, indicated (o). As we reach 10^{-30}s these particles are no longer produced and we have, instead, their decay products quarks (q) and anti-quarks $(\bar{q})$. The quarks are favored by a number of the order of $(N(q) - N(\bar{q}))/N(q) = 10^{-9}$. Most of the anti-quarks are eventually annihilated into photons with a tiny faction of quarks surviving to yield ordinary matter (protons and neutrons). The numbers of particles in the figure is not to scale.

Chapter 10

A Brief Introduction to Cosmology

In this chapter we are going to discuss recent developments in cosmology with emphasis on those aspects related to particle physics. As we will see the word, derived from the Greek word $\kappa o' \sigma \mu o \varsigma$, which means beautiful in the Greek view of the universe, indeed describes our current model of the universe 3000 years later.

10.1 Cosmological principles

Before proceeding with consideration of the facts and observations of cosmology, we are going to examine the basic underlying principles needed for a better understanding and organization of these facts. These are:

i. First principle: The universe appears the same to all observers.
ii. The universe is homogeneous and isotropic (at large enough scale)
iii. The universe can be understood in physical terms, i.e. in terms of the laws governing the events in our microscopic laboratories and the theories explaining them. In particular in terms of the general theory of relativity.

The first and the last principle are more or less obvious. One would like a picture of the universe, which is not observer dependent, especially in the absence of the ability to transform such information from one observer to the next. It is also obvious that one would like to have a picture of the universe, which incorporates as firm an information as our knowledge of the laws of nature permits.

The second, principle, however seems to be counter-intuitive. The world around us, the planetary system, the structure of the galaxy and generally a look through the sky indicates that the density of matter and energy in the universe is not homogeneous and isotropic. Even the furthest segments we have been able to see are not quite isotropic and homogeneous, see Fig. 10.1. And yet we accept this principle, crucial as it is in understanding the evolution of the universe, and expect it to hold, if we will be able to see even further.

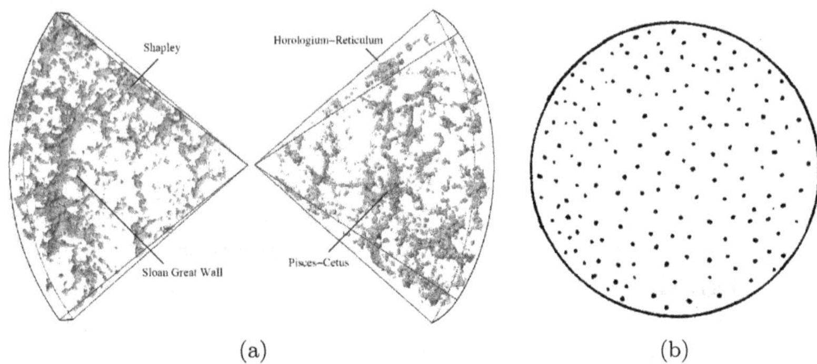

Fig. 10.1: Even at the scale of 1000 Mpc, approximately 2×10^9 light years the universe is not completely homogeneous and isotropic as seen by the 2df Galaxy Redshift Survey of the galaxy cluster Coma. One still sees regions of high density and voids extending up 50 Mpc or 160×10^6 light years, which are among the largest seen in the universe (a). A view of what we expect a homogeneous and isotropic universe to be is shown in (b).

10.2 The expanding universe; the Big Bang scenario

From various observations it is clear that the our universe was born out of a huge explosion, **Big Bang**, which took place about 14 billion years ago. This is not like the explosions we see in everyday life, in which things are moving away from some center. In the ig bang scenario there is no such center. As we will see this simply means that the distance between any two objects, which are not gravitationally bound, increases by a time dependent scale factor. For a finite system this can be described by the blowing up of a balloon. General theory of relativity, which, at present, provides the best framework for the study of the evolution of the universe, accepts solutions which do not have a meaning for $t \leq 0$.

After the Big Bang the universe started expanding uniformly with a rate which depends on time. This is supported mainly by the following observations:

(1) The receding of the galaxies.
(2) The existence of the background microwave radiation.
(3) The relic abundance of the $^4_2$e in the universe.

We will now examine each these observations.

10.2.1 *The receding of galaxies*

The receding of galaxies is perhaps the greatest discovery in cosmology. It is not clear when it was proposed for the first time and by whom. The prevailing view is that it was first conceived by Lemaitre in 1927 and it was formulated by Hubble in 1929, when he resolved Andromeda constellation into stars. He then discovered that in the Andromeda spirals there exist variable stars, which resemble the variable stars of our galaxy known Cepheids. This was indeed very important, since such stars can be used as prototype candles, in the sense that their absolute luminosity (power output) is known. This is because their exists a relationship between the absolute luminosity and their period. The absolute luminosity was calibrated from the Cepheids that exist in our galaxy, since their distance can be determined by the triangulation method. Thus, by measuring the relative luminosity (energy flux) in the laboratory one can determine the distance of these stars, which are much further away. He concluded that the Andromeda cloud was about 400000 light years away[1]. It is clear that cosmology can be viewed as a race in measuring large distances. For really large distances two prototypes are used.

- Prototype candles.
  These are objects whose absolute luminosity (power output) is known. Then, as in the case of Cepheids, the "volometric" distance can be determined by measuring the energy flux (relative luminosity) in the laboratory. Fortunately there exist in nature such prototypes. These are supernovae of type Ia. They are double stars, in which one member, usually a white dwarf, eats up its companion, e.g. a red giant. When its size reaches 1.44 solar masses (Chandrashekar limit) the supernova explosion occurs, with an output associated with this well defined mass. Fortunately

[1]Subsequent more precise measurements indicated that the actual distance was a bit larger, 2000000 light years. This method of determining distances was indeed a breakthrough for cosmology.

these rather rare supernovae can be distinguished from the sea of other supernovae by their characteristic spectrum and light curve. This is the currently used method for determining distances up to 1/10 of what is believed to be the size of the universe.

- Prototype lengths.
  If the size of an object, perpendicular to the viewer, is known, its distance can be determined by measuring the angle it suspends. Unfortunately not many such prototypes exist. Furthermore measuring small angles is rather tricky.

Having measured distances with the Cepheid candles Hubble proposed in 1932 **Hubble's law**, that is the idea that all galaxies are receding from us with a velocity proportional to their distance, i.e.:

$$v = Hr, \tag{10.1}$$

where v is the relative velocity of the observed galaxy with respect to ours, r its distance from us and is a function of time, but the same for all observers, known as Hubble's parameter.
The Hubble parameter today, $H(t_0) = H_0$, extracted from data like those of 10.2, is usually given as follows:

$$H_0 = (73.4 \pm 2.8) \text{ km/sMpc}^{-1} \Leftrightarrow \tag{10.2}$$
$$H_0 = 100h \text{ km/sMpc}^{-1}, \quad 0.706 < h < 0.762 \Leftrightarrow \tag{10.3}$$
$$\text{or} \tag{10.4}$$
$$H_0 = 100h \text{ km/sMpc}^{-1}, \quad h = 0.734 \pm 0.028. \tag{10.5}$$

This is the most important parameter of Cosmology. It has recently been measured by the satellite experiments WMAP, see e.g. [Spergel *et al.* (2003)] and [Tegmark *et al.* (2004)] and PLANCK [Ade *et al.* (2015)]. The obtained results are:

$$H_0 = (70.4 \pm 1.4) \text{ km/sMpc}^{-1}{}_{(\text{WMAP})},\ H_0 = (67.8 \pm 0.8) \text{ km/sMpc}^{-1}{}_{(\text{PLANCK})}$$

It is worth noting that Hubble's law admits an expansion of the universe without the need of a center (see Fig. 10.3 below).

The determination of the receding velocity was easier thanks to the red shift of the spectrum due to the motion of the source (Doppler effect). In fact in the context of special relativity:

$$1 + z = \frac{\lambda}{\lambda_0} = \left(\frac{1 + v/c}{1 - v/c}\right)^{1/2}, \tag{10.6}$$

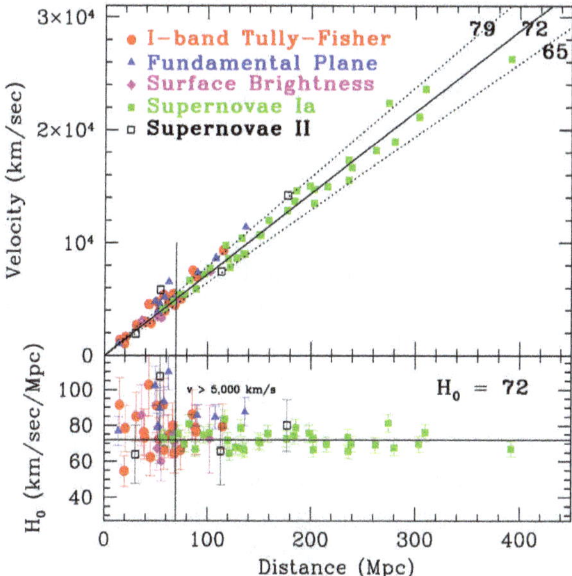

Fig. 10.2: Experimental verification of Hubble's law (upper panel) and the extracted value of H_0 (lower panel). The diagrams are marked by the corresponding five possible ways of measuring the distances.

where λ_0 is the wavelength produced by the source and λ the wavelength seen by the observer. Such in effect exists in the context of the general theory of relativity (GTR) as well. We will see later that in the context of GTR all lengths expand. So do the wavelengths. From Eq. (10.6) we can see that for a receding galaxy $\upsilon > 0$, i.e $z > 0$. The observed shift, $\lambda = \lambda - \lambda_0$, is given by:

$$z = \frac{\Delta\lambda}{\lambda_0} = \frac{\upsilon}{c} \times \begin{cases} 1, & \upsilon \ll c \\ \gamma, \quad \gamma = (1 - \upsilon^2/c^2)^{-1/2}, & \upsilon \leq c. \end{cases} \tag{10.7}$$

Thus

$$z = \frac{H}{c} r, \quad \upsilon \ll c \quad \text{for not too large distances.} \tag{10.8}$$

The deeper we look into the universe, the larger the parameter z!

The experimental verification of Hubble's law is shown in Fig. 10.2. It is impressive that Hubble's equation is a mathematical consequence of the first cosmological principle in the context of classical mechanics.

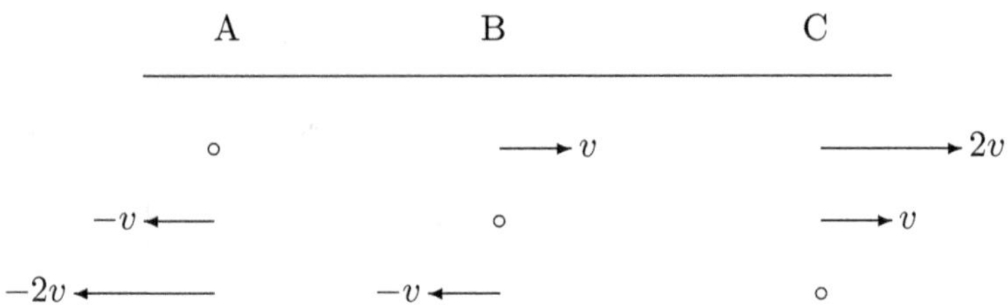

Fig. 10.3: In the context of classical mechanics equivalence of all observers leads to such a picture. In the first line we show the velocities of the two observers as seen by A. In the second as seen by B and in the last as seen by C. Exactly as Hubble's law implies.

We have already mentioned that the evolution of the universe is governed by GTR. We will see later that some aspects of it can be understood by classical mechanics. Suppose, e.g., that all lengths involving not gravitationally bound systems are changed by a scale factor $a(t)$. Suppose, e.g., that such a system with mass m has a dimension $\tilde{r}$ and after some time it becomes r. Then

$$r = a(t)\tilde{r} \rightarrow v = \frac{dr}{dt} = \dot{R}\tilde{r} = \frac{\dot{a}}{a}r = Hr. \tag{10.9}$$

The dependence of and a on time will be examined later. We should mention here that it is an expansion of the space itself, not an enlargement of the size of bound objects. The velocity of expansion can exceed the velocity of light, i.e. it can become superluminal, without violating the special theory of relativity. It is not possible to exchange signals between observers, who are so far away[2]. The expansion of the universe does not affect gravitationally bound systems (galaxies, planetary systems etc). If it were operating on all scales, we would not have matter condensation to yield stars etc. It is believed that initial quantum mechanical fluctuations lead to classical matter fluctuations, which under the influence of the gravitational interaction resulted in the structures observed today.

10.2.2 *The background microwave radiation*

It is clear that initially the temperature of the universe was very high. It soon started dropping like the temperature of any rapidly expanding

[2]You can imagine bugs moving on the surface of a balloon with a velocity 10 cm/s. Imagine now that the balloon is blown up with a high expansion rate. Any two of the bags are moving way from each other with a high speed, but their velocity relative to the balloon still is 10 cm/s!

system. At the end of the first second the temperature is 10^{10}K, which corresponds to an energy of a few MeV. at such a temperature, however, one cannot have stable atoms. Quite a bit later, when the universe was about 400000 years old its temperature was about 3000 K which, as we will discuss below, corresponds to about 0.25 eV. Then the first neutral atoms began to form. At this point the interaction of light and matter became much weaker than that with a plasma (a collection of charged particles). So essentially light becomes decoupled from the rest of the universe. It simply gets cooler and cooler and today it should be around as black body radiation with a characteristic temperature of about 3K. These ideas had in fact been proposed by Gamow, but few paid attention and nobody believed in them.

In 1964 two physicists, A. Penzias and R. Wilson, working on radars, caught with their antennas a "noise". It was in fact microwave radiation, coming pretty uniformly from all directions of observation, independently of the site of the experiment and the part of the year[3]. They could not understand it, but with these characteristics of homogeneity and isotropy it could not have originated from a single source of electromagnetic radiation. Anyway the energy density of this radiation was quite high, about 300 photons/cm^3 or about 0.25 eV/cm^3. This is about 10 times larger than that of inner galactic radiation.

It is not surprising that that this had to wait till the temperature fell so low, even though the binding energy of the electron in the hydrogen atom is 13.6 eV. The reason is that even at lower temperatures the photons could break the hydrogen atom due to those photons that have energy in the distribution above, above, say, the energy of the first excited state of 10.2 eV (if that is done another photon can come and kick the electron off the atom). This is known as the Saha effect. The portion of photons with that energy is about 10^{-10} (see Fig. 10.4). Note, however, that the number of photons in the universe is about 10^{10} larger, since a tiny fraction of matter survived when most of it was annihilated with the disappearance of antimatter. The spectrum is that of a black body radiation. It is in fact the best such spectrum ever observed (see Fig. 10.5). Before radiation was decoupled from matter, however, it took a picture of the universe. This, in fact, has been seen in the recent satellite data of WMAP, see e.g. [Spergel *et al.* (2003)] and [Tegmark *et al.* (2004)] and PLANCK [Ade *et al.* (2015)].

---

[3]It is an ironic coincidence that at that time an experiment was about to start in Princeton following the ideas of Gamow. The experimental paper was published together with a theoretical one [Weinberg (2008.)], [Weinberg (1972.)].

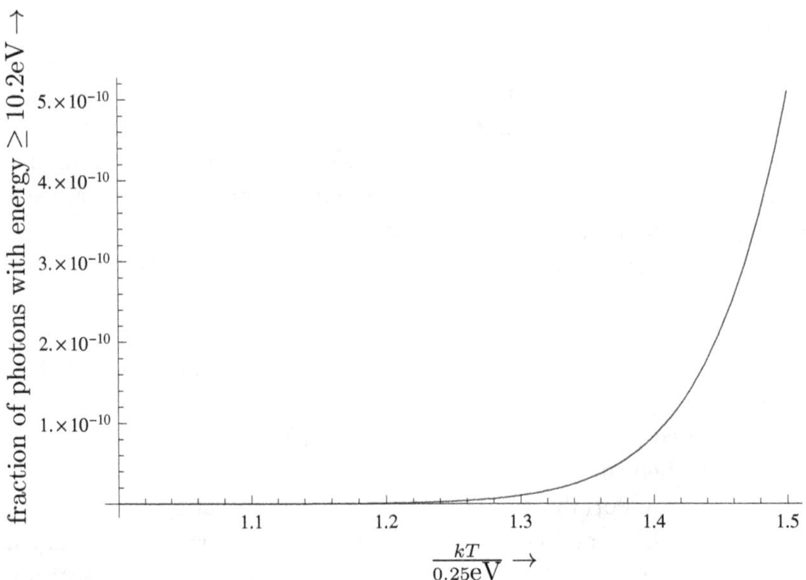

Fig. 10.4: The fraction of photons which can excite the hydrogen atom with kT in units 0.25 eV.

In these pictures we see the universe the way it had been when it was only about 400000 old.

These findings will be discussed in detail later.

10.2.3 *The abundance of primordial ^4_2He and other light elements in the universe*

Various observations indicate that the matter of the universe is mainly composed by hydrogen and $^4_2$He, in a ratio 3:1. We will see below that this much helium could not have been produced by fusion of hydrogen inside the stars. The stars are too young to have produced so much helium.

To understand the primordial nucleosynthesis we have to understand the availability of neutrons. This is not because of the fact that they decay. At time scales we are interested, the age of the universe being 1 min old, this life time is sufficiently long to consider the neutron as stable. What matters is the fact that there existed fewer neutrons than protons due to their larger mass. In other words the then existing ratio of (n/p) becomes crucial. From statistical mechanics we know at the then prevailing temperatures the

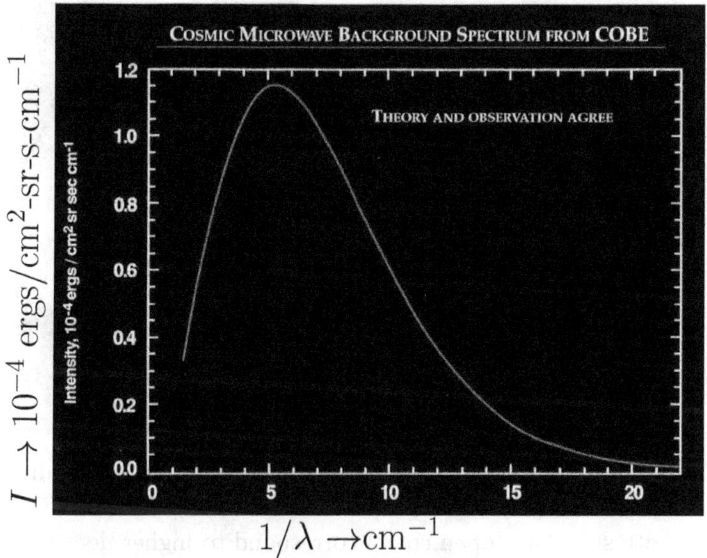

Fig. 10.5: The spectrum of the black body radiation as seen by FIRAS of COBE satellite. It is, perhaps, the best black body distribution ever observed. The errors are too small to be visible (from slide 36, of COBE Slide Set). The maximum is located at $1/\lambda \approx 5.92\text{cm}^{-1}$, which corresponds to a temperature of $= 2.728\text{K}$.

Fermi-Dirac distribution becomes Maxwell-Boltzmann, i.e this ratio is given by $(n/p) = e^{-\frac{\Delta m}{kT}}$, i.e. it depends on the neutron proton mass difference. Δm of about 0.8 MeV, and the then temperature of the universe. Thus from $(n/p) = 1/7$ we obtain $kT \approx$0.41MeV. This value is implied by the fact that the the observed mass of He to proton is $1/3$, i.e. the number of He atoms is $1/12$ of that of protons.

Indeed for understanding he helium abundance the following quantities are employed:

$$X_4 = \frac{N_{^4_2\text{He}}}{N_{\text{Hy}}} = \frac{1}{2}\frac{n}{p}/\left(1 - \frac{n}{p}\right). \tag{10.10}$$

The factor $(1 - \frac{n}{p})$ represents the fraction of protons and, for given ratio, the number of He atoms is by a factor of two less than the number of protons. The faction of the mass carried by He is given by

$$Y_4 = \frac{4X_4}{1 + 4X_4} = 2\frac{n}{p}/\left(1 + \frac{n}{p}\right), \tag{10.11}$$

where n and p the initial number of protons and neutrons respectively. From observations we know that $Y_4 \approx 1/4$ and $X_4 = 1/12$ or a mass ratio of He

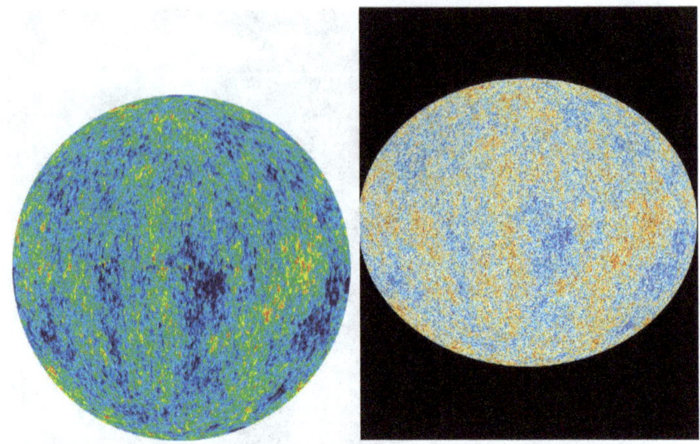

Fig. 10.6: The inhomogeneity of the microwave background radiation, as observed by WMAP (2001-2007) (a) and more recently by PLANCK (1012) (b). The hottest points (open color) correspond to higher density.

to proton to be $4X_4$ i.e 1 to 3. From these we find n/p =1/7. Furthermore from the evolution of the universe we find that the kT $\approx$0.41MeV was achieved when the universe was 1 min old.

Thus at the end of the first minute the fraction of He mass is Y_4, i.e. we have 75% in protons and 25% in He. Basically, the hydrogen-helium abundance helps us to model the expansion rate of the early universe. If it had been faster, there would be more neutrons and more helium. If it had been slower, more of the free neutrons would have decayed before the deuterium stability point and there would be less helium.

Let us make a detour to see how energy is produced in the sun. Eddington first proposed 4p $\rightarrow^4_2$He as an energy source for our sun

$$4p \rightarrow^4_2 \text{He} + 2e^+ + 2\nu_e \tag{10.12}$$

The mass difference for this reaction is about 25.7 MeV, while the typical kinetic energy carried off by neutrinos about 0.4 MeV. Therefore about 25 MeV is released per four protons consumed. This is the energy that keeps the electron gas in our sun hot: energy is produced in the solar core at the required rate, just about balancing the energy that is carried off the sun by the photons emitted from the photosphere. Thus we can estimate the rate of fusion in the solar core from the measured solar constant. The solar constant, i.e. the light energy flux on earth is $\Phi = 0.033$ cal/sec/cm^2. The sun-Earth distance is $r = 1.49 \times 10^{13}$cm. Thus the power output is

$$4\pi r^2 \Phi \approx 0.92 \times 10^{26} \text{cal/s} = 2.4 \times 10^{39} \text{MeV/s}.$$

Since, however, 4 protons are consumed for every 25 MeV produced, the sun consumes about 4×10^{38} protons/s. Now the sun has a mass equal to that of about 1.19×10^{57} protons. Let us assume a solar age of 5 billion years (b.y.) (the sun is estimated to be about 4.6 b.y. in age) and it is burning at the current power level over that period. Then we can estimate the number of protons consumed over that lifetime 5×10^9y $= 3.156 \times 10^7 \times 5 \times 10^9 = 1.57 \times 10^{17}$s. Thus

$$1.57 \times 10^{17} \text{s} \times 4 \times 10^{38} \text{ protons/s} \approx 0.6310^{56} \text{ protons}.$$

So this is 0.63 /11.9 = 5:3% of the sun's mass, Thus only about 5% of protons converted in 5 b.y. Furthermore, of course, this ${}^4_2$He is locked in the core of our sun, not in places like the inter- stellar medium where it could be counted by those interested in determining abundances. In addition many protons are not residing in stars. Thus the tentative conclusion is that stellar burning contributes to, but cannot account for all, of the ${}^4_2$He observed in the universe. In fact, this conclusion can be put on much firmer ground by looking at the ${}^4_2$He abundance as a function of stellar metallicity, which is kind of a "clock" for the galaxy. Very metal poor stars presumably were formed very early (before supernovae and novae created many of the metals). The surfaces of such stars should not know about the ${}^4_2$He synthesis in the core, but rather be representative of the star at its birth. So even if the surface shows a large ${}^4_2$He abundance, one would conclude that that ${}^4_2$He was primordial. We know today more about 4p$\to{}^4_2$He. It was first described by Bethe and Critchfield in 1939 and was further developed by Salpeter and by Burbidge, Burbidge, Fowler, and Hoyle in the 1950's.

An equilibrium between protons and neutrons is established via the reactions:

$$n + p \leftrightarrow d + \gamma$$

$$p + e^- \leftrightarrow n + \nu_e$$

$$n + e^+ \leftrightarrow p + \bar{\nu}_e$$

$$n \to p + e^- + \bar{\nu}_e$$

Thus at the end of the first minute we have 75% protons and 25% neutrons. Up to this point the neutron, with a mean lifetime of $\tau = 885. \pm 0.9$s, is considered as stable. Basically, the hydrogen-helium abundance helps us to model the expansion rate of the early universe. If it had been faster, there would be more neutrons and more helium. If it had been slower, more of the free neutrons would have decayed before the deuterium stability point and there would be less helium.

Stable nuclei of ${}^4_2\text{He}$ begin to form in the first three minutes of the age of the universe.

Before proceeding further with the Big Bang Nucleosynthesis (BBN) let us elaborate a bit on what we know today about the fusion of hydrogen to helium in stars. In fact we know that there exist two main processes:

- The pp cycle.

$${}^1_1\text{H} + {}^1_1\text{H} \rightarrow {}^2_2\text{He} + \gamma$$

$${}^2_2\text{He} \rightarrow {}^2_1\text{H} + e^+\nu_e$$

$${}^2_1\text{H} + {}^1_1\text{H} \rightarrow {}^3_2\text{He} + \gamma\,(5.49\text{ MeV})$$

$${}^3_2\text{He} + {}^3_2\text{He} \rightarrow {}^4_2\text{He} + 2\,{}^1_1\text{H} + 12.86\text{ MeV}$$

  The above process is quite slow, since the second reaction is very slow.

- The CNO cycle. It was independently proposed by Carl von Weizsäcker and Hans Bethe in 1938 and 1939, respectively.

$${}^{12}_6\text{C} + p \rightarrow {}^{13}_7\text{N} + \gamma$$

$${}^{13}_7\text{N} \rightarrow {}^{13}_6\text{C} + e^+ + \nu_e$$

$${}^{13}_6\text{C} + p \rightarrow {}^{14}_7\text{N} + \gamma$$

$${}^{14}_4\text{N} + p \rightarrow {}^{15}_8\text{O} + \gamma$$

$${}^{15}_8\text{O} \rightarrow {}^{15}_7\text{N} + e^+ + \nu_e$$

$${}^{15}_7\text{N} + p \rightarrow {}^{16}_8\text{N}^* \rightarrow {}^{12}_6\text{C} + {}^4_2\text{He}$$

  So ${}^{12}_6\text{C}$ acts as a catalyst. The net result the same as in the pp cycle. Four protons are consumed to produce two positrons, two neutrinos and an α particle plus energy.

  There exist some slight variations of the above cycles (e.g. CNO-I, CNO-II etc)

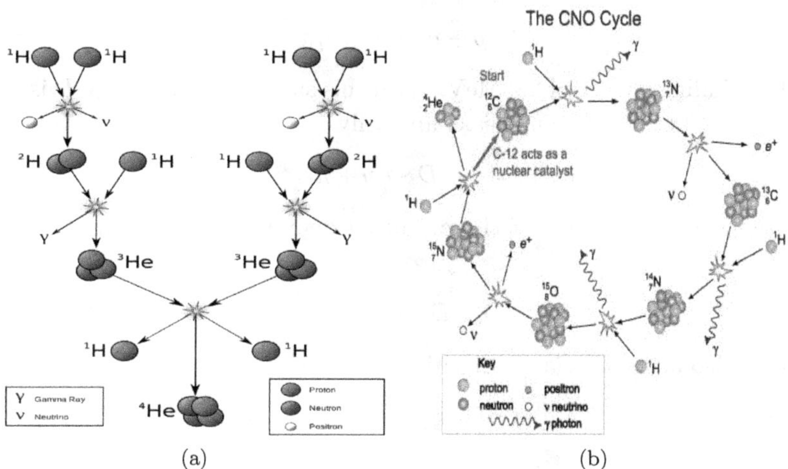

(a) (b)

Fig. 10.7: The fusion of hydrogen to He in stars. The pp cycle (a) and the CNO cycle (b)

The above reactions are shown pictorially in Fig. 10.7. Additional energy is produced by the pair annihilation of the positrons:

$$e^+ + e^- \rightarrow 2\gamma \leftrightarrow (1.05 \text{ MeV})$$

Theoretical models suggest that the CNO cycle is the dominant source of energy in stars more massive than about 1.3 times the mass of the Sun.The proton-proton chain is more important in stars the mass of the Sun or less. This difference stems from temperature dependent differences between the two reactions; pp-chain reactions start occurring at temperatures around 4×10^6K (4 megakelvins), making it the dominant energy source in smaller stars [Salaris and Cassisi (2005)], A self-maintaining CNO chain starts occurring at approximately 15 MK, but its energy output rises much more rapidly with increasing temperatures. At approximately 17 MK, the CNO cycle starts becoming the dominant source of energy.[3] The Sun has a core temperature of around 15.7 MK, and only 1.7% of $^4_2$4He nuclei being produced in the Sun are born in the CNO cycle.

Elements like deuterium, D, and ^{7}Li are much rarer. Also Oxygen, carbon, iron etc are quite a bit less abundant for reasons that will be given below.

At about the same time the first deuterium, D, nuclei begin to form:

$$p + n \rightarrow D + \gamma, \tag{10.13}$$

with a binding energy of 2.2 MeV. With this small binding energy D is very fragile, i.e. the deuterium breaks up easily,

$$\gamma + D \rightarrow p + n. \tag{10.14}$$

Furthermore quite a bit of the produced deuterium is lost, since it fuses yielding ${}^4_2\mathrm{He}$

$$D + D \rightarrow {}^4_2\mathrm{He} + \gamma. \tag{10.15}$$

This is also achieved via the reactions

$$D{+}p \rightarrow {}^3_2\,\mathrm{He}{+}\gamma, \quad D{+}n \rightarrow {}^3_1\mathrm{H}{+}\gamma, \quad D{+}D \rightarrow {}^3_2\mathrm{He}{+}n, \quad D{+}D \rightarrow {}^3_1\mathrm{H}{+}p, \tag{10.16}$$

$${}^3_1\mathrm{H} + p \rightarrow {}^4_2\mathrm{He} + \gamma, \quad {}^3_1\mathrm{H} + n \rightarrow {}^4_2\mathrm{He} + \gamma, \tag{10.17}$$

$${}^3_1\mathrm{H} + D \rightarrow {}^4_2\mathrm{He} + n, \quad {}^3_2\mathrm{H} + D \rightarrow {}^4_2\mathrm{He} + p. \tag{10.18}$$

With a much slower rate heavier elements are also produced

$${}^4_2\mathrm{He}{+}D \rightarrow {}^6_3\mathrm{Li}{+}\gamma, \quad {}^4_2\mathrm{He}{+}{}^3_1\mathrm{H} \rightarrow {}^7_3\mathrm{Li}{+}\gamma, \quad {}^4_2\mathrm{He}{+}{}^3_2\mathrm{He} \rightarrow {}^7_4\mathrm{B}{+}\gamma. \tag{10.19}$$

So the deuterium plays a crucial role in the nucleosynthesis and some of it survives as primordial element.

The process ends up here since time is running out and there no stable nuclei with 8 nucleons. Indeed the most suitable is the process:

$${}^4_2\mathrm{He} +{}^4_2\,\mathrm{He} \rightarrow {}^8_4\mathrm{Be} + \gamma\,. \tag{10.20}$$

The produced nucleus, however, has a mean life time $\tau = 2 \times 10^{-16}$s.

We should stress that the above elements (see Fig. 10.8) were produced primordially. They began to form in the first three minutes of the age of the universe. We have seen that the majority of the stars are very young and they did not have time to burn so much hydrogen . So their abundance constitutes evidence of the Big Bang scenario.

10.3 Evolution of the universe

The evolution of the universe can truly be described in terms of the General Theory of Relativity (GTR). Some conclusions, however, can be extracted classically.

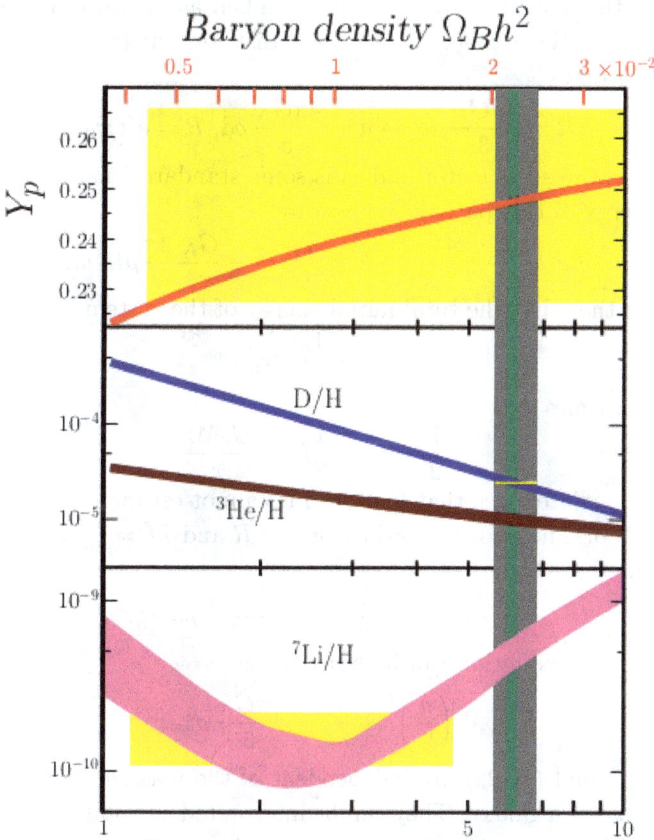

Fig. 10.8: The fraction of the abundance of the elements of primordial nucleosynthesis, relative to hydrogen, as a function of the baryon density, in gr/cm^3 on the bottom axis and as in units of the critical density at the top. The acceptable density region is also marked. The horizontal rectangular (yellow) bands indicate the measured error bars in $Y_p = Y_4, Y_7$ etc. One can see that Y_4 is quite uncertain and Y(D/H) is the most accurate. The narrow vertical band (green) is the CMB measure, while the wider (gray) band is the concordance range with BBN deuterium (95%CL).

10.3.1 *A classical prelude*

Let us consider a particle of mass m moving radially in the gravitational field generated by another mass M. Then the equation of motion is

$$m\ddot{r} = -\frac{G_N M}{r^2} m \rightarrow \ddot{r} = -\frac{G_N M}{r^2} \rightarrow \ddot{r} = -\frac{4\pi G_N}{3} \rho \frac{R^3}{r^2} \tag{10.21}$$

where ρ is the density of the source M. Then assuming that the motion takes place on the surface of the mass distribution ($r = R$) the above equation becomes

$$\ddot{R} = \frac{4\pi G_N}{3}\rho R \rightarrow \ddot{a} = \frac{4\pi G_N}{3}\rho a,\ R = a(t)r_0, \tag{10.22}$$

where $a(t)$ is the scale factor and r_0 is some standard length.

The energy of this particle is given by

$$E = \frac{1}{2}m\dot{r}^2 - \frac{G_N M}{r^2}m \rightarrow E = \frac{1}{2}m\dot{r}^2 - \frac{G_N}{r^2}\frac{4\pi}{3}\rho R^3 m. \tag{10.23}$$

Suppose further that the total initial energy of the system is given by

$$E = -\frac{1}{2}\kappa m r_0^2 \tag{10.24}$$

Then for all times

$$-\frac{1}{2}\kappa m r_0^2 = \frac{1}{2}\dot{r}^2 - \frac{G_N M}{r}. \tag{10.25}$$

We can immediately see that for $\kappa > 0$ it cannot escape to infinity. It will do so if $\kappa \leq 0$. The above equation for $r = R$ and $M = \frac{4\pi\rho}{3}R^3$ can be cast in the form

$$-\frac{1}{2}\kappa m r_0^2 = \frac{1}{2}\dot{a}^2 r_0^2 - \frac{4\pi G_N}{3}\rho a^2 r_0^2. \tag{10.26}$$

i.e. the previous equation can be written as:

$$\left(\frac{\dot{a}}{a}\right)^2 + \frac{\kappa}{a^2} = \frac{8\pi G_N}{3}\rho. \tag{10.27}$$

Eqs. (10.27) and (10.22) are independent of the mass m, which could be, e.g., the mass of a galaxy. They can be interpreted as applying to the whole universe with ρ being the density of the universe. We are gratified that, at least in the case of matter, they agree with the exact equations derived in the context of GTR, Eqs (10.42) and (10.44) below, known as Friedman equations.

Introducing now the critical density ρ_c by

$$\frac{8\pi G_N}{3}\rho_c = H^2,\ H = \frac{\dot{a}}{a} \tag{10.28}$$

we find:

$$\kappa = (Ha)^2\left(\frac{\rho}{\rho_c} - 1\right) \tag{10.29}$$

In the study of the evolution of the universe instead of κ one finds more convenient to use the quantity Ω defined by:

$$\Omega = 1 + \frac{\kappa}{(aH)^2} \tag{10.30}$$

Classically $\Omega = \frac{\rho}{\rho_c}$.

From this we see that applying the above results to the universe we find:

- if $\kappa > 0 \leftrightarrow \rho > \rho_c$ the universe will remain finite. Geometrically we say that it has positive curvature (closed geometry).
- if $\kappa < 0 \leftrightarrow \rho < \rho_c$ the universe will expand for ever. Geometrically we say that it has positive curvature (open geometry).
- if $\kappa = 0 \leftrightarrow \rho = \rho_c$ the universe will expand for ever. Geometrically we say that it has zero curvature (flat geometry).

The above results are, however, obtained classically and they do not include any other form of energy. This can only be done in the context of GTR.

The critical density is changing with time as H changes with time. Its present day value is:

$$\rho_c(t_0) = \frac{3H_0^2}{8\pi G_N} = 1.87 \times 10^{-29}\text{gr/cm}^{-3}h_0^2 = 11.2h_0^2 \text{ nucleons/m}^3, \quad (10.31)$$

where

$$0.706 \leq h_0 \leq 0.752\,.$$

We will see later that the actual density of protons is much smaller than this critical density. Thus the number of protons must be compared with the 300 photons per cm^3 or 3×10^8 per m^3 observed in the microwave background radiation.. This is another indication that the number of photons in the universe 1s about 10^{10} than that of protons.

The relative distance R(t) of two points in space is related to their co-moving distance r_0 via the scale factor $a(t)$. The latter gives the expansion rate of the universe.

$$R(t) = a(t)r_0, \quad (10.32)$$

with r_0 constant in time.

10.3.2 *The Robertson-Walker metric in the context of GTR*

The equations, describing the evolution of the universe, can be obtained by the **Robertson-Walker metric** in the context of GTR. This metric can be derived taking into account the demand of homogeneity and isotropy of space- time (see [Weinberg (2008.)]).

$$ds^2 = c^2dt^2 - d\ell^2, \quad d\ell^2 = a^2(t)\left(\frac{dx^2}{1-\kappa(x/R_0)^2} + x^2[(d\theta)^2 + \sin^2\theta(d\phi)^2]\right), \quad (10.33)$$

where κ is the curvature, κ=1,0,-1 for closed, flat and open universe respectively. This metric is sometimes written in a different form by defining:

$$dr = \frac{dx}{\sqrt{1 - \kappa(x/R_0)^2}}. \tag{10.34}$$

Integrating this equation we obtain:

$$x = S_\kappa \left(r/R_0\right) = \begin{cases} R_0 \sin \dfrac{r}{R_0}, & \kappa = 1 \\ r, & \kappa = 0 \\ R_0 \sinh \dfrac{r}{R_0}, & \kappa = -1. \end{cases} \tag{10.35}$$

Thus the metric can be written as

$$ds^2 = c^2 dt^2 - d\ell^2, \quad d\ell^2 = a^2(t) \left(dr^2 + S_\kappa \left(r/R_0\right)^2 [(d\theta)^2 + \sin^2\theta (d\phi)^2], \right) \tag{10.36}$$

$$\begin{aligned} \kappa =& 1 \Leftrightarrow \text{Closed (curved) universe}, \\ \kappa =& 0 \Leftrightarrow \text{flat universe}, \\ \kappa =& -1 \Leftrightarrow \text{Open (curved) universe}. \end{aligned}$$

Einstein's equation is given by:

$$R_{\mu\nu} - \frac{1}{2} R g_{\mu\nu} = -8\pi G_N T_{\mu\nu}, \tag{10.37}$$

where $R_{\mu\nu}$ is Ricci tensor, $R = g^{\mu\nu} R_{\mu\nu}$, and $T_{\mu\nu}$ is the energy momentum tensor.

For a homogeneous and isotropic fluid it takes the form:

$$T_{\mu\nu} = -\frac{P}{c^2} g_{\mu\nu} + \left(\frac{u}{c^2} + \frac{P}{c^2} \hat{u}_\mu \hat{u}_\nu \right), \quad \hat{u} = (1, 0, 0, 0). \tag{10.38}$$

In the co-moving reference frame one gets:

$$T_{\mu\nu} = \text{diag} \left(\frac{u}{c^2}, \frac{P}{c^2}, \frac{P}{c^2}, \frac{P}{c^2} \right), \tag{10.39}$$

where u is the energy density ($u = \rho c^2$ for matter) and P the pressure. One can show (see [Weinberg (2008.)])

$$R_{00} = 3 \frac{\ddot{a}(t)}{a(t)}, \; R_{11} = R_{22} = R_{33} = - \left(\frac{\ddot{a}(t)}{a(t)} + 2 \left(\frac{\dot{a}}{a} \right)^2 + \frac{2\kappa}{a^2} \right), \tag{10.40}$$

$$R = 6 \left(\frac{\ddot{a}(t)}{a(t)} + \left(\frac{\dot{a}}{a} \right)^2 + \frac{\kappa}{a^2} \right). \tag{10.41}$$

Thus the time-time component of Einstein's equation gives:

$$\left(\frac{\dot{a}}{a}\right)^2 + \frac{\kappa}{a^2} = \frac{8\pi}{3c^2} G_N u \tag{10.42}$$

and any space-space component of Einstein's equation yields:

$$-2\frac{\ddot{a}}{a} + \left(\frac{\dot{a}}{a}\right)^2 + \frac{\kappa}{a^2} = \frac{8\pi}{c^2} G_N 3P. \tag{10.43}$$

Combining the last two equations we get:

$$\frac{\ddot{a}}{a} = -\frac{4\pi}{3} G_N \left(\frac{u}{c^2} + 3\frac{P}{c^2}\right). \tag{10.44}$$

Eqs. (10.42) and (10.44) are known as Friedman equations.

To proceed further we need one more equation to connect the pressure and the energy density. It is of the form:

$$P = wu, \quad w = \text{constant} \quad \text{(Equation of state)}. \tag{10.45}$$

Taking the time derivative of Eq. (10.42) and then eliminating $\ddot{a}$ via Eq.(10.44) after the use of (10.45) we obtain:

$$\dot{u} + 3\frac{\dot{a}}{a}(1+w)u = 0 \Rightarrow \frac{du}{u} = -3(w+1)\frac{da}{a} \Rightarrow u = \frac{u_0}{a^{3+3w}}. \tag{10.46}$$

Thus

$$\Omega_i = \frac{\Omega_{i_0}}{a^{3+3w_i}} \quad \text{(for any type of energy } i\text{)}. \tag{10.47}$$

Putting this in the Friedman equation, Eq. (10.42), we obtain a differential equation for $a(t)$, which describes the evolution of the universe.

In the case of a classical fluid (k here is Boltzmann's constant) we obtain:

$$P = \frac{N}{V}kT = \frac{\rho}{m}kT = \frac{u}{mc^2}\frac{1}{3}m < v^2 > = u\frac{< v^2 >}{3c^2}, \tag{10.48}$$

where $u = \rho c^2$. Thus

$$\frac{P}{u} = \frac{< v^2 >}{3c^2} \ll 1 \Rightarrow w = 0\,. \tag{10.49}$$

In the case of (black body) radiation we have[4]:

$$u_r = \tilde{a}T^4, \quad P_r = \frac{1}{3}\tilde{a}T^4 \Rightarrow P_r = \frac{1}{3}u_r \Rightarrow w = \frac{1}{3}. \tag{10.50}$$

Thus

$$\Omega_r = \frac{\Omega_{r_0}}{a^4}, \tag{10.51}$$

[4]We have used here $\tilde{a}$ instead of the usual notation a to avoid possible confusion with the scale factor.

which must be compared with the expression for matter:

$$\Omega_m = \frac{\Omega_{mo}}{a^3}. \tag{10.52}$$

It is clear that even if radiation is irrelevant, i.e. too small, today ($a(t_0) = 1$), it must have dominated the universe in in the past ($a(t) << 1$).

Using now the relations

$$\Omega_{0i} = \frac{u_{0i}}{u_{crit}} = u_{0i}\frac{8\pi G_N}{3c^2 H_0^2} \rightarrow u_{0i}\frac{8\pi G_N}{3c^2} = \Omega_{0i} H_0^2,$$

where the critical density today is $u_{crit} = (3H_0^2 c^2)/(8\pi G_N)$, we can cast the Friedman equation into the form:

$$\left(\frac{\dot{a}}{a}\right)^2 + \frac{\kappa}{a^2} = H_0^2 \sum_i \frac{\Omega_{0i}}{a^{3(1+w_i)}} \tag{10.53}$$

One instead of the curvature κ we use the parameter Ω_0 defined by:

$$\Omega_0 = 1 + \frac{\kappa}{H_0^2} \tag{10.54}$$

(see Eq. 10.30 and note that $a(t_0) = 1$). Then the Friedman equation for two components matter and radiation becomes:

$$\frac{\dot{a}}{a} = H_0 \sqrt{\frac{\Omega_{mo}}{a^3} + \frac{\Omega_{ro}}{a^4} + \frac{1 - \Omega_0}{a^2}} \tag{10.55}$$

The solution for the scale factor can be cast into the implicit form:

$$\int_0^a \frac{dx}{\sqrt{\Omega_{mo}/x + \Omega_{ro}/x^2 + (1 - \Omega_0)}} = H_0 t, \; a(0) = 0. \tag{10.56}$$

The age of the universe t_0, taking the present scale to be unity, $a(t_0) = 1$, is given by:

$$t_0 = \frac{1}{H_0} \int_0^1 \frac{dx}{\sqrt{\Omega_{mo}/x + \Omega_{ro}/x^2 + (1 - \Omega_0)}} \tag{10.57}$$

The scale is set by the Hubble constant!

We can see now what happens if there is no matter, radiation in the universe, which we call an empty universe. then $\Omega_0 < 1$, i.e. $\kappa < 0$, so the space must be negatively curved. Then the scale factor is

$$a(t) = H_0 \sqrt{1 - \Omega_0} t,$$

i.e. the rate of expansion is constant. We call it an empty universe, which, however, evolves!

It is also easy to find the solution for one component and flat universe.

$$a(t) = \left(\frac{t}{t_0}\right)^{2/3}, \quad t_0 = \frac{2}{3}\frac{1}{\sqrt{\Omega_{m0}}}\frac{1}{H_0} \qquad \text{matter.}$$

$$a(t) = \left(\frac{t}{t_0}\right)^{1/2}, \quad t_0 = \frac{1}{2}\frac{1}{\sqrt{\Omega_{r0}}}\frac{1}{H_0} \qquad \text{radiation.} \tag{10.58}$$

10.3.3 *Presence of matter and curvature*

We will distinguish two cases

i) Negative curvature ($\Omega_0 < 1$).
Now the solution of Eq. (10.56) is a bit more complicated one finds:

$$\frac{1}{\sqrt{\Omega_{m_0}}}\frac{1}{s^{3/2}}\left(\sqrt{as(1-as)}-\tan^{-1}\left(\sqrt{\frac{1}{as}-1}\right)\right)=H_0 t \quad (10.59)$$

where $s=(1-\Omega_0)/\Omega_{m_0}$. The age of the universe is given by:

$$t_0=\frac{1}{H_0}\frac{1}{\sqrt{\Omega_{m_0}}}\frac{1}{s^{3/2}} \quad (10.60)$$

The scale factor $a(t)$ is exhibited in Fig. 10.9.

ii) Positive curvature $\Omega_0 > 1$.
In this case one finds:

$$\frac{\cos^{-1}\left(\sqrt{\frac{a(1-\Omega_0)+\Omega_{m_0}}{\Omega_{m_0}}}\right)\Omega_{m_0}-\sqrt{a\,(\Omega_0-1)}\sqrt{a\,(1-\Omega_0)+\Omega_{m_0}}}{(\Omega_0-1)^{\,3/2}}$$
$$=H_0 t. \quad (10.61)$$

In this case $a(t)$ reaches a maximum

$$a_{max}=\frac{\Omega_{m_0}}{\Omega_0-1} \quad (10.62)$$

at time:

$$t_1=\frac{\pi\Omega_{m_0}}{2H_0\,(\Omega_0-1)^{\,3/2}}. \quad (10.63)$$

For $t > t_1$ the Friedman equation must be solved with the initial condition $a(t_1)=a_{max}$. This way we find that $a(t)$ becomes zero at time $2t_1$. Subsequently the above cycle is repeated at infinitum. The age of the universe is given by

$$t_0=\frac{1}{H_0}\left(\frac{\cos^{-1}\left(\sqrt{\frac{-\Omega_0+\Omega_{m_0}+1}{\Omega_m}}\right)\Omega_{m_0}}{(\Omega_0-1)^{\,3/2}}-\frac{\sqrt{-\Omega_0+\Omega_{m_0}+1}}{\Omega_0-1}\right). \quad (10.64)$$

The scale factor $a(t)$ is now exhibited in Fig. 10.10.

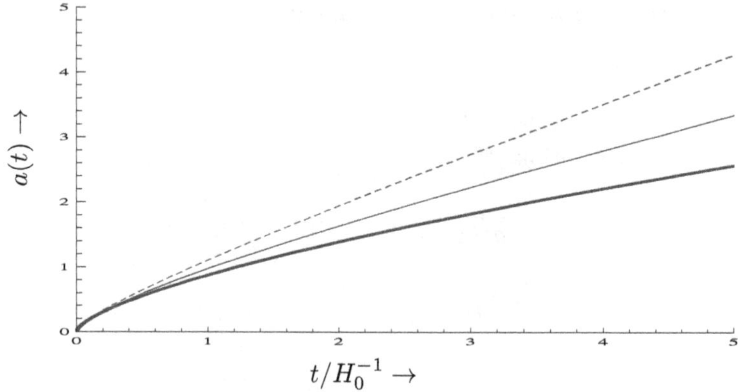

Fig. 10.9: The scale factor $a(t)$ as a function of time in units H_0^{-1}) for $\Omega_{m_0} = 0.3$. The thick solid, the thin solid and the dashed line correspond to flat universe, ($\Omega_0 = 1.0$), $\Omega_0 = 0.8$ and $\Omega_0 = 0.5$ respectively. The age of the universe is $\frac{2}{3H_0}$, $\frac{0.567}{H_0}$ and $\frac{0.481}{H_0}$ respectively.

10.3.4 *Presence of radiation and curvature (the early years)*

From the observation of microwave background radiation (MBR) we know that

$$T_0 = 2.724 \pm 0.001\text{K}, \quad u_r = aT_0^4 = 4.17 \times 10^{-14}\text{Jm}^{-3} = 0.260\text{MeV-m}^{-3} \tag{10.65}$$

Using the expression for the critical energy density

$$u_c = \frac{3H_0^2c^2}{8\pi G_N} = (8.3{\pm}1.7){\times}10^{-10}\,\text{Jm}^{-3}\,h_0^2 = 5.2{\times}10^3\,\text{MeV}\,\text{m}^{-3}\,h_0^2. \tag{10.66}$$

we obtain

$$\Omega_{r_0} = \frac{u_r}{u_c} = \frac{0.260\text{MeV-m}^{-3}}{5.2 \times 10^3\text{MeV-m}^{-3}} = 5.0 \times 10^{-5} \tag{10.67}$$

This density appears today quite small, but, at some point back in time, it was dominant. Indeed using the current value of Ω_{r_0} and $\Omega_{m_0} = 0.3$ we find

$$\frac{\Omega_r}{\Omega_m} = \frac{5.0 \times 10^{-5}}{0.3}\frac{1}{a} = 1.7 \times 10^{-4}\frac{1}{a}. \tag{10.68}$$

Thus when the scale factor satisfied the condition $a < 1.7 \times 10^{-4}$ ($a = 1$ today), radiation was dominant.

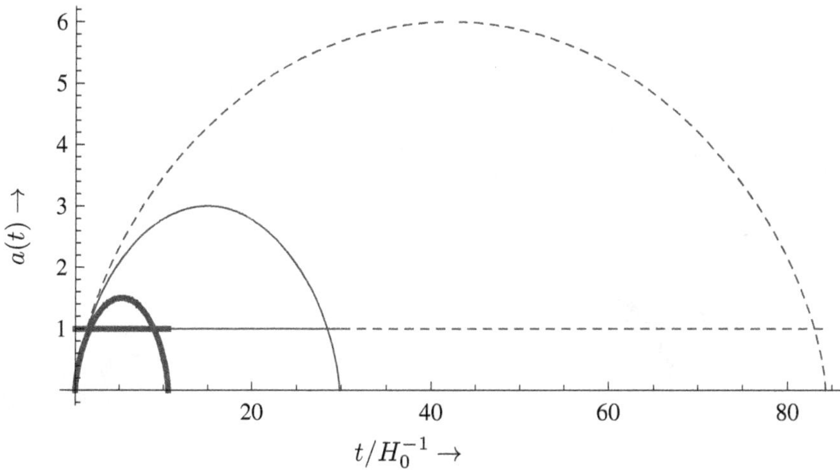

Fig. 10.10: The same as in Fig. 10.9 for $\Omega_{m_0} = 0.3$. The thick solid, the thin solid and the dashed line correspond to $\Omega_0 = 1.2$, $\Omega_0 = 1.1$ and $\Omega_0 = 1.05$ respectively. Since the universe is currently expanding, the age of the universe is at the first intersection of the corresponding curve with the line $a(t) = 1$. One numerically finds $\frac{1.6}{H_0}$, $\frac{1.4}{H_0}$ and $\frac{1.3}{H_0}$ respectively. It is not far from the origin. The collapse will take a long time!

The Friedman equation for our two component system yields

$$\int_0^a \frac{xdx}{\sqrt{\Omega_{r_0} + (1 - \Omega_0)x^2}} = H_0 t, \quad a(t_0) = 1. \tag{10.69}$$

Once this equation is inverted one finds that the Hubble parameter in terms of Ω_{r_0} and Ω_0 is given by

$$H(t) = H_0 \frac{\sqrt{\Omega_{r_0} + (1 - \Omega_0)a^2}}{a^2}. \tag{10.70}$$

We will now consider the following possibilities

i) Flat universe $\Omega_0 = 1$.

$$a(t) = \sqrt{2H_0 t \Omega_{r_0}^{1/2}} \Rightarrow t_0 = \frac{1}{2} \frac{1}{\Omega_{r_0} H_0} \Rightarrow$$

$$a(t) = \sqrt{\frac{t}{t_0}}. \tag{10.71}$$

ii) Closed curved universe $\Omega_0 > 1$. Now

$$a = \sqrt{\xi \left(-\Omega_0 \xi + \xi + 2\sqrt{\Omega_{r_0}} \right)}, \quad \xi = H_0 t. \tag{10.72}$$

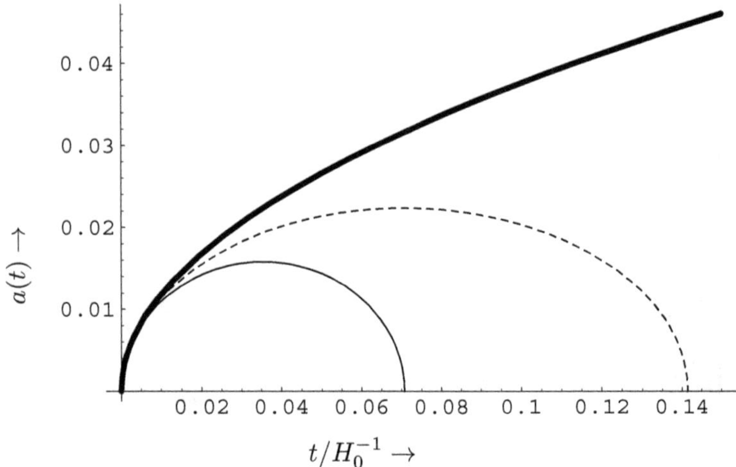

Fig. 10.11: The scale factor $a(t)$ as a function of time (in units of H_0^{-1}) in a universe dominated by radiation for $\Omega_{r_0} = 5.0 \times 10^{-5}$. The thick solid the fine solid correspond to a flat universe $\Omega_0 = 1.0$), $\Omega_0 = 1.2$ and $\Omega_0 = 1.1$ respectively.

The scale factor will attain a maximum at t_1:

$$t_1 = \frac{1}{H_0} \frac{\sqrt{\Omega_{r_0}}}{\Omega_0 - 1} \Rightarrow a_{max} = \sqrt{\frac{\Omega_{r_0}}{\Omega_0 - 1}}. \tag{10.73}$$

After that the system will go through a contraction stage as is exhibited in Fig. 10.11.

iii) Open curved universe, $\Omega_0 < 1$. Now

$$a = \sqrt{\xi \left(-\Omega_0 \xi + \xi + 2\sqrt{\Omega_{m_0}} \right)}, \quad \xi = H_0 t. \tag{10.74}$$

It is thus seen that for sufficiently large times the scale factor increases linearly. The main points are shown in Figs. 10.12-10.13.

10.4 The cosmological constant; dark energy

The Friedman equations lead to a non static universe. At the time of Einstein this was a defect. To cure this Einstein introduced in his GTR a new term, the cosmological constant. This lead him to a new classical repulsive gravitational force, which increases with distance:

$$F = \frac{1}{3} \Lambda m a, \quad \text{(gravitational repulsion a la Einstein)}. \tag{10.75}$$

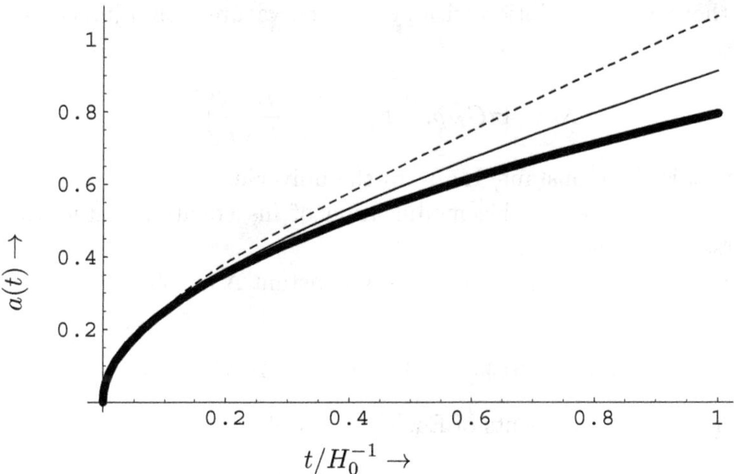

Fig. 10.12: The same as in Fig. 10.11. Now the curves correspond to $\Omega_0 = 1.0$, $\Omega_0 = 0.8$ and $\Omega_0 = 0.5$.

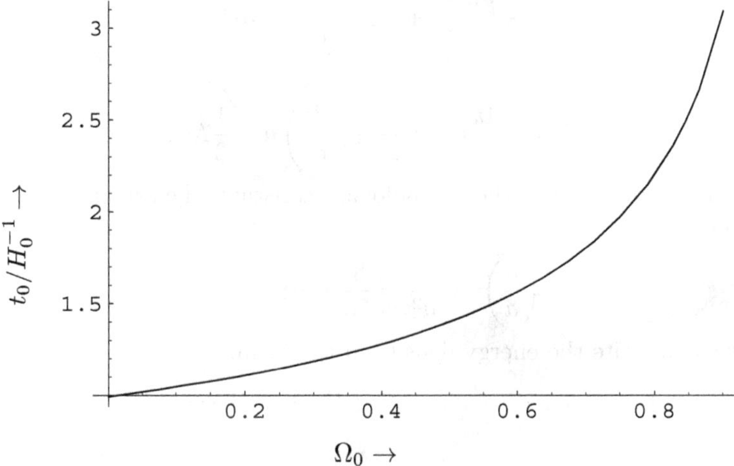

Fig. 10.13: The time (in units of H_0^{-1}) needed to get to the current size, $a = 1$, in an open universe dominated by radiation for $\Omega_{r_0} = 5.0 \times 10^{-5}$.

The condition for a static universe is:

$$\dot{a} = 0, \quad \ddot{a} = 0, \tag{10.76}$$

which leads to

$$-\frac{4\pi}{3} G_N \rho R_c + \frac{1}{3}\Lambda R_c = 0, \quad \frac{8\pi}{3} G_N \rho R_c^2 + \frac{1}{3}\Lambda R_c^2 = -\kappa. \tag{10.77}$$

Thus there exists a solution with positive curvature with a judicious choice $\Lambda = \Lambda_c$:

$$\Lambda_c = 4\pi G_N \rho, \quad R_c = \left(\frac{\kappa}{4\pi G_N \rho}\right)^{1/2}, \tag{10.78}$$

where R_c is the (constant) radius of the universe.
Einstein had to refute this modification of his equations later when the universe was found to expand.

Einstein introduced a cosmological constant Λ as follows:

$$R_{\mu\nu} - \frac{1}{2} R g_{\mu\nu} = -8\pi G_N T_{\mu\nu} - \Lambda g_{\mu\nu}, \quad \Lambda > 0. \tag{10.79}$$

The time-time components of Eq. (10.79) yields:

$$\left(\frac{\dot{a}}{a}\right)^2 + \frac{\kappa}{a^2} = \frac{8\pi}{3} G_N \frac{u}{c^2} + \frac{1}{3}\Lambda, \tag{10.80}$$

while any space-space component of Eq. (10.79) gives:

$$-2\frac{\ddot{a}}{a} + \left(\frac{\dot{a}}{a}\right)^2 + \frac{\kappa}{a^2} = \frac{8\pi}{c^2} G_N 3P - \frac{1}{3}\Lambda. \tag{10.81}$$

Thus

$$\ddot{a} = -\frac{4\pi}{3} G_N \left(\frac{u}{c^2} + 3\frac{P}{c^2}\right) a + \frac{1}{3}\Lambda a. \tag{10.82}$$

After the introduction of the Cosmological constant, the Friedman equation becomes:

$$\left(\frac{\dot{a}}{a}\right)^2 + \frac{\kappa}{a^2} = \frac{8\pi}{3c^2} G_N u + \frac{1}{3}\Lambda. \tag{10.83}$$

Now we can write the energy density in the form:

$$u = u_m + u_r + \cdots + u_\Lambda, \quad P = P_m + P_r + \cdots + P_\Lambda, \tag{10.84}$$

where

$$u_m = \rho c^2 \;, \quad u_r = \tilde{a} T^4, \quad u_\Lambda = \frac{c^2}{8\pi G_N}\Lambda. \tag{10.85}$$

We thus find:

$$\left(\frac{\dot{a}}{a}\right)^2 + \frac{\kappa}{a^2} = \frac{8\pi}{3c^2} G_N \, u, \tag{10.86}$$

$$\ddot{a} = -\frac{4\pi}{3c^2} G_N \left(u + 3P\right) a. \tag{10.87}$$

In order to proceed further we need an equation of state:

$$P_i = w_i u_i, \quad i = m, r, \cdots, \Lambda, \tag{10.88}$$

where

$$w_m = 0, \quad w_r = \frac{1}{3}, \cdots, \quad w_\Lambda = -1. \tag{10.89}$$

Thus Eq. (10.86) can be cast in the form of of Eq (10.53) with the addition in the sum of the term:

$$\Omega_\Lambda = \frac{u_\Lambda}{u_{crit}}$$

A component of the universe with an equation of state of the form of Eq. (10.45) with $w < -1/3$ is called dark energy.

$$P = wu, \quad w < -1/3 \quad \text{(dark energy)}. \tag{10.90}$$

For the special case of the cosmological constant we have $w = -1$, i.e.

$$P_\Lambda = -u_\Lambda. \tag{10.91}$$

from Eq. (10.46) we see that the density associated with the cosmological constant does not change with time.

Eq. (10.86) can now be written in the form:

$$H^2 + \frac{\kappa}{a^2} = H^2 \sum_i \Omega_i, \sum_i \Omega_i = \Omega_m + \Omega_r + \Omega_\Lambda$$

or

$$\kappa = (aH)^2 \left(\sum_i \Omega_i - 1 \right) \tag{10.92}$$

Now we can express the condition for a curved universe in terms of its matter content:

$$\kappa = 1 \Leftrightarrow \sum_i \Omega_i > 1 \Leftrightarrow \text{(closed curved universe)},$$

$$\kappa = 0 \Leftrightarrow \sum_i \Omega_i = 1 \Leftrightarrow \text{(flat universe)},$$

$$\kappa = -1 \Leftrightarrow \sum_i \Omega_i < 1 \Leftrightarrow \text{(open curved universe)}.$$

Returning to the multi-component universe we have:

$$\left(\frac{\dot{a}}{a}\right)^2 = \frac{H_0^2}{a^2}\left[\sum_i \Omega_{i_0} a^{-1-3w_i} + 1 - \Omega_0\right], \quad \Omega_{i_0} = \frac{u_{i_0}}{u_c}, \quad \Omega_\Lambda = \Omega_{\Lambda_0}. \tag{10.93}$$

and

$$\frac{\ddot{a}}{a} = -\frac{1}{2}H_0^2 \sum_i (1 + 3w_i)\Omega_{i_0} a^{-3-3w_i}. \tag{10.94}$$

Eq. (10.55) can now be rewritten as:

$$\frac{\dot{a}}{a} = H_0 \sqrt{\frac{\Omega_{mo}}{a^3} + \frac{\Omega_{ro}}{a^4} + \frac{1 - \Omega_0}{a^2} + \Omega_\Lambda}. \tag{10.95}$$

Integrating Eq. (10.95) we obtain the the scale factor $a(t)$ implicitly written as:

$$\int_0^a \frac{dx}{\sqrt{\Omega_{m_0}/x + \Omega_{r_0}/x^2 + \Omega_\Lambda x^2 + (1 - \Omega_0)}} = H_0 t. \tag{10.96}$$

From this, if solved for a=a(t), one obtains the Hubble function:

$$H(t) = H_0 \frac{\sqrt{\Omega_{m_0} a + \Omega_{r_0} + \Omega_\Lambda a^4 + (1 - \Omega_0) a^2}}{a}. \tag{10.97}$$

From these we obtain the observationally important deceleration factor:

$$q = -\frac{a\ddot{a}}{\dot{a}^2}, \tag{10.98}$$

The age of the universe is:

$$\frac{1}{t_0} = H_0 \int_0^1 \frac{dx}{\sqrt{\Omega_{m_0}/x + \Omega_{r_0}/x^2 + \Omega_\Lambda x^2 + (1 - \Omega_0)}}. \tag{10.99}$$

Thus it can be obtained without integrating the differential equation (assuming $a(t_0) = 1$).

Specializing the Friedman equation for for $u_m = u_r = 0$ we find that the universe obeys the equation:

$$\int \frac{da}{\sqrt{\Omega_\Lambda a^2 + (1 - \Omega_0)}} = H_0 t + C_1, \tag{10.100}$$

where C_1 chosen so that today $a(t_0) = 1$. Now the equation

$$\int_0^a \frac{dx}{\sqrt{\Omega_\Lambda x^2 + (1 - \Omega_0)}} = H_0 t, \quad \Omega_\Lambda = \frac{u_\Lambda}{u_c}, \quad a(t_0) = 1. \tag{10.101}$$

is not always available since the solution may not vanish at $t = 0$

Hubble's function is given by:

$$H(t) = H_0 \frac{\sqrt{\Omega_\Lambda a^2 + (1 - \Omega_0)}}{a}. \tag{10.102}$$

One thus finds $a(t)$ as a function of the parameters Ω_Λ and Ω_0. Then one finds the Hubble function H(t). The condition $H = H_0$ for $a = 1$ implies

$\Omega_\Lambda = \Omega_0$. Thus:

- For flat universe ($\Omega_\Lambda = \Omega_0 = 1$):
$$a = e^{H_0(t-t_0)},\ H(t) = H_0. \tag{10.103}$$
In this case the scale factor increases exponentially (inflation!), but no solution can be zero for $t = 0$. Something else must have preceded the inflation. No expression for the age of the universe in terms of H_0.
- Positive curvature $\Omega_\Lambda = \Omega_0 > 1$. Then:
$$a = \sqrt{\frac{\Omega_\Lambda - 1}{\Omega_\Lambda}} \cosh H_0 t, \quad t_0 = \frac{1}{H_0} \cosh^{-1}\left(\sqrt{\frac{\Omega_\Lambda}{\Omega_\Lambda - 1}}\right), \tag{10.104}$$
where
$$\sqrt{1 - \frac{1}{\Omega_\Lambda}} \le a \le 1 \Rightarrow t_1 \le t \le t_0,\ t_1 = \frac{1}{H_0} \cosh^{-1}(1).$$
- Negative curvature $\Omega_\Lambda = \Omega_0 < 1$. Then:
$$a = \sqrt{\frac{1 - \Omega_\Lambda}{\Omega_\Lambda}} \sinh H_0 t, \quad t_0 = \frac{1}{H_0} \sinh^{-1}\left(\frac{\Omega_\Lambda}{\sqrt{1 - \Omega_\Lambda}}\right). \tag{10.105}$$

Summarizing the above results in the special cases of one component universe we find:

$$a(t) = \left(\frac{t}{t_0}\right)^{2/3}, \quad t_0 = \frac{2}{3} \frac{1}{\sqrt{\Omega_{m0}}} \frac{1}{H_0} \qquad \text{matter.}$$
$$a(t) = \left(\frac{t}{t_0}\right)^{1/2}, \quad t_0 = \frac{1}{2} \frac{1}{\sqrt{\Omega_{r0}}} \frac{1}{H_0} \qquad \text{radiation.}$$
$$a(t) = e^{H_0(t-t_0)}, t_0 \text{ unspecified} \qquad \text{cosmological constant } \Omega_\Lambda = 1$$

$$\begin{aligned}
H(t) &= \tfrac{2}{3t}, \quad q = -\frac{\ddot{a}(t)a(t)}{(\dot{a}(t))^2} = \frac{1}{2} \text{ (matter).} \\
H(t) &= \tfrac{1}{2t}, \quad q = -\frac{\ddot{a}(t)a(t)}{(\dot{a}(t))^2} = 1 \text{ (radiation).} \\
H(t) &= H_0, \quad q = -\frac{\ddot{a}(t)a(t)}{(\dot{a}(t))^2} = -1 \text{ (cosmological constant).}
\end{aligned} \tag{10.106}$$

10.5 The standard cosmological model

Observationally we know that:

$$\Omega_{m_0} = 0.3, \quad \Omega_{r_0} = 5.0 \times 10^{-5}, \quad \Omega_\Lambda = 0.7.$$

These parameters define the standard cosmological model (SCM).

- At the beginning of the universe the scale factor is small and radiation is dominant, even though Ω_{r_0} is small. Furthermore we believe that all particles were then massless.
- As the universe expands, there will come a time that matter begins to dominate. This will happen when

  $$\Omega_{m_0} a \geq \Omega_{r_0} \Rightarrow a \geq \Omega_{r_0}/\Omega_{m_0} \approx 5.0 \times 10^{-5}/0.3 = 1.7 \times 10^{-4}.$$

  The age of the universe is then:

  $$t_1 = t_0 a^2 = 13.6 \times 10^9 \text{y} \times (1.7 \times 10^{-4})^2 \approx 400\text{y}.$$

- Sometime later, as the universe gets bigger and bigger, dark energy will take over. This will happen when:

  $$\Omega_\Lambda a^3 \geq \Omega_{m_0} \Rightarrow a \geq (\Omega_{m_0}/\Omega_\Lambda)^{1/3}$$

  or

  $$a \approx (0.3/0.7)^{1/3} = 0.75 \Rightarrow t_2 = t_0 (0.75)^{3/2} \approx 9 \text{ billion years.}$$

  This has happened already. The rate of acceleration is increasing. We are eventually going to get a new inflation.

The evolution of the universe in the SM of cosmology is given in Fig. 10.14.

10.6 The proper length and the horizon

We will suppose that the system is homogeneous and isotropic, so that it is described by the Robertson-Walker metric. It is evident that, as the system expands, the distance between two points is increasing. The proper length is defined by:

$$d\ell^2 = (a(t))^2 \left[dr^2 + S_\kappa \left(r/R_0 \right)^2 d\Omega \right]. \tag{10.107}$$

where r, θ, ϕ are the coordinates of a point in a co-moving frame. In the direction of observation $d\Omega = 0$, i.e.

$$d\ell = a(t) dr. \tag{10.108}$$

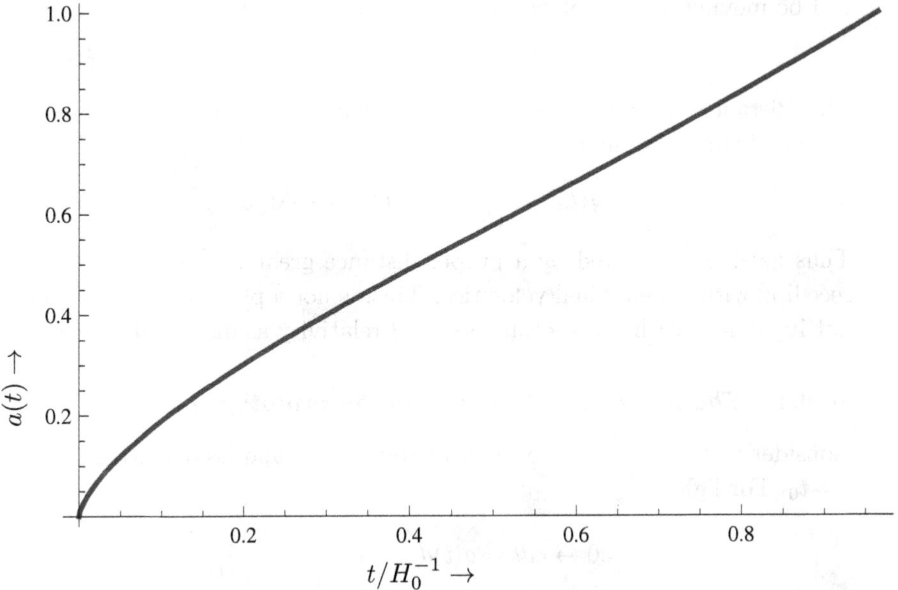

Fig. 10.14: The scale factor a(t) as a function of time (in units of H_0^{-1}=13.5 billion years) for a flat universe with $\Omega_m = 0.3$ and $\Omega_\Lambda = 0.7$.

Thus the proper distance becomes:

$$d_P(t,r) = a(t)r \Rightarrow d_P(t,x) = \begin{cases} a(t)R_0 \sin^{-1}\dfrac{x}{R_0}, & \kappa = 1 \\ a(t)x, & \kappa = 0 \\ a(t)R_0 \sinh^{-1}\dfrac{x}{R_0}, & \kappa = -1. \end{cases} \tag{10.109}$$

Since $d_P(t,r) = a(t)r$ and in the co-moving system r =constant, we find:

$$\dot{d}_P(t,r) = \dot{a}(t)r = \frac{\dot{a}(t)}{a(t)} d_P(t,r) = H d_P(t,r). \tag{10.110}$$

Omitting for simplicity of notation the indication r we find:

$$\upsilon_P(t_0) = H_0 d_P(t_0), \quad \upsilon_P(t_0) = \dot{d}_P(t_0). \tag{10.111}$$

i.e. we recover Hubble's law.

This means that points, which are separated by a proper distance greater than a critical distance

$$d_H(t_0) = \frac{c}{H_0}, \tag{10.112}$$

will be moving with a velocity greater than that of light:

$$v_P = \dot{d}_P > c. \tag{10.113}$$

The distance $d_H(t_0)$ is called Hubble distance. Using $H_0 = (70 \pm 7)$ kms^{-1}Mpc^{-1} we find:

$$d_H(t_0) = \frac{c}{H_0} = (4300 \pm 400)\text{Mpc}. \tag{10.114}$$

Thus galaxies separated by a proper distance greater than 4300 Mpc are receding with supeluminal velocities. This, is not a physical velocity and is not in conflict with the special theory of relativity as discussed elsewhere.

10.6.1 *The proper length entering the prototype candles*

Consider now the emission of light at time $t = t_e$ and its detection at time $t = t_0$. For light

$$d\ell = 0 \to cdt = a(t)dr \to r = c\int_{t_e}^{t_0} \frac{dt}{a(t)}. \tag{10.115}$$

This implies

$$d_P(t_e, t_0, r) = ca(t_0)\int_{t_e}^{t_0} \frac{dt}{a(t)}. \tag{10.116}$$

Expanding $a(t)$ around t_0 and noting that $a(t_0) = 1$, we get:

$$\begin{aligned} a(t) &= a(t_0) + \left.\frac{da}{dt}\right|_{t_0}(t - t_0) + \frac{1}{2}\left.\frac{d^2a}{dt^2}\right|_{t_0}(t - t_0)^2 + \cdots \\ &\approx 1 + H_0(t - t_0) - \frac{1}{2}q_0 H_0^2(t - t_0)^2, \end{aligned} \tag{10.117}$$

$$\frac{1}{a(t)} \approx \frac{1}{a(t_0)}\left[1 - H_0(t - t_0) + H_0^2(1 + \frac{1}{2}q_0)(t - t_0)^2\right], \tag{10.118}$$

$$d_P(t_e, t_0, r) \approx c\left(t_0 - t_e + \frac{1}{2}H(t_0)\,(t_e - t_0)^2\right). \tag{10.119}$$

Recall that

$$1 + z = \frac{1}{a(t_e)}. \tag{10.120}$$

Hence

$$z \approx \frac{1}{2}H_0^2\,(q_0 + 2)\,(t_0 - t_e)^2 + H_0\,(t_0 - t_e)\,. \tag{10.121}$$

From this we get

$$t_0 - t_e \approx \frac{1}{H_0} z \left(1 - (1 + q_0/2) z\right). \tag{10.122}$$

Thus we obtain the useful relation

$$d_P(z) \approx \frac{c}{H_0} z \left(1 - \frac{1+q_0}{2} z\right). \tag{10.123}$$

For a flat universe it is easy to show that:

$$d_P(z) = \frac{c}{H_0} \begin{cases} 2\left(1 - \frac{1}{\sqrt{1+z}}\right) & \text{(matter only)} \\ \frac{z}{1+z} & \text{(radiation only)} \\ z & \text{(cosmological constant only)} \end{cases} \tag{10.124}$$

10.6.2 *The horizon*

In cosmology as, of course, in special relativity, the visible universe must be within a region called **horizon**. From the Robertson-Walker metric for observing light, $ds = 0$, and $d\Omega = 0$, we find:

$$dr = c\frac{dt}{a(t)} \Rightarrow r = c\int_0^t \frac{dt'}{a(t')}. \tag{10.125}$$

The horizon is defined as:

$$d_H(t) = a(t) r = c a(t) \int_0^t \frac{dt'}{a(t')}. \tag{10.126}$$

At the epoch of radiation dominance for a flat universe we have seen that the scale factor is proportional to $t^{1/2}$. At some time t_r, the scale factor will be $a_r = a(t_r)$, while the horizon will be:

$$d_H(t_r) = c a_r \int_0^{t_r} \frac{dt}{a_r (t/t_r)^{1/2}} = 2ct_r. \tag{10.127}$$

At the end of radiation dominance we will have

$$2ct_r = 2 \times 3 \times 10^8 \text{m/s} \times 400\text{y} \approx 2 \times 3 \times 10^8 \text{m/s} \times 400 \times 3.156 \times 10^7 \text{s} = 7.6 \times 10^{18} \text{m} \Rightarrow$$

$$d_H(t_r) \approx 250\text{pc},$$

which is quite small.

Quite generally for a universe with one component with equation of state $P = wu$ the horizon after some time t_1 is:

$$d_H(t_1) = c a_1 \int_0^{t_1} \frac{dt}{a_1 (t/t_1)^{2/(3+3w)}} = \frac{3+3w}{1+3w} c t_1, \quad w \neq -1, \tag{10.128}$$

that is the horizon increases linearly with time. On the contrary, if $w = -1$, we have:

$$a(t) = a_1 e^{H_0(t-t_1)\sqrt{\Omega_\Lambda}} \Rightarrow d_H(t_1) = c a_1 \int_0^{t_1} \frac{dt}{a_1 e^{H_0(t-t_1)\sqrt{\Omega_\Lambda}}}, \tag{10.129}$$

$$d_H(t_1) = \frac{c}{H_0\sqrt{\Omega_\Lambda}} e^{H_0 t_1 \sqrt{\Omega_\Lambda}}. \tag{10.130}$$

The horizon now increases exponentially.

The situation gets a bit complicated wen the dominance passes from one constituent to the other. Thus passing, e.g., from radiation to matter dominance we have:

$$d_H(t_m) = c a_m \int_0^{t_m} \frac{dt}{a(t)} = c a_m \left[\frac{2ct_r}{ca_r} + \int_{t_r}^{t_m} \frac{dt}{a(t)} \right], \tag{10.131}$$

$$a_m(t) = \left(a_r^{3/2} + \frac{3}{2}\left(H_0 t - H_0 t_r\right)\sqrt{\Omega_0} \right)^{2/3}. \tag{10.132}$$

Thus

$$d_H(t_m) = \frac{c}{H_0}\frac{a_m}{a_r} H_0 t_r + \frac{c}{H_0} a_m$$

$$2^{2/3}\left(\sqrt[3]{2a_r^{3/2} + 3t_m H_0\sqrt{\Omega_0} - 3H_0 t_r\sqrt{\Omega_0}} - \sqrt[3]{2a_r^{3/2} - 3H_0 t_r \sqrt{\Omega_0}} \right), \tag{10.133}$$

where $a_m = a(t_m)$. The first term makes a little contribution. For $\Omega_{m_0} = 0.3$ we get the picture of 10.15.

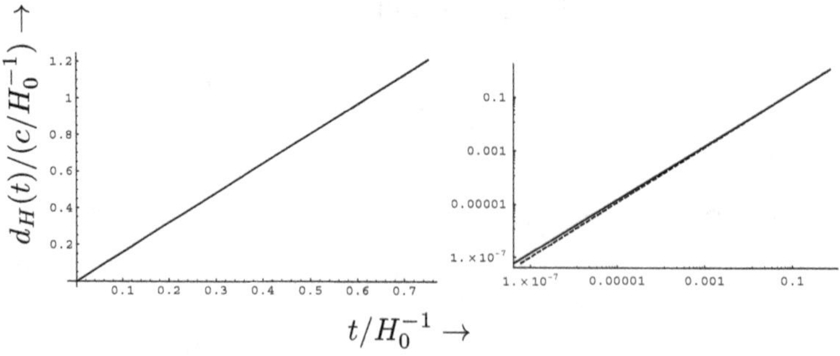

Fig. 10.15: The horizon corresponding to $\Omega_{m_0} = 0.3$ in a linear scale on the left and a logarithmic on the right. The horizon during the radiation dominance is not visible. The dashed line is associated with the second term of Eq. (10.133) by itself.

10.7 Evolution of temperature

From the scale factor one can obtain the temperature assuming that the evolution of the universe is adiabatic (reversible adiabat $\Leftrightarrow$ isentropic). All we need to know is an expression for the entropy.

Suppose that the system is radiation dominated. Then for the black body radiation

$$S = \frac{4}{3}\tilde{a}T^3 V = \frac{4}{3}\tilde{a}T^3 \frac{4}{3}\pi a^3(t). \tag{10.134}$$

with

$$\tilde{a} = \frac{g_s}{2}\frac{\hbar}{\pi^2 c^3}\left(\frac{k}{\hbar}\right)^4 \int_0^\infty \frac{x^3 dx}{e^x - 1} = \frac{\hbar}{c^3}\left(\frac{k}{\hbar}\right)^4 \frac{\pi^2}{15}, \tag{10.135}$$

for the photon $g_s = 2$, the degeneracy (the number of polarizations). Its numerical value is: $\tilde{a} = 7.56 \times 10^{-16}$ Nt-m^{-2} (K)$^{-4}$. Then

$$S = \text{constant} \Leftrightarrow a(t)T = \text{constant} \Leftrightarrow \frac{T}{T_0} = \frac{a(t_0)}{a(t)} \Rightarrow T = T_0 \frac{1}{a(t)} \tag{10.136}$$

or

$$T = T_0 \sqrt{\frac{t_0}{t}} = 2.725\text{K}\sqrt{\frac{t_0}{t}}. \tag{10.137}$$

For matter

$$PV^\gamma = \text{constant}, \quad \gamma = 5/3 \Leftrightarrow VT^{1/(\gamma-1)} = \text{constant}.$$

Hence

$$a(t) = \left(\frac{T_0}{T}\right)^{1/(3(\gamma-1))} = \left(\frac{T_0}{T}\right)^{1/2} \tag{10.138}$$

From the known expression for a(t) we get

$$\frac{T}{T_0} = \left(\frac{t_0}{t}\right)^{4/3}. \tag{10.139}$$

Returning to the radiation dominated universe, we can obtain the thermal energy of the universe:

$$T = 10^{10}\left(\frac{t}{1/s}\right)^{-1/2} \text{K for } g_s = 2, \tag{10.140}$$

$$< E >= kT = \left(\frac{t}{1/s}\right)^{-1/2} \text{MeV for } g_s = 2. \tag{10.141}$$

This relation will be extended to other relativistic particles. We only need replace g_s with a statistical factor $n(T)$, which is a function of the temperature. More specifically:

$$n(T) = \sum_{i}^{N(T)} g_s(i), \tag{10.142}$$

which represents the number of particles coexisting in equilibrium at temperature T.

For the determination of $n(T)$ we observe the following:

- For massless bosons $g_s = 2$. For massive bosons $g_s = 2s + 1$.
- Every degree of freedom of a fermion contributes 7/8 of that of the photon.
- The neutrinos are left handed. If they are of Majorana type, the particle coincides with its antiparticle, $g_s = 2$. Otherwise $g_s = 1$.

We will illustrate this we the following examples:

- For the γ, e^+, e^- after the neutrinos are decoupled.
$$n(T) = g_s(\gamma) + \frac{7}{8}\left(g_s(e^+) + g_s(e^-)\right) = 2 + \frac{7}{8}(2+2) = \frac{11}{2}. \tag{10.143}$$
- For the neutrino subsystem ν_e, ν_μ, ν_τ:
$$n(T) = \frac{7}{8}(3 \times 2) = \frac{21}{4}. \tag{10.144}$$
- For all of them $\gamma, e^+, e^-, \nu_e, \nu_\mu, \nu_\tau$, before of neutrino decoupling, but for temperatures $kT < m_\mu c^2$
$$n(T) = \frac{11}{2} + \frac{21}{4} = \frac{43}{4}. \tag{10.145}$$
- For the case of the last system, enriched with muons and temperatures $m_\mu c^2 < kT < m_\pi c^2$, we find $n(T) = \frac{57}{4}$.

Generally speaking the parameter $n(T)$ becomes smaller as the temperature drops, since some particles have decayed. The situation is exhibited in Table 10.1.

10.7.1 *The relic neutrinos*

In addition to the microwave background radiation one expects relic neutrinos, neutrinos produced at some point in the history of the Unverse and now freely streaming around us. At the beginning the neutrinos were

Table 10.1: The quantity $4n(T)$ is shown for various temperatures together with the coexisting particles. For the quarks and gluons the picture is not so clear since it depends on the de-confinement energy E_{dc}. In any case the color degree of freedom has been included. In the GUT $SU(5)$ model, we have 12 additional gauge bosons, 26 Higgs particles. In the case of supersymmetry the number of particles must be doubled for the age of the universe before the time of their decay.

Temperature	New particles	$4n(T)$
$0 < kT < m_e c^2$	$\gamma, \nu_e, \nu_\mu, \nu_\tau$	29
$m_e c^2 < kT < m_\mu c^2$	e^+, e^-	43
$m_\mu c^2 < kT < m_\pi c^2$	μ^+, μ^-	57
$m_\pi c^2 < kT < E_{qg}$	π^+, π^-, π^0	69
$m_\mu c^2 < kT < m_W c^2$	$\tau, \bar{\tau}$	73
$E_{\rm dc} < kT < m_s c^2$	$u, \bar{u}, d, \bar{d}, g_i$	185
$m_s c^2 < kT < m_c c^2$	$s, \bar{s}$	227
$m_s c^2 < kT < m_c c^2$	$c, \bar{c}$	271
$m_c c^2 < kT < m_b c^2$	$b, \bar{b}$	313
$m_b c^2 < kT < m_t c^2$	$t, \bar{t}$	355
$m_W c^2 < kT$	W^+, W^-, Z^0, H^0	395
$SU(5)$, $kT \approx m_{\rm GUT}$	$X, Y, H^{(24)}, H^{(5)}, H^{*(5)}$	611

strongly interacting with matter. The number of interactions was later on decreased partly because matter was diluted as the universe expanded and partly because the interaction became what we know it to be today, i.e. weak interaction. Eventually the neutrinos were decoupled from the rest of the universe and they simply cooled as the universe expanded. This happened when the temperature reached $kT \approx 1$ MeV, i.e. when the universe was 1s old. We expect the neutrinos to be a bit colder than the photons, since the latter continued to have interactions, which ended up heating the photon component. The most important such reaction was the interaction of photons with the plasma of electrons and positrons:

$$e^+ + e^- \rightleftharpoons \gamma + \gamma \tag{10.146}$$

So long as the rate of this reaction is faster than the expansion rate of the universe the above system is in equillibrium. Thus the chemical potentials satisfy:

$$\mu(e^+) + \mu(e^-) = \mu(\gamma) \rightarrow \mu(e^+) + \mu(e^-) = 0 \tag{10.147}$$

Table 10.2: The main points in the history of the universe.

time	energy	temperature	content	forces
$0 < t \leq 10^{-43}$s	10^{19}GeV	$T = 10^{32}$ K	only energy	unification
10^{-43}s $\leq t \leq 10^{-35}$s	10^{19}GeV $\leq E \leq 10^{14}$GeV	10^{32} K $\leq T \leq 10^{27}$ K	gauge bosons, inflation	gravity separates
10^{-35}s $\leq t \leq 10^{-12}$s	10^{14}GeV $\leq E \leq 10^{2}$GeV	10^{27} K $\leq T \leq 10^{15}$ K	quark, gluon and lepton plasma	strong and EW separate
10^{-12}s $\leq t \leq 10^{-6}$s	10^{2}GeV $\leq E \leq 1$ GeV	10^{15} K $\leq T \leq 10^{13}$ K	quark gluon and lepton plasma	EM and weak separate
10^{-6}s $\leq t \leq 1$ s	1 GeV $\leq E \leq 1$ MeV	10^{13} K $\leq T \leq 10^{10}$ K	hadrons form neutron still lives	
1 s $\leq t \leq 3$ s	1 MeV $\leq E \leq 0.6$ MeV	10^{10} K $\leq T \leq 0.6 \times 10^{10}$ K	ν's decouple, ν-rad nuclei are formed	
3 s $\leq t \leq 100$ s	0.6 MeV $\leq E \leq 0.1$ MeV	0.6×10^{10} K $\leq T \leq 1.0 \times 10^{9}$ K	$e^+ + e^- \to$ photons	
$t \approx 4 \times 10^4$y	$E \approx 0.25$ eV	$T \approx 3.0 \times 10^3$ K	first atoms formed	
4×10^4 y $\leq t \leq 0.5 \times 10^9$y			dark ages opaque universe	
0.5×10^9 y $\leq t \leq 1.0 \times 10^9$y			first stars form reonization	
$t \geq 9 \times 10^9$y			Uni. acceleration	DE dominance

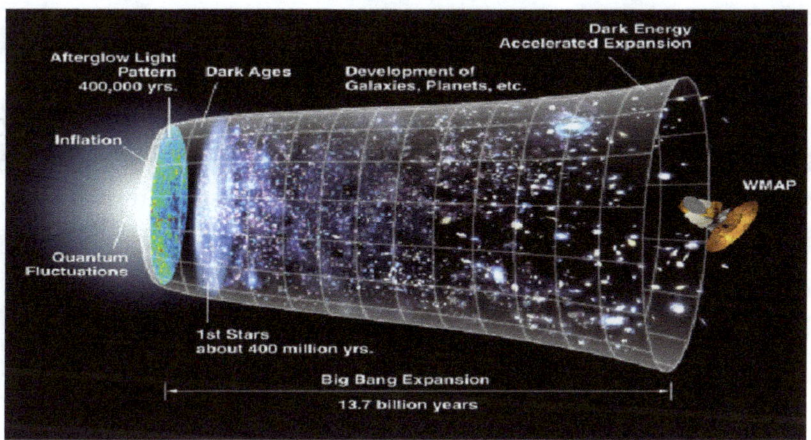

Fig. 10.16: Panoramic view of the history of the universe. Note that there appear two axes: the time and the size of the universe (see p. 1, colored section).

The chemical potential of the photons is zero, since the photonic system is not characterized by a fixed number of photons. On the other hand the fermions obey Fermi-Dirac distribution $\propto 1/(e^{(\epsilon-\mu_{\pm})/kT}+1)$, while charge conservation requires that the average number of electrons to be the same with that of positrons, $\langle(e+)\rangle = \langle(e+)\rangle$. This implies that $\mu_+ = \mu_-$, i.e. $\mu(e^+) = \mu(e^-)$. In other words all chemical potentials are zero ($\mu(e^+) = \mu(e^-) = \mu(\gamma) = 0$). This means that the electrons and the positrons follow the same distribution $\propto 1/(e^{\epsilon/kT}+1)$, while the photons follow the Bose-Einstein distribution $\propto 1/(e^{\epsilon/kT}-1)$. The process is, therefore, isentropic[5]

[5]The useful relations, known from statistical mechanics for relativistic particles with degeneracy g_s, are:

- The particle density:

$$\frac{dn_{\pm}}{d\epsilon} = 4\pi g_s \frac{\epsilon^2}{h^3} f_{\pm}(\nu, x), \quad f_{\pm}(\nu, x) = \frac{1}{x^{-\nu} \pm 1}, \quad x = \frac{\epsilon}{kT}, \quad \nu = \frac{\mu}{kT}, \tag{10.148}$$

  with μ the chemical potential.

- Energy density is

$$\frac{du_{\pm}}{d\epsilon} = 4\pi g_s \frac{\epsilon^3}{h^3} f_{\pm}(\nu, x), \tag{10.149}$$

- The specific entropy (entropy per unit volume) for $\mu = 0$ is

$$s_{\nu} = \frac{4}{3}\frac{7}{8}\frac{\pi^4}{15} 4\pi g_s \frac{(kT)^3}{h^3} k, \quad s_{\gamma} = \frac{4}{3}\frac{\pi^4}{15} 4\pi g_s \frac{(kT)^3}{h^3} k \tag{10.150}$$

we find:

$$s_1 = s_2 \rightarrow \left(4 \times \frac{7}{8} + 2\right) T_1^3 = 2T_2^3 \rightarrow \frac{11}{2} T_1^3 = 2T_2^3 \rightarrow T_2 = \left(\frac{11}{4}\right)^{1/3} T_1 \tag{10.152}$$

Thus when the temperature becomes sufficiently low, $kT \leq 2m_ec^2$, the reaction proceeds only to the left and the temperatures are related by

$$T_\gamma = \left(\frac{11}{4}\right)^{1/3} T_\nu.$$

This relation continues to hold thereafter. Today

$$T_{0\gamma} = \left(\frac{11}{4}\right)^{1/3} T_{0\nu} \rightarrow T_{0\nu} = \left(\frac{4}{11}\right)^{1/3} 2.723\text{K} = 1.95\text{K}$$

Since the energy density is proportional to the fourth power of the temperature and, noting that the ratio of the energy density of neutrinos is scaled up by 7/8 compared to that of the photons, we find:

$$u_\nu = 3 \times \frac{7}{8} \left(\frac{11}{4}\right)^{4/3} u_{\text{rad}} = 0.68 u_{\text{rad}}$$

Thus today

$$\Omega_\nu = 0.68 \Omega_\gamma \tag{10.153}$$

From neutrino oscillations (see chapter 11) we know that the neutrinos are massive, but we do not know the absolute scale of their mass. From the available neutrino oscillation data we know, however, the the sum of the neutrino masses satisfy the following limit $\sum_i m_{\nu i} \geq \sqrt{\Delta m^2_{\text{atm}}} + \sqrt{\Delta m^2_{\text{sun}}} =$ 0.1 eV. This indeed larger than the mean kinetic energy $3kT_0 \approx 5.5 \times 10^{-4}$ eV. We conclude that the relic neutrinos today are non relativistic. Thus:

$$\langle n_\nu(T) \rangle >= 3.6 \frac{1}{(2\pi)^2} \left(\frac{kT}{\hbar c}\right)^3 \tag{10.154}$$

- The corresponding average energy densities are:

$$\langle u_\nu \rangle = 4\pi g_s \left(\frac{kT}{h}\right)^3 kT \frac{7}{8} \frac{\pi^4}{15}, \quad \langle u_\gamma \rangle = 4\pi g_s \left(\frac{kT}{h}\right)^3 kT \frac{\pi^4}{15} \tag{10.151}$$

We have used of the following relations:

$$I_1 = \int_0^\infty \frac{x^2}{e^x + 1} = \frac{3}{2}\zeta(3), \quad \int_0^\infty \frac{x^2}{e^x - 1} = 2\zeta(3), \quad \zeta(3) \approx 1.2$$

$$I_2 = \int_0^\infty \frac{x^3}{e^x + 1} = \frac{7}{8} \frac{\pi^4}{15}, \quad \int_0^\infty \frac{x^3}{e^x - 1} = \frac{\pi^4}{15}, \quad \frac{I_2}{I_1} \approx 3$$

which hold for $\mu = 0$.

In other words for, $T_0 = 1.95\,\mathrm{K}$, we find

$$\langle n_\nu(T_0)\rangle >= 57\mathrm{cm}^{-3}. \tag{10.155}$$

The neutrino mass density is given by:

$$\Omega_m(\nu) = \Omega_\nu \frac{\left(\sum m_\nu c^2\right)/3}{3kT_0} = \frac{\sum m_\nu c^2}{45\mathrm{eV}} \tag{10.156}$$

10.7.2 *The re-ionization era*

The diagram of Fig. 10.17 below, provides a good graphical representation of the history of the universe. After the Big Bang, the universe was hot, but quickly cooling soup of fundamental particles. After a few hundred thousand years, things cooled enough so that protons and electrons could combine to form neutral hydrogen. This was a rather sudden event, and allowed the thermal glow of the fireball plasma, as it existed immediately before the hydrogen formation event, to radiate throughout the universe unimpeded by constant interactions with the charged particles of the now-absent plasma. This glow, red-shifted by a factor of 1100 or so, is what we now observe as the Cosmic Microwave Background (CMB) in all directions. The CMB carries a frozen imprint of the density fluctuations in the early universe, the study of which, by the observational cosmology community, is intense and sustained.

After the universe became neutral, it became unobservable across much of the electromagnetic spectrum. Any short wavelength radiation that might have been emitted was quickly absorbed by the atomic gas, and a long interval known as the Dark Ages began. Slowly, gravitational collapse of over-dense regions, the same regions we can see in the CMB imprint from earlier times, led to the formation of more and more pronounced structure in the neutral medium, and eventually the first stars, galaxies and quasars started to form. The exact mechanism and nature of this formation, poorly constrained by observation, is a topic of much research and great importance. We know what the universe looked like at the time of the CMB, and we know what it looks like now, but how did it get from one to the other?

As the collapse of structures proceeded, temperature variations developed. Gradually, energetic radiation emitted by the first sources caused local heating, and then ionization of the hydrogen in the universe. It will have started with "bubbles" of ionized plasma surrounding the most energetic sources. As the bubbles grew and became more numerous, they started to overlap, and more and more of the neutral medium became exposed to the harsh ionizing radiation, which travels unimpeded through ionized regions. The final phase of re-ionization of the universe may have occurred swiftly. As soon as the bulk of the universe was re-ionized, light at many wavelengths could escape from the early galaxies and quasars, revealing the distant universe that we see today with optical and infrared telescopes.

Re-ionization was complete about 1 billion years after the Big Bang, corresponding to a redshift of $z \approx 6.5$. Before that time, observations rapidly become more difficult. By and large, one must hope to find isolated, very luminous objects whose radiation in one form or another manages to reach us through the increasingly neutral medium. Perhaps the best hope for a more general and comprehensive probe of these early epochs is the 21cm hyperfine transition line of neutral hydrogen, red-shifted to frequencies below 200 MHz. Sensitive observations of emission and absorption in this line can probe deeply into the reheating and re-ionization epochs, and give us a detailed view of the density, temperature and velocity field of the material. We would get a view, not just of isolated luminous objects and the material which happens to lie in front of them, but of large volumes of the universe at the target redshifts. Such a view would yield a treasure of information from which to deduce the early history of structure formation, and the origin of the stars, galaxies, clusters and quasars that we see today. So today we the nice WMAP and PLANCK pictures of around 400000 years after the Big Bang. After that we do not have any observations till we again recover a luminous universe after re-ionization began at the age of about 500 million years. This is exhibited in Fig. 10.17.

10.8 Some problems with the Standard Cosmological Model

In spite of its successes the Standard Cosmological Model (SCM) suffers from a few problems. The most serious are:

(1) The horizon problem.
We know that

$$d_P(t_s, t_0) = c \int_{t_s}^{t_0} \frac{dt}{a(t)} \tag{10.157}$$

$$d_H(t_0) = ca(t_0) \int_0^{t_0} \frac{dt}{a(t)}. \tag{10.158}$$

From $t_s \ll t_0$ we conclude that the proper length at the time of decoupling about the same with today's horizon $P(t_s, t_0) \approx d_H(t_0)$,

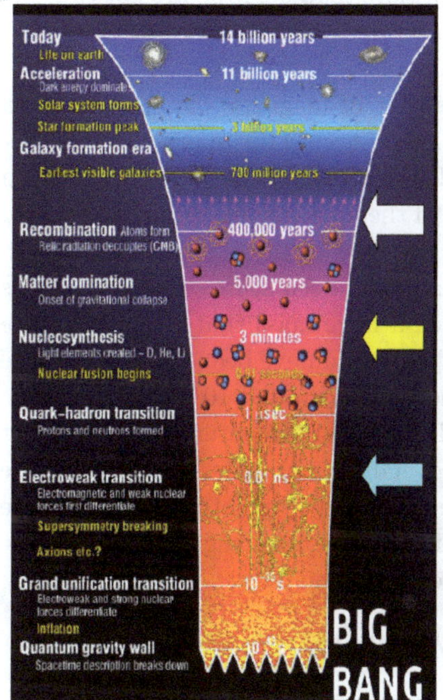

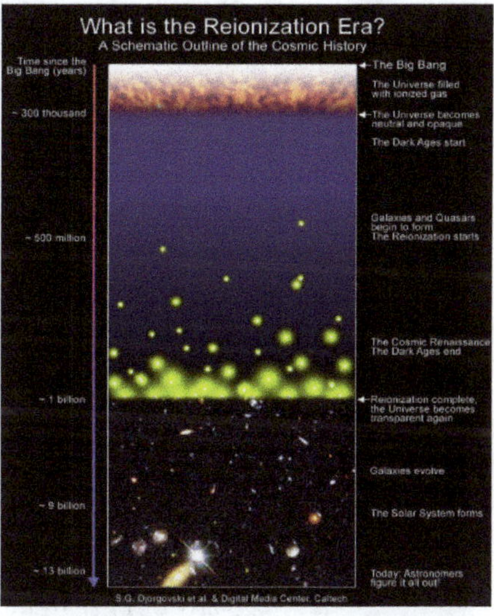

Fig. 10.17: The early history of the universe (left) and its real story, as observed by optical and infrared telescopes, after re-ionization (right).

or more precisely $d_P(t_s, t_0) = 0.98 d_H(t_0)$. This implies that two anti-diametrical points on the surface of last scattering are separated, as seen by an observer on Earth, by a distance $1.96 d_H(t_0)$. On the other hand points at a distance larger than the horizon are not causally connected. Therefore they did not have adequate to communicate. How come they have been in thermal equilibrium? And yet the temperature of the universe appears to be uniform with accuracy 1 to10^5. How did this happen?

We should also note that temperature variations represent variations in density of matter and energy existing at that time. In the SCM at the time of decoupling of photons the universe was dominated by matter. In this case for a flat universe:

$$d_H(t_s) = 2\frac{c}{H(t_s)} = \frac{t_s}{t_0} 2\frac{c}{H_0} \approx \frac{0.35 \times 10^6}{1.36 \times 10^10} 2\frac{c}{H_0} \approx 0.2\,\text{Mpc}\,. \tag{10.159}$$

As a result points separated by more than 0.2 Mpc are not causally connected. But we observe structures of 13.8 Mpc.

(2) The problem of magnetic monopoles.
Magnetic monopoles, i.e. the unit of magnetic charge, discussed originally by Dirac, are predicted to exist in many theories, beyond the SM, e.g. Grand Unified Theories (GUT). Their detection would have been easy because of the huge magnetic flux that would follow their motion. And yet in spite of the extensive hunt with all kinds of detectors they have not been seen. The way out is to imagine that they have such a huge mass that they cannot be produced in the laboratory. They could have been produced in the universe when the average energy was around 10^{14} m_p. Since the magnetic charge is, like the electric charge, conserved, the lightest monopole should have survived. This is a problem since, during the GUT phase transition following the GUT symmetry breaking, there appear topological effects which behave like magnetic monopoles. The number of such monopoles would be huge:

$$n_m(t_{\mathrm{GUT}}) = 10^{82}\,\mathrm{m}^{-3} \Rightarrow$$

$$\epsilon_m(t_{\mathrm{GUT}}) = m_{\mathrm{GUT}} n_m(t_{\mathrm{GUT}}) = 10^{94}\,\mathrm{TeV\,m}^{-3}\,.$$

In spite of the their huge energy density they were not dominant, since the radiation energy density is even higher,

$$\epsilon_{\mathrm{rad}}(t_{\mathrm{GUT}}) = aT^4_{\mathrm{GUT}} = 10^{104}\,\mathrm{TeV\,m}^{-3}\,.$$

Eventually, however, they would have become non relativistic and they would dominate since their energy would follow the rule a^{-3}.

(3) The flatness problem.
We know that the curviture is given by:

$$1 - \Omega = -\frac{\kappa c^2}{a^2 R_0^2 H^2} \Rightarrow 1 - \Omega(t) = \frac{H_0^2}{H^2(t) a^2(t)}(1 - \Omega_0). \quad (10.160)$$

The Friedman equation yields:

$$\frac{H^2}{H_0^2} = \frac{\Omega_{r_0}}{a^4} + \frac{\Omega_{m_0}}{a^3} + \Omega_{\Lambda_0}. \quad (10.161)$$

Thus

$$1 - \Omega(t) = \frac{a^2(1 - \Omega_0)}{\Omega_{r_0} + \Omega_{m_0} a + \Omega_{\Lambda_0} a^4}. \quad (10.162)$$

From observations we find $|1-\Omega_0| < 0.2$, which is consistent with SCM:

$$\Omega_{r_0} = 8.4 \times 10^{-4}, \quad \Omega_{m_0} = 0.28 \pm 0.05, \quad \Omega_{\Lambda_0} = 0.72 \pm 0.05 \tag{10.163}$$

(see 10.3 for more precise values). In the far distant past, i.e. during the period of radiation dominance, we have:

$$1 - \Omega(t) = \frac{a^2}{\Omega_{r_0}}(1-\Omega_0) = \frac{a_{rm}^2}{\Omega_{r_0}} \frac{t}{t_{rm}}(1-\Omega_0) \tag{10.164}$$

where a_{rm} is the scale factor at the end of this period, t_{rm}. Consequently if $\Omega = 1$, the universe has always been flat. However

$$|1-\Omega_0| \leq 0.2 \Rightarrow$$

$$|1-\Omega_{\,\mathrm{rm}}| \leq 2\times 10^{-2}, \quad |1-\Omega_{\,\mathrm{nuc}}| \leq 3\times 10^{-14}, \quad |1-\Omega_{\,\mathrm{P}}| \leq 1\times 10^{-60}$$

for the radiation era, the period of deuterium synthesis and the Planck time respectively. This requires a real fine tuning, which could have occured after God's intervention, but it is not acceptable by a bona fide physical theory.

To the above problems we should add some particle physics issues:

- How, when and why matter dominated over antimatter?
- Why does the cosmological constant have the value that it does? We know that Ω_Λ is not changing with time. In fact

$$\Omega_\Lambda = \frac{u_\Lambda}{u_c} \Rightarrow \Lambda = 3\Omega_\Lambda H_0^2 \tag{10.165}$$

(see Eqs (10.85) and (10.66). Thus

$$\Omega_\Lambda = 3 \times 0.7 \times (1.4 \times 10^{10} \times 3.156 \times 10^7)^{-2}\,\mathrm{s}^{-2} = 1.1 \times 10^{-35}\,\mathrm{s}^{-2}\,. \tag{10.166}$$

On the other hand

$$\mathrm{s}^{-1} = 5.4 \times 10^{-44}\,\mathrm{m_P}$$

where $\mathrm{m_P}$ is the Planck mass. Thus finally:

$$\Omega_\Lambda = 3.1 \times 10^{-122}\,\mathrm{m_P^2}\,. \tag{10.167}$$

How can one understand such a small value?

10.9 The period of inflation

Up to now the evolution of the universe is characterized by the following periods:

- Initially radiation dominance.
- Matter domination epoch, which lasted during most of the age of the universe.
- The dominant role of dark energy today.
- The expected complete domination of dark energy eventually.

The problems of the SCM mentioned above can be solved by the inflationary scenario proposed by Guth and Linde in 1981.

During the initial period all particles are massless and radiation is dominant. Suppose that at some time t_i during this period the cosmological constant appears and becomes dominant till a time t_f. Then we have the evolution of the unverse is described by the scale factor:

$$a(t) = \begin{cases} a_i \left(\dfrac{t}{t_i}\right)^{1/2}, & t \leq t_i \\ a_i e^{H_i(t-t_i)}, & t_i \leq t \leq t_f \\ a_i e^{H_i(t_f-t_i)} \left(\dfrac{t}{t_f}\right)^{1/2}, & t_f \leq t \leq t_{rm} \end{cases} \tag{10.168}$$

where t_{rm} is the time that matter begins to play a role. During the period $t_i \leq t \leq t_f$ the scale factor increased by a factor e^N, $N = 100$, with $H(t) = H_i$. At the moment t_{GUT}, i.e. when gravity becomes different than the other interactions, we have:

$$t_i = t_{\text{GUT}} = 10^{-36}\,\text{s}, \quad t_f = (N+1)t_i = 10^{-34}\,\text{s},$$

$$H_i = H_0\sqrt{\Omega_{\Lambda_i}} \rightarrow \Omega_{\Lambda_i} = \left(\frac{H_0^{-1}}{H_i^{-1}}\right)^2 = \left(\frac{4.4\times 10^{17}\,\text{s}}{10^{-36}\,\text{s}}\right)^2 = 2\times 10^{107},$$

which should be compared to the present value $\Omega_{\Lambda_0} = 0.7$. The corresponding energy density had then also been tremendous:

$$u_{\Lambda_i} = \Omega_{\Lambda_i} u_c = 2\times 10^{107} \times 5.2\times 10^3\,\text{MeV m}^{-3} \approx 10^{101}\,\text{MeV m}^{-3} = 10^{105}\,\text{TeV m}^{-3}$$

(see (10.66)).

With the introduction of the inflationary scenario the above problems are solved. In particular:

(1) There is no flatness problem.

The relation:

$$|1-\Omega(t)| = \frac{c^2}{R_0^2 a(t)^2 H(t)^2} \tag{10.169}$$

during the inflationary period, as we have seen, gives:

$$|1-\Omega(t)| = \frac{c^2}{R_0^{22}} e^{-2H_i(t-t_i)}. \tag{10.170}$$

Thus comparing the times t_i and $t_f = (N+1)t_i$ we get

$$|1-\Omega(t_f)| = e^{-2N}|1-\Omega(t_i)| \tag{10.171}$$

$$|1-\Omega(t_i)| = \kappa \rightarrow |1-\Omega(t_f)| = e^{-2N}\kappa = e^{-100}\kappa = 10^{-87}\kappa \approx 0. \tag{10.172}$$

In other words even if at the beginning of inflation the universe had a reasonable curvature, even $\kappa = e^{120}$, at the end of inflation is practically flat. In other words the present limit $|1-\Omega_0| \leq 0.2$ can be achieved even with the huge value $-1.5 \times 10^{86} \leq \Omega(t_i) \leq 1.5 \times 10^{86}$.

(2) There is no horizon problem.
We already know (see section 10.6.2) that

$$d_H t_i = 2ct_i \tag{10.173}$$

$$\begin{aligned} d_H(t_f) &= a_i c e^N \left(\int_0^{t_i} \frac{dt}{a_i(t/t_i)^{1/2}} + \int_{t_i}^{t_f} \frac{dt}{exp(H_i(t-t_i))} \right) \\ &= a_i c e^N \left(2\frac{t_i}{a_i} + \frac{1}{a_i H_i} \left(1 - e^{-Hi(t_f - t_i)} \right) \right) \\ &= ce^N \left(2t_i + H_i^{-1} \right) \end{aligned} \tag{10.174}$$

$$N = 100, \quad t_i = H_i^{-1} = 10^{-36}\,\text{s}.$$

In other words:

$$d_H(t_i) = 2ct_i = 6 \times 10^{-28}\,\text{m} \qquad \text{(before inflation)} \tag{10.175}$$

$$d_H(t_f) = e^N 3ct_i \approx 2 \times 10^{16}\,\text{m} = 0.8\,\text{pc} \qquad \text{(after inflation)}, \tag{10.176}$$

i.e. it increased by a factor of 3.3×10^{43}. Thus, if, as we have seen, during the radiation era the size of the universe had been 0.5 pc, after inflation it became 1.3×10^{43} pc, large enough to accommodate the whole universe. No horizon problem.

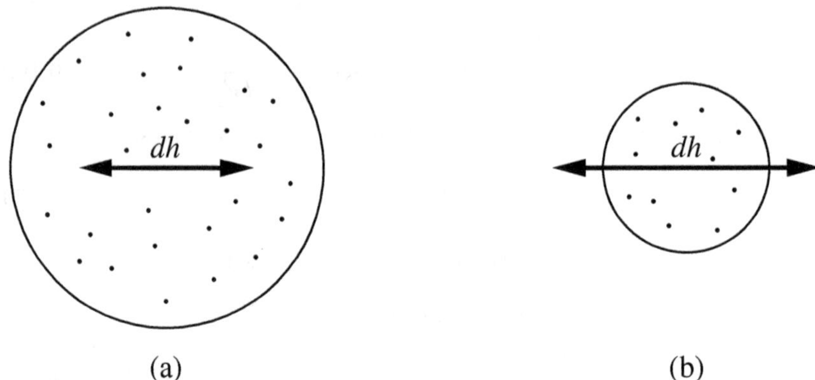

Fig. 10.18: The horizon problem is exhibited. In the SCM the universe had a dimension much larger than the horizon (a) while after inflation the universe arises from the exponential growth of a small region, which is smaller than the horizon (b). So the various parts of this domain are causally connected. The scale is not real, but adopted for illustration purposes.

Let us illustrate the picture going back in time. At the time of the decoupling of radiation, $z = 1100$, the proper length is

$$d_P(1100) = \frac{3c}{H_0} = \frac{34 \times 10^8}{70 \times 10^3}\,\text{Mpc} = 1.3 \times 10^4\,\text{Mpc}$$

(see Eq. (10.124)). The scale factor is:

$$a(t_d) = \left(\frac{t_d}{t_0}\right)^{2/3} = \left(\frac{0.35 \times 10^6}{1.36 \times 10^{10}}\right)^{2/3} = 8.7 \times 10^{-4},$$

i.e.

$$a(t_f) = a_{t_d}\left(\frac{t_f}{t_d}\right)^{1/2} = 8.7 \times 10^{-4}\left(\frac{10^{-34}}{0.35 \times 10^6\ 3.156 \times 10^7}\right)^{1/2}$$

$$= 2.6 \times 10^{-27}$$

$$d_P(t_f) = a(t_f) d_P(1000) = 0.9 \times 10^{-24}\,\text{Mpc} = 1.0\,\text{m}\,.$$

The whole energy of the universe is inside a sphere of radius 1 m!. Before inflation the universe was enclosed in radius given by

$$d_P(t_i) \approx e^{-100} d_P(1000) = 4.0 \times 10^{-44}\,\text{m}\,. \tag{10.177}$$

In spite of the fact that the horizon is small (Eq. (10.175)), it is 1.5×10^{16} larger than the then radius of the universe. This is exhibited in Fig. 10.18.

Two points separated by a distance lrger than the horizon, are not causally connected. Indeed the Robertson-Walker metric for a flat universe yields:

$$ds^2 = (cdt)^2 - (a(t)dx)^2 . \tag{10.178}$$

Thus, even if one uses the light as asignal the communication between two points separated dx, $ds = 0$, one finds:

$$d\tau^2 = dx^2, \quad d\tau = \frac{cdt}{a(t)} \quad \text{(light cone ,} \tag{10.179}$$

In other words only points inside this cone can exchange information. The situation is is exhibited in Figs. 10.19 and 10.20.

(3) There is no magnetic monopole problem.
Assuming that the magnetic monopoles have masses of the order of $10^{14}\mathrm{m}_p$, they must have been produced before inflation, corresponding to a monopole density $n_M(t_{\mathrm{GUT}}) \approx 10^{82}\,\mathrm{m}^{-3}$. Let us suppose that their number did not change during the inflation. Thus their density after inflation has greatly been reduced:

$$n_M(t_f) \approx \left(e^{-100}\right)^3 10^{82}\,\mathrm{m}^3 = 5 \times 10^{-49}\,\mathrm{m}^{-3} = 15\,\mathrm{pc}^{-3} . \tag{10.180}$$

and after the subsequent expansion, $a(_f) = 1.0 \times 10^{-27} \to a(t_0) = 1$,

$$n_M(t_0) \approx 15 \times 10^{-81}\,\mathrm{pc}^{-3} = 1.5 \times 10^{-62}\,\mathrm{Mpc}^{-3} . \tag{10.181}$$

Thus the probability of detecting even a single monopole is astronomically small.

10.10 Particle physics and the inflationary scenario

In this section we will make a simple effort to crudely understand inflation from the side of particle physics. We will begin with the time 10^{-43} s, which correspond GUTs with temperature 10^{27}K. This is the symmetric phase, i.e. the symmetry has not been broken spontaneously. This is not going to last long, since the scalar particles are anxious to break the symmetry as soon as the temperature drops below some critical value T_c, sometime before some time 10^{-30}s. The energy density is then given by:

$$u = u + u_H = u + \frac{1}{2}\dot{\Phi}^2 + V(\Phi) \tag{10.182}$$

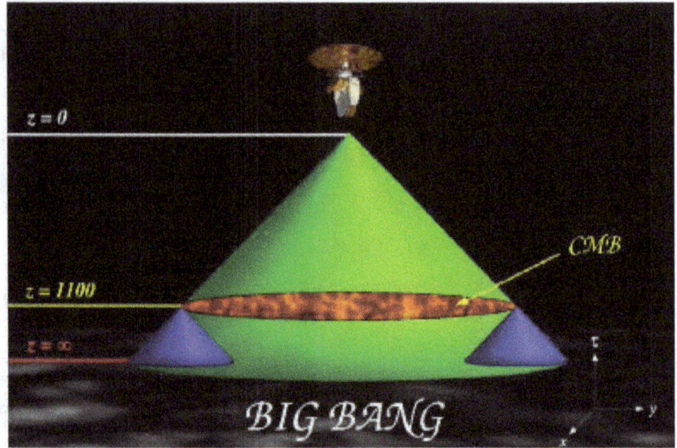

Fig. 10.19: The horison without inflation. Only signals inside the light cone can reach the observer at the top of the cone ($z = 0$). The two blue cones, which end at the edge of the universe at $z = 1100$ they do not intersect at the base they do not intersect in the base of the large cone (Big Bang). As a reult two points at the top of the blue cones cannot be causally correlated. WMAP, however, finds them correlated.

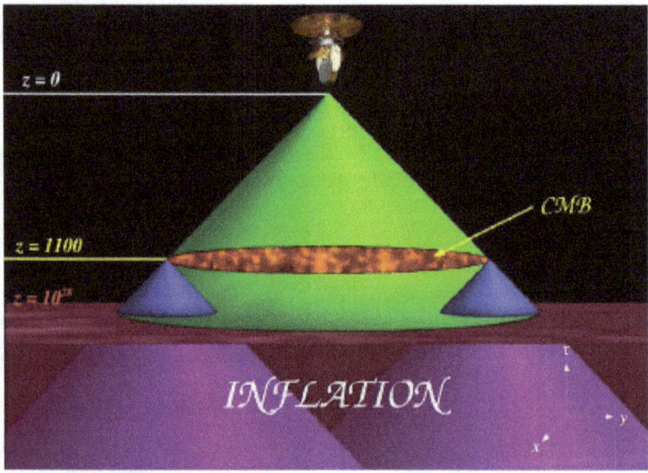

Fig. 10.20: This figure is similar to Fig. 10.19, but in the case the inflationary scenario is operative. Now the two blue cones intersect at the base (Big Bang) and, thus, the edges of the universe at $z = 1100$ can be in causal contact.

where $V(\Phi)$ is the potential of the scalar field Φ, which we will now call **inflaton**. For simlicity we will take it to be a simple complex scalar field with a potential of the form:

$$V(\Phi) = -\frac{1}{2}\mu^2\Phi^2 + \frac{\lambda}{4}|\Phi|^4 + \frac{1}{4}\frac{\mu^4}{\lambda}, \quad \lambda > 0, \tag{10.183}$$

see section 4.1.1. We have found that we have a maximum at $|\Phi| = 0$ and a minimum at $|\Phi| = \upsilon$ Then we have found that by setting $\mathrm{Re}\,\Phi = \eta + \upsilon$, $\mathrm{Im}\,\Phi = \xi$ the potential can be cat in the form:

$$V = \frac{\lambda\eta^4}{4} + \sqrt{\lambda}\mu\eta^3 + \mu^2\eta^2 + \frac{\lambda\xi^4}{4} + \left(\frac{\lambda\eta^2}{2} + \sqrt{\lambda}\mu\eta\right)\xi^2. \tag{10.184}$$

Thus the field η attained a mass μ, while ξ remains massless. The potential has in the η direction a maximum at $Re\Phi = 0$ and a minimum at $\mathrm{Re}\,\Phi = \upsilon$. In the direction ξ the symmetry remains unbroken and we have a minimum at $\xi = 0$. This is exhibited in Fig. 10.21.

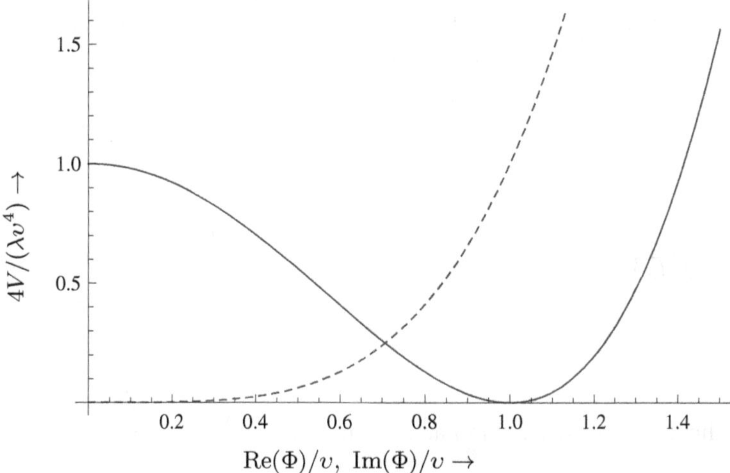

Fig. 10.21: The potential is exhibited after the symmetry breaking in the direction $\mathrm{Re}(\Phi)/\upsilon$ (solid line) and $\mathrm{Im}(\Phi)/\upsilon$ (dashed line).

It clear that if the inflaton is found near the origin, see Fig. 4.2, it will roll down towards the minimum. The time required for this depends on the potential. Suppose now that the radiation energy density $u = aT^4$ is much higher than than the value of the potential $V(|\Phi|)|_{\Phi=0} = V_0$. Then the Friedman equations remain unaffected by the presence of the inflaton.

So inflation is expected to begin at a temperature T_i given by:

$$T_i \approx \left(\frac{V_0}{a}\right)^{1/4} \approx 2 \times 10^{28}\,\mathrm{K}\left(\frac{V_0}{10^{105}\,\mathrm{TeV\,m^{-3}}}\right)^{1/4} \tag{10.185}$$

and the corresponding time will be:

$$t_i \approx \left(\frac{c^2}{G_N V_0}\right)^{1/2} \approx 3 \times 10^{-36}\mathrm{s}\left(\frac{V_0}{10^{105}\,\mathrm{TeV\,m^{-3}}}\right)^{1/2}. \tag{10.186}$$

The corresponding quantities after inflation are still given be Eqs. (10.185–10.186) with the replacement $V_0 = V(|\Phi|)|_{\mathrm{Re}(\Phi)=\upsilon}$ The above results hold at zero temperature.

What happens at high temperatures? At higher temperatures instead of the potential one has to minimize the free energy. We will consider a simple expression of the form:

$$f(T, \Phi) = V(\Phi) + \frac{1}{24} m_\Phi^2 T^2 - \frac{\pi^2}{90}{}^4. \tag{10.187}$$

where the temperature is given in units of energy. In such cases where the minimum occurs depends on the temperature (see section 4.1.4, Fig. 4.3). Let us examine it a little bit further.
We have seen that

$$m_\Phi^2 = \frac{1}{2}\frac{\partial^2 V}{\partial \Phi^2}\bigg|_{\Phi=0} = \mu^2 \tag{10.188}$$

Hence

$$f[T, V] = \frac{1}{2}\mu_{\text{eff}}^2(T)|\Phi|^2 + \frac{1}{4}|\Phi|^4 - \frac{1}{24}\mu^2 T^2 - \frac{\pi^2}{90}{}^4 \tag{10.189}$$

where

$$\mu_{\text{eff}}^2(T) = \frac{\lambda}{4}T^2 - \mu^2. \tag{10.190}$$

From the above analysis one has a minimum if $\mu_{\text{eff}}^2(T) < 0$, i.e.

$$T < T_c, \quad T_c = 2\sqrt{\frac{\mu^2}{\lambda}} = 2\upsilon, \quad T_c = \text{ (critical temperature)}. \tag{10.191}$$

Notice that $\mu_{\text{eff}}(T_c) = 0$. The minimum occurs

$$|\Phi| = \sqrt{\frac{-\mu_{\text{eff}}^2(T)}{\lambda}} = \sqrt{\upsilon^2 - \frac{T^2}{4}}. \tag{10.192}$$

On the other hand for $> T_c$ the minimum occurs at $|\Phi| = 0$. This situation is depicted in Figs. 10.22 and 10.23. Above the critical temperature the inflatons, even though they interact with the other particles, they sit in the

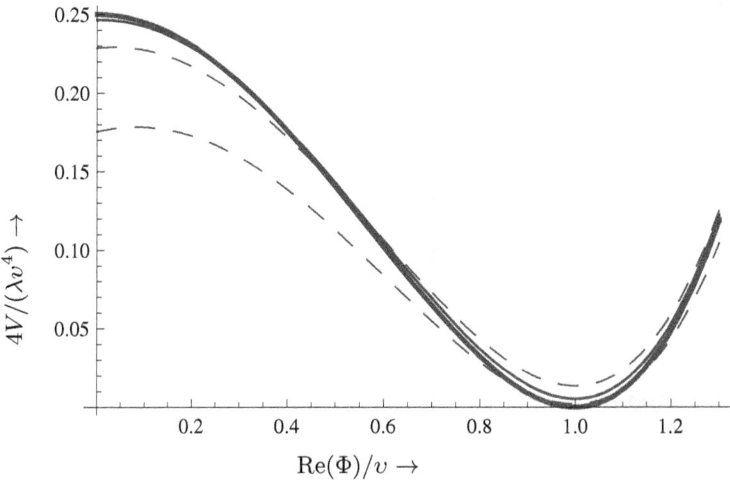

Fig. 10.22: The scalar potential is exhibited after the symmetry breaking as a function of $\mathrm{Re}(\Phi)/v$ for $T < T_c$. The thick solid, the solid, the short dash and the long dash correspond to $T/(2v) = 0.0, 0.1, 0.2$ and 0.5 respectively.

pseudo-vacuum at $\Phi = 0$. When the temperature drops they roll to the real vacuum at $\Phi = v$, breaking the symmetry spontaneously.

If the rolling from $\Phi = 0$ to $\Phi = v$ had happened very fast, we would encounter the situation of the SCM, where the phase transition is adiabatic. In the inflationary scenario, however, this rolling happened relatively slowly so that the temperature of the universe fell far below the critical temperature. As a result, even though the temperature is below 10^{27}K, the symmetry has not yet been broken. In other words one encounters a situation like the supercooling of the liquid phase of water, i.e. it remains a liquid in spite of the fact that its temperature is below the freezing point (in fact in the case of water some concentration centers ignite the freezing).

It is known that the inflationary scenario requires

$$p = -u = -\rho c^2 \tag{10.193}$$

which can be achieved only via quantum fluctuations. As the temperature drops and the potential curve approaches the lowest curve of Fig. 10.22, some quantum fluctuations of the field Φ show up. As a result the space of the false vacuum breaks up in areas, inside which the fields Φ can be considered uniform. These area are called bubbles or fluctuation bubbles. The space now is filled with such bubbles contained with walls, not necessarily spherical, but irregular. Their size however, is smaller than the horizon.

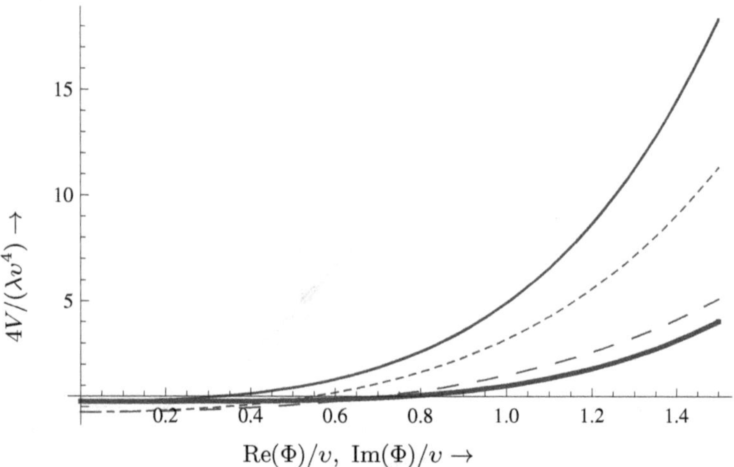

Fig. 10.23: As in Fig. 10.22 for $T > T_c$. The thick solid, the solid, the short dash and the long dash correspond to $T/(2v) = 1.0, 1.1, 1.3$ and 1.5. Since the value of the potential at $\Phi = 0$ depends on the temperature this value was adjusted so that they coincide with the values seen in Fig. 10.22.

Due to he fluctuations inside the bubbles, the field Φ moves away from the origin. Since the potential near the origin is almost horizontal the rolling at he beginning is quite slow, the time this slow rolling lasts is called rolling period τ. For our estimates we will use

$$\tau = 60H_i = 4 \times 10^{-33}\mathrm{s}\ , \tag{10.194}$$

where H_i gives the rate of expansion of the universe during the period of inflation (section 10.9).

Let us now return to the quantum fluctuation bubbles. Evidently they will follow the evolution of the fields Φ inside them. The bubble density is constant, independent of the radius. As a result the expansion is exponential, i.e the size of the bubble will increase many orders of magnitude.

$$D_{\text{initial}} = H^{-1} = 10^{-24}\text{cm} \Rightarrow D_{\text{final}} = H^{-1}e^{Ht_i} = 10^{28}H_i^{-1} = 10^6 \tag{10.195}$$

(see section 10.9). In other words the size of the a given bubble has increased by a factor 10^{28}. It thus appears that our universe may have started out as one of these bubbles.

From the moment the field Φ gets the value $\Phi = v$ its starts oscillating around this minimum. The radiation and matter density that resides initially inside the bubble at the end of inflation is insignificant. The huge amount of energy, acquired by the inflaton as kinetic energy as it reaches the minimum of the potential, makes it oscillate around this minimum. Thus energy is transferred to the matter and energy since the oscillations are damped. Furthermore at this stage the inflaton is unstable and decays to other particles and radiation. Now point the universe has enough energy in the form of matter and radiation.

In fact the large amount of energy coming from the kinetic energy of the inflaton lead to an increase in the temperature. As a result the universe was reheated above the $10^{27}\,\text{K}$, near the GUT phase transitions. At this point many very interesting processes occurred, like the asymmetry baryon-antibaryon etc.

From then on our universe evolves as predicted by the SCM.

Finally it is tempting to assume that the source of dark energy is the scalar field and, for this reason, the name, the fifth essence, quintesense. The role of dark energy increases as the universe gets larger and eventually dark energy will dominate leading to a new inflation.

The inflationary scenario is very appealing. It should, however, be better connected with a successful particle model, equipped with a precise particle symmetry beyond the standard model. This will illuminate the nature of the dark energy and dark matter. More specifically such a model, among other things, will give us the form of the scalar potential $V(\Phi)$, required by the necessary to understand the slow roll motion, and the symmetry breaking at the GUT.

Cosmology may aid particle physics more in the future and particle physics has undoubtly aided cosmology. The coupling between them has been a great triumph for science.

10.11 The role of dark matter

We have seen that matter represents about 30% of the energy content of the universe. From meaurements of the promordial abundance of light elements (see Fig. 10.8) the usual baryonic matter (consisting of stars, intergalactic dust, black holes etc) cannot exceed 5%. The vast majority is composed of **dark matter**, i.e. matter that cannot possibly radiate[6]. The presence of dark matter can directly be verified by its gravitational effects.

- The rotational velocities of stars around the center of the galaxy they belong to.

[6]Some of these effecta can probably be explained by a modified gravity. No such modification, however, has been able to expain all phenomena.

- gravitational lensing.

10.11.1 *The rotational velocities*

Dark matter can be inferred by its gravitational effects around galaxies. Astronomers believe that dark matter was first hypothesised nearly 100 years ago, as a result of the observations of galactic rotation by Dutch astronomer Jacobus Kapteyn. Studies of galaxies revealed that there was a significant amount of matter missing from the universe. It was put on more solid basis by the observed rotational curves of stars far way from the center of their galaxies by Zwicky in 1937. Let us see how this is derived from Newton's laws.

From Classical mechanics one finds that given a mass m in a gravitational field its tangential velocity υ is given by Newton's law:

$$\frac{\upsilon^2(r)}{r} = \frac{G_N}{r^2} M(r). \tag{10.196}$$

Strictly speaking this is true for a spherically symmetric matter distribution. Regardless of the shape of the distribution, however, for distances r much larger than the size of the matter distribution one finds that $\upsilon^2 \propto 1/r$ For a uniform mass distribution of mass M within a spherical galaxy of radius a

$$\rho(r) = \begin{cases} \dfrac{3M}{4\pi a^3}, & r < a \\ 0, & r > a \end{cases} \tag{10.197}$$

we find:

$$M(r) = \begin{cases} M\dfrac{r^3}{a^3}, & r < a \\ M, & r > a \end{cases} \tag{10.198}$$

that is:

$$\upsilon^2(r) = \begin{cases} \dfrac{G_N M}{r}, & r > a \\ \dfrac{G_N M}{a}\dfrac{r^2}{a^2}, & r < a. \end{cases} \tag{10.199}$$

In other words for small distances the velocity is increasing (here linearly), while for large distances always falls $1/\sqrt{r}$.

Alternatively for a given mass distribution ρ we can obtain the potential Φ by solving Poisson's equation. Then the tangential velocity is given by:

$$\frac{\upsilon^2}{r} = -\frac{d\Phi}{dr}, \quad \Phi = \text{ the potential.} \tag{10.200}$$

In case the density is of the form

$$\rho(r) = \rho_0 \left(1 + \frac{r^2}{a^2}\right)^{-5/2}. \tag{10.201}$$

We can obtain the solution analytically:

$$\Phi(r) = \frac{4\pi G_N \rho_0}{3} \left(1 + \frac{r^2}{a^2}\right)^{-1/2}. \tag{10.202}$$

We demand that the mass enclosed up to a distance $r = a$ to be practically the total mass M of the galaxy we find:

$$M = \frac{4\pi}{3} \frac{1}{2\sqrt{2}} a^3 \rho_0 \Rightarrow \Phi(r) = \frac{G_N M}{a} \frac{2\sqrt{2}}{\sqrt{1 + (r/a)^2}}. \tag{10.203}$$

Hence

$$\upsilon^2(r) = \frac{MG_N}{a} 2\sqrt{2} \frac{(r/a)^2}{(1 + (r/a)^2)^{3/2}}. \tag{10.204}$$

We know that some galaxies, including our own, are essentially flat with a surface density $\sigma(\mathbf{r})$. Finding the potential is now much tougher. The behavior, however, is crudely the same. We will consider here the density:

$$\sigma = \sigma_0 \frac{1}{(1 + x^2)^{5/2}}, \quad x = \frac{r}{a}. \tag{10.205}$$

The obtained results are shown in Fig. 10.24. The constant σ_0 was related to ρ_0 so that we get the same potential at the origin. Generally speaking initially the rotational velocity increases, but eventually it falls as $1/\sqrt{r}$.

It is impressive that a measurement of the radius of a galaxy and the rotational velocity of "suburban" galaxies determines the mass of the galaxy (Newton can find not only the weight of stars in our galaxy, but also the weight of the galaxies!):

$$\upsilon^2(r = a) = \frac{MG_N}{a} \Rightarrow M = \frac{a}{G_N} \upsilon^2(r = a). \tag{10.206}$$

For galaxy NGC6503 $\upsilon(r = a) = 100$km/s and $a = 1.5$kpc we get $\approx 10^8$ solar masses.

The actual observations are shown in Fig. 10.25. A somewhat artistic picture of our galaxy with the dark matter halo is shown in Fig. 10.26.

A very simple dark matter density profile is given by:

$$\rho = \frac{\rho_0}{1 + (r/a)^2}, \quad \rho_0 = \frac{3M}{4\pi a^3}, \tag{10.207}$$

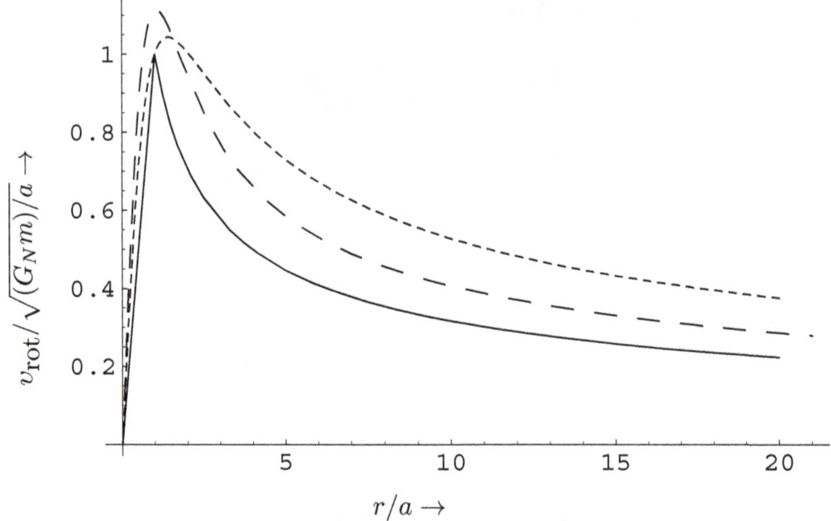

Fig. 10.24: The rotational velocity in units of $\sqrt{(G_N m/a)}$ as a function of r/a for a uniform distribution (dotted line), for the density of Eq. (10.201) (solid line) and for the density of Eq. (10.205) (long dashed line).

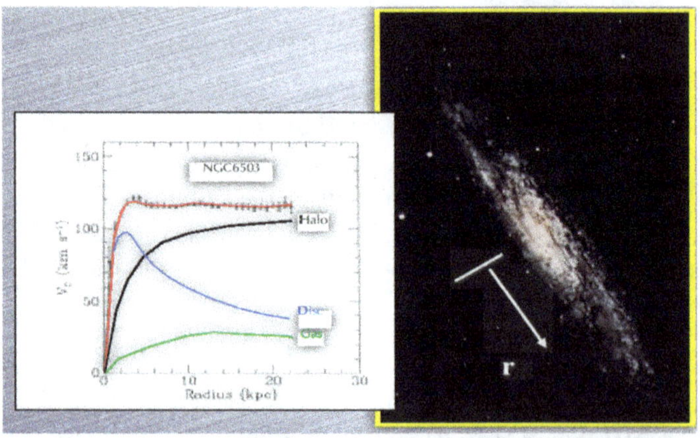

Fig. 10.25: Newton weighted galaxies! The observed rotational velocities around a typical galaxy, NGC6503. Note the disagreement between experiment and expectations outside the luminous galaxy. The predictions are based on the reasonable assumption that the luminosity of a segment of the galaxy is proportional to its mass. Note that 10 kpc $\approx$ 30000 light years.

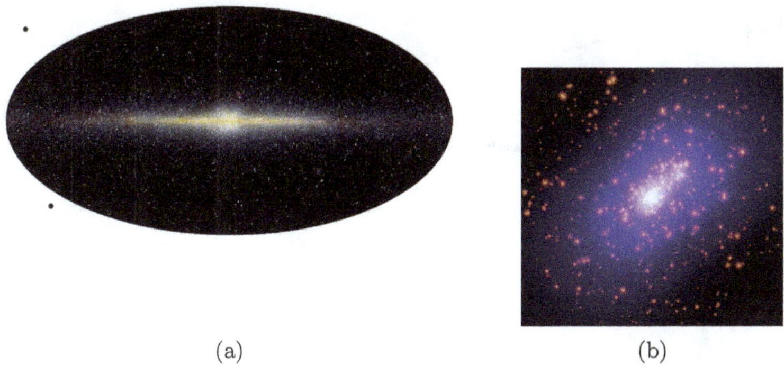

(a) (b)

Fig. 10.26: A panoramic view of the luminous part of our galaxy with the dark matter halo (left) and galaxy clusters (shown as dots) embedded in dark matter exhibited in blue (right).

where ρ_0 has been fit so that at $r = 0$ gives a density, which is the same with a galaxy of constant luminosity. Then it is easy to show that

$$v^2(x) = \frac{MG_N}{a}\left(1 - \frac{\tan^{-1}(x)}{x}\right), \quad x = \frac{r}{a} \tag{10.208}$$

and

$$v^2(x \to \infty) = \frac{MG_N}{a}. \tag{10.209}$$

In reality we have both luminous and dark matter, with a potential, which is the sum of the two (the equation is linear!). Then

$$v(r) = \sqrt{v^2_{\text{lum}}(r) + v^2_{\text{dark}}(r)}. \tag{10.210}$$

The obtained results are shown in Fig. 10.27. It is seen that the agreement with the data, Fig. 10.25, is not bad. The rotational velocities also may involve galaxies moving around the center of a galaxy cluster. These velocities cannot be accounted for by the visible matter alone.

Another evidence for the existence of dark matter is the temperature of very hot X-ray emitting gas. Galaxy clusters contain large amounts of such gas. The temperature of this hot gas measures the clusters mass, because the gas has to be held in the cluster by gravity (hydrostatic equilibrium).

10.11.2 *Gravitational lensing*

Dark matter can be observed some other gravitational effect known **gravitational lensing**. It is due to the fact that in the context of general

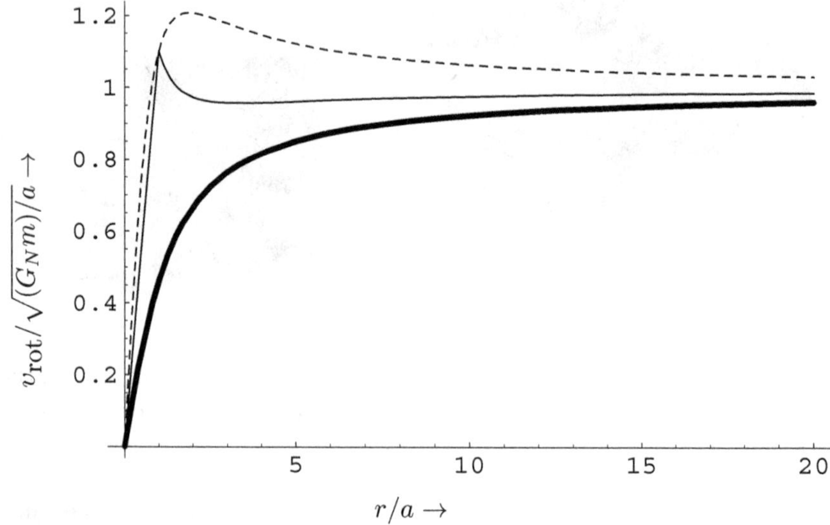

Fig. 10.27: The rotational velocity in units of $\sqrt{(G_N m/a)}$ as a function of r/a in the presence of only dark matter, Eq. (10.207) (thick line), a combination of dark matter and ordinary matter with uniform distribution (dashed line) and with the density of Eq. (10.201) (solid line).

theory relativity light can be bent if it passes through a gravitational field (see Fig. 10.28). This bending, leading to a number of circularly arranged images of an otherwise invisible object, can be caused as efficiently by dark matter as ordinary matter. The presence and the amount of dark matter can be inferred, if any luminous matter in front of the object cannot alone explain the observed pattern of images (for a recent review see [Massey *et al.* (2010)]).

Gravitational lensing has excluded the possibility of the non luminous mass being MACHOS (MAssive Compact Halo ObjectS), i.e. dead or failed stars in the halos of galaxies (brown dwarfs, white dwarfs, small black holes). Too few such lensings.

10.11.3 *The observation of bullet cluster*

In an announcement of NASA in 2006 it was reported the observation of two cluster galaxies known as 1E0657-56. These two clusters, known as bullet cluster, 3.5 billion light years away from us, went through each other and they are now 2 million years apart. The stars and the galaxies

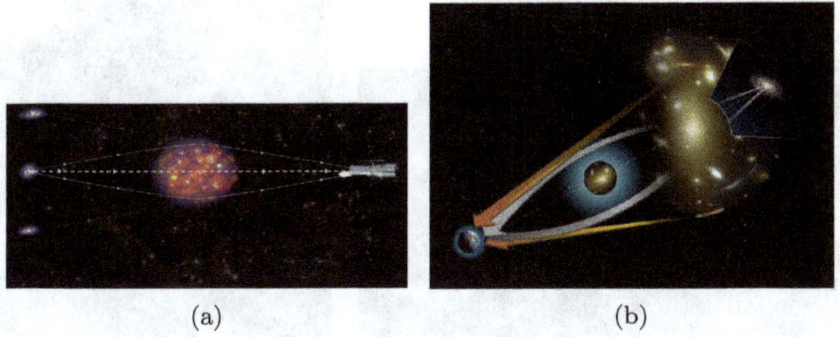

(a) (b)

Fig. 10.28: The idea of light bending in a gravitational field (a). This "elevates" the image of the object so it can be seen. In fact we see not just one but a ring of images arranged symmetrically around the lens in a plane perpendicular to the line connecting the observer with the focusing object. An ideal but rather rare arrangement (b). In practice the geometry is not so simple. So the more laborious method called weak lensing is employed [Massey *et al.* (2010)].

remained essentially intact[7]. The gas in the galaxies, however, were really shaken up resulting in intense X-ray emission. The dark matter inferred from gravitational lensing also remained intact. The gases, however, like a comet followed their host galaxy. This is exhibited in a rather artistic form in Fig. 10.30.

Another strong evidence is the existence of filament between huge colliding galaxies (Fig. 10.31).

Before concluding this section we note that the rotational curves permit the determination of dark matter in the suburbs of a galaxy. Thus at $a =$ 20kpc, where dark matter is dominant the rotational velocity is $\approx$ 300km/s. Combining Eqs. (10.207) and 10.209 we find:

$$\rho_0 = \frac{3v^2}{4\pi G_N a^2} \approx 1.0 \times 10^{-21}\text{kg m}^{-3}.$$

In our vicinity, $r_s = 5$kpc,

$$\rho(r_s) \approx \rho_0/1.16 = 8.8 \times 10^{-22}\text{kg m}^{-3} \Rightarrow n = \frac{8.8 \times 10^{-22}\text{kg m}^{-3}}{1.6710^{-27}\text{kg}} =$$

$$0.53 \times 10^6 \text{ nucleons m}^{-3} = 0.53 \text{ nucleons cm}^{-3}.$$

A more detail analysis yields

$$\rho(a) \approx 0.3 \text{ nucleons cm}^{-3} \quad \text{in our vicinity.}$$

[7]The galaxies are essentially empty, in the sense that the stars occupy a small fraction of the volume, just like the nucleus occupies a small fraction of the volume of the atom ($d_{\text{galaxy}}/d_{\text{sun}} = 3 \times 10^7$).

(a) (b)

Fig. 10.29: The ring is convincing about gravitational lensing (a), even if the focusing mass is not visible. In the case of dark matter the focusing mass is quite extensive.Strong gravitational lensing around galaxy cluster CL0024+17 (b), demonstrating at least three layers projected onto a single 2D image. Of interest to us are the elongated blue objects, which are much more distant galaxies, physically unassociated with, and lying behind, the cluster. Gravitational lensing has distorted their apparent images into a series of tangential arcs.

Which is huge compared with the average dark matter density in the universe:

$$\Omega_{\text{galaxy}} = \frac{\rho}{\rho_c} = \frac{(0.3/0.53) \times 8.8 \times 10^{-22}\text{kg m}^{-3}}{9.2 \times 10^{-27}\text{kg m}^{-3}} \approx 5 \times 10^4!$$

Dark matter exists and it played a crucial role in structure formation. Billions of such particles pass through our bodies per second. But there is no reason to worry. Only a few nuclei in our body may be knocked off. They interact very weakly, if at all! We do not know the nature of dark matter constituents. We know, however, that:

- They must be stable or have a lifetime longer than the age of the universe.
- They constitute cold dark matter, i.e. they are non relativistic. This is necessary to explain the large structure formation in the universe.
- They posses gravitational interaction (from rotational curves).
- They do not feel the strong interaction, since otherwise they would have been detected in the laboratory.
- They are electrically neutral. Otherwise they would have formed atoms etc.

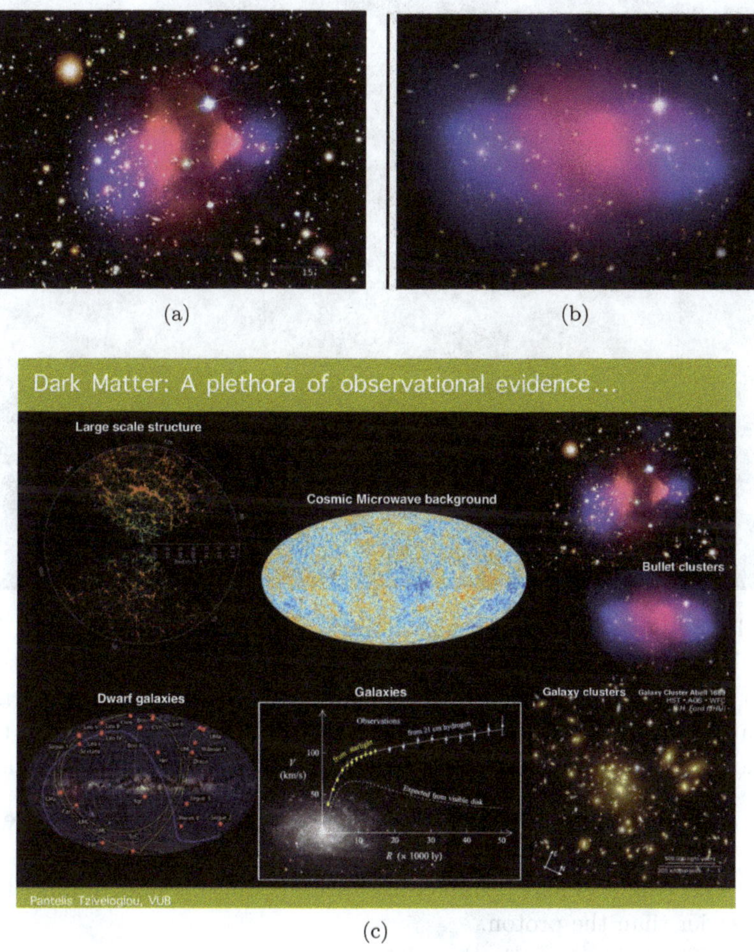

(a) (b)

(c)

Fig. 10.30: A panoramic view of the bullet cluster or 1E0657-56 (a). White or orange color indicates ordinary luminous galaxies, magenta the X-rays and blue the dark matter inferred from gravitational lensing. The baby bullet MACSJ0025.4-1222 (b). Plenty of dark matter observed at all scales (c).

Their nature can be determined if they are found in the ongoing underground WIMP (Weakly Interacting Massive Particles) searching experiments[8]. Possible theoretical candidates are:

[8]The discovery of exotic particles in accelerators may, however, give a clue about the nature of the candidate.

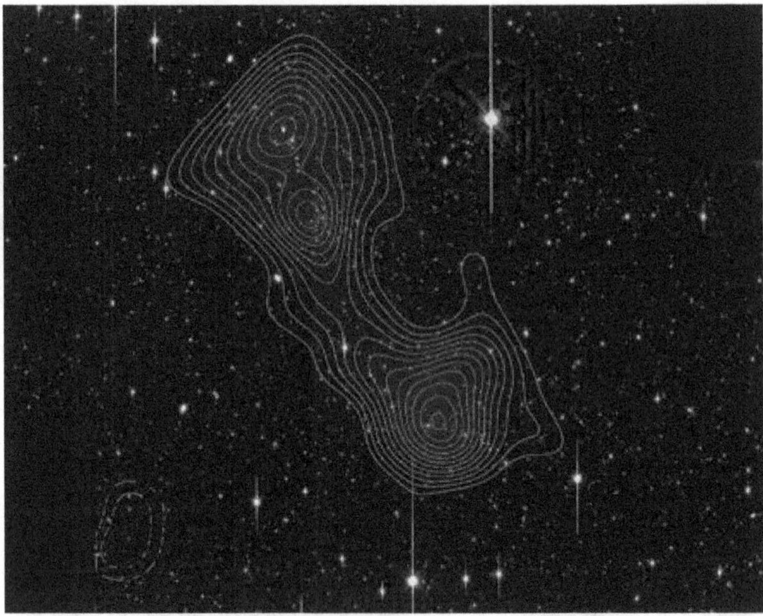

Fig. 10.31: Observations of huge colliding galaxies, like Abel 222/223, (Science News, July 14 , 2012) indicate the exchange of dark matter between them and a channel (filament) between them. By comparing X-rays, a manifestation of ordinary matter, and mass inferred by gravitational lensing (blue in the figure) one concludes that ordinary matter is only 9% of the total

- The axion, a scalar particle with a mass in the range 10^{-6} eV $< m_a < 10^{-3}$ eV.
- The lightest supersymmetric particle (LSP) , at least 10 times heavier than the proton.
- The lightest exotic aluza-Klein particle of theories in more than 3 space time dimensions, which is expected to more than 500 times heavier than the proton.

The experiments aiming at its detection are of the type:

- Indirect detection.
  In this case one tries to detect particles produced when two WIMPs are destroyed (mainly photons and neutrinos of high energy) in regions of high WIMP density, e.g. in the sun or the center of the galaxy. If they happen to be fermions the collision occurs between a particle and its antiparticle or between two particles, if they happen

to be of the Majorana variety.

- Direct detection.

  a) The axion is detected via its probability to be converted into a photon via the Primakoff effect. Thus the axion penetrates into the detector and a photon is detected in the presence of a magnetic field. Such experiments have been going on, e.g. the ADMX [Asztalos *et al.* (2010)] and more recently looking for special signatures [Vergados and Semertzidis (2016); Zioutas *et al.* (2017)].

  b) The others, after they collide with the nucleus, can be detected by measuring the energy of the recoiling nuclei. Since the WIMP energy is very low, ≤ 100 keV, the probability of getting the nucleus excited is extremely low. Te energy of the recoil can be converted into light or into heat or it can cause ionization destroy superconductivity. The experiments are hard, since the counting rate is very small and the signal cannot be distinguished from the background. The background is annoying, even if the experiment is done underground. Such experiments were first attempted long time ago [Ahlen *et al.* (1987)] and have been going on by many experimental groups scattered all over the globe. Nothing has yet been found, only rather stringent limits on the WIMP-nucleon cross sections have been set for the coherent nucleon cross section, e.g. by the XENON100 collaboration [Aprile *et al.* (2011)] $\sigma \leq 10^{-8}\text{pb} = 10^{-20}\text{b} = 10^{-44}\text{cm}^2$. This amounts to less than 1 event per Kg target per year!

10.12 Dark energy

The presence of dark energy is manifested by the increase in the expansion rate of the universe, i.e. its acceleration. This, unfortunately cannot be measured directly. So it is inferred from the behavior of the apparent luminosity of the receding galaxies compared to that of an empty universe, i.e. a universe in which the receding velocity remains constant. If it is brighter the universe is decelerating, if it is dimmer it is accelerating.

10.12.1 *Observations with standard candles*

The standard candles are those that have very well defined energy output, i.e. absolute luminosity, so they can be used to measure large distances. The following steps are needed:

- A population of standard candles with absolute luminosity L.
- Measurement of the energy flux I associated with a given shift z in the spectrum.
- Extraction of the length $d_L(z) = \sqrt{L/4\pi I}$ from the standard candle.
- Plot of the function $d_L(z)$.
- Determination of the Hubble constant H_0 (from the slope of the previous curve for low z).
- Plot the apparent magnitude m and the absolute magnitude M of the source (typically a logarithmic plot).
- Comparison with cosmological models.

The first step is the hardest (one has to catch the rabbit first!). Fortunately there exist prototype candles. For large distances the best are supernovae of type Ia. The exploding star is the white dwarf member of double star, which explodes when it reaches the Chandrashekar limit, after a process of eating up mass from its companion. So it has a very well defined mass of about 1.44 solar masses and, thus, very well defined energy output. Its identity is revealed by its spectrum and the light curve. This type of supernova is a bit rare, one per century in our galaxy, but plenty of them occur in the universe. They are quite bright, $L/L_{\text{sun}} = 4 \times 10^9$, exceeding the luminosity of a whole galaxy. Since the observation of galaxies at $z = 1$ is fairly easy, such supernovae are ideal prototype candles.Two such projects were recently executed, the Supernova Cosmology Project (SCP) and the High-z Supernova Search Team (HSST). They obtained results consistent with each other.

Before proceeding with the analysis of the experiments let us briefly discuss the bolometric distance of prototype candles. Suppose t_e and t_0 are the times of emission and observation of a photon respectively. Then the the change in the wavelengths yields:

$$E_e = \frac{hc}{\lambda_e}, \quad \lambda_e = \frac{\lambda_0}{1+z} \Rightarrow E_0 = \frac{E_e}{1+z}. \tag{10.211}$$

Consider now two photons. Suppose that at the emission are timewise separated δt_e, while at the observation by δt_0. The corresponding proper times are related by:

$$\delta t_0 = \delta t_e (1+z).$$

Which affects the observed power. We thus find:

$$I = \frac{L}{4\pi \left(S_\kappa(r/R_0)(1+z)\right)^2} \Rightarrow \tag{10.212}$$

$$d_L = S_\kappa(r/R_0)(1+z). \tag{10.213}$$

So the bolometric distance is related to the proper distance by

$$d_L(z) = d_P(z)(1+z). \tag{10.214}$$

Before proceeding further we will discuss the magnitude of apparent and absolute luminosity, used by astronomers since the time of Hipparchus (second century BC). It involves a logarithmic scale, not only because the eye works this way, but because the quantities involved cover a wide range. It is only this way the various models can be distinguished from each other. We define the magnitude of bolometric flux as follows:

$$m = -2.5\log_{10}\left(\frac{I}{I_s}\right) \quad \text{(relative magnitude)}, \quad I_s = 2.53\times 10^{-8}\text{Wm}^{-2}. \tag{10.215}$$

and the magnitude of the absolute luminosity:

$$M = -2.5\log_{10}\left(\frac{L}{L_s}\right) \quad \text{(absolute magnitude)}, \quad L_s = 78.7L_{\text{sun}}. \tag{10.216}$$

with I_s as seen at 10 pc. From these we find the useful relations:

$$M = m - 5\log_{10}\left(\frac{d_L}{10\text{pc}}\right) \tag{10.217}$$

or

$$M = m - 5\log_{10}\left(\frac{d_L}{\text{Mpc}}\right) - 25. \tag{10.218}$$

The reduced magnitude, distance modulus, $m - M$, is defined as:

$$m - M = 5\log_{10}\left(\frac{d_L}{1\text{Mpc}}\right) + 25. \tag{10.219}$$

For $z \ll 1$

$$d_P(z) \approx \frac{c}{H_0}z\left(1 - \frac{1+q_0}{2}z\right), \quad d_L(z) = d_P(z)(1+z) \approx \frac{c}{H_0}z\left(1 - \frac{1-q_0}{2}z\right). \tag{10.220}$$

(see sec. 10.6). Thus

$$m - M = 43.17 - 5\log_{10}\left(\frac{H_0}{70\text{km/s}^{-1}\text{Mpc}^{-1}}\right) + 5\log_{10} z + 1.086(1 - q_0 z). \tag{10.221}$$

It is better if we plot these functions relatively to an empty universe, i.e. a universe in which the galaxies are receding with constant velocity Fig. 10.32. The observations of the SCP (HSP06) teams are shown in Figs. 10.33–10.35.

From these and other data the existence of dark energy was established and as result the Nobel in physics for 2011 was awarded. This has been confirmed by the WMAP observations.

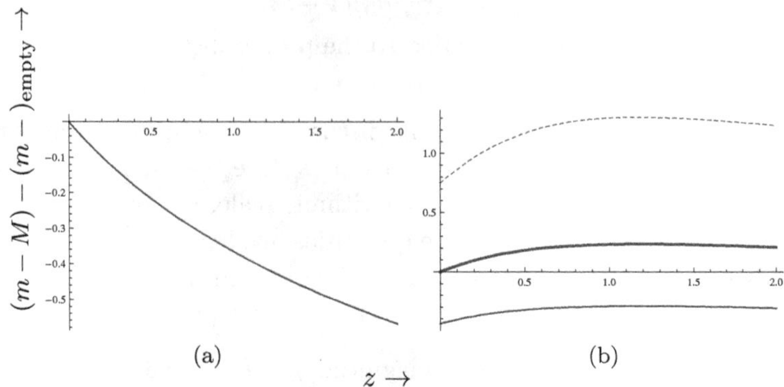

Fig. 10.32: The reduced magnitude $(m - M)$ relative to that of an empty universe as a function of z in the case of $(\Omega_{m_0}, \Omega_{\Lambda_0}) = (1.0, 0.0)$ (a). The quantity is negative, which means that in the presence of matter the apparent luminosity is greater compared to that of an empty universe. The case of $(\Omega_{m_0}, \Omega_{\Lambda_0}) = (0.3, 0.7)$ is exhibited with thick solid, fine solid and dashed lines for $\Omega_0 = 1$, $\Omega_0 = 0.5$ and $\Omega_0 = 1.5$ respectively (b). Clearly, due to the cosmological constant, the situation has dramatically changed compared to that of case (a) (except in the case of negative curvature).

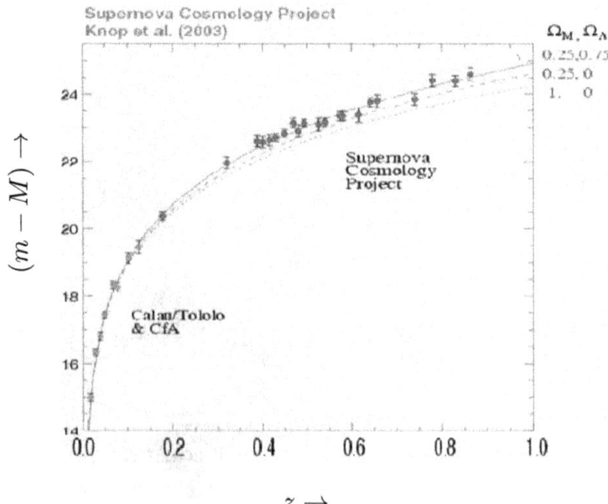

Fig. 10.33: The reduced magnitude $(m - M)$ in various models for the universe as a function of z, together with the data of the SCP team.

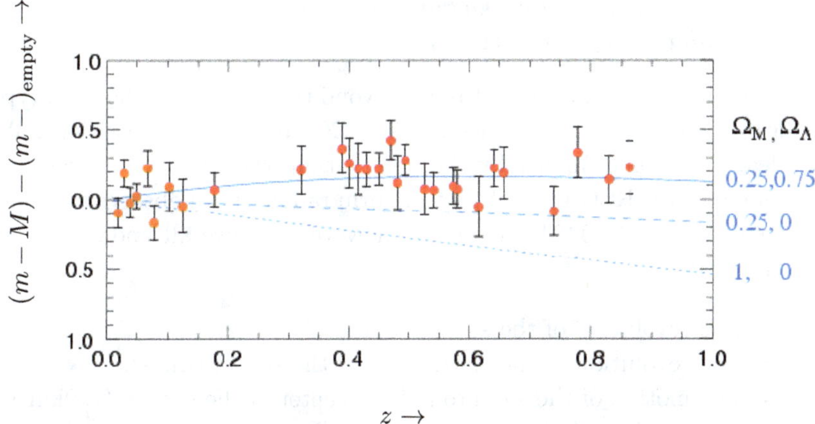

Fig. 10.34: The quantity $(m - M) - (m - M)_{\text{empty}}$. For the rest as in Fig. 10.33.

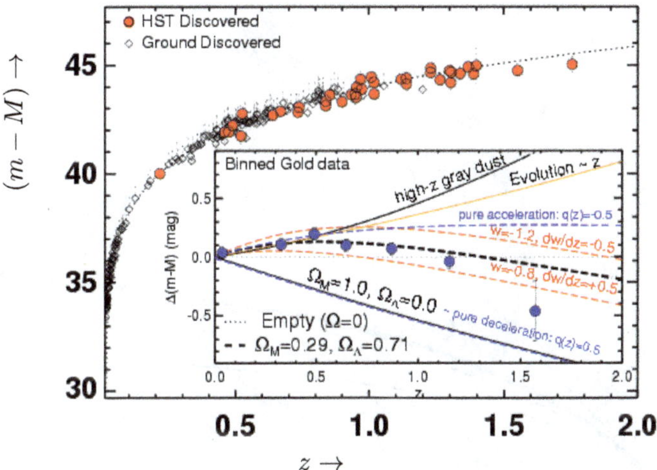

Fig. 10.35: The reduced magnitude $(m - M)$, alongside with the standard cosmological model, and $\Delta(m -) = (m - M) - (m -)_{\text{empty}}$ as a function of z, as they were obtained by appeared Hubble Space Telescope (HSP06), compared with various models.

10.12.2 *The microwave background radiation; the earliest picture of the universe*

At present it seems very difficult to go beyond the $z = 7$ with the prototype candles. We may be able to reach $z = 15 - 20$ with γ-ray observations. We can clearly go much further by the observation of microwave background radiation of 2.724 K. On top of that uniform radiation we observe asymmetries at the level of 10^{-5} K. We must, however, remove all known sources of asymmetry:

- The revolution of the satellite (8km/s).
- The revolution of the Earth around the sun (30km/s).
- The motion of the sun around the center of the galaxy (220km/s).
- The motion of the galaxy around the center of the local group (89km/s).
- The motion of the local group around the hydra complex (630 ± 20km/s).

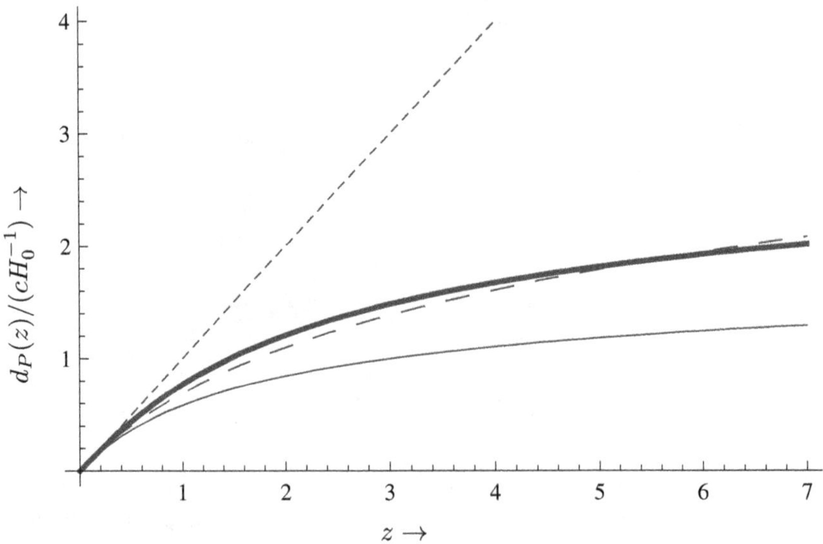

Fig. 10.36: The proper length for a flat universe for matter only (fine line), Cosmological constant only (short dashed line) and the standard cosmological model (thick solid line). With long dashed line we indicate the corresponding quantity for an empty universe. Only relatively small z are included (near objects).

After that there still remains some asymmetry seen by the COBE (1992) and more recently WMAP (2001-2007). The latter are shown in Fig. 10.38. The angular resolution is $\delta\theta = 1^\circ = 0.0174$ radians. The angular diameter in universe dominated by matter is

$$d_A(z) = \frac{2c}{H_0}\left(1 - \frac{1}{1+z}\right)\frac{1}{(1+z)}. \tag{10.222}$$

For the standard cosmological model for $z \gg 1$:

$$d_A(z) = d_P(z)/(1+z) = \frac{3c}{H_0}\frac{1}{(1+z)} \tag{10.223}$$

(See Fig. 10.37). Hence at the time of decoupling

$$d_A(z = 1000) = \frac{3 \times 3 \times 10^8}{70 \times 10^3}\text{Mpc}\,10^{-3} = 13.8\text{Mpc}.$$

Thus we find $\ell = \delta\theta d_A = 0.22$Mpc then, that is $\ell = 220$Mpc today, which resembles the size of large galaxy clusters.

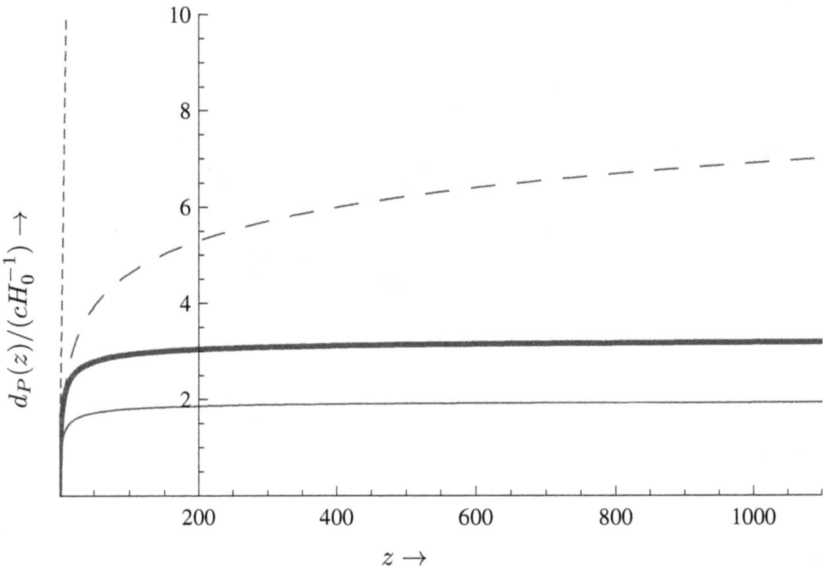

Fig. 10.37: The same as in Fig. 10.36, but up to $z \approx 1100$.

These inhomogeneities started as quantum fluctuations, which eventually became classical, i.e. in matter density variations described by the Sachs-Wolfe theory. This way variations in gravitational potential became

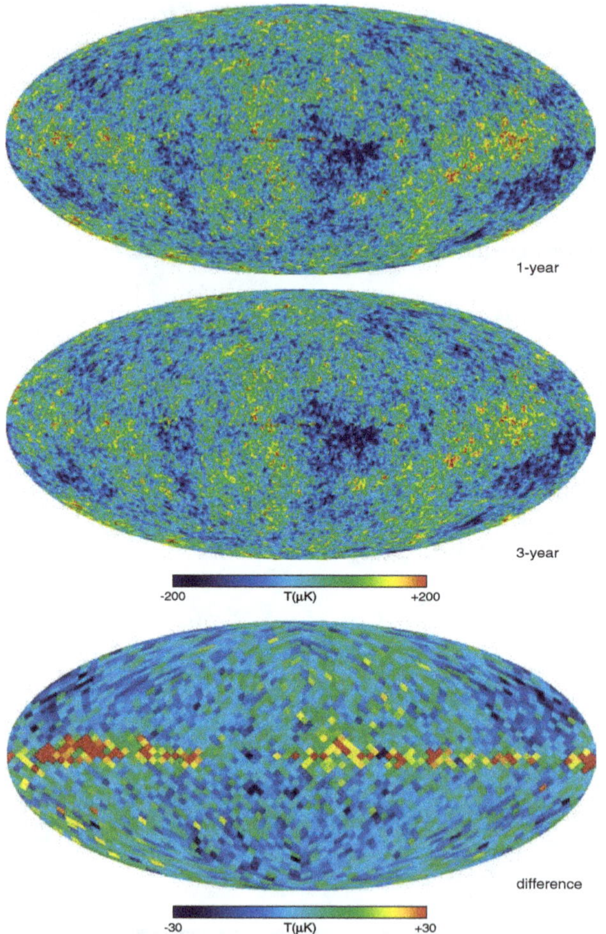

Fig. 10.38: The inhomogeneity of the microwave background radiation, as observed by WMAP (2001-2007). Shown are the data of the first and third year measurements and the difference between them. The hottest points (open color) correspond to higher density. The angular resolution is $\delta\theta = 1^\circ = 0.0174$ radians. Given that the distance of angular diameter is $d_A = 13.8$Mpc, we find $\ell = \delta\theta d_A = 0.22$Mpc then, that is $\ell = 150$Mpc today, comparable to that of galaxy clusters.

temperature variations and survived till the present. Thus:

- When a photon exits from a gravitational potential smaller than the one it entered, it is blue shifted (its energy increases).

- When a photon exits from a gravitational potential greater than the one it entered, it is red shifted (its energy decreases).

It can be shown that:

$$\frac{\delta T}{T} = \frac{1}{3}\frac{\delta \Phi}{\Phi} \quad \text{(Sachs-Wolfe effect)}, \tag{10.224}$$

where Φ is the gravitational potential. Thus the black body radiation registered the variations in the gravitational potential prevailing before decoupling, appearing to us as temperature variations. These photons took a photograph the surface of last scattering. More than that it registered that the the usual matter had been attracted by the prevailing dark matter as follows:

- The coupled system of baryons and photons tends to move to the minimum of the dark matter potential. Thus the pressure increases and it recoils back.
- As it moves away the pressure decreases and it moves backwards.
- Such changes in pressure lead to density oscillations, which were registered as acoustic waves, which reached to us.
- The light we observe is red shifted, if the fluid at the time of decoupling was receding from us; it is blue shifted, if it were moving towards us.
- The position of the maxima depend on the cosmological model.
- The location of the absolute maximum corresponds to the maximum pressure.

We note that we have no such information long before decoupling because the universe was then opaque. Only after that the registered information reached us.

10.13 The WMAP and PLANCK observations

These are satellite experiments analyzing the anisotropy of the microwave background radiation and they are going to be discussed below.

The results of the WMAP observations [Hinshaw *et al.* (2008)] are sown in Fig. 10.39. That data represent an impressive confirmation of the standard cosmological model. We hope that the future observations at smaller angles will give information about structures at $\ell < 150$Mpc today.

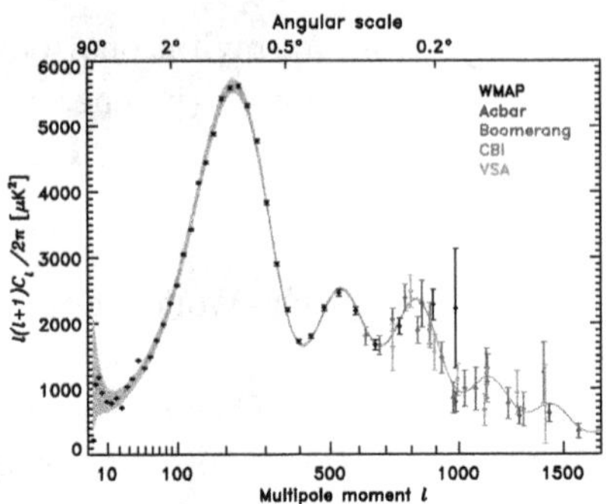

Fig. 10.39: The three year WMAP spectrum (in black) alongside other observations not discussed here. At the top the angular scale is shown, while at the bottom the multiplicity $\ell = \pi/\theta$. The agreement is good. It is an impressive confirmation of the standard cosmological model for large angles (continuous line). We hope in the near future we will have data for small angles corresponding to smaller scales.

The basic cosmological parameters, as the result of the various recent experiments, are shown in table 10.3. We usually write, e.g.,

$$\begin{aligned} H_0 &= (73.4 \pm 2.8)\text{kms}^{-1}\text{Mpc}^{-1} \Leftrightarrow \\ H_0 &= 100h \text{ km s}^{-1}\text{Mpc}^{-1}, \quad 0.706 < h < 0.762 \Leftrightarrow \\ H_0 &= 100h \text{ km s}^{-1}\text{Mpc}^{-1}, \quad h = 0.734 \pm 0.028. \end{aligned}$$

for the definition of the parameters see [Spergel *et al.* (2007)]. We find a remainder $1-\Omega_m-\Omega_b = 0.72\pm0.04$, which is characterized with an equation of state with $w = -0.97^{+0.07}_{-0.09}$, indicating that the universe is dominated by dark energy or cosmological constant within the experimental errors, $\Omega_\Lambda = 0.72 \pm 0.04$. From these data one concludes that the universe is, flat $\Omega = 1.00 \pm 0.04$. This Ω_Λ will eventually lead to a new inflation, since this fraction will not change with time, but the one associated with matter will keep getting smaller as the universe expands.

During the last five years the PLANCK collaboration released through NASA a map of Microwave Background Radiation with a resolution 2.5 times better than WMAP. The situation is exhibited in Figs. 10.40 and 10.41. A complete analysis of the PLANCK data indicates that the universe:

- It is 100 million years older than we thought at 13.82 billion years old.

Table 10.3: The Λ-CDM model combining the probabilities and the errors of the indicated experiments.

parameter	WMAP only	WMAP + CBI +VSA	WMAP+ ACBAR+ BOOMERang	WMAP+ 2dFGRS
$100\Omega_b h^2$	$2.233^{+0.072}_{-0.091}$	$2.212^{+0.066}_{-0.084}$	$2.231^{+0.070}_{-0.088}$	$2.223^{+0.066}_{-0.083}$
$\Omega_m h^2$	$0.1268^{+0.007}_{-0.010}$	$0.1233^{+0.007}_{-0.009}$	$0.1259^{+0.008}_{-0.010}$	$0.1262^{+0.005}_{-0.006}$
h	$0.734^{+0.028}_{-0.038}$	$0.743^{+0.027}_{-0.037}$	$0.739^{+0.028}_{-0.038}$	$0.732^{+0.018}_{-0.025}$
A	$0.801^{+0.043}_{-0.054}$	$0.796^{+0.042}_{-0.052}$	$0.798^{+0.046}_{-0.054}$	$0.799^{+0.042}_{-0.051}$
τ	$0.088^{+0.028}_{-0.034}$	$0.088^{+0.027}_{-0.033}$	$0.088^{+0.030}_{-0.033}$	$0.083^{+0.027}_{-0.031}$
n_s	$0.951^{+0.015}_{-0.019}$	$0.947^{+0.014}_{-0.017}$	$0.951^{+0.015}_{-0.020}$	$0.948^{+0.014}_{-0.018}$
σ_8	$0.744^{+0.050}_{-0.060}$	$0.722^{+0.043}_{-0.053}$	$0.739^{+0.047}_{-0.059}$	$0.737^{+0.033}_{-0.045}$
Ω_m	$0.238^{+0.030}_{-0.041}$	$0.226^{+0.026}_{-0.036}$	$0.233^{+0.029}_{-0.041}$	$0.236^{+0.016}_{-0.024}$

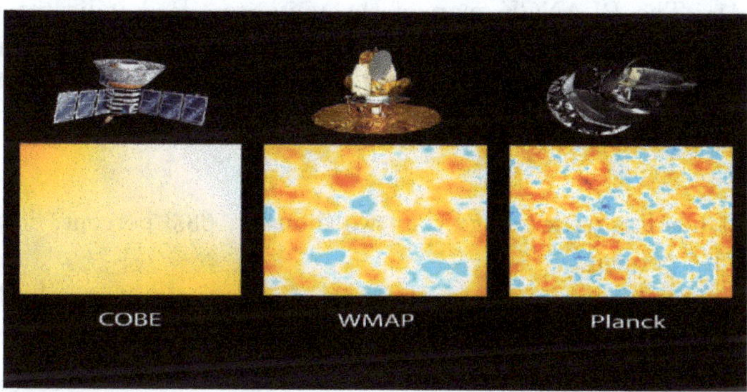

Fig. 10.40: A comparison of the resolution in the COBE, WMAP and PLANCK spectra.

- It is expanding slower than we thought: 67.15 kilometers/second/megaparsec. A megaparsec is roughly 3 million light-years.
- Contains more dark matter than we thought: 26.8 percent, up from 24 percent.

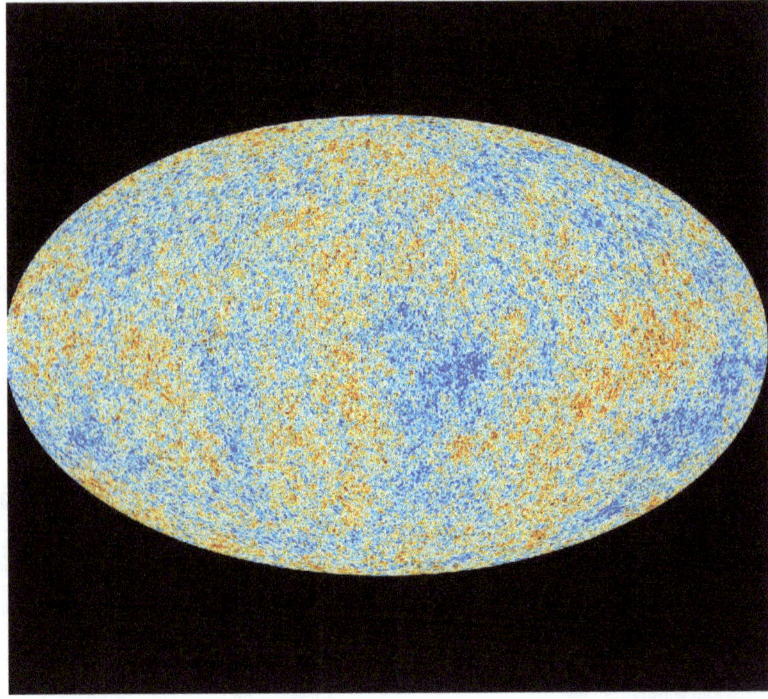

Fig. 10.41: The PLANCK map of the universe. Red indicates warmer (denser) spots, while blue means colder (less dense) parts of the universe, when its age was 370000 years old. Compare this with the WMAP (see Fig. 10.38).

- It has less dark energy than we thought: 68.3 percent, down from 71.4 percent.
- It has more "normal matter", i.e. in the form of standard model matter (atoms etc). It is now 4.9 percent, up from 4.6 percent.

Chapter 11

Aspects of Neutrino Physics; Neutrino Oscillations

11.1 The elusive neutrino

From its discovery neutrino has been elusive, but it does not really escape in neutron decay:

$$n \to p + e^- + (?)$$

its presence was inferred by Pauli (1930,1932), since without it one has violation of energy and angular momentum.Thus they proposed[1]:

$$n \to p + e^- + \tilde{\nu}_e$$

Soon afterward Fermi (1934) proposed a theory of beta decay as a contact interaction.

Bethe (1939) finds for it an important role in our world. Without it our sun and the stars cannot shine. Symbolically:

$$4p \to \text{He} + 2\nu_e + 2e^+ + \text{light}$$

(proton fusion). Its direct discovery was made much later by Reiness and Cowan (1953)

$$\tilde{\nu}_e + p \to n + e^+$$

The fact the the neutrino produced in weak interaction is different from its anti-neutrino was shown by Davis (1955) looking at the reactions:

$$\nu_e +^{37} \text{C}\ell \to^{37} \text{Ar} + e^- (\text{yes!}),$$

$$\tilde{\nu}_e +^{37} \text{C}\ell \to^{37} \text{Ar} + e^- (\text{No!}),$$

[1] They actually wrote ν, at that time they did not about the existence the antineutrino or of the presence three neutrino generations.

Neutrino has been a real revolutionary; it violates sacred laws: The fall of parity Lie & Yang and Wu *et al.* (1956) Furthermore the neutrino is very prejudiced: always left handed! As shown by Goldhaber *et al.* (1958). A new neutrino is born in 1947, named ν_μ, since it always accompanies μ^-, with its own antineutrino found at about the same time. This neutrino was different from its brother since it was shown that $\nu_\mu \neq \nu_e$, Brookhaven experiments (1962). One more neutrino, named ν_τ, was later born. A neutral fermion that appears always together with the charged lepton τ^- found at SPEAR and SLAC (1976). It was later shown that $\nu_\tau \neq \nu_\mu \neq \nu_e$ at Fermilab (1997).

The neutrino is the only known particle that does not interact electromagnetically or strongly. Its properties were the cornerstone in the development of the standard model by GSW (1967-1971), as we have seen in chapter 3.

11.2 Lepton flavor conservation

In weak interactions it appears that the neutrino appears in 3 distinct generations as it is manifested by the fact that it is always accompanied by its charged partner. This behavior is understood if we associate with each lepton pair, neutrino and charged particle, an additive quantum number called flavor, i.e. $(L_e, L_\mu, L_\tau$ and postulate that it is conserved. The antineutrinos have an opposite lepton number and, as we have seen above, they are distinct from the neutrinos. A consequence of lepton flavor violation is that processes like:

$$\mu^- \to e^- + \gamma, \tau^- \to e^- + \gamma, \tau^- \to \mu^- + \gamma \tag{11.1}$$

$$\mu^- \to e^- + e^- + e^+, \tau^- \to e^- + e^- + e^+ \text{ etc} \tag{11.2}$$

$$\mu_b^- + (A,Z) \to e^- + (A,Z)^* \text{ (bound muon conversion in the nucleus)} \tag{11.3}$$

are forbidden. Experimentalists have, in fact, been searching for such processes even long before the advent of gauge theories, but no such signals have been observed. So rather stringent experimental limits on the observed branching ratios (BR) exist:

$$\text{BR}(\mu^- \to e^- + \gamma) \leq 5.6 \times 10^{-13}, \text{ MEG (2012)}$$

$$\text{BR}(\tau^- \to \mu^- + \gamma) \leq 3.8 \times 10^{-8}, \text{ B-factories (2011)}$$

$$\text{BR}(\tau^- \to e^- + \gamma) \leq 4.8 \times 10^{-8}, \text{ B-factories (2011)}$$

$$\text{BR}(\mu_b^- + \text{Au} \to e^- + \text{Au}) \le 7 \times 10^{-13} \text{ SINDRUM II (2006)}$$

Lepton flavor conservation implies that the total lepton number $L = L_e + L_\mu + L_\tau$ is also conserved. So neutrinoless double beta decay

$$(A, Z) -> (A, Z + 2) + e^- + e^-$$

is forbidden. In fact it has been searched for over a half a century, but has not yet been observed. This important process will be discussed below.

Lepton flavor conservation implies that a neutrino of a given flavor cannot be transformed into one of a different flavor. Neutrino oscillation experiments persisted, however, and, as we will see below, they were finally successful.

11.3 History of neutrino oscillations

It is known from quantum mechanics that if a system is found in a state, which is not stationary, i.e. it is not an eigenstate of the Hamiltonian, it oscillates with time between the various eigenstates. Many such oscillations are known, e.g. in molecules like NH_3 with the possibility of different arrangements of the atoms in space, which are degenerate to zeroth order, but the degeneracy may be removed by some perturbation. Another system, known to exhibit oscillations since the middle of the 20th century, which we will study in detail in chapter 12, is that of the neutral kaons, K^0 and $\bar{K}^0$, which are eigenstates of the strong interactions, i.e they are produced as such. However K^0 and $\bar{K}^0$ are not eigenstates of the symmetry CP. The eigenstates of CP are the K_1 and K_2, i.e.

$$K_1 = \frac{1}{\sqrt{2}} \left(K^0 + \bar{K}^0\right), \quad K_2 = \frac{1}{\sqrt{2}} \left(K^0 - \bar{K}^0\right). \tag{11.4}$$

Thus we encounter the transitions:

$$|\langle K^0(t)|K^0(0)\rangle|^2 = \cos^2 \omega t, \quad |\langle \bar{K}^0(t)|K^0(0)\rangle|^2 = \sin^2 \omega t, \quad \omega = \frac{\delta m c^2}{2\hbar}. \tag{11.5}$$

CP symmetry, however, is not conserved in weak interactions and we have further transitions from K_1 and K_2 to the K_L (K-long) and K_S (K-short), see chapter 12. Can a similar situation exist in another neutral system, like the neutrinos? This idea was frst put forward by Pontecorvo (1958). The neutrinos, $(\nu_e, \nu_\mu, \nu_\tau)$ are produced in weak interactions, which are flavor conserving. Thus they have a definite flavor associated with that of their charged partners, e^-, μ^-, τ^-. Flavor, however, is only a global symmetry and there is no reason to expect it to be absolutely conserved. If it is not, one expects to see neutrino oscillations. In such a case we need to go beyond the Standard Model of Particle Physics.

The first hints of neutrino oscillations came from the deficit of solar neutrinos, Davis (1960-1970). The first evidence of neutrino oscillations came from observations of atmospheric neutrinos Super Kamiokande (1998), the observations of solar neutrino oscillations, CHOOZE (1999), and the Sudbury Neutrino Observatory (SNO) (2001). The settlement of important outstanding neutrino oscillation issues were settled with reactor neutrinos, KamLAND (2003). More precise measurement of the oscillation parameters followed, culminating with the constraint of the small θ_{13}, providing upper and lower limits, T2K (2011), and the actual measurement of θ_{13}, Daya Bay (march 2012).

11.4 Aspects of neutrino masses and mixing

In the discussion of the SM we have seen that the neutrinos cannot acquire a mass like all the other fermions because the right handed neutrino does not exist. If the right handed neutrino exists then the neutrino can acquire a mass through the Ykakawa coupling, similarly to the up quarks:

$$\mathcal{L}_D = y_D(\bar{\nu}_L, \bar{e}_L) \begin{pmatrix} \phi^{*0} \\ -\phi^- \end{pmatrix} \nu_R + HC = y_D \left(\bar{\nu}_L \nu_R \phi^{*0} - \bar{e} \nu_R \phi^- \right) + HC \tag{11.6}$$

In the unitary gauge this can be written as

$$\mathcal{L}_D = y_D \bar{\nu}_L \nu_R \frac{\eta + \upsilon}{\sqrt{2}} + HC = \bar{\nu}_L \mathcal{M}_D \nu_R + \frac{\mathcal{M}_D}{\upsilon} \bar{\nu}_L \eta \nu_R + HC \tag{11.7}$$

where

$$\mathcal{M}_D = \frac{\upsilon}{\sqrt{2}} y_D \tag{11.8}$$

is called the Dirac mass. This can be generalized to a Dirac mass matrix. This matrix can be non diagonal, if there exist lepton flavor violating interactions. Anyway, if this is all there is, the above matrix can be diagonalized a la up quarks yielding the neutrino eigenstates and eigenmasses. In this case the mass eigenstates are of the Dirac type (the charge conjugate (antiparticle) is different from its parent state). This scenario is not accepted by most physicists, since it does not explain why the neutrinos are so much lighter than the other fermions. Thus in extensions of the standard model, like Grand Unified Theories, the neutrino mass ought to be as large as the up quark masses. In such theories as well as in string theories or theories in extra dimensions there also appear other types of mass as we will see immediately.

11.4.1 *The Majorana neutrino mass*

Since the neutrino is a neutral particle it is possible to consider mass terms, which do not involve the right handed neutrino, but the existing right handed anti-neutrino, i.e. of the type

$$\mathcal{L}_{mass} = m_{m_\nu} \bar{\nu}^c_R \nu_L$$

These terms violate lepton number, but this conservation is the consequence of local and not gauge symmetry and, thus they are not sacred. In fact Weinberg (PRD 22 (1980) 1694) has shown that this can be achieved via two collaborating scalar isodoublets. The most economic way to achieve Majorana mass matrix is via the introduction of an isotriplet scalar $T^{I=1}_{I_3}$ particle, which can acquire a vacuum expectation value, i.e. a scalar that possess an electrically neutral component. If we choose it to be an $Y = 2$ object its neutral component is $T^0 = T^1_{-1}$. The other two components are $T^+ = T^1_0$ and $T^{++} = T^1_1$. Then the Majorana mass terms are of the type:

$$\mathcal{L}_{m_\nu} = f_{m_\nu} \left[\sum_{q=-1,0,1} \left(\bar{\ell}^c \otimes \ell \right) \right]^1_q (-1)^q T^1_q \text{ or}$$

$$\mathcal{L}_{m_\nu} = f_{m_\nu} \left(- \left[\bar{\ell}^c \otimes \ell \right]^1_{+1} T^0 + \left[\bar{\ell}^c \otimes \ell \right]^1_0 T^+ - \left[\bar{\ell}^c \otimes \ell \right]^1_{-1} T^{++} \right) + HC$$

where

$$\left[\bar{\ell}^c \otimes \ell \right]^1_{+1} = -\bar{\nu}^c_R \nu_L,\ \left[\bar{\ell}^c \otimes \ell \right]^1_{-1} = \bar{e}^c_R e_L,\ \left[\bar{\ell}^c \otimes \ell \right]^1_0 = \frac{1}{\sqrt{2}} \left(\bar{\nu}^c_R e_L - \bar{e}^c_R \nu_L \right)$$

This can be cast in convenient form

$$\mathcal{L}_{m_\nu} = f_{m_\nu} \left(\bar{\nu}^c_R, -\bar{e}^c_R \right) \begin{pmatrix} T^0 & \frac{1}{\sqrt{2}} T^+ \\ \frac{1}{\sqrt{2}} T^+ & T^{++} \end{pmatrix} \begin{pmatrix} \nu_L \\ e_L \end{pmatrix} + HC,$$

or in the form most commonly used

$$\mathcal{L}_{m_\nu} = f_{m_\nu} \left(\bar{\nu}_L, \bar{e}_L \right) \begin{pmatrix} T^0 & \frac{1}{\sqrt{2}} T^- \\ \frac{1}{\sqrt{2}} T^- & T^{--} \end{pmatrix} \begin{pmatrix} \nu^c_R \\ -e^c_R \end{pmatrix} + HC \tag{11.9}$$

Now if the isotriplet acquires a vacuum expectation value υ_T we get the majorana mass matrix:

$$\mathcal{L}_{m_\nu} = \bar{\nu}_L \mathcal{M}_{m_\nu} \nu^c_R + HC,\ \mathcal{M}_{m_\nu} = f_{m_\nu} \upsilon_T \tag{11.10}$$

The above expression is the celebrated light neutrino Majorana mass matrix. Again as in the case of the other fermion masses, the models cannot predict the numerical values of the entries of this matrix. One can arrange them to fit the data, if the vacuum expectation value is sufficiently small. This is all fine, except that neither the isotriplet has been found nor other effects attributed to it have yet been observed.

11.4.2 *Transformation properties under C*

The reader should be familiar with the content of section 2.6 before proceeding further.

11.4.2.1 *Transformation of currents*

$$J_\mu = \bar{\psi}_{aL}\gamma_\mu\psi_{bL} = -\left(\psi^c_{aR}\right)^T C^{-1}\gamma_\mu C\left(\bar{\psi}^c_{bR}\right)^T = \left(\psi^c_{aR}\right)^T\left(\gamma_\mu\right)^T\left(\bar{\psi}^c_{bR}\right)^T$$
$$= -\bar{\psi}^c_{bR}\gamma_\mu\psi^c_{aR} \tag{11.11}$$

(the last sign change entered since the fermion fields anti-commute).

11.4.2.2 *Symmetry relation of the Dirac mass term*

$$\mathcal{M}^D \equiv m^D_{a,b}\bar{\psi}_{aL}\psi_{bR} = -m^D_{a,b}\left(\psi^c_{aR}\right)^T C^{-1}C\left(\bar{\psi}^c_{bL}\right)^T = -\,m^D_{a,b}\left(\psi^c_{aR}\right)^T\left(\bar{\psi}^c_{bL}\right)^T$$
$$= m^D_{a,b}\bar{\psi}_{bL}\psi_{aR} \tag{11.12}$$

(again the last sign changed entered since the fermion fields anti-commute). But

$$\mathcal{M}^D \equiv m^D_{b,a}\bar{\psi}_{bL}\psi_{aR} \Rightarrow m^D_{b,a} = m^D_{a,b}.$$

Thus the Dirac mass matrix is symmetric.

11.4.2.3 *Symmetry relation of the Majorana mass term*

We distinguish two possibilities:

- A term connecting a left handed neutrino with a right handed antineutrino:

$$\mathcal{M}_\nu = m^M_{a,b}\bar{\psi}_{aL}\psi^c_{bR} = -m^M_{a,b}\left(\psi^c_{aR}\right)^T C^{-1}C\left(\bar{\psi}_{bL}\right)^T$$

$$= -m^M_{a,b} \left(\psi^c_{aR}\right)^T \left(\bar{\psi}_{bL}\right)^T = m^M_{a,b} \bar{\psi}_{bL} \psi^c_{aR}$$

(again the last sign changed entered since the fermion fields anti-commute).

$$\mathcal{M}_{nu} \equiv m^M_{b,a} \bar{\psi}_{bL} \psi^c_{aR} \Rightarrow m^M_{b,a} = m^M_{a,b}$$

- A term connects a left handed anti-neutrino with a right handed neutrino (if they exist)

$$\mathcal{M}_N = M^M_{a,b} \bar{\psi}^c_{aL} \psi_{bR} = M^M_{a,b} \bar{\psi}^c_{bL} \psi_{aR} \equiv M^M_{b,a} \bar{\psi}^c_{bL} \psi_{aR}.$$

Both types of Majorana mass matrices are symmetric. The full matrix is:

$$\left(\begin{array}{c|cc} & \nu^{0c}_R & \nu^0_R \\ \hline \bar{\nu}^0_L & \mathcal{M}_\nu & m^D \\ \bar{\nu}^{0c}_L & \left(m^D\right)^T & \mathcal{M}_N \end{array} \right).$$

11.4.3 *The see-saw mechanism*

Once the right handed neutrinos have have been introduced, one does not need the isotriplet to yield a Majorana mass $\mathcal{M}_{m_\nu}$ term. One can construct an effective such term involving the left handed neutrinos by introducing a Majorana mas term for the right handed neutrinos, which can easily be done since they do not carry any SM quantum numbers (they are isosinglets). This will be of the form:

$$\mathcal{L}_{m_N} = \bar{\nu}^c_L \mathcal{M}_{m_N} \nu_R + HC,\ \mathcal{M}_{m_N} = f_{m_N} \upsilon_S$$

Combining this with the Dirac Mass we obtain the 6×6 matrix:

$$\mathcal{L}_m = (\bar{\nu}_L\, \bar{\nu}^c_L) \begin{pmatrix} 0 & \mathcal{M}_D \\ \mathcal{M}^T_D & \mathcal{M}_{m_N} \end{pmatrix} \begin{pmatrix} \nu^c_R \\ \nu_R \end{pmatrix} \tag{11.13}$$

Assuming now that the Majorana mass has eigenvalues which are all much larger than the entries of the Dirac mass we obtain an effective light Majorana mass of the form

$$(\mathcal{M}_{m_\nu})_{eff} = -\mathcal{M}^T_D \mathcal{M}^{-1}_N \mathcal{M}_D \tag{11.14}$$

similar in form with that obtained in the case of the isotriplet. In fact one can see that even if the Dirac mass entries are of the order of 100 MeV, the resulting neutrinos can be quite light, if the isosinglet neutrinos are extremely heavy, e.g. $\geq 10^{10}$MeV. E.g.:

$$m_\nu \approx \frac{m_D^2}{M_N} = \frac{(100\text{MeV})^2}{10^{10}\text{MeV}} = 10^{-6}\text{MeV} = 1\text{eV},$$

which is of the order expected for neutrino oscillations. This is the celebrated see saw mechanism. The resulting egenstates are necessarily Majorana particles, i.e. $\nu_j^c = e^{-i\alpha_j}\nu_j$, with α_j real parameters called Majorana phases.

The question of predicting the fermion masses remains open. Many efforts in this direction over the years have goven some interesting models, but a real theory is missing.

The form of the unitary matrix U, which expresses the original weak eigenstates, the neutrino flavor states, in terms of the mass eigenstates is known as U_{PMNS} (Pontecorvo-Maki-Nakagawa-Sakata matrix). It is commonly parametrized in terms of three elementary rotations, one of which involves a phase, times a diagonal matrix of the Majorana phases, out of which only two are independent. It is expressed as follows:

$$U = \begin{pmatrix} 1 & 0 & 0 \\ 0 & c_{23} & s_{23} \\ 0 & -s_{23} & c_{23} \end{pmatrix} \begin{pmatrix} c_{13} & 0 & s_{13}e^{-i\delta} \\ & 0\,1 & 0 \\ -s_{13}e^{i\delta} & 0 & c_{13} \end{pmatrix} \begin{pmatrix} c_{12} & s_{12} & 0 \\ -s_{12} & c_{12} & 0 \\ 0 & 0 & 0 \end{pmatrix} \begin{pmatrix} 1 & 0 & 0 \\ & e^{i\alpha_{21}/2} & 0 \\ 0 & 0 & e^{i\alpha_{31}/2} \end{pmatrix} \tag{11.15}$$

with

$$s_{ij} = \sin\theta_{ij}, \quad c_{ij} = \cos\theta_{ij}, \; \alpha_{21} = \alpha_2 - \alpha_1, \; \alpha_{31} = \alpha_3 - \alpha_1 \tag{11.16}$$

or

$$U = \begin{pmatrix} c_{12}c_{13} & c_{13}s_{12} & e^{i\delta}s_{13} \\ -c_{23}s_{12} - e^{-i\delta}c_{12}s_{13}s_{23} & c_{12}c_{23} - e^{-i\delta}s_{12}s_{13}s_{23} & c_{13}s_{23} \\ s_{12}s_{23} - e^{-i\delta}c_{12}c_{23}s_{13} & -e^{-i\delta}c_{23}s_{12}s_{13} - c_{12}s_{23} & c_{13}c_{23} \end{pmatrix}$$
$$\begin{pmatrix} 1 & 0 & 0 \\ & e^{i\alpha_{21}/2} & 0 \\ 0 & 0 & e^{i\alpha_{31}/2} \end{pmatrix}. \tag{11.17}$$

The Majorana phases do not appear in neutrino oscillations or in any other lepton flavor changing processes. They appear only in lepton number violating processes, neutrinoless double beta decay etc.

11.5 Neutrino oscillations in the case of two generations

Let us consider two types of neutrinos ν_α and ν_β produced in weak interactions, say the electron and the muon neutrinos. Suppose however that these neutrinos are not eigenstates of the world Hamiltonian, in other words the Hamiltonian in flavor space is non diagonal, represented by a Hermitian matrix. The eigenstates are say ν_1 and ν_2, with masses m_1 and m_2 respectively. Then we know that:

$$\nu_1 = \cos\theta\nu_\alpha + \sin\theta\nu_\beta,\ \nu_2 = -\sin\theta\nu_\alpha + \cos\theta\nu_\beta$$

where the mixing angle θ depends on the interaction. We could as well have written:

$$\nu_\alpha = \cos\theta\nu_1 - \sin\theta\nu_2,\ \nu_\beta = \sin\theta\nu_1 + \cos\theta\nu_2 \tag{11.18}$$

The above expressions hold, say, at $t = 0$. Clearly these states are orthogonal, $\langle\nu_\alpha(0)|\nu_\beta(0)\rangle = 0$, but they are not stationary. They evolve according to the laws of quantum mechanics, e.g. at a later time:

$$\nu_\beta(t) = \sin\theta\nu_1 e^{-iE_1t} + \cos\theta\nu_2 e^{-iE_2t} \tag{11.19}$$

and

$$\langle\nu_\alpha(0)|\nu_\beta(t)\rangle = \cos\theta\sin\theta\left(e^{-iE_1t} - e^{-iE_2t}\right) \tag{11.20}$$

So these states are no longer orthogonal. Suppose we now define

$$\Delta_{12} = \frac{E_1 - E_2}{2}, \tag{11.21}$$

We get

$$P(\nu_\alpha(0) \to \nu_\beta(t)) = |\langle\nu_\alpha(0)|\nu_\beta(t)\rangle|^2 = 2\cos\theta\sin\theta\,(1 - \cos(2\Delta_{12})$$

$$= \sin^2 2\theta \sin^2 \Delta_{12}t$$

and

$$P(\nu_\alpha(0) \to \nu_\alpha(t)) = 1 - P(\nu_\alpha(0), \nu_\beta(t)) = \sin^2 2\theta\sin^2\Delta_{12}t$$

Suppose further that the beam is characterized by momentum p, while the neutrino masses involved are much smaller than the momentum. Then

$$\Delta_{12} = \frac{1}{2}(E_1 - E_2) = \frac{1}{2}(\sqrt{p^2 + m_1^2} - \sqrt{p^2 + m_1^2})$$

$$\approx \frac{1}{2}\left((p(1 + \frac{m_1^2}{1p^2}) - p(1 + \frac{m_2^2}{2p^2})\right) = \frac{m_1^2 - m_2^2}{4E}$$

We thus find:

$$P(\nu_\alpha(0) \to \nu_\beta(t)) = \sin^2 2\theta \sin^2 \pi \frac{L}{\ell_{12}},$$
$$P(\nu_\alpha(0) \to \nu_\alpha(t)) = 1 - \sin^2 2\theta \sin^2 \pi \frac{L}{\ell_{12}} \tag{11.22}$$

With

$$\ell_{12} = \frac{4\pi E_\nu}{\Delta m_{12}^2}, \; \Delta m_{12}^2 = m_1^2 - m_2^2 \tag{11.23}$$

So a measurement at distance $L = ct$ will detect the appearance of a new neutrino with flavor β with amplitude $\sin^2 2\theta$ and an **oscillation length** ("period") ℓ_{12}. It will also detect a diminishing flux of neutrinos with flavor α (disappearance). The situation is exhibited in Fig. 11.1. Note that in the standard experiments one does not see the full waveform, but essentially one point if $\ell_{ij} \gg$ detector size

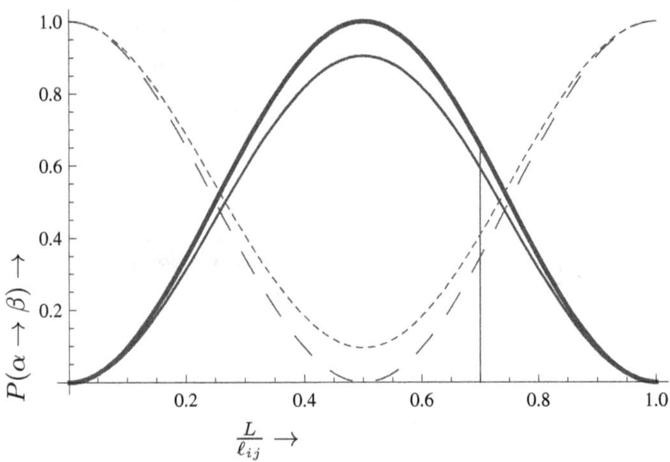

Fig. 11.1: Transition probability in the case of two flavors. The solid line describes the oscillation to one flavor, say β, from another one, say α. The thick line corresponds to mixing $\theta = \pi/4$, while the fine one to $\theta = 0.2\pi$. We will see that the first corresponds to atmospheric neutrinos, while the latter to solar neutrinos. With long (short) dashed line the neutrino disappearance is shown for $\theta = \pi/4$ ($\theta = 0.2\pi$) respectively. The intersection of the vertical line with the graphs indicates the values expected, if the detector is placed at $L = 0.7\ell_{ij}$. This value is, of course, above what is expected at $L = 0$, i.e. without neutrino oscillations.

11.6 The three generation formalism

Let us assume that a neutrino flavor ν_α is produced at time $t = 0$. What is the probability to appear as flavor ν_β at a later time? We have

$$\nu_\alpha(t) = \sum_j U_{\alpha_j} e^{-iE_j t}|\nu_j\rangle, \quad \nu_\beta(0) = \sum_j U_{\beta_j}|\nu_j\rangle, \tag{11.24}$$

with $\langle \nu_k|\nu_j\rangle = \delta_{ij}$. Thus we get:

$$P(\nu_\alpha, \nu_\beta) = |\langle \nu_\beta(0)|\nu_\alpha(t)\rangle|^2, \quad \langle \nu_\beta(0)|\nu_\alpha(t)\rangle = \sum_j U^*_{\beta_j} U_{\alpha_j} e^{-iE_j t}, \tag{11.25}$$

$$P(\nu_\alpha \to \nu_\beta) = |\langle \nu_\beta(0)|\nu_\alpha(t)\rangle|^2 = \sum_{j,k} U^*_{\beta_j} U_{\alpha_j} U_{\beta_k} U^*_{\alpha_k} e^{2i\Delta_{jk}L},$$

$$\Delta_{jk} = \frac{E_j - E_k}{2}, \quad L = ct. \tag{11.26}$$

Noting that the states $\nu_\alpha(0)$ and $\nu_\beta(0)$ are orthogonal we get $\sum_{j,k} U^*_{\beta_j} U_{\alpha_j} U_{\beta_k} U^*_{\alpha_k} = \delta_{\alpha\beta}$, so that the above equation becomes:

$$P(\nu_\alpha \to \nu_\beta) = \delta_{\alpha\beta} - \sum_{j,k} U^*_{\beta_j} U_{\alpha_j} U_{\beta_k} U^*_{\alpha_k} (1 - e^{-2i\Delta_{jk}L}), \tag{11.27}$$

which can be written as

$$P(\nu_\alpha \to \nu_\beta) = \delta_{\alpha\beta} - 2\mathrm{Re}\left\{\sum_{j<k} U^*_{\beta_j} U_{\alpha_j} U_{\beta_k} U^*_{\alpha_k} (1 - e^{-2i\Delta_{jk}L})\right\}. \tag{11.28}$$

For light neutrinos $E_j = \sqrt{p^2 + m_j^2} \approx p + m_j^2/2p$ with the momentum p independent of j. Thus

$$\Delta_{jk} L \approx \frac{m_j^2 - m_k^2}{4E_\nu} L = \pi \frac{L}{\ell_{jk}}, \quad \ell_{jk} = \frac{4\pi E_\nu}{m_j^2 - m_k^2} \tag{11.29}$$

or

$$\begin{aligned} P(\nu_\alpha \to \nu_\beta) &= \delta_{\alpha\beta} - 4\mathrm{Re}\left\{\sum_{j<k} U^*_{\beta_j} U_{\alpha_j} U_{\beta_k} U^*_{\alpha_k} \sin^2 \pi \frac{L}{\ell_{jk}}\right\} \\ &\quad -2\,\mathrm{Im}\left\{\sum_{j<k} U^*_{\beta_j} U_{\alpha_j} U_{\beta_k} U^*_{\alpha_k} \sin\left(2\pi \frac{L}{\ell_{jk}}\right)\right\}. \end{aligned} \tag{11.30}$$

Note, however, that if the phase δ of the standard U_{PMNS} matrix vanishes we have:

$$\mathrm{Im}\left\{\sum_{j<k} U^*_{\beta_j} U_{\alpha_j} U_{\beta_k} U^*_{\alpha_k}\right\} = 0. \tag{11.31}$$

We should also mention that this term comes with opposite sign in the case of antineutrinos. Ignoring this term we get:

$$P(\nu_\alpha \to \nu_\beta) = \delta_{\alpha\beta} - 4\mathrm{Re}\left\{\sum_{j<k} U^*_{\beta_j} U_{\alpha_j} U_{\beta_k} U^*_{\alpha_k} \sin^2 \pi \frac{L}{\ell_{jk}}\right\}, \tag{11.32}$$

with ℓ_{jk} conveniently written as:

$$\begin{aligned} \ell_{jk} &= 2.476\mathrm{km}\frac{E_\nu/1\mathrm{GeV}}{|\Delta m^2_{jk}|/1\mathrm{eV}^2} = 2.476\mathrm{m}\frac{E_\nu/1\mathrm{MeV}}{|\Delta m^2_{jk}|/1\mathrm{eV}^2} \\ &= 2.476\mathrm{m}\frac{E_\nu/1\mathrm{keV}}{|\Delta m^2_{jk}|/10^{-3}\mathrm{eV}^2}. \end{aligned} \tag{11.33}$$

In the special case of two generations we recover the formulas obtained above: Note there exist 3 values of $|\Delta m^2_{jk}|$, with two of them independent. If these 3 values are different, one can isolate only one mode by a judicious choice of the source detector distance L. One can thus treat it as a two generation problem with an effective mixing angle θ_{eff}. In the presence of three generations in the case of ν_e disappearance we get

$$\begin{aligned} P(\nu_e \to \nu_e) = 1 - 4 \Bigg[&\cos^4\theta_{13}\cos^2\theta_{12}\sin^2\theta_{12}\sin^2\left(\pi\frac{L}{\ell_{21}}\right) \\ &\cos^2\theta_{13}\sin^2\theta_{12}\sin^2\theta_{13}\sin^2\left(\pi\frac{L}{\ell_{32}}\right) \\ &\cos^2\theta_{13}\cos^2\theta_{12}\sin^2\theta_{13}\sin^2\left(\pi\frac{L}{\ell_{31}}\right)\Bigg]. \end{aligned} \tag{11.34}$$

In the special case of $\ell_{31} \approx \ell_{32}$ and $\theta_{13} << 1$, we find

$$P(\nu_e \to \nu_e) = 1 - \left[\sin^2 2\theta_{12}\sin^2\left(\pi\frac{L}{\ell_{21}}\right) + \sin^2 2\theta_{13}\sin^2\left(\pi\frac{L}{\ell_{32}}\right)\right], \tag{11.35}$$

i.e. we have to deal with the superposition of two oscillations, one due to θ_{12} and the other due to θ_{13}.

For three generations we find that some useful relations which are going to be useful in the study of CP violation in the leptonic sector (see the problem in the end of this chapter).

11.7 Neutrino oscillation experiments

Depending on the neutrino source we have the following possibilities:

- Solar neutrinos.

They are neutrinos ν_e originating from the sun during the proton fusion. Since their energy is known their oscillation can be detected as neutrino disappearance provided that the original flux is known.

- Reactor neutrinos.
  These are primarily anti-neutrinos produced in β-decay. Their oscillation cannot be detected by transition to other flavors, since their energy is low. So only disappearance of electron anti-neutrinos may be seen.
- Accelerator neutrinos.
  In this case all neutrino flavors are produced. Transitions to all flavors can be detected.
- The atmospheric neutrinos.
  These are produced in the atmosphere from cosmic rays producing π^+ and π^-, which eventually produce neutrinos.

$$\pi^- \to \mu^- + \tilde{\nu}_\mu, \quad \mu^- \to e^- + \nu_\mu + \bar{\nu}_e, \tag{11.36}$$

$$\pi^+ \to \mu^+ + \nu_\mu, \quad \mu^+ \to e^+ + \tilde{\nu}_\mu + \nu_e. \tag{11.37}$$

(see Fig. 11.2). It is seen from the last expression that the number of muon type to electron type neutrinos is 2:1. What is the ratio in the laboratory? Note that the ratio of the initial neutrino flux is adequate.

11.7.0.1 *The atmospheric neutrinos*

The following indication were chronologically observed:

(1) A difference in the muon neutrino to electron neutrino ratio.
The measurement of the ratio:

$$R_{\mu,e} = \frac{N_\mu}{N_e}, \tag{11.38}$$

indicated a value deviating from the value 2. Indeed, if :

$$fR_{\mu,e} = \frac{R_{\mu,e}(\text{experiment})}{R_{\mu,e}(\text{no oscillation})}, \tag{11.39}$$

it was found that [Fukuda *et al.* (1998a)]

$$fR_{\mu,e} = 0.63 \pm .03(\text{statistical}) \pm .05(\text{systematic}). \tag{11.40}$$

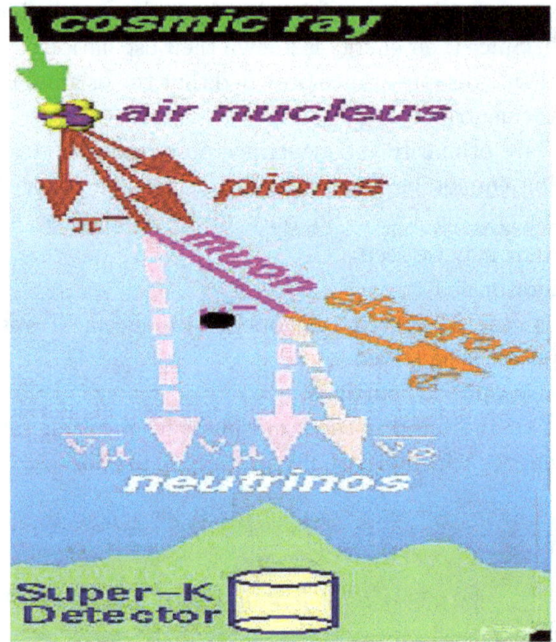

Fig. 11.2: The neutrinos produced from the decay of π^-. Note that from cosmic rays the muon like neutrinos are twice as many as the electron type.

(2) A difference in the behavior of neutrinos, depending whether they were coming from above (near zenith) or below (from the opposite side of the Earth), see Fig. 11.3. This could not be attributed to the magnetic field of the Earth, since the neutrino energy was high.

One defines the asymmetry:

$$A = \frac{N(\text{above}) - N(\text{below})}{N(\text{below}) + N(\text{above})}. \tag{11.41}$$

Then after a detailed analysis of the data [Fukuda *et al.* (1998b)] the obtained results are exhibited in Fig. 11.4. From the data of 11.4, one finds

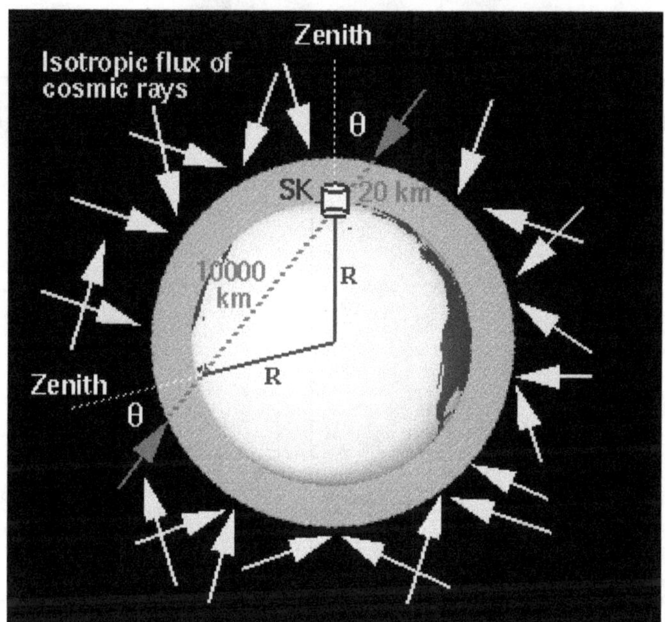

Fig. 11.3: There is no preferred direction for neutrinos from cosmic rays above 10 GeV, since the Earth's magnetic field is negligible.

$A = 0.296 \pm 0.048(\text{statistical}) \pm 0.01(\text{systematic})$. One finds:

$$\theta \approx \frac{\pi}{4}, \quad \Delta m^2 \approx 2.5 \times 10^{-3}(\text{eV}/\text{c}^2)^2. \tag{11.42}$$

Furthermore they claim that it is an oscillation of the type $\nu_\mu \to \nu_\tau$.

11.7.0.2 *Solar neutrinos*

The detection of solar neutrino oscillations is extremely difficult for the following reasons:

- The energy of neutrinos with large flux is extremely low, i.e. below the energy threshold of the standard detectors. On the other hand the flux of the higher energy neutrinos is not adequate to give detectable events (see Fig. 11.5).
- The oscillation, manifested as a reduction of the neutrino flux in the detector, presupposes knowledge of the corresponding flux in the surface of the sun. This cannot be measured, but only estimated based on an assumed solar model. The existing solar models, even

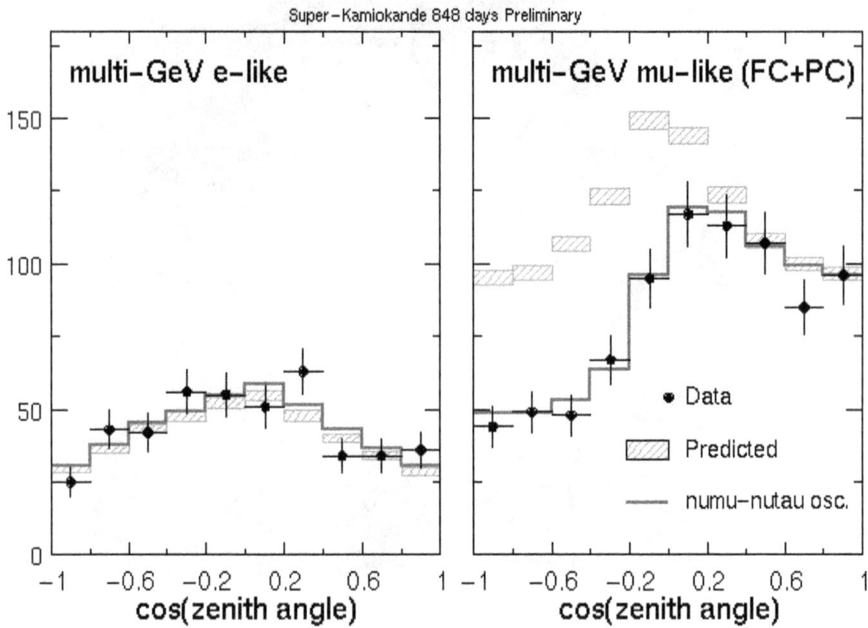

Fig. 11.4: The distribution of muon-like (on the left) and electron-like events (on the right) exhibits a characteristic asymmetry. Note that $\cos\theta_{zenith} < 0$ corresponds to upward events.

though they were considered otherwise reliable, could not exclude the possibility of an overestimate of the neutrino flux.

- The neutrinos traverse a long distance to travel to Earth. The oscillation could take place inside the sun. Then its description is quite complicated, due to the presence of electrons, whose distribution also depends on the solar model. This very interesting scenario, however, is not going to be discussed here.

For all these reasons the question of solar neutrino oscillations was finally resolved with the aid of reactor neutrino experiments. The most important stages are:

(1) Solar neutrinos
   - Cl experiments, which began in 1968, known as Davies

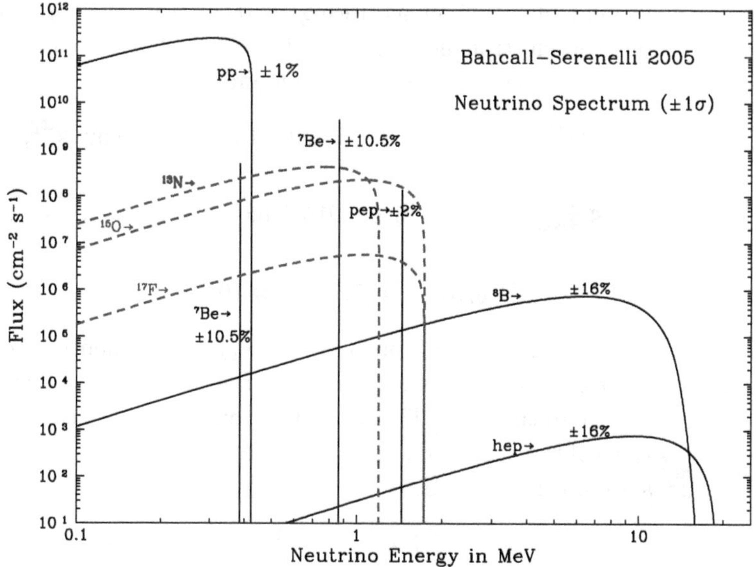

Fig. 11.5: The spectrum of solar neutrinos.

experiments:

$$\nu_e +^{37} C\ell \longrightarrow e^- +^{37} Ar \ , \ (E_{th} = 0.814 MeV).$$

It is remarkable that Davies was able to extract a dozen of Ar atoms from tons of $CC\ell_4$, which he himself put in to check the efficiency of the apparatus, before the actual measurement.

- The Gallium experiment, GNO, SAGE, GALLEX:

$$\nu_e +^{71} Ga \longrightarrow e^- +^{71} Ge \ , \ (E_{th} = 0.523 MeV).$$

- The Sudbury experiment, SNO.
  It involved 1000 tons D_2O , $E_\nu \geq$5 MeV utilizing the reactions:

$$\nu_e + d \longrightarrow e^- + p + p \ (\text{ cc}),$$

$$\nu_\alpha + d \longrightarrow \nu_\alpha + p + n \ , \ (\alpha = e, \mu, \tau), \ (\text{nc})$$

followed by: $n + p \longrightarrow d + \gamma$. Also the reaction:

$$\nu_\alpha + e^- \longrightarrow \nu_\alpha + e^-, \quad (\alpha = e, \mu, \tau) \ (\text{ES}).$$

In neutrino electron scattering (ES) all electrons contribute, but the electron neutrinos ν_e dominate. For the the boron neutrinos (see Fig. 11.5) it was found that

$$\Phi^{cc}_{SNO}(\nu_e) = (1.68^{+0.08}_{-0.09} \pm 0.015) \times 10^{-6} \ \ \mathrm{cm}^2 \ \mathrm{s}^{-1},$$

$$\Phi^{ES}_{SNO}(\nu_x) = (2.35 \pm 0.015 \pm 0.012) \times 10^{-6} \ \mathrm{cm}^2 \ \mathrm{s}^{-1},$$

$$\Phi^{nc}_{SNO}(\nu_x) = (5.2^{+0.7}_{-0.8}) \times 10^{-6} \ \mathrm{cm}^2 \ \mathrm{s}^{-1}.$$

Clearly, since the nc process is insensitive to neutrino oscillations, this gives us a measure of the initial flux, which was measured this way! This value was consistent with the predictions of the standard solar model (SSM) as well as the particle standard model (PSM):

$$\Phi^{nc}_{SSM} = 5.7 \times 10^{-6} \ \mathrm{cm}^2 \mathrm{en} \ \mathrm{s}^{-1} \ \ \text{neutrinos} \ .$$

The cc process gives a measure of neutrino disappearance via:

$$\Phi^{cc}_{SNO} = \langle P_{ee} \rangle \Phi^{nc}_{SNO}, \ \ \langle P_{ee} \rangle \approx 0.34^{0.07}_{-0.06} \ . \tag{11.43}$$

Furthermore

$$\Phi^{ES}_{SNO} = [\langle P_{ee} \rangle \Phi^{nc}_{SNO} + r_\sigma (1 - \langle P_{ee} \rangle)] \, \Phi^{nc}_{SNO}, \ \ r_\sigma \approx 0.15 \ , \tag{11.44}$$

where $r_\sigma$is the contribution of ν_μ and ν_τ.

(2) The reactor experiments, KamLAND.

$$\tilde{\nu}_e + p \longrightarrow e^+ + n \ , \ (E_{th} = 1.8MeV),$$

followed by: $n + p \longrightarrow d + \gamma$. Here, of course, anti-neutrinos are involved, but, as far as oscillations are concerned, their behavior is the same. These experiments [Araki *et al.* (2005)] not only removed the ambiguities of solar neutrinos, but they also improved the accuracy of the atmospheric neutrino measurements, yielding

$$\tan^2\theta \approx 0.45, \Delta m^2 = \Delta m^2_{12} \approx (5.0 - 7.5) \times 10^{-5} (\ \mathrm{eV/c^2})^2, \theta \approx \theta_{12} \tag{11.45}$$

Even though these experiments are ν_e disappearance experiments, a more detailed analysis reveals that it is an oscillation of the type $\nu_e \to \nu_\mu$.

(3) The NOSTOS experiment [Vergados *et al.* (2010)].

This employs monochromatic neutrinos resulting from electron capture, with energies around 10 keV which subsequently interact with electrons:

$$\nu_e + e^-_{\text{bound}} \longrightarrow \nu_e + e^- .$$

The evaluation of the oscillation probability is a bit tricky, since all neutrino flavors interact with electrons. We have seen in the discussion of the standard model that the ν_e interact both via the charged and the neutral current, while ν_μ and ν_τ do so only via the neutral current. Thus the total electron neutrino cross section in the presence of oscillations is given by:

$$\begin{aligned}\sigma_{\text{tot}} &= P(\nu_e \to \nu_e)\sigma + \sum_{\alpha=\mu,\tau} P(\nu_e \to \nu_\alpha)\sigma^{'} \\ &= \sigma - \left(\sigma - \sigma^{'}\right) \sum_{\alpha=\mu,\tau} P(\nu_e \to \nu_\alpha), \end{aligned} \quad (11.46)$$

where the last equation followed from unitarity. This equation can be rewritten as

$$\begin{aligned}\sigma_{\text{tot}} &= \sigma \left\{1 - \left(1 - \frac{\sigma^{'}}{\sigma}\right) \sum_{\alpha=\mu,\tau} P(\nu_e \to \nu_\alpha)\right\} \\ &= \sigma \left\{1 - \left(1 - \frac{\sigma^{'}}{\sigma}\right) \left[\sin^2 2\theta_{12} \sin^2\left(\pi \frac{L}{\ell_{21}}\right) + \sin^2 2\theta_{13} \sin^2\left(\pi \frac{L}{\ell_{32}}\right)\right]\right\}. \end{aligned} \quad (11.47)$$

In the last step we have used Eq. (11.35). Thus the effective oscillation probability takes the form:

$$P(\nu_e \to \nu_e) = 1 - \chi(E_\nu) \left[\sin^2 2\theta_{12} \sin^2\left(\pi \frac{L}{\ell_{21}}\right) + \sin^2 2\theta_{13} \sin^2\left(\pi \frac{L}{\ell_{32}}\right)\right], \quad (11.48)$$

with

$$\chi(E_\nu) = 1 - \frac{\sigma^{'}}{\sigma} \quad (11.49)$$

(the cross sections are functions of the neutrino energy). We see that, if $\sigma^{'} = \sigma$, there is no oscillation, reminiscent of the absence of neutrino oscillations induced by the hadronic neutral current.

Since in such experiments the oscillation length ℓ_{32} can be a few meters, one can see the full oscillation or a major part of it (see Fig. 11.6), with a spherical TPC detector of radius about 5 meters, as opposed to the standard experiments, which can only see one point (indicated by a vertical bar in Fig. 11.1).

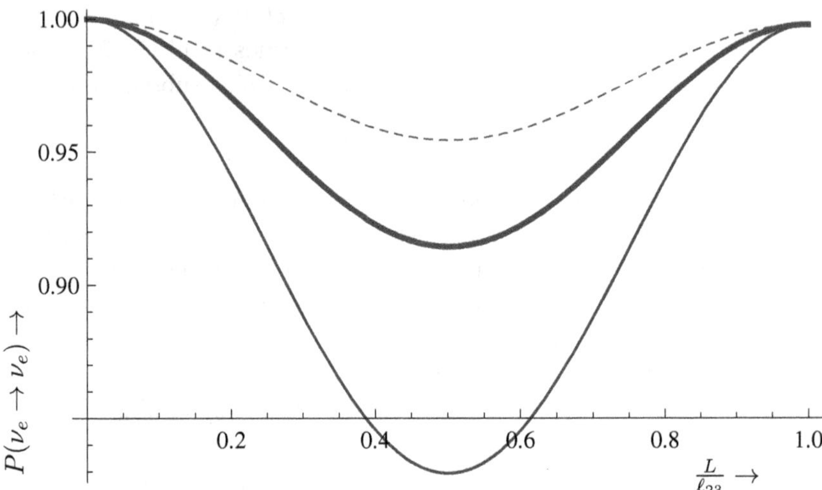

Fig. 11.6: The oscillation probability for the disappearance of electron neutrinos due to the small mixing angle θ_{13}. ℓ_{23} is the small oscillation length, which corresponds to the large mass squared difference Δm^2_{23}. The fine solid, thick solid and dashed lines correspond to $\sin^2 2\theta_{13} = 0.170, 0.085$ and 0.045. At the time the experiment was designed, $\sin^2 2\theta_{13}$ was not known. The values employed here are consistent what was found later (see text). The source is at the center of the spherical detector.

We note that in the reactor experiments one can also observe not only the strong oscillation due to the large angle θ_{12}, but also the weak one with the small oscillation length associated with the small angle θ_{13} (see Eq. (11.35)).

In summary with the discovery of neutrino oscillations quite a lot of information regarding the neutrino sector has become available (e.g., for a recent review see [Vergados *et al.* (2012)]). More specifically we know:

- The mixing angles θ_{12}, θ_{23} and θ_{13}.
- The mass squared differences:
$$\Delta^2_{\text{SUN}} = \Delta^2_{12} = m_2^2 - m_1^2 \quad \text{and } \Delta^2_{\text{ATM}} = |\Delta^2_{23}| = |m_3^2 - m_2^2|,$$
entering the solar and atmospheric neutrino oscillation experiments. Note that we do not know the absolute scale of the neutrino mass and the sign of Δ^2_{23}.

For determination of an absolute scale of the neutrino mass the relevant neutrino oscillation parameters are the MINOS value $\Delta m^2_{\text{ATM}} = (2.43 \pm$

$0.13) \times 10^{-3}$ eV^2, the global fit value $\Delta m^2_{\text{SUN}} = (7.65^{+0.13}_{-0.20}) \times 10^{-5}$ eV^2, the solar-KamLAND value $\tan^2\theta_{12} = 0.452^{+0.035}_{-0.033}$ and the recent Daya Bay [An *et al.* (2012)] observation $\sin^2 2\theta_{13} = 0.092 \pm 0.016$ (stat) ± 0.005 (syst) with a significance of 5.2 standard deviations. We note that non-zero value of mixing angle θ_{13} was already observed also by the T2K ($0.04 < \sin^2 2\theta_{13} < 0.34$), the DOUBLE CHOOZ ($\sin^2 2\theta_{13} = 0.085 \pm 0.029$ (stat) ± 0.042 (syst) (68% CL)) and RENO ($\sin^2 2\theta_{13} = 0.103 \pm 0.013$ (stat) ± 0.011 (syst)) collaborations.

A more recent analysis of the 3-generation data is given in table 11.1 (see [Capozzi *et al.* (2014)]).

At this point we would like to exhibit the essential features of the K-M (Kobayashi-Maskava) charded mixing matrix for hadrons, which is almost diagonal, and the PMNS (Pontecorvo-Maki-Nakagawa-Sakata) matrix for leptons, which is characterized by large off diagonal elements.

$$\text{K-M} \approx \begin{pmatrix} 0.975 & 0.221 & 0.003 \\ 0.221 & 0.975 & 0.040 \\ 0.009 & 0.039 & 0.999 \end{pmatrix}, \ \text{PMNS} \approx \begin{pmatrix} 0.82 & 0.55 & 0.15 \\ -0.49 & 0.52 & 0.70 \\ 0.30 & -0.65 & 0.70 \end{pmatrix} \tag{11.50}$$

The matrix PMNS is not very well understood. It involves physics beyond the standard model. Horizontal, i.e. in flavor discreet symmetries, to leading order predict the tri-bimaxial mixing matrix:

$$\text{PMNS} \approx \begin{pmatrix} -2/\sqrt{6} & 1/\sqrt{3} & 0 \\ 1/\sqrt{6} & 1/\sqrt{3} & 1/\sqrt{2} \\ 1/\sqrt{6} & 1/\sqrt{3} & -1/\sqrt{2} \end{pmatrix} \tag{11.51}$$

Based on the above we have the following scenarios:

- Normal Spectrum (NS), $m_1 < m_2 < m_3$:

$$\Delta m^2_{\text{SUN}} = m_2^2 - m_1^2\ , \ \Delta m^2_{\text{ATM}} = m_3^2 - m_1^2,$$

$$m_0 = m_1,\ m_2 = \sqrt{\Delta m^2_{\text{SUN}} + m_0^2}\ ,\ m_3 = \sqrt{\Delta m^2_{\text{ATM}} + m_0^2}.$$

- Inverted Spectrum (IS), $m_3 < m_1 < m_2$:

$$\Delta m^2_{\text{SUN}} = m_2^2 - m_1^2\ , \ \Delta m^2_{\text{ATM}} = m_2^2 - m_3^2,$$

$$m_0 = m_3,\ m_2 = \sqrt{\Delta m^2_{\text{ATM}} + m_0^2},$$

$$m_1 = \sqrt{\Delta m^2_{\text{ATM}} - \Delta m^2_{\text{SUN}} + m_0^2}.$$

The flavor content and the order of the eigenstates is exhibited in Fig. 11.7.

Table 11.1: Results of the global 3ν oscillation analysis, in terms of best-fit values and allowed 1, 2 and 3σ ranges for the 3ν mass-mixing parameters. We remind that Δm^2 is defined herein as $m_3^2 - (m_1^2 + m_2^2)/2$, with $+\Delta m^2$ for Normal Hierarchy (NH) and $-\Delta m^2$ for Inverted Hierarchy(IH). The CP violating phase is taken in the (cyclic) interval $\delta/\pi \in [0, 2]$. The overall χ^2 difference between IH and NH is insignificant ($\Delta\chi^2_{\text{I-N}} = -0.3$). The notion of NH and IH is given below in the text.

Parameter	Best fit	1σ range	2σ range	3σ range
$\delta m^2/10^{-5}$ eV^2 (NH or IH)	7.54	7.32 – 7.80	7.15 – 8.00	6.99 – 8.18
$\sin^2\theta_{12}/10^{-1}$ (NH or IH)	3.08	2.91 – 3.25	2.75 – 3.42	2.59 – 3.59
$\Delta m^2/10^{-3}$ eV^2 (NH)	2.43	2.37 – 2.49	2.30 – 2.55	2.23 – 2.61
$\Delta m^2/10^{-3}$ eV^2 (IH)	2.38	2.32 – 2.44	2.25 – 2.50	2.19 – 2.56
$\sin^2\theta_{13}/10^{-2}$ (NH)	2.34	2.15 – 2.54	1.95 – 2.74	1.76 – 2.95
$\sin^2\theta_{13}/10^{-2}$ (IH)	2.40	2.18 – 2.59	1.98 – 2.79	1.78 – 2.98
$\sin^2\theta_{23}/10^{-1}$ (NH)	4.37	4.14 – 4.70	3.93 – 5.52	3.74 – 6.26
$\sin^2\theta_{23}/10^{-1}$ (IH)	4.55	4.24 – 5.94	4.00 – 6.20	3.80 – 6.41
δ/π (NH)	1.39	1.12 – 1.77	0.00 – 0.16 $\oplus$ 0.86 – 2.00	—
δ/π (IH)	1.31	0.98 – 1.60	0.00 – 0.02 $\oplus$ 0.70 – 2.00	—

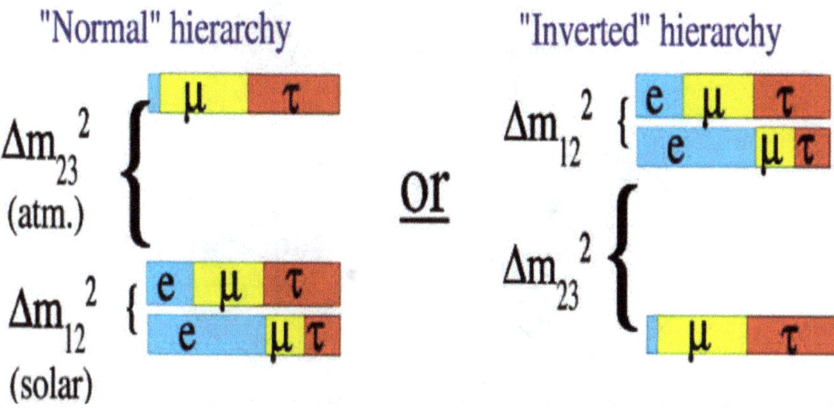

Fig. 11.7: The ordering and the flavor content of the neutrino mass eigenstates. The absolute scale m_{ν_1} in the normal hierarchy and m_{ν_3} in the inverted one is not known.

11.8 The elusive absolute scale of the neutrino mass

For determination of an absolute scale of the neutrino mass the relevant neutrino oscillation parameters are the MINOS value $\Delta m^2_{\text{ATM}} = (2.43 \pm 0.13) \times 10^{-3}$ eV^2 [Vergados *et al.* (2012)], the global fit value $\Delta m^2_{\text{SUN}} = (7.65^{+0.13}_{-0.20}) \times 10^{-5}$ eV^2 [Vergados *et al.* (2012)], the solar-KamLAND value $\tan^2\theta_{12} = 0.452^{+0.035}_{-0.033}$ and the recent Daya Bay observation $\sin^2 2\theta_{13} = 0.092 \pm 0.016$ (stat) $\pm\, 0.005$ (syst) with a significance of 5.2 standard deviations [Vergados *et al.* (2012)]. We note that non-zero value of mixing angle θ_{13} was already observed also by the T2K $(0.04 < \sin^2 2\theta_{13} < 0.34)$ [Vergados *et al.* (2012)]. The absolute scale m_0 of neutrino mass can in principle be determined by the following observations:

- Neutrinoless double beta decay.

  There exist nuclei, which are absolutely stable against ordinary weak decays, but they are unstable against double beta decay (see Fig. 11.8). This can occur with the emission of two neutrinos, which is not exotic, but a very slow second order weak interaction. This has in fact been observed in many systems with life

Nuclear Femto Laboratory
for ββ experiments

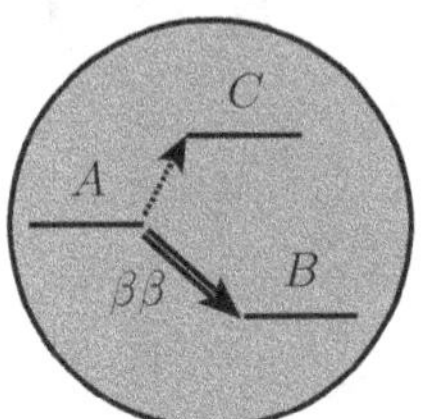

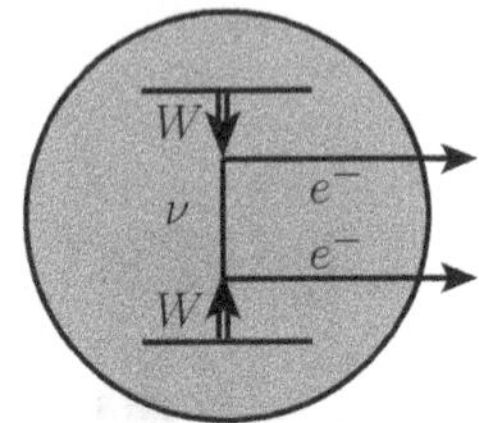

Fig. 11.8: Schematic diagrams of $\beta\beta$ decays in nuclear femto(10^{-15}m) laboratories, where single β-decay is not allowed and neutrinoless Double Beta Decay (DBD) is much enhanced. Left hand side: DBD decay scheme. Right hand side: Neutrinoless DBD with the Majorana ν exchange between two nucleons.

times in the range $10^{19} - 10^{24}$y (the longest time ever measured). If, however, there exist lepton number violating interactions and the neutrinos happen to be Majorana particles with an appreciable mass, it can also proceed without emission of neutrinos, $0\nu\beta\beta$ decay as shown in Fig. 11.9. It is somewhat favored kinematically against the standard $2\nu\beta\beta$ decay. It is suppressed by the smallness of the effective Majorana neutrino mass [Vergados *et al.* (2012)], [Vergados *et al.* (2016)].

The effective light neutrino mass $\langle m_\nu \rangle$ extracted in such experiments is given, e.g. in [Vergados *et al.* (2012)], as follows[2]:

$$\langle m_\nu \rangle = \sum_k^3 \left(U_{ek}\right)^2 m_k$$
$$= c_{12}^2 c_{13}^2 m_1 + c_{13}^2 s_{12}^2 e^{i\alpha_{21}} m_2 + s_{13}^2 e^{i\alpha_{31}} m_3. \quad (11.52)$$

The relative phases, $\alpha_{31} = 2\left(\alpha_3 - \alpha_1\right)$, $\alpha_{21} = 2\left(\alpha_2 - \alpha_1\right)$, result from both the Majorana phases α_k, $k = 12, 2$ of the neutrino and the CP violating phase of the matrix U_{PMNS}.

[2]An updated version [Vergados *et al.* (2016)] appeared while this book was in its final stage.

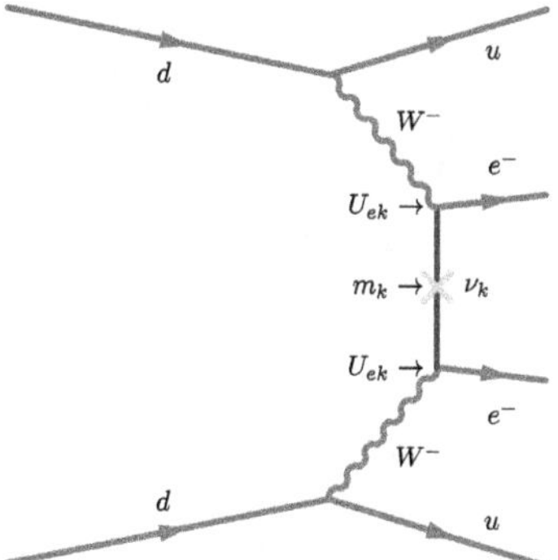

Fig. 11.9: A Feynman diagram leading to neutrinoless double beta decay at the quark level. The chiralites are such that only the mass term in the propagator $\approx \left(m_k + \gamma_\lambda p_\nu^\lambda\right)/p_\nu^2$ contributes, weighted by the square of the amplitude for neutrino production, i.e. U_{ek}^2, with the Majorana phases absorbed in it.

- The neutrino mass extracted from ordinary beta decay, e.g. from triton decay [Vergados *et al.* (2012)].

$$\langle m_\nu \rangle_{\text{decay}} = \sqrt{\sum_k^3 |U_{ek}|^2 m_k^2}$$
$$= \sqrt{c_{12}^2 c_{13}^2 m_1^2 + c_{13}^2 s_{12}^2 m_2^2 + s_{13}^2 m_3^2}. \tag{11.53}$$

  assuming, of course, that the three neutrino states cannot be resolved.
- From astrophysical and cosmological observations (see, e.g., the recent summary[Vergados *et al.* (2012)]).

$$m_\nu = \sum_k^3 m_k \le m_{\text{astro}}. \tag{11.54}$$

  The current limit m_{astro} depends on the type of observation[Vergados *et al.* (2012)]. Thus CMB primordial gives 1.3 eV, CMB+distance 0.58 eV, galaxy distribution and and lensing of galaxies 0.6 eV. On the other hand the largest photometric red shift survey yields 0.28 eV [Vergados *et al.* (2012)]. For purposes of illustration we will take a world average of $m_{\text{astro}} = 0.71$ eV.

The above mass combinations can be written as follows:

(1) Normal Hierarchy (NH), $m_1 \ll m_2 \ll m_3$.
In this case we set $m_0 = m_1$ and

$$\Delta m^2_{\rm SUN} = m_2^2 - m_1^2 \ , \ \Delta m^2_{\rm ATM} = m_3^2 - m_1^2.$$

Thus:

- Triton decay.

$$\langle m_\nu \rangle_{\rm decay} = \sqrt{c_{12}^2 c_{13}^2 m_0^2 + s_{12}^2 c_{13}^2 \left(\Delta m^2_{\rm SUN} + m_0^2\right) + s_{13}^2 (\Delta m^2_{\rm ATM} + m_0^2)}. \quad (11.55)$$

- Cosmological bound:

$$m_\nu = m_0 + \sqrt{\Delta m^2_{\rm SUN} + m_0^2} + \sqrt{(\Delta m^2_{\rm ATM} + m_0^2)}. \quad (11.56)$$

- $0\nu\beta\beta$ decay:

$$\langle m_\nu \rangle = c_{12}^2 c_{13}^2 m_0 + s_{12}^2 c_{13}^2 e^{i\alpha_{21}} \sqrt{\Delta m^2_{\rm SUN} + m_0^2} + s_{13}^2 e^{i\alpha_{31}} \sqrt{(\Delta m^2_{\rm ATM} + m_0^2)}, \quad (11.57)$$

Note the presence of the Majorana phases as we have mentioned earlier.

By assuming NH, i.e., that m_0 is negligibly small ($m_1 \ll \sqrt{\Delta m^2_{\rm SUN}}$, $m_2 \simeq \sqrt{\Delta m^2_{\rm SUN}}$, and $m_3 \simeq \sqrt{\Delta m^2_{\rm ATM}}$), we obtain

$$|\langle m_\nu \rangle| \simeq |c_{13}^2 s_{12}^2 \sqrt{\Delta m^2_{\rm SUN}} + s_{13}^2 \sqrt{\Delta m^2_{\rm ATM}} e^{-2i\alpha_2}| \leq 4 \cdot 10^{-3} \text{ eV}. \quad (11.58)$$

(2) Inverted Hierarchy (IH), $m_3 \ll m_1 < m_2$:
In this case we set $m_0 = m_3$ and

$$\Delta m^2_{\rm SUN} = m_2^2 - m_1^2 \ , \ \Delta m^2_{\rm ATM} = m_2^2 - m_3^2.$$

- Triton decay.

$$\langle m_\nu \rangle_{\rm decay} = \left(s_{13}^2 m_0^2 + s_{12}^2 c_{13}^2 (\Delta m^2_{\rm ATM} + m_0^2) + c_{12}^2 c_{13}^2 (\Delta m^2_{\rm ATM} - \Delta m^2_{\rm SUN} + m_0^2)\right)^{1/2}. \quad (11.59)$$

- Cosmological bound:

$$m_\nu = m_0 + \sqrt{(\Delta m^2_{\rm ATM} + m_0^2)} + \sqrt{(\Delta m^2_{\rm ATM} - \Delta m^2_{\rm SUN} + m_0^2)}. \quad (11.60)$$

- $0\nu\beta\beta$ decay:

$$\langle m_\nu \rangle = s_{13}^2 e^{i\alpha_{31}} m_0 + e^{i\alpha_{21}} s_{12}^2 c_{13}^2 \sqrt{(\Delta m^2_{\rm ATM} + m_0^2)} + c_{12}^2 c_{13}^2 \sqrt{(\Delta m^2_{\rm ATM} - \Delta m^2_{\rm SUN} + m_0^2)}. \quad (11.61)$$

Since in IH scenario m_0 is negligibly small ($m_1 \simeq m_2 \simeq \sqrt{\Delta m^2_{\rm Atm}}$ and $m_0 \ll \sqrt{\Delta m^2_{\rm ATM}}$), we find

$$|\langle m_\nu \rangle| \simeq \sqrt{\Delta m^2_{\rm Atm}} c_{13}^2 (1 - \sin^2 2\,\theta_{12} \sin^2 \alpha_{12})^{\frac{1}{2}}, \quad (11.62)$$

The phase difference α_{12} is the only unknown parameter in the expression for $|\langle m_\nu \rangle|$. From (11.62) we obtain the following inequality [Vergados *et al.* (2012)]

$$1.5 \cdot 10^{-2}\ \text{eV} \leq |\langle m_\nu \rangle| \leq 5.0 \cdot 10^{-2}\ \text{eV}. \quad (11.63)$$

(3) Quasi-degenerate (QD) spectrum, $m_0 = m_1 \simeq m_2 \simeq m_3$. Then

- Triton decay and Cosmology:

$$\langle m_\nu \rangle_{\rm decay} = m_0, \quad m_\nu = 3m_0. \quad (11.64)$$

- $0\nu\beta\beta$ decay:
The effective Majorana mass is relatively large in this case and for both types of the neutrino mass spectrum is given by the expression

$$m_0 |1 - 2c_{13}^2 c_{12}^2| \leq |\langle m_\nu \rangle| \leq m_0. \quad (11.65)$$

The above results are exhibited in Fig. 11.10 for the tritium β-decay and cosmological limits as a function of the lowest neutrino mass and in Fig. 11.11 for the case of the $0\nu\beta\beta$-decay both for the NS and the IS scenarios. The allowed range values of $|\langle m_\nu \rangle|$ as a function of the lowest mass eigenstate m_0 is exhibited. For the values of the parameter $\sin^2 2\theta_{13}$ new Double Chooz data are used [Vergados *et al.* (2012)]. The IH allowed region for $|\langle m_\nu \rangle|$ is presented by the region between two parallel lines in the upper part of Fig. 11.11. The NH allowed region for $|\langle m_\nu \rangle| \approx$ few meV is compatible with m_0 smaller than 10 meV. The quasi-degenerate spectrum can be determined, if m_0 is known from future β-decay experiments KATRIN [Vergados *et al.* (2012)] and MARE or from cosmological observations. The lowest value for the sum of the neutrino masses, which can be reached in future cosmological measurements, is about (0.05-0.1) eV. The corresponding values of m_0 are in the region, where the IS and the NS predictions for $|\langle m_\nu \rangle|$ differ significantly from each other.

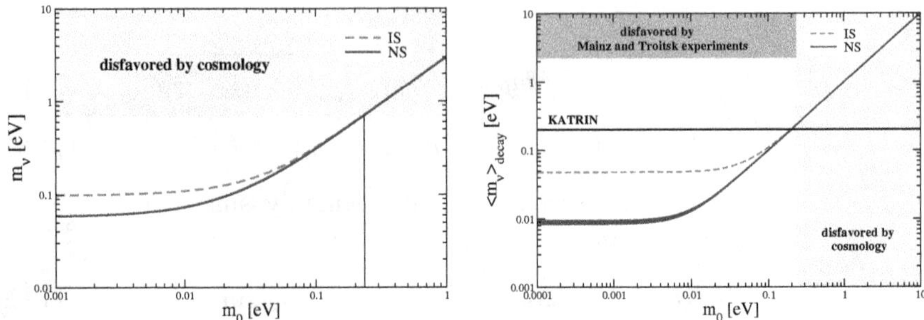

Fig. 11.10: (Color online) The neutrino mass limits in eV as a function of mass of the lowest eigenstate m_0 also in eV, extracted from cosmology (left panel) triton decay (right panel). From the current upper limit of 2.2 eV of the Mainz and Troitsk experiments we deduce a lowest neutrino mass of 2.2 eV both for the NS and IS. From the astrophysical limit value of 0.71 eV the corresponding neutrino mass extracted is about 0.23 eV for the NS and IS. It is assumed [Vergados *et al.* (2012)]: $\Delta m^2_{\rm ATM} = (2.43 \pm 0.13) \times 10^{-3}$ eV^2, $\Delta m^2_{\rm SUN} = (7.65^{+0.13}_{-0.20}) \times 10^{-5}$ eV^2 , $\tan^2\theta_{12} = 0.452^{+0.035}_{-0.033}$ and $\sin^2 2\theta_{13} = 0.092 \pm 0.016$.

From the most precise experiments on the search for $0\nu\beta\beta$-decay [Vergados *et al.* (2012)] by using of nuclear matrix elements the following stringent bounds were inferred:

$$\begin{aligned} |\langle m_\nu \rangle| &< (0.20 - 0.32)\ eV \quad (^{76}\mathrm{Ge}), \\ &< (0.33 - 0.46)\ eV \quad (^{130}\mathrm{Te}), \\ &< (0.17 - 0.30)\ eV \quad (^{136}\mathrm{Xe}). \end{aligned} \tag{11.66}$$

In future experiments [Vergados *et al.* (2012)] CUORE, EXO, MAJORANA, SuperNEMO, SNO+, KamLAND-Zen and others, a sensitivity of

$$|\langle m_\nu \rangle| \simeq \text{a few } 10^{-2} \text{ eV} \tag{11.67}$$

is expected to be reached, which is the region of the IH of neutrino masses. In the case of the normal mass hierarchy $|\langle m_\nu \rangle|$ is too small in order to be probed in the $0\nu\beta\beta$-decay experiments of the next generation.

11.9 Neutrinos as probes

We have seen the very interesting physics that are associated with neutrinos. Furthermore in recent years neutrinos are beginning to be thought of

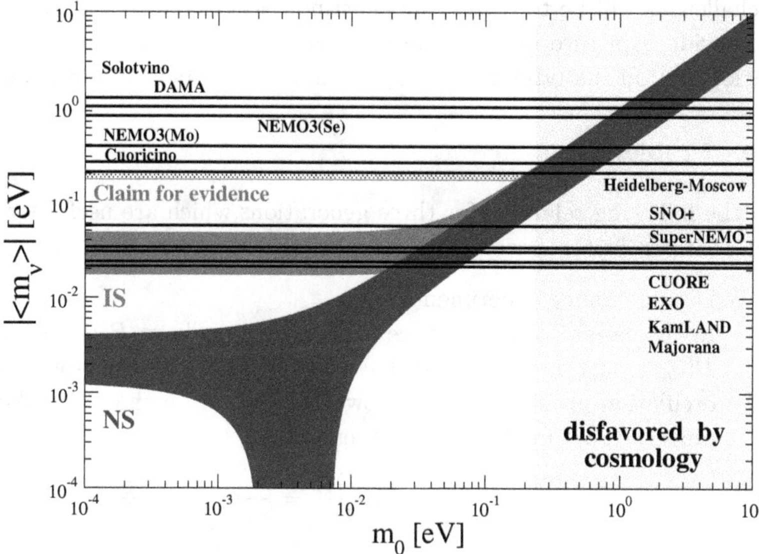

Fig. 11.11: (Color online) We show the allowed range of values for $|\langle m_\nu \rangle|$ as a function of the lowest mass eigenstate m_0 using the three standard neutrinos for the cases of normal (NS, $m_0 = m_1$) and inverted (IS, $m_0 = m_3$) spectrum of neutrino masses. Also shown are the current experimental limits and the expected future results [Vergados *et al.* (2012)] (QRPA NMEs with CD-Bonn short-range correlations and $g_A^{eff} = 1.25$ are assumed). Note that in the inverted hierarchy there is a lower bound, which means that in such a scenario the $0\nu\beta\beta$-decay should definitely be observed, if the experiments reach the required level. The same set of neutrino oscillation parameters as in Fig. 11.10 is considered.

as probes for studying other systems, in particular the interior of large objects, like the interior of stars in their last stages (supernova explosions), the interior of the Earth (geoneutrinos) and the deep sky (neutrino telescopes). The essential reason is that they can travel undistorted large distances with the speed of light. They are not affected by magnetic fields or other objects on their way and they, thus, carry undistorted information about their source. The remote control and/or surveillance of reactors is another useful application. In particular, these possibilities can be fully exploited, if low threshold detectors involving the neutral current interaction are employed, since these are not affected by neutrino oscillations. One of the real challenges will be to detect the neutrino background radiation, which may provide a picture of the Universe before the microwave black body radiation. Detailed study of such objects is, however, beyond our goals.

11.10 Problems

Show the following relations for three generations,which are useful for CP violation in the leptonic sector:

i) Disappearance experiments.
In the case of disappearance experiments we have $P(\nu_\alpha \to \nu_\alpha) = P(\bar{\nu}_\alpha \to \bar{\nu}_\alpha)$, i.e. we do not have CP violation. The disappearance oscillation probability may, however, depend on the CP violating phase δ. For simplicity we will use

$$\Delta_{ij} = \frac{\Delta_{ij}^2}{4E_\nu} = \frac{\pi}{\ell_{ij}}$$

Thus:

$$\begin{aligned} P(\nu_e \to \nu_e) = 1 - 2 \big(&\sin^2(\theta_{13})\cos^2(\theta_{13}) \\ &\big(\sin^2(\Delta_{23})\sin^2(\theta_{12}) + \sin^2(\Delta_{13})\cos^2(\theta_{12})\big) + \\ &\sin^2(\Delta_{12})\sin^2(\theta_{12})\cos^2(\theta_{12})\cos^4(\theta_{13})\big) \end{aligned} \tag{11.68}$$

In the case of ν_e disappearance the oscillation probability is independent of δ.

$$\begin{aligned} &P(\nu_\mu \to \nu_\mu) = 1 \\ &-2\sin^2(\Delta_{12})\cos^2(\theta_{23})\big(\sin^2(\theta_{12})\sin^2(\theta_{13}) + \cos^2(\theta_{12})\big) \\ &\big(\sin^2(\theta_{12})\sin^2(\theta_{23}) + \sin^2(\theta_{13})\cos^2(\theta_{12})\cos^2(\theta_{23})\big) \\ &+\cos(\delta)\sin^2(\Delta_{12})\sin(2\theta_{12})\sin(\theta_{13})\cos^2(\theta_{23}) \\ &\big(\sin^2(\theta_{12})\big(\sin(2\theta_{23})\sin^2(\theta_{13}) + 2\sin^2(\theta_{23})\big) \\ &+\cos^2(\theta_{12})\big(\sin(2\theta_{23}) + 2\sin^2(\theta_{13})\cos^2(\theta_{23})\big)\big) \\ &-\cos^2(\delta)\sin^2(\Delta_{12})\sin^2(2\theta_{12})\sin^2(\theta_{13})\sin(2\theta_{23})\cos^2(\theta_{23}) \\ &-2\sin^2(\Delta_{13})\big(\sin^2(\theta_{12})\cos^2(\theta_{23}) + \sin^2(\theta_{23})\big(\cos^2(\theta_{13}) \\ &+\sin^2(\theta_{13})\cos^2(\theta_{12})\big)\big) \\ &-\cos(\delta)\sin^2(\Delta_{13})\sin(2\theta_{12})\sin(\theta_{13})\sin(2\theta_{23}) \\ &-2\sin^2(\Delta_{23})\big(\cos^2(\theta_{23})\big(\sin^2(\theta_{12})\sin^2(\theta_{13}) + \cos^2(\theta_{13})\big) \\ &+\sin^2(\theta_{23})\cos^2(\theta_{12})\big) \\ &-\cos(\delta)\sin^2(\Delta_{23})\sin(2\theta_{12})\sin(\theta_{13})\sin(2\theta_{23}) \end{aligned} \tag{11.69}$$

$$
\begin{aligned}
&P(\nu_\tau \to \nu_\tau) = 1 \\
&-2\sin^2(\Delta_{12})\cos^2(\theta_{23})\left(\sin^2(\theta_{12})\sin^2(\theta_{13}) + \cos^2(\theta_{12})\right) \\
&\left(\sin^2(\theta_{12})\sin^2(\theta_{23}) + \sin^2(\theta_{13})\cos^2(\theta_{12})\cos^2(\theta_{23})\right) \\
&+\cos(\delta)\sin^2(\Delta_{12})\sin(2\theta_{12})\sin(\theta_{13})\cos^2(\theta_{23}) \\
&\left(\sin^2(\theta_{12})\left(\sin(2\theta_{23})\sin^2(\theta_{13}) + 2\sin^2(\theta_{23})\right) + \right. \\
&\left.\cos^2(\theta_{12})\left(\sin(2\theta_{23}) + 2\sin^2(\theta_{13})\cos^2(\theta_{23})\right)\right) \\
&-2\cos^2(\delta)\sin^2(\Delta_{12})\sin^2(2\theta_{12})\sin^2(\theta_{13})\sin(\theta_{23})\cos^3(\theta_{23}) \\
&-2\sin^2(\Delta_{13})\left(\sin^2(\theta_{12})\sin^2(\theta_{23})\right. \\
&\left.+\cos^2(\theta_{23})\left(\cos^2(\theta_{13}) + \sin^2(\theta_{13})\cos^2(\theta_{12})\right)\right) \\
&+\cos(\delta)\sin^2(\Delta_{13})\sin(2\theta_{12})\sin(\theta_{13})\sin(2\theta_{23}) \\
&-2\sin^2(\Delta_{23})\left(\cos^2(\theta_{23})\left(\sin^2(\theta_{12})\sin^2(\theta_{13}) + \cos^2(\theta_{13})\right)\right. \\
&\left.+\sin^2(\theta_{23})\cos^2(\theta_{12})\right) \\
&-4\cos(\delta)\sin^2(\Delta_{23})\sin(\theta_{12})\sin(\theta_{13})\sin(\theta_{23})\cos(\theta_{12})\cos(\theta_{23})
\end{aligned}
\tag{11.70}
$$

ii) CP violation.
The expression of interest in the same spirit is going to be:

$$
r = \frac{P(\nu_\alpha \to \nu_\beta) - P(\bar{\nu}_\alpha \to \bar{\nu}_\beta)}{P(\nu_\alpha \to \nu_\beta) + P(\bar{\nu}_\alpha \to \bar{\nu}_\beta)} \tag{11.71}
$$

We find

$$
\begin{aligned}
&\frac{1}{2}\left(P(\nu_e \to \nu_\mu) + P(\bar{\nu}_e \to \bar{\nu}_\mu)\right) = \\
&\sin^2(\Delta_{12})\sin^2(2\theta_{12})\cos^2(\theta_{13})\left(\sin^2(\theta_{13})\cos^2(\theta_{23}) - \sin^2(\theta_{23})\right) \\
&+\cos(\delta)\sin^2(\Delta_{12})\sin(4\theta_{12})\sin(\theta_{13})\sin(2\theta_{23})\cos^2(\theta_{13}) \\
&+8\sin^2(\Delta_{13})\sin^2(\theta_{13})\sin^2(\theta_{23})\cos^2(\theta_{12})\cos^2(\theta_{13}) \\
&+8\cos(\delta)\sin^2(\Delta_{13})\sin(\theta_{12})\sin(\theta_{13})\sin(\theta_{23})\cos(\theta_{12}) \\
&\cos^2(\theta_{13})\cos(\theta_{23}) \\
&+2\sin^2(\Delta_{23})\sin^2(\theta_{12})\sin^2(2\theta_{13})\sin^2(\theta_{23}) \\
&-8\cos(\delta)\sin^2(\Delta_{23})\sin(\theta_{12})\sin(\theta_{13})\sin(\theta_{23}) \\
&\cos(\theta_{12})\cos^2(\theta_{13})\cos(\theta_{23})
\end{aligned}
\tag{11.72}
$$

$$
\begin{aligned}
&\frac{1}{2}\left(P(\nu_e \to \nu_\tau) + P(\bar{\nu}_e \to \bar{\nu}_\tau)\right) = \\
&2\sin^2(\Delta_{12})\sin^2(2\theta_{12})\cos^2(\theta_{13})\left(\sin^2(\theta_{23}) - \sin^2(\theta_{13})\cos^2(\theta_{23})\right) \\
&-\cos(\delta)\sin^2(\Delta_{12})\sin(4\theta_{12})\sin(\theta_{13})\sin(2\theta_{23})\cos^2(\theta_{13}) \\
&+8\sin^2(\Delta_{13})\sin^2(\theta_{13})\cos^2(\theta_{12})\cos^2(\theta_{13})\cos^2(\theta_{23}) \\
&-8\cos(\delta)\sin^2(\Delta_{13})\sin(\theta_{12})\sin(\theta_{13})\sin(\theta_{23}) \\
&\cos(\theta_{12})\cos^2(\theta_{13})\cos(\theta_{23}) \\
&+8\sin^2(\Delta_{23})\sin^2(\theta_{12})\sin^2(\theta_{13})\cos^2(\theta_{13})\cos^2(\theta_{23}) \\
&8\cos(\delta)\sin^2(\Delta_{23})\sin(\theta_{12})\sin(\theta_{13})\sin(\theta_{23})\cos(\theta_{12}) \\
&\cos^2(\theta_{13})\cos(\theta_{23})
\end{aligned} \tag{11.73}
$$

$$
\begin{aligned}
&\frac{1}{2}\left(P(\nu_\mu \to \nu_\tau) + P(\bar{\nu}_\mu \to \bar{\nu}_\tau)\right) = \\
&\sin^2(\Delta_{12})\left(\sin^2(\theta_{13})\cos^2(\theta_{12})\right. \\
&\left(-2\cos(4\theta_{23})\cos^2(\theta_{12}) - 5\cos(2\theta_{12}) + 7\right) \\
&\left.-\left(\sin^2(2\theta_{12})\left(\sin^4(\theta_{13}) + 1\right) - 4\sin^4(\theta_{12})\sin^2(\theta_{13})\right)\sin^2(2\theta_{23})\right) \\
&+\cos(\delta)\sin^2(\Delta_{12})\sin(4\theta_{12})\sin(\theta_{13})\left(\sin^2(\theta_{13}) + 1\right)\sin(4\theta_{23}) \\
&+8\cos^2(\delta)\sin^2(\Delta_{12})\sin^2(\theta_{12})\sin^2(\theta_{13})\cos^2(\theta_{12})\cos(4\theta_{23}) \\
&+\sin^2(\Delta_{13})\sin^2(2\theta_{23})\cos^2(\theta_{13}) \\
&\left(2\cos(2\theta_{13})\cos^2(\theta_{12}) - 3\cos(2\theta_{12}) + 1\right) \\
&-2\cos(\delta)\sin^2(\Delta_{13})\sin(2\theta_{12})\sin(\theta_{13})\sin(2\theta_{23})\cos^2(\theta_{13}) \\
&+\sin^2(\Delta_{23})\sin^2(2\theta_{23})\cos^2(\theta_{13}) \\
&\left(3\cos(2\theta_{12}) + 2\sin^2(\theta_{12})\cos(2\theta_{13}) + 1\right) \\
&+2\cos(\delta)\sin^2(\Delta_{23})\sin(2\theta_{12})\sin(\theta_{13})\sin(4\theta_{23})\cos^2(\theta_{13})
\end{aligned} \tag{11.74}
$$

$$
\begin{aligned}
&\frac{1}{2}\left(P(\nu_e \to \nu_\mu) - P(\bar{\nu}_e \to \bar{\nu}_\mu)\right) = \\
&2\sin(\delta)\left(\sin(2\Delta_{12}) - \sin(2\Delta_{13}) + \sin(2\Delta_{23})\right)\sin(2\theta_{12}) \\
&\sin(\theta_{13})\sin(2\theta_{23})\cos^2\theta_{13}
\end{aligned} \tag{11.75}
$$

$$
\begin{aligned}
&\frac{1}{2}\left(P(\nu_e \to \nu_\tau) - P(\bar{\nu}_e \to \bar{\nu}_\tau)\right) = \\
&-2\sin(\delta)\left(\sin(2\Delta_{12}) - \sin(2\Delta_{13}) + \sin(2\Delta_{23})\right) \\
&\sin(2\theta_{12})\sin(\theta_{13})\sin(2\theta_{23})\cos^2(\theta_{13})
\end{aligned} \tag{11.76}
$$

$$\frac{1}{2}\left(P(\nu_\mu \to \nu_\tau) - P(\bar{\nu}_\mu \to \bar{\nu}_\tau)\right) =$$
$$2\sin(\delta)\left(\sin(2\Delta_{12}) - \sin(2\Delta_{13}) + \sin(2\Delta_{23})\right)$$
$$\sin(2\theta_{12})\sin(\theta_{13})\sin(2\theta_{23})\cos^2(\theta_{13}) \tag{11.77}$$

One can, of course, rewrite the above expressions using only two osculation lengths by, e.g., eliminating the Δ_{32}, since $\Delta_{32} = \Delta_{31} - \delta_{21}$ We thus get

$$\sin 2\Delta_{21} - \sin\Delta_{31} + \sin\Delta_{32} = 2(-\sin^2\Delta_{21}\sin 2\Delta_{31} + \sin^2\Delta_{31}\sin 2\Delta_{21})$$

We prefer to leave it as it is, however, to exhibit the behavior of the long oscillation length separately, since this is much longer than that relevant to current experiments. The factor which contains the relevant mixing angles is known as the Jarlskog invariant [Jarlskog (1985)].

Chapter 12

Discrete Symmetries: *C*, *P*, *T* and All That

In particle physics of special interest are groups with two elements T and E, with $T^2 = E$, where E the identity element. These are like reflection symmetries. Examples: i) space reflection (P), ii) time inversion (T), iii) charge conjugation (C) iv) A combination of any two of them v) The product of all three. We will examine these symmetries and their possible violation. We will also discuss the three historic experiments, namely parity and CP non conservation as well as the determination of the neutrino helicity.

12.1 Space inversion P

$$x \to -x, \qquad y \to -y, \qquad z \to -z, \, t \to t. \tag{12.1}$$

The action of P on some physical quantities is given as follows:

$$\begin{aligned}
&i) && \text{Position operator} \\
& && \vec{r}\,' = P\vec{r}P^{-1} = -\vec{r} \\
&ii) && \text{Momentum operator} \\
& && \vec{p}\,' = P\vec{p}P^{-1} = -\vec{p} \\
&iii) && \text{angular momentum operator} \\
& && \vec{L}' = P\vec{L}P^{-1} = -P(\vec{r}\times\vec{p})P^{-1} = (-\vec{r})\times(-\vec{p}) = \vec{L} \\
&iv) && \text{helicity operator} \\
& && h' = PhP^{-1} = -P(\vec{\sigma}\cdot\hat{p})P^{-1} = \vec{\sigma}(-\vec{p}) = -h
\end{aligned} \tag{12.2}$$

The helicity operator is defined by $\vec{h} = \boldsymbol{\vec{\sigma}} \cdot \hat{p}$ where $\boldsymbol{\sigma}$ is the vector with components the Pauli matrices $\sigma_1, \sigma_2, \sigma_3$, and $\hat{p}$ is a unit vector in the direction of momentum. From Eqs. (12.2) we see that the position and momentum operators transform as vectors while the angular momentum and helicity as pseudovector and pseudoscalar respectively. One can see that under P a right handed coordinate system is transformed into a left handed one.

The operator P, defined as above, induces on the space of functions $\psi(\mathbf{r},t)$ an operator O_P defined by:

$$\psi(\vec{r},t) \to \psi(-\vec{r},t) = O_P\psi(\vec{r},t), \tag{12.3}$$

with O_P a suitable operator[1] related to P A function $\Psi(\vec{r})$ is even if:

$$\psi(-\vec{r}) = \psi(\vec{r}) \to P\psi(\vec{r}) = \psi(\vec{r}), \tag{12.4}$$

while it is odd if

$$\psi(-\vec{r}) = -\psi(\vec{r}) \to P\psi(\vec{r}) = -\psi(\vec{r})\,. \tag{12.5}$$

On the other hand any function $\psi(\vec{r})$ can be written as the sum of an even and and an odd function:

$$\psi(\vec{r}) = \psi_e(\vec{r}) + \psi_o(\vec{r}), \quad \psi_e(\vec{r}) = \frac{\psi(\vec{r}) + \psi(-\vec{r})}{2}, \quad \psi_o(\vec{r}) = \frac{\psi(\vec{r}) - \psi(-\vec{r})}{2}\,. \tag{12.6}$$

Thus we can select as a basis a set of even and odd functions so that the operator P takes the form:

$$P = \left(\begin{array}{c|c} (E) & 0 \\ \hline 0 & -(E) \end{array}\right), \tag{12.7}$$

where E is the identity operator. Then it is easy to see that, acting on the space of functions $\psi_e(\vec{r})$, P is Hermitian. Indeed:

$$\begin{aligned}\langle\phi(\vec{r})|P\psi(\vec{r})\rangle &= \int_V d^3\vec{r}\phi^*(\vec{r})P\psi(\vec{r}) = \int_V d^3\vec{r}\phi^*(\vec{r})\psi(-\vec{r}) \\ &= \int_V d^3\vec{r}\phi^*(-\vec{r})\psi(\vec{r}) = \int_V d^3\vec{r}P\phi^*(\vec{r})\psi(\vec{r}) \\ &= \langle P\phi(\vec{r})|\psi(\vec{r})\rangle.\end{aligned}$$

Thus the operator P is both unitary and hermitian:

$$O_P^+ = O_P = O_P^{-1}\,. \tag{12.8}$$

The relations (12.4) and (12.5) imply that the possible eigenvalues of P are $\pi = \pm 1$ (parity).

[1]For simplicity of notation we will often write P instead of the correct O_P.

The operator O_P cannot be written as $O_P = e^{iA}$, with A a Hermitian operator, see section 1.3. In fact it is an improper orthogonal transformation (or an improper Lorentz transformation if in addition we consider a time transformation $t \to t$), not continuously connected to the identity. This means that parity is a multiplicative not an additive quantum number.

Parity is a conserved quantity, if the Hamiltonian of the system H commutes with P:

$$[P, H] = 0 \tag{12.9}$$

Then P and H have a common set of eigenfunctions $|E, \pi\rangle$, that is

$$H|E,\pi\rangle = E|E,\pi\rangle, \quad P|E,\pi\rangle = \pi|E,\pi\rangle\,. \tag{12.10}$$

In polar coordinates relations (12.1) become:

$$r \to r,\ \theta \to \pi - \theta,\ \phi \to \pi + \phi. \tag{12.11}$$

Thus using Eq. (12.11) and the known properties of spherical harmonics we find:

$$Y^{\ell}_{m}(-\hat{r}) = Y^{\ell}_{m}(\pi - \theta, \pi + \phi) = (-1)^{l} Y^{\ell}_{m}(\theta, \phi)$$

and

$$PY^{\ell}_{m}(\hat{r}) = (-1)^{\ell} Y^{\ell}_{m}(\hat{r})\,. \tag{12.12}$$

In other words the eigenfunctions of angular momentum have definite parity, i.e. ($\pi = +1$, if $\ell =$ even and $\pi = -1$, if $\ell =$ odd). Let us now consider a system of two particles with wave function (wf):

$$|\psi_{12}\rangle = |\psi_1\rangle|\psi_2\rangle|\psi_{rel}\rangle, \tag{12.13}$$

where $|\psi_i\rangle$, is the wf particle i, $i = 1, 2$, and $|\psi_{rel}\rangle$ describes their relative motion. Then, ignoring the center of mass motion, we have:

$$P|\psi_{12}\rangle = \big(P|\psi_1\rangle\big)\big(P|\psi_2\rangle\big)\big(P|\psi_{rel}\rangle\big)\,. \tag{12.14}$$

with $|\psi_1\rangle$ and $|\psi_2\rangle$ eigenfunction of P with parities π_1 and π_2 . Then

$$P|\psi_1\rangle = \pi_1 P|\psi_1\rangle, \quad P|\psi_2\rangle = \pi_2 P|\psi_2\rangle, \quad P\psi_{rel} = (-1)^{\ell_{12}}\psi_{rel}. \tag{12.15}$$

Consider now the process:

$$a + b \to c + d,$$

proceeding via parity conserving process. Then

$$\pi_a \cdot \pi_b (-1)^{l_{ab}} = \pi_c \cdot \pi_d (-1)^{l_{cd}}\,. \tag{12.16}$$

Before proceeding further we should mention that it is not always possible to determine the relative parity of two particles via a parity conserving process if the particles differ in one of the internal quantum numbers ($Q, L_i, , S, C$, etc), which is conserved in this process, e,g, the relative parity of proton and neutron, of $\pi^{\pm}$ relative π^0, of a quark with $Q = 2/3$ relative to that of one with $Q = -1/3$. We, thus, arbitrarily set the relative quark parities such that:

$$\pi_u = \pi_d = \pi_s = \pi_c = \pi_b = \pi_t \,. \tag{12.17}$$

Then assuming that the relative motion of the quarks in the nucleon is characterized by angular momentum zero we get:

$$\pi_p = \pi(uud) = \pi_u \pi_u \pi_d = \pi_d,$$
$$\pi_n = \pi(ddu) = \pi_u \pi_d \pi_d = \pi_u \Rightarrow \pi(p) = \pi(n)$$

Thus, selecting arbitrarily $\pi(p) = +1$, we find

$$\pi_u = \pi_d = \pi_s = \pi_c = \pi_b = \pi_t = +1 \,.$$

Furthermore Dirac's theory predicts that for particles with spin $s = 1/2$, the particle and the antiparticle have opposite parities. In fact if $\psi(\mathbf{x}, t)$ is a solution of the Dirac equation so is $P\psi(\mathbf{x}, t) = \gamma_0 \psi(-\mathbf{x}, t)$. Thus

$$\begin{aligned} &(i\partial^\mu \gamma_\mu - m)\, \gamma_0 \psi(-\mathbf{x}, t) = \gamma_0 \left(i\gamma_0 \partial^0 - i\gamma_k \partial^k - m\right) \psi(-\mathbf{x}, t) \\ &= \gamma_0 \left(i\gamma_0 \partial^0 + i\gamma_k \partial^k - m\right) \psi(\mathbf{x}, t) = \gamma_0 \left(i\gamma_\mu \partial^\mu - m\right) \psi(\mathbf{x}, t) = 0, \end{aligned} \tag{12.18}$$

$$P\psi^c = PC\bar{\psi}(\mathbf{x}, t)^T = PCP^{-1} P\bar{\psi}(\mathbf{x}, t)^T = -\gamma_0 \bar{\psi}(-\mathbf{x}, t)^T = -\psi^c. \tag{12.19}$$

This has in fact been confirmed by experiments studying $e^+ - e^-$ annihilation, which showed that $\pi(e^+) = -\pi(e^-)$.

Generally speaking we accept the following rules:

The elementary fermions have parity $+1$.
The elementary anti-fermions have parity -1.
The vector bosons have parity -1.
The Higgs scalars have parity $+1$.

From these rules the parities of all composite particles can be obtained. Thus the mesons $q\bar{q}$ have internal parity -1. Thus if the relative angular momentum of the two quarks is ℓ we find:

$$\begin{aligned} &\pi_M = \pi_{q_1} \pi_{\bar{q}_2} (-1)^\ell = (-1)^{\ell+1}, \\ &\pi_{\bar{M}} = (-\pi_{\bar{q}_1})\,(-\pi_{q_2})\,(-1)^\ell = \pi_{q_1} \pi_{\bar{q}_2} (-1)^\ell = (-1)^{\ell+1} \Rightarrow \pi_{\bar{M}} = \pi_M \end{aligned} \tag{12.20}$$

The π mesons have angular momentum zero, i.e. $\ell = 0$, and their parity is -1. They are pseudoscalar. The ρ mesons have angular momentum (spin) 1, $\ell = 0$, and parity -1, i.e. they transform like vectors.

The internal parity of all baryons is $+1$ and of all anti-baryons is -1. The total parity, of course, depends on the relative motion of the quarks. The spin J and parity π are usually designated as J^π

12.1.1 *Parity conservation; parity violation*

If the Hamiltonian is invariant under parity its eigenstates can be chosen to have a definite parity[2]. On the contrary, if (12.9) is not satisfied, the wf ψ can be cast in the form:

$$|\psi\rangle = \alpha|\psi_e\rangle + \sqrt{1 - |\alpha|^2}|\psi_0\rangle, \tag{12.21}$$

with

$$P|\psi_e\rangle = \underbrace{|\psi_e\rangle}_{\text{even}} \quad \text{and } P|\psi_0\rangle = -\underbrace{|\psi_0\rangle}_{\text{odd}} .$$

Then

$$P|\psi\rangle = \alpha|\psi_e\rangle - \sqrt{1 - |\alpha|^2}|\psi_0\rangle \neq \pi|\psi\rangle . \tag{12.22}$$

Let us now suppose that the dominant component of ψ is even. Then the quantity

$$|f|^2 = \frac{1 - |\alpha|^2}{|\alpha|^2} = \frac{1}{|\alpha|^2} - 1 \tag{12.23}$$

is a measure of parity violation. The maximum, $|f| = 1$, happens when $\alpha = 1/\sqrt{2}$.

Many experimental tests had been made for parity conservation in electromagnetic and strong interactions, affirming the "obvious", i.e. parity conservation. Consider, e.g. the strong interaction. A consequence of parity conservation is that the decay:

$$\rho^0 \to \pi^+ + \pi^- \tag{12.24}$$

is allowed. Indeed the ρ mesons have spin-one and the pions have spin zero. Thus the the 2-pion final state must have $\ell = 1$. The initial state has negative parity and the pions also have intrinsic parity -1. Thus the

[2]The eigenfunctions must have a definite parity except in the unlikely case that two states of opposite parity are degenerate. Then we can choose them to have a definite parity.

parity of the final state is $(-1)^1 = -1$ and the decay is allowed by strong interactions. On the other hand the process

$$\rho^0 \to \pi^0 + \pi^0 \tag{12.25}$$

is not allowed. The reason is that the final state of two identical mesons must be symmetric. This means that ℓ =even. Therefore the parity of the final state is $+1$ and, if parity is conserved, the process is not allowed. Experimentally the branching ratio for the second process is very small.

So, based on theoretical prejudices, parity conservation was elevated to a dogma. Not even the so-called τ, θ puzzle[3], succeeded in questioning this dogma.

This was the situation when, to everyone's surprise, two then Young Chinese, T. D. Lee and C. N. Yang, in a detailed study showed that nobody had tested parity non conservation in weak interactions. They even proposed various experiments to perform such tests. In fact the group of professor Wu at Colombia University [Wu *et al.* (1957)] produced the earth shaking news: parity was not conserved! To get an idea of what this meant then, see C. N. Yang's book [Yang (1962)]. It is amusing to show in Fig. 12.1 a telegram attributed to Pauli.

The main idea of the experiment [Wu *et al.* (1957)] is exhibited in Fig. 12.2, involving the beta decay of the polarized target ^{60}Co, with a ground state with $J^\pi = 5^+$:

$$^{60}\text{Co}(5^+, \text{polarized}) \to ^{60}\text{Ni}(4^+) + e^- + \bar{\nu_e}\,.$$

The intensities I_1 and I_2 of the produced electrons are measured in two directions, namely the electrons with momenta $\vec{p}_1$, and $\vec{p}_2$. Under the parity transformation we have:

$$\vec{x} \to -\vec{x}, \quad \vec{p} \to -\vec{p}, \quad \vec{J} \to \vec{J},$$

i.e. the two intensities are interchanged, Fig. 12.2. If P is a good symmetry, the two situations are not distinguishable. In other words, then, even though the nucleus is polarized the electron intensities are the same in both directions $\vec{J}$ and $-\vec{J}$ ($I_1 = I_2$). The actual results[4] are shown in Fig. 12.3. We notice the asymmetry with respect to the two directions $\vec{B}$. We also

[3]It involved two particles that they were identical in mass, charge, spin etc, except that one of them decayed into positive parity and the other into negative parity states. Today we know that it is one particle, known as K^+, decaying as $K^+ \to \pi^+\pi^+\pi^-$ (decay of type τ) and $K^+ \to \pi^+\pi^0$ (decay of type θ). Since the K^+ has zero spin none of the final products is allowed to have any angular momentum. Thus the two pion decay has positive parity and is excluded by parity conservation, while the three particle has parity -1 and is allowed, since the initial meson has parity -1.

[4]It is much harder to move the apparatus. The researchers, therefore, chose to change the direction of the magnetic field.

Pauli's Reaction to the Downfall of Parity

Pauli's Reaction to the Downfall of Parity

Es ist uns eine traurige Pflicht, bekannt zu geben, daß unsere langjährige ewige Freundin	It is our sad duty to announce that our loyal friend of many years
PARITY	PARITY
den 19. Januar 1957 nach kurzen Leiden bei weiteren experimentellen Eingriffen sanfte entschlafen ist.	went peacefully to her eternal rest on the nineteenth of January 1957, after a short period of suffering in the face of further experimental interventions.
Für die hinterbliebenen	For those who survive her,
$e \ \mu \ \nu$	

Fig. 12.1: The fall of parity shocked the world.

notice that the asymmetry vanishes after a time of 8 minutes, since the sample has been heated up and the polarization has vanished due to the interactions between the dipoles.

We should mention that the asymmetry should correspond to a pseudoscalar quantity. Since the neutrino escapes observation and the momentum of the final nucleus is negligible the only available pseudoscalar quantity is of the form $(\vec{J} \cdot \vec{p_e})$, with p_e the electron momentum, that is

$$\vec{J} \cdot \vec{p_e} \xrightarrow{P} -\vec{J} \cdot \vec{p_e}\,.$$

Thus the left of Fig. 12.2 corresponds to $\vec{J} \cdot \vec{p_e}$, while the right to $-\vec{J} \cdot \vec{p_e}$. The second is obtained from the first by changing either $\vec{p_e}$ or $\vec{J}$. The experiment was quite difficult with the then available cryostatic technology (temperatures of 0.01 degree ${}^0K$ obtained with adiabatic nuclear demagnetization etc).

It is interesting to remark that parity violation allows one to define a right handed system. Suppose that an observer here on Earth and someone in the Andromeda galaxy compare their Cartesian systems and in particular they want to check whether they are right handed. So each one of them selects an x-axis and an y-axis perpendicular to it in any way they like. They only need compare the direction of their z-axis. To this they put a polarized ^{60}Co so that the direction of its spin corresponds to the rotation

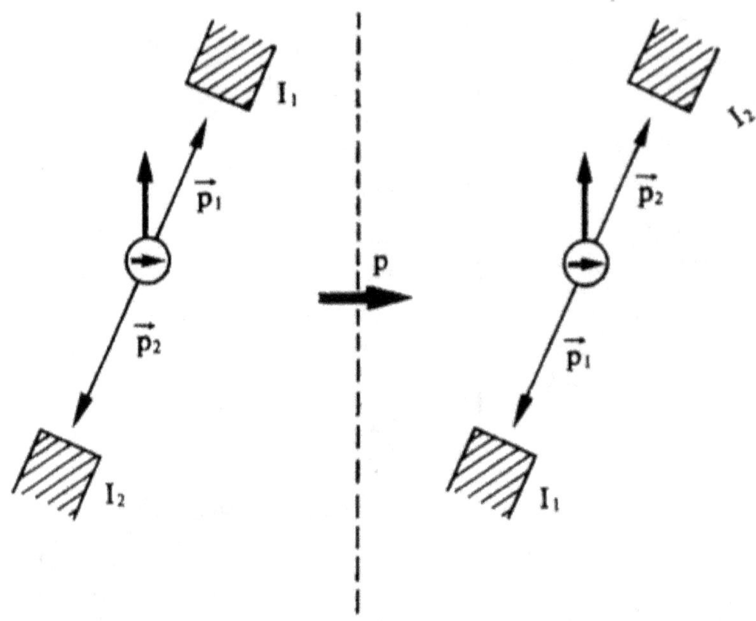

Fig. 12.2: The main idea of the experiment of Wu *et al* [Wu *et al.* (1957)]. A polarized nucleus ^{60}Co with spin parity $J = 5^+$ emits electrons in all directions. We detect those in the directions of momenta $\vec{p}_1$ and $\vec{p}_2$. The initial situation is shown on the left and the one after the inversion P on the right. Invariance implies that the two ought to be identical and experimentally indistinguishable. In the experiment, however, the direction $\vec{p}_1||\vec{J}$ corresponds to small intensity, while the direction $\vec{p}_2$ (opposite to **J**) is the one with the strong intensity.

$x \to y$. Then if the direction of the minimum electron density is in the z-direction, the coordinate system is right-handed. On the other hand if they observe the maximum intensity in the z-direction, the coordinate system is left handed. They only need a radar to convey the needed information in physics terms (see Fig. 12.4).

The above conclusions depend, of course, the fact that both observers live in a world with matter, as seems to be the case in our Universe. If that is not the case, they need to know that they deal with cobalt and not anti-cobalt, since a cobalt source in a left handed system behaves like anti-cobalt in a right handed system. So they need another violation to settle the issue

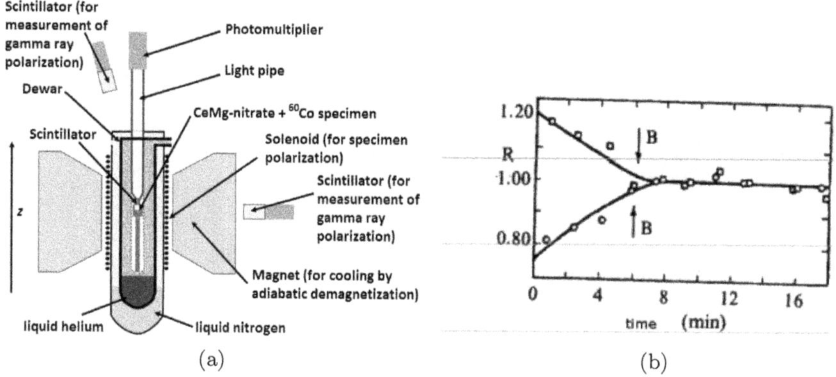

Fig. 12.3: (a) An apparatus detecting electrons originating from a polarized nucleus is shown on the left. (b) On the right we show the experimental results of the now classic experiment Wu *et al.* R gives the ratio of the event rate for the polarized sample divided by that associated with the sample not polarized, for two different magnetic fields. After some time R becomes unity since the system is depolarized, when it is warmed up.

of matter vs antimatter. This, as we will see below, is accomplished due to the violation of charge conjugation symmetry in weak interactions (See Fig. 12.10 below).

12.2 The helicity of the neutrinos

The concept of helicity was introduced in section 12.1. For any particle with non zero spin and momentum $\mathbf{p}$ one can define a quantity called helicity, as seen in Fig. 12.5. If the particle has spin 1/2, there exist two spin orientations, $m_s = \pm 1/2$. If the spin of the particle is parallel to its momentum the particle is right handed or is characterized by helicity +1. If its spin is opposite to its momentum, the particle is left handed or it is characterized by helicity −1 (see Fig. 12.6).

The helicity concept is useful, strictly speaking, only in the case of massless particles, which move with the velocity of light. For massive particles the helicity is frame dependent. This means that a particle can have helicity +1 for one observer and −1 for another. For massless particles the helicity is frame independent, i.e. a property of the particle itself.

It is well known that a massless spin one boson (photon, graviton) has two degrees of freedom, transverse polarizations, i.e. helicities (see 3.4.1).

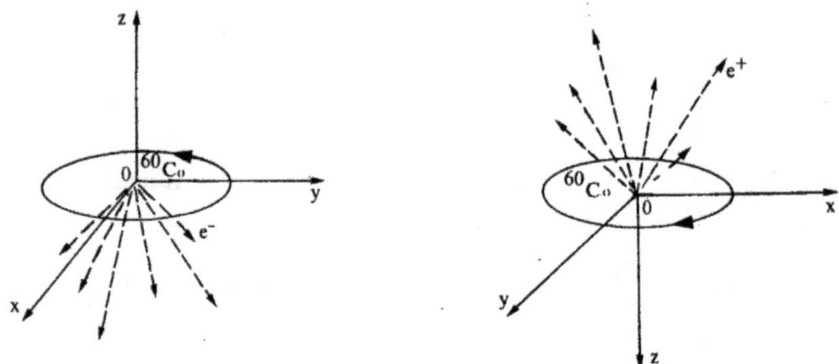

Fig. 12.4: Placing polarized nuclei ^{60}Co at the intersection of two perpendicular axes x and y one can determine the polarity of the z axis, which coincides with the direction of polarization, by comparing it with the direction of the highest electron intensity. It is opposite to the z-axis and the frame is by agreement a right handed one. The situation can, however, be confused if one of them uses anti-cobalt as on the left. Now most anti-electrons are emitted again in a direction opposite to the z-axis, but the system is now left handed. Thus they can settle the issue, if they somehow know whether they deal with matter or antimatter. This, in fact, can be achieved due to charge conjugation violation.

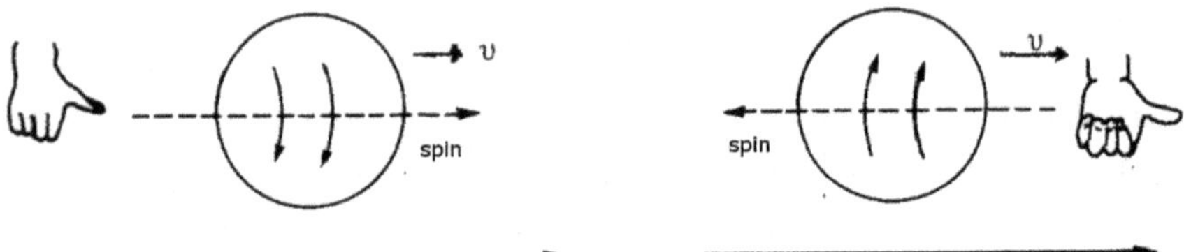

Fig. 12.5: The helicity of a particle is defined by the direction of its spin relative to its momentum. The helicity is +1, right handed particle, if the two directions coincide (left panel). The helicity is −1, left handed particle, if their directions are opposite (right panel).

Since with the then evidence the neutrino was massless[5] the question posed in the middle of last century was: What is the helicity of the neutrino? Is the neutrino left handed (ν_L) or right handed (ν_R)?

[5]Today we know that it is non zero, but at most 1 eV/c^2.

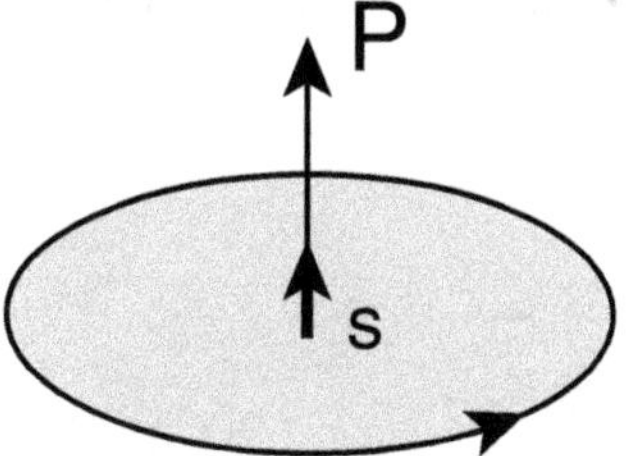

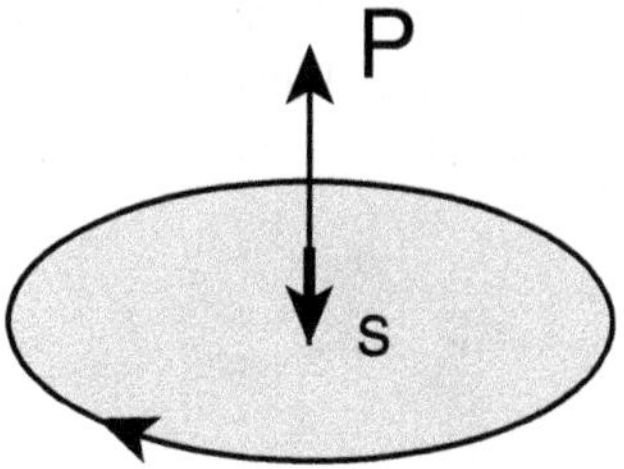

Fig. 12.6: The particle is right handed (helicity +1) is the spin and the momentum have the same direction (left panel). It is left handed (helicity -1) if they are in opposite directions (right panel).

12.2.1 *The determination of the helicity of neutrino*

This question was answered by an experiment performed by the team of . Goldhaber and collaborators in 1958 by studying the process [Goldhaber *et al.* (1958)]:

$$e^- +^{152} \mathrm{Eu}(0^+) \rightarrow^{152} \mathrm{Sm}^*(1^+) + \nu_e, \quad {}^{152}\mathrm{Sm}^*(1^+) \rightarrow^{152} \mathrm{Sm}(0^+) + \gamma. \tag{12.26}$$

The measurement consisting in detecting and measuring the polarization of the photons produced in the de-excitation of the ${}^{152}\mathrm{Sm}^*(1^+)$ nucleus. The electron is captured from the $1s$ orbit and the initial nucleus is characterized by spin and momentum, which are zero. Thus the final nucleus and the neutrino have opposite spins. Thus if we know the helicity of the nucleus we can determine the helicity of the neutrino, since they are equal. The helicity of the nucleus can be determined by measuring the helicity of the emitted photons, in particular those that moved in the direction of motion of the nucleus $\mathrm{Sm}^*(1^+)$. Thus, in the reaction:

$$e^- +^{152} \mathrm{Eu}(0^+) \rightarrow^{152} \mathrm{Sm}(0^+) + \nu_e + \gamma,$$

the momenta of the three produced particles are co-linear.

To be specific let us take as a quantization axis the spin direction of ${}^{152}\mathrm{Sm}^*(1^+)$. Then the photon and the neutrino have opposite momenta and an angular momentum such that its projection equal to that of the absorbed electron. In other words:

- The electron and the photon have the same helicity (Fig. 12.7).
- The photon helicity can be determined by its absorption in magnetized iron.
- The only photon helicity measured was -1.

As a result the produced neutrinos are always left handed [Goldhaber *et al.* (1958)]. Thus the right handed neutrinos either need not exist or, if they do, they do not interact (sterile neutrinos).

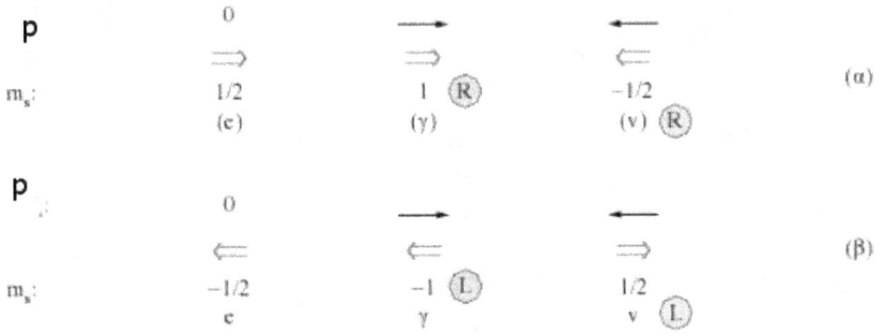

Fig. 12.7: The possible helicities consistent with momentum and angular momentum conservation. In case (α) the photon and the neutrino each have helicity $+1$, while in (β) the opposite is true.

In an analogous experiment performed later in the β decay of ^{203}Hg, in which anti-neutrinos and photons are produced, revealed that the anti-neutrinos are right-handed. This is consistent with the theory of Dirac. Thus:

Neutrinos are always left handed.
anti-neutrinos are always right handed

A pretty picture concerning muon decay is exhibited in Fig. 12.8.

12.2.2 *Weak interaction and handedness*

Since only left handed neutrinos have been detected it is natural to assume that the weak interaction, the only one, in addition to gravity, that the neutrinos feel, must involve neutrinos with left handed chirality. This lead to a weak interaction that is of the form $V - A$. For anti-neutrinos a right handed chirality is involved. The other fermions also feel the weak interaction. They are, however, massive and they can appear in both chiralities. They were classified according to their chralities., e.g. $f_L = (1/2)(1 - \gamma_5)f$, $f_R = (1/2)(1 + \gamma_5)f$, $f = e$, q, etc. Glashow, Weinberg and Salam (GSW)

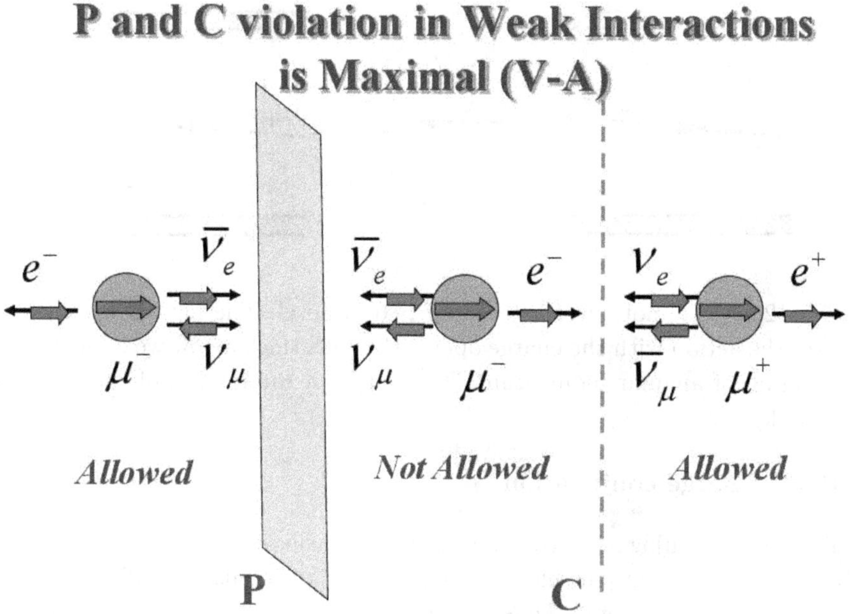

Fig. 12.8: The possible helicities, consistent with momentum and angular momentum conservation, involving the products of polarized muon decay.

with a stroke of genius, incorporated into a theory the results of the above two celebrated experiments and postulated that only left handed fermions participate in weak interactions, see chapter 3. Thus the celebrated Standard Model (SM)of elementary particles emerged. We recall that in this model the weakly interacting particles are members of an isodoublet with $SU(2)$ symmetry:

$$\ell_L = \begin{pmatrix} \nu_e \\ e \end{pmatrix}_L, \begin{pmatrix} \nu_\mu \\ \mu \end{pmatrix}_L \begin{pmatrix} \nu_\tau \\ \tau \end{pmatrix}_L, \quad q_L = \begin{pmatrix} u \\ d \end{pmatrix}_L, \begin{pmatrix} c \\ s \end{pmatrix}_L \begin{pmatrix} t \\ b \end{pmatrix}_L, \quad (12.27)$$

with the interaction understood to act vertically. Their right handed partners

$$e_R, \mu_R, \tau_R, u_R, d_R, c_R, s_R, t_R, b_R \quad (12.28)$$

are weak isospin singlets and they do not participate in weak interactions. Notice in the SM the absence of the right handed neutrino.

In this formulation we see explicitly that the weak interactions violate parity maximally (V-A).

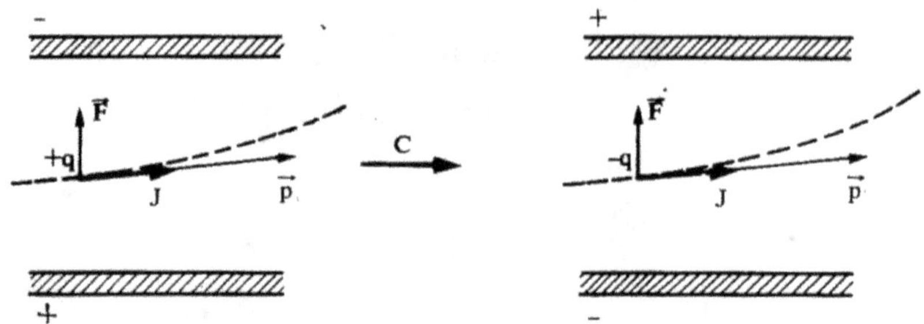

Fig. 12.9: The motion of a charge $q > 0$ in an electric field **E** before and after the action with the charge operator. With thick arrow we indicate the direction of angular momentum. No change in the motion of the charged particle is seen.

12.3 Charge conjugation

The charge conjugation transformation C involves the changing of every particle by its antiparticle. This means that it changes all the internal quantum numbers. Thus if a given state is characterized by the set $|q, B, S, L_i, \ldots\rangle$ with $i = e, \mu, \tau$, we have:

$$C|q, B, S, L_i, \ldots\rangle = \lambda| - q, -B, -S, -L_i, \ldots\rangle \,. \tag{12.29}$$

In particular the action of C on the charge Q and baryon B etc operators is given by:

$$CQC^{-1} = -Q, \quad CBC^{-1} = -B, \quad CSC^{-1} = -S, \quad CL_iC^{-1} = -L_i \,. \tag{12.30}$$

C acts on the internal space of a particle leaving unchanged all variables depending on space time.

At the classical level charge conjugation is obvious. It is clear that Maxwell's equations do not change under the transformation:

$$q \to -q, \quad \vec{E} \to -\vec{E}, \quad \vec{B} \to -\vec{B} \,.$$

Thus for a particle moving in an electric field we get the picture of Fig. 12.9. The situation is analogous for the case of the magnetic field. Since the charge operator C affects only the internal quantum numbers we have:

$$\langle \phi | C\psi \rangle = \int_v \phi^* C\psi d^3\vec{r} = \int_v C\phi^* \psi d^3\vec{r} = \langle C\phi | \psi \rangle \,.$$

Thus C is both unitary and hermitian and $C^2 = E$. Thus its possible eigenvalues are $\lambda = \pm 1$. It has, however, eigenvalues only if all the internal quantum numbers are zero. This is the result of Eq. (12.29). This happens, even if it commutes with the Hamiltonian, since it does not commute with the operators Q, B etc. Consider, e.g., the charge operator and an eigenstate $|q\rangle$ of Q with charge q. Then

$$Q|q\rangle = q|q\rangle, \quad C|q\rangle = \lambda|-q\rangle\,.$$

But

$$CQ|q\rangle = qC|q\rangle = \lambda q|-q\rangle,$$
$$QC|q\rangle = \lambda Q|-q\rangle = -\lambda q|-q\rangle,$$

where $\lambda = \pm 1$, i.e.

$$[CQ, QC]|q\rangle = 2\lambda q|-q\rangle = 2CQ|q\rangle$$

or

$$[CQ, QC] = 2CQ\,. \tag{12.31}$$

The operators C and Q do not commute. Thus the eigenstates of Q with $q \neq 0$ cannot be eigenstates of C. The same arguments apply in the case of the other operators S (or Y), etc.

For the truly neutral particles, however, we have:

$$C|q = 0, B = 0, Y = 0, c = 0, b = 0, \cdots\rangle =$$
$$\lambda|q = 0, B = 0, Y = 0, c = 0, b = 0, \cdots\rangle$$

with $\lambda = \pm 1$. λ is called eigenvalue of the charge conjugation operator or charge parity. For the reasons we discussed in the case of parity, the charge parity is a multiplicative quantum number. Thus in any process respecting the charge conjugation symmetry we have

$$\prod_i \lambda_i = \text{constant.} \tag{12.32}$$

Truly neutral particles are the photon, all mesons composed of a quark and its anti-quark and their spin and radial excitations like $\pi^0, \eta_c, \Psi/J, \Psi', Y \ldots$. what is their eigenvalue λ?

As is well known the photon is described by the vector potential, which under C changes sign. Therefore

$$C|\gamma\rangle = -|\gamma\rangle, \quad \text{that is} \quad \lambda(\gamma) = -1\,. \tag{12.33}$$

The meson π^0 decays electromagnetically into two photons:

$$\pi^0 \to 2\gamma\,.$$

Thus assuming charge parity conservation in electromagnetic interactions we get

$$\lambda(\pi^0) = |\lambda(\gamma)|^2 = 1$$

or

$$C|\pi^0\rangle = |\pi^0\rangle\,. \tag{12.34}$$

From Eq. (12.29) we get:

$$C|e^-\rangle = \lambda_e|e^+\rangle, \tag{12.35}$$

where $\lambda_e^2 = 1$ or $\lambda_e = \pm 1$. Similarly for a given quark q we have:

$$C|q\rangle = \lambda_q|\bar{q}\rangle, \tag{12.36}$$

where again $\lambda_q^2 = 1$ or $\lambda_q = \pm 1$. From these relations we find

$$C|e^+\rangle = \frac{1}{\lambda_e}C^2|e^-\rangle = \frac{1}{\lambda_e}|e^-\rangle = \lambda_e|e^-\rangle \tag{12.37}$$

and

$$C|\bar{q}\rangle = \lambda_q|q\rangle\,. \tag{12.38}$$

We can thus select λ_e and λ_q so that $\lambda_e = \lambda_q = 1$.

12.3.1 *Charge conjugation for 4-spinors*

The reader may want to be familiar with the content of section 2.6 before proceeding further.

In addition to changing the sign of the internal quantum numbers, the charge conjugation in Dirac theory is defined by:

$$\psi \to \psi^c = C(\bar{\psi})^T = C\gamma_0\psi^*, \text{ (under charge conjugation transformation).}$$

C satisfies the conditions:

$$C^{-1}\gamma_\mu C = -\left(\gamma_\mu\right)^T,$$

$$C^{-1} = -C = C^+ = C^T.$$

A possible choice is:

$$C = i\gamma_2\gamma_0.$$

Furthermore:

$$\bar{\psi} \to \bar{\psi}^c = -\psi^T C^{-1} = \psi^T \text{ (under charge conjugation transformation).}$$

The chiral states transform analogously:

$$\psi_{L,R} \to \psi^c_{L,R} = P_{L,R}\psi^c, \quad P_L = \frac{1}{2}(1-\gamma_5), \quad P_R = \frac{1}{2}(1+\gamma_5).$$

Note, however, that:

$$(\psi_R)^c = C\left(\bar{\psi}_R\right)^T = C\left(\bar{\psi}\frac{(1-\gamma_5)}{2}\right)^T$$

$$= C\frac{(1-\gamma_5)}{2}\left(\bar{\psi}\right)^T = \frac{(1-\gamma_5)}{2}C\left(\bar{\psi}\right)^T = \psi^c_L.$$

Similarly:

$$(\psi_L)^c = \psi^c_R.$$

We have only two independent chiral combinations. Usually these are taken to be:

$$(\psi_L, \psi_R) \text{ or } (\psi_L, (\psi_L)^c).$$

We will now examine some important consequences.

12.3.2 *The charge conjugation of composite systems*

The relation (12.34) is consistent with Eq. (12.36). Indeed consider the meson:

$$M = |q\bar{q}Is\ell\rangle.$$

$$C|(q\bar{q})Is\ell\rangle = |(\bar{q}q)Is\ell\rangle = (-1)^{I_1+I_2-I+\ell+s}|(q\bar{q})Is\rangle, \tag{12.39}$$

where I_1, I_2 and are the isotopic spins of $q, \bar{q}$ and , s the spin of the meson and ℓ the relative angular momentum. The phase in Eq. (12.39) results from the change in the order of the two quarks. $(-1)^\ell$ results from the spherical harmonic associated with the relative motion, $(-1)^{I_1+I_2-I}$ from the change in the order of the coupling of isospin and $(-)^{-1+s}$ from the spin, related to the symmetry property of the relevant Glebsch-Gordan coefficient. We get an extra $(-)$, which results from the change in the order of two fermions q and $\bar{q}$, which anti-commute.

Thus in the case of the π meson we get:

$$I_1 + I_2 - I + 1 + s = \frac{1}{2} + \frac{1}{2} - 1 + 0 + 0 = 0,$$

which yields Eq. (12.34). In the special case of mesons with $\ell = 0$ we get:

$$C|\pi^0\rangle = |\pi^0\rangle,\ C|J/\Psi\rangle = |J/\Psi\rangle,\ C|\omega^0\rangle = |\omega^0\rangle,\ C|\phi\rangle = |\phi\rangle \to C = +1,$$

$$C|\rho^0\rangle = -|\rho^0\rangle,\ C|\eta\rangle = -|\eta\rangle,\ C|\eta_c\rangle = -|\eta_c\rangle \to C = -1\,.$$

From Eq. (12.39) we also find:

$$C|\pi^+\rangle = C|(u\bar{d})I = 1, s = 0\rangle = |(d\bar{u})I = 1, s = 0\rangle = |\pi^-\rangle,$$

$$C|\rho^+\rangle = -|\rho^-\rangle,$$

$$C|K^+\rangle = C|(u\bar{s})\rangle = |\bar{u}s\rangle = (-1)^{1/2+0-1/2+1/2+1/2-0+1}|s\bar{u}\rangle = |K^-\rangle,$$

$$C|K^{*+}\rangle = -|K^{*-}\rangle,$$

$$C|K^0\rangle = |\bar{K}^0\rangle,\quad C|D^+\rangle = |D^-\rangle\,.$$

The other flavors are treated similarly, e.g.

$$C|Y\rangle = |Y\rangle\,.$$

The same principle applies in the case of systems of many particles, e.g. the (e^+, e^-) and (π^+, π^-).

$$C|(e^+, e^-)l, s\rangle = |(e^-, e^+)\ell, s\rangle = (-1)^{\ell+s}|(e^+, e^-)\ell, s\rangle,$$

$$C|(\pi^+, \pi^-)\ell\rangle = (-1)^{\ell}|(\pi^+, \pi^-)\ell\rangle,$$

where ℓ characterizes the relative angular momentum.

The 0^- state of positronium has $\ell = 0$ and spin $s = 0$. It is thus characterized by $C = 1$. It can thus decay to two photons (the photon has $C = -1$, the same with the vector potential). It cannot be produced by a single (virtual) photon. The positronium state 1^- has spin $s = 1$, i.e. $C = -1$. It can thus decay to 3 photons and it can easily be produced by a single (virtual) photon. This explains why it has a much longer life time:

$$0^- \to \gamma\gamma,\ \tau = 1.25 \times 10^{-10}\ \text{s},\ 1^- \to \gamma\gamma\gamma,\ \tau = 1.40 \times 10^{-7}\ \text{s}$$

A list of the masses and the quantum numbers of some quarkonium mesons of the type $c\bar{c}$ and $b\bar{b}$ are shown in 12.1. Except for the mass scale these spectra are similar (they remain similar even if states with net charm and beauty are included. They are also similar to those of the positronium. This is firm evidence of the existence of the quarks, even though they have not been found free (quantum chromodynamics predicts that they never will).

Table 12.1: The mass and the quantum numbers of some mesons of the type $c\bar{c}$ and $b\bar{b}$. The G-parity is not shown, since for these systems, it coincides with C.

state	$c\bar{c}$ (MeV)	J^{PC}	$b\bar{b}$ (MeV)	J^{PC}
$1s$	$\eta_c(2980)$	0^{-+}	$\eta_b(9388)$	0^{-+}
$1s$	$J/\psi(3097)$	1^{--}	$\Upsilon(9460)$	1^{--}
$1p$	$\chi_{c0}(3415)$	0^{++}	χ_{b0}	0^{++}
$1p$	$\chi_{c1}(3511)$	1^{++}	χ_{b1}	1^{++}
$1p$	$\chi_{c2}(3556)$	2^{++}	χ_{b2}	2^{++}
$2s$	$\eta'_c(3680)$	1^{-+}	$\eta'_b(?)$	1^{-+}
$2s$	$\psi(3680)$	1^{--}	$\Upsilon(10023)$	1^{--}
$1d$	$\psi(3773)$	2^{--}	$\Upsilon(10161)$	2^{--}

We have seen above the eigenstates of C are truly neutral. For particles with non-zero charge the charge operator simply sends a state to its charge conjugate. On the other hand the isospin[6] rotation operator $e^{iI_2\pi}$, associated with the isospin operator I_2 through via an angle π, changes a state $|I,m\rangle$ to $|I,-m\rangle$, up to a phase. In fact the matrix elements of this operator are the Wigner $d^I_{m_1,m_2}(\pi)$ functions (see Eq. (12.51) below). Thus, if the I-multipole contains both charges, it is possible to construct eigenstates of the operator $G = Ce^{iI_2\pi}$. This is known as **G-parity operator**.

It is clear that for isoscalar particles the G-parity does not give anything new. G in this case coincides with the C. In particular this is true for $s\bar{s}$, $c\bar{c}$ and $b\bar{b}$ resonances. So we will examine $I \neq 0$ cases.

Using the C matrices constructed above we find: i) for the π meson

$$d^1(\pi) = \begin{pmatrix} 0 & 0 & -1 \\ 0 & 1 & 0 \\ -1 & 0 & 0 \end{pmatrix}, C = \begin{pmatrix} 0 & 0 & 1 \\ 0 & -1 & 0 \\ 1 & 0 & 0 \end{pmatrix} \Rightarrow G = Cd^1(\pi) = \begin{pmatrix} -1 & 0 & 0 \\ 0 & -1 & 0 \\ 0 & 0 & -1 \end{pmatrix} \tag{12.40}$$

i.e. the π mesons have a G-parity -1,i.e. $I^G = 1^-$. Similarly for n π mesons we have $G|n\pi\rangle = (-1)^n|n\pi\rangle$

i) for the ρ meson

[6]It is clear that here we consider the strong, not the weak, isospin of the states

$$d^1(\pi) = \begin{pmatrix} 0 & 0 & -1 \\ 0 & 1 & 0 \\ -1 & 0 & 0 \end{pmatrix}, C = \begin{pmatrix} 0 & 0 & -1 \\ 0 & 1 & 0 \\ -1 & 0 & 0 \end{pmatrix} \Rightarrow G = Cd^1(\pi) = \begin{pmatrix} 1 & 0 & 0 \\ 0 & 1 & 0 \\ 0 & 0 & 1 \end{pmatrix} \tag{12.41}$$

$I^G = 1^+$.

For the isoscalar mesons $G = C$. Thus for the ω^0 meson $|q\bar{q} = I = 0, s = 1\rangle$ $G = -1$. Similarly for the η meson $|q\bar{q} = I = 0, s = 0\rangle$ and ϕ^0 meson $|S\bar{S}, s = 0\rangle$ we find $G = -1$. Strong interactions conserve G-parity. Thus the ρ meson decays strongly to two ions, but it cannot decay to 3π. Similarly ω^0, ϕ^0 and η cannot decay to 2π but they can decay to 3π. The following processes

$$\rho \to 2\pi, \quad \omega^0 \to 3\pi, \quad \phi \to 3\pi$$
$$f^0 \to 2\pi, \quad \eta(549) \to 3\pi, \quad \eta'(958) \to 5\pi$$

have been observed in strong interactions as expected.

It is amusing to find the transformation under G of the case of quarks. Now $d^{1/2} = i\tau_2$. Thus

$$G = Ci\tau_2 = C\begin{pmatrix} 0 & 1 \\ -1 & 0 \end{pmatrix}.$$

Thus

$$Gq = C\begin{pmatrix} 0 & 1 \\ -1 & 0 \end{pmatrix}\begin{pmatrix} u \\ d \end{pmatrix} = C\begin{pmatrix} d \\ -u \end{pmatrix} = \begin{pmatrix} \bar{d} \\ -\bar{u} \end{pmatrix} = \bar{q}$$

$$G\bar{q} = C\begin{pmatrix} 0 & 1 \\ -1 & 0 \end{pmatrix}\begin{pmatrix} \bar{d} \\ -u \end{pmatrix} = C\begin{pmatrix} -\bar{u} \\ -\bar{d} \end{pmatrix} = -\begin{pmatrix} u \\ d \end{pmatrix} = -q\,.$$

Thus

$$\begin{aligned} G|(q\bar{q})\ell, I, s\rangle &= |(-\bar{q}q)\ell, I, s\rangle = -(-1)^{1+\ell+I+s}|(q\bar{q})\ell, I, s\rangle \\ &= (-1)^{\ell+I+s}|(q\bar{q})\ell, I, s\rangle \end{aligned} \tag{12.42}$$

In the same spirit we can treat the nucleon-antinucleon system, since the nucleon (p, n) has the same quantum numbers as the quark (u, d). Thus

$$\begin{aligned} G|(N\bar{N})\ell, I, s\rangle &= |(-\bar{N}N)\ell, I, s\rangle = -(-1)^{1+\ell+I+s}|(N\bar{N})\ell, I, s\rangle \\ &= (-1)^{\ell+I+s}|(N\bar{N})\ell, I, s\rangle \end{aligned} \tag{12.43}$$

12.3.3 *How good a symmetry is the charge conjugation?*

We have already seen that Dirac's theory predicts that it is a good symmetry. What is the experimental situation? In answering this question we will examine each of the interactions separately.

- The electromagnetic interaction.
  Let us begin with a process which is forbidden, if the symmetry holds:
  $$\pi^0 \quad \rightarrow \quad 3\,\gamma$$
  Experimentally the branching ratio is:
  $$R = \left(\frac{\pi^0 \rightarrow 3\gamma}{\pi^0 \rightarrow 2\gamma}\right) < 4 \times 10^{-7}\,.$$
  We conclude that the above experimental limit is satisfactory, since, considering effects of phase space alone one finds $R \approx 10^{-5}$. So the symmetry C is good in EM interactions.
- The strong interaction
  $$p + \bar{p} \rightarrow \pi^+\pi^-\pi^0\,. \tag{12.44}$$
  The initial state is invariant under C, if we assume that the relative angular momentum ℓ and the spin are zero. C conservation implies, therefore, that, due to the fact that $C|\pi^\pm\rangle = |\pi^\pm\rangle, C|\pi^0\rangle = |\pi^0\rangle$, we expect the distribution of π^+ and π^- mesons in Eq. (12.44) to be the same, since their relative angular momentum must be zero. Based on this and similar studies one my conclude that the strong interaction conserves C at least at the level of 1%.
- The weak interaction.
  Here C is not conserved. A characteristic example is the decay of μ^- and μ^+:
  $$\mu^- \rightarrow e^- + \bar{\nu}_e + \nu_\mu, \quad \mu^+ \rightarrow e^+ + \bar{\nu}_\mu + \nu_e\,.$$
  We focus our attention on the angular distribution of the produced e^- and e^+ relative to the direction of the polarization $\vec{s}$ of the initial particle. We obtain the picture of Fig. 12.10. It is clear that C is not conserved, since if it were, the two diagrams ought to have been the same. We still see that the two diagrams are mirror images of one another with respect to a plane perpendicular to $\vec{s}$. The reason is that the symmetry CP is conserved.

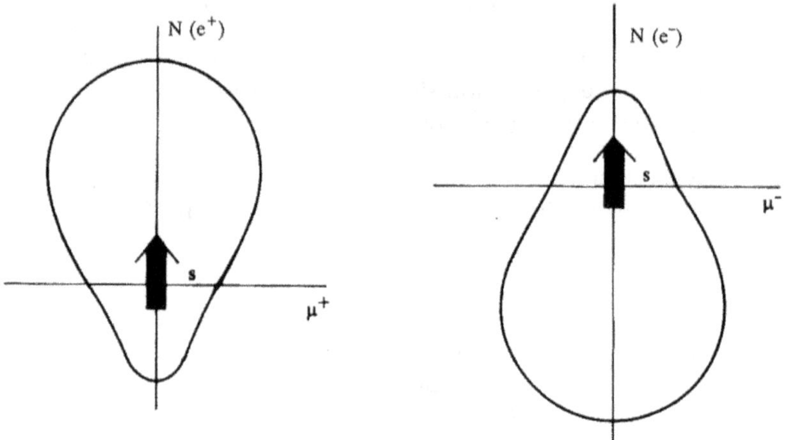

Fig. 12.10: Polar diagrams of the $e^-(e^+)$ distributions resulting from the decay of polarized $\mu^-(\mu^+)$. The positrons tend to be produced in the direction of the spin polarization $\vec{s}$, while the electrons in the opposite direction.

12.4 Time reversal

We will now examine the behavior of the physical systems under the transformation of time reversal T:

$$T:\; t \to t' = -t\,. \tag{12.45}$$

Time reversal is not as dramatic a procedure as it seems at first sight. We do not have to go backwards in time.

12.4.1 *Time reversal in classical mechanics*

In classical mechanics it can be simply formulated as follows: Suppose that a given particle at time t_0 is found in position $\vec{r}_0 = \vec{r}(t_0)$ with velocity $\vec{v}_0 = \vec{v}(t_0)$. Suppose next that, under the laws of classical mechanics, at time t_1 it is in position $\vec{r}_1 = \vec{r}(t_0+t_1)$ with velocity $\vec{v}_1 = \vec{v}(t_0+t_1)$. Suppose further that a similar particle at time t_0+t_1 is found at position $\vec{r}_1$ with velocity $\vec{v}_0' = -\vec{v}_1$. What will happen after time t_1, i.e. at time t_0+2t_1? The answer is one of the two:

(*i*) either it will be found at the position $\vec{r}_0$ with velocity $-\vec{v}_0$.
(*ii*) or it will not be found at the position $\vec{r}_0$ with velocity $-\vec{v}_0$.

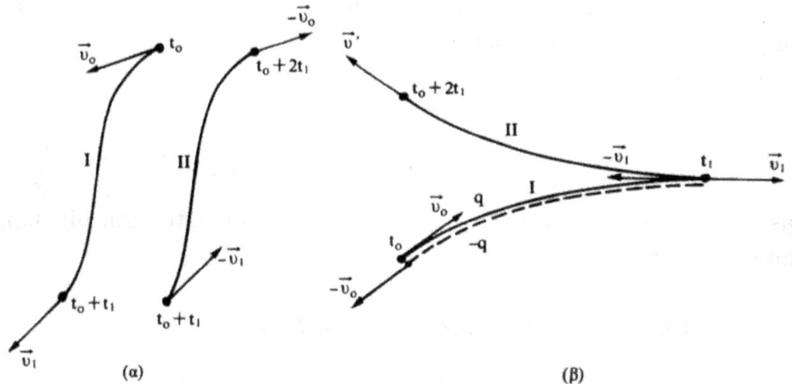

Fig. 12.11: Two time reversed orbits. In case (α) the two time reversed orbits coincide (they were drawn a little apart to be visible). The dots indicate the position and the arrows the velocities. The case (β) corresponds to the motion of a charge Q in a magnetic field. Now the time reversed orbits do not coincide. To make them coincide one must reverse the charge.

In the case (i) we wll say that the laws of mechanics are invariant with respect to time reversal, while in the case (ii) we will claim that they are not. The two cases are outlined in Fig. 12.11. This can be understood as follows: The basic law of mechanics is written

$$m\frac{d^2\vec{r}}{dt^2} = \vec{F}(\vec{r},\vec{v},t) \quad \text{or} \quad m\vec{a} = \vec{F}(\vec{r},\vec{v},t)\,.$$

Under time reversal T the acceleration does not change, while the velocity changes sign. Thus Newton's law is written

$$m\frac{d^2\vec{r}}{dt^2} = F(\vec{r},-\vec{v},-t)\,.$$

Thus the laws of mechanics do not change provided that:

$$\vec{F}(\vec{r},-\vec{v},-t) = \vec{F}(\vec{r},\vec{v},t)\,. \tag{12.46}$$

This holds for forces which depend only on the coordinates. The Lorentz force

$$\vec{F} = q\,\vec{v}\times\vec{B}$$

does not obey this condition, since

$$T:\ \vec{F}\to\vec{F}' = -\vec{F}$$

We have assumed, of course, that the external magnetic field remains unchanged. Obviously the equations remain unchanged if the charge is also reversed, that is

$$T:\ t\to -t,\quad q\to -q,\quad \vec{F}\to\vec{F},$$
$$\vec{v}\to -\vec{v},\quad \vec{r}\to\vec{r},\quad \vec{a}\to\vec{a}\,.$$

Classical non relativistic physics, however, does not automatically supply solutions of opposite charge.

12.4.2 *Time reversal in quantum mechanics*

Wigner has shown that the only transformations S acting on a Hilbert space, which preserve the length of any vector of the space are either unitary, $S = U$, $U^+ = U^{-1}$ or anti-unitary $S = UK$, where K performs complex conjugation acting on any function of the space. It turns out that the time reversal operator T cannot be unitary, since in that case invariance of the expression e^{iEt} under time reversal, would imply a change of the energy to its opposite, which is not acceptable, something possible for the momentum (see below). Time reversal therefore is anti-unitary and it can be cast in the form:

$$T = UK \tag{12.47}$$

with U and K as above ($K\alpha = \alpha^*$, α=complex number), such that:

$$K^{-1} = K.$$

This is necessary to guarantee $T^2 = E$. Note that

$$T^{-1} = K^{-1}U^{-1} = KU^{\dagger},$$

with the conjugation operation performed last.

Quite generally we have

$$T\big[\lambda_1|\psi_1\rangle + \lambda_2|\psi_2\rangle\big] = \lambda_1^* U|\psi_1^*\rangle + \lambda_2^* U|\psi_2^*\rangle\,. \tag{12.48}$$

Furthermore due to the non unitarity we also have

$$\begin{aligned}\langle T\Psi|T\phi\rangle &= \langle UK\Psi|UK\phi\rangle = \langle UU^{\dagger}K\Psi|K\phi\rangle \\ &= \langle K\Psi|K\phi\rangle = \langle\phi|\Psi\rangle \neq \langle\Psi|\phi\rangle\,.\end{aligned} \tag{12.49}$$

Suppose now that $|\vec{x}'\rangle$ and $|\vec{p}'\rangle$ are the eigenfunctions of the position and momentum operators. Then under the time inversion T we have

$$T|\vec{p}'\rangle = |-\vec{p}'\rangle,\quad T|\vec{x}'\rangle = |\vec{x}'\rangle\,.$$

Noting that

$$\vec{p}|-\vec{p}'\rangle = -\vec{p}'|-\vec{p}'\rangle \text{ (eigenfunction of momentum operator)}$$

and applying to it the operator T, we find:

$$(T\vec{p}T^{-1})T|-\vec{p}'\rangle = -\vec{p}'T|-\vec{p}'\rangle$$

or

$$(T\vec{p}T^{-1})|\vec{p}'\rangle = -\vec{p}'|-\vec{p}'\rangle\,.$$

Thus

$$T\vec{p}T^{-1} = -\vec{p}.$$

Proceeding analogously we get:

$$T\vec{x}T^{-1} = \vec{x}\,.$$

These define the action of on any function of $\vec{x}$ and $\vec{p}$. In the case of the angular momentum L, e.g., one finds:

$$T\vec{L}T^{-1} = T\vec{x}\times\vec{p}T^{-1} = T\vec{x}T^{-1}\times T\vec{p}T^{-1} = \vec{x}\times(-\vec{p}) = -\vec{L}\,. \tag{12.50}$$

Since the time reversal operator is anti-unitary we cannot associate with it a Hermitian operator. This implies that there is no quantum number associated with it. Thus any possible eigenvalues of T are not measurable quantities.

Before proceeding further let us examine the transformation properties of angular momentum eigenstates under T. Under the time reversal transformation T the angular momentum operator changes sign, $T^{-1}\mathbf{J}T = -\mathbf{J}$, which means that T and $\mathbf{J}$ anti-commute. In particular $TJ_z = -J_zT$. Also the rotation around the y-axis by π, indicated by $d^j_{m,m'}(\pi)$ or simply $\left(d^j(\pi)\right)$ also anticommutes with J_z,

$$\left(d^j(\pi)\right)(J_z) = -(J_z)\left(d^j(\pi)\right)$$

Some simple expressions are:

$$d^{1/2}(\pi) = \begin{pmatrix} 0 & 1 \\ -1 & 0 \end{pmatrix},\ d^1(\pi) = \begin{pmatrix} 0 & 0 & -1 \\ 0 & 1 & 0 \\ -1 & 0 & 0 \end{pmatrix},\ d^{3/2}(\pi) = \begin{pmatrix} 0 & 0 & 0 & 1 \\ 0 & 0 & -1 & 0 \\ 0 & 1 & 0 & 0 \\ -1 & 0 & 0 & 0 \end{pmatrix} \tag{12.51}$$

or in general

$$\left(d^j(\pi)\right)_{k,\ell} = \begin{cases} (-1)^{j+1-\ell}, & k+\ell=0 \\ 0, & \text{otherwise} \end{cases},\ k,\ell = j,\, j-1,\cdots,\, -j+1, -j \tag{12.52}$$

in the indicated order of magnetic sub-states and

$$(J_z) = \text{diagonal}(j, j-1, \cdots, -j+1, -j)$$

Thus one can show that the operator $T\left(d^j(\pi)\right)$ commutes with (J_z) and $\mathbf{J}^2$, so that all three can be diagonalized simultaneously. Indeed:

$$[T\left(d^j(\pi)\right), (J_z)] = T\left(d^j(\pi)\right)(J_z) - (J_z)T\left(d^j(\pi)\right) =$$

$$T\left(d^j(\pi)\right)(J_z) + T(J_z)\left(d^j(\pi)\right) = T\left(d^j(\pi)\right)(J_z) - T\left(d^j(\pi)\right)(J_z) = 0$$

Similarly for $\mathbf{J}^2$. In other words one can choose the eigenvectors so that

$$T\left(d^j(\pi)\right)|j, m> = -|j, m>$$

Noting now that, see Eq. (12.52),

$$\left(d^j(\pi)\right)|j, m> = -(-1)^{j-m}|j, -m>, \tag{12.53}$$

we find

$$T|j, m> = (-1)^{j+m}|j, -m>, \tag{12.54}$$

a well known relation.

Thus invariance under T cannot be checked by traditional observables. Fortunately, however, there exist other measurable quantities, like the transition rate for a given process must be the same with that of its inverse. This is the celebrated principle of detailed balance. Based on this, the cross section for the process

$$A + B \rightarrow C + D,$$

must be the the same with that of the process

$$C + D \rightarrow A + B\,.$$

Many experimental tests of this kind have been done, but no T violation has been found. Such experiments are not easy and that it is why the upper limits are relatively poor, $10^{-2} - 10^{-3}$. We should mention, however, that CP violation (the product of C and P operations) has been found in nature. Since on quite general grounds the symmetry CPT is believed to hold, CP violation implies T violation. At this point we present in table 12.2 the transformation behavior of a number of common physical observables under the influence P and T. From this table we conclude that an elementary particle must necessarily have zero electric dipole moment.

Indeed if a particle is truly elementary the only direction that can be defined is that of the angular momentum (spin) $\vec{J}$. Thus the electric $\vec{\mu}_E$ as

Table 12.2: Transformation properties of some common and useful physical observables under the operations of space inversion P and time reversal T.

variable	P	T	name
$\vec{x}$	$-\vec{x}$	$\vec{x}$	polar vector (position)
$\vec{p}$	$-\vec{p}$	$-\vec{p}$	polar vector (momentum)
$\vec{J}$	$\vec{J}$	$-\vec{J}$	axial vector (angular momentum)
$\vec{A}$	$-\vec{A}$	$-\vec{A}$	vector potential
ϕ	ϕ	ϕ	scalar potential
$\vec{E}$	$-\vec{E}$	$\vec{E}$	$\vec{E} = -\partial\vec{A}/\partial t - \vec{\nabla}\phi$ (electric field)
$\vec{B}$	$\vec{B}$	$-\vec{B}$	$\vec{B} = \vec{\nabla} \times \vec{A}$ (magnetic field)
$\vec{J} \cdot \vec{B}$	$\vec{J} \cdot \vec{B}$	$\vec{J} \cdot \vec{B}$	magnetic dipole moment
$\vec{J} \cdot \vec{E}$	$-\vec{J} \cdot \vec{E}$	$-\vec{J} \cdot \vec{E}$	electric dipole moment
$\vec{J} \cdot \vec{P}$	$-\vec{J} \cdot \vec{P}$	$\vec{J} \cdot \vec{P}$	$\vec{P} =$ longitudinal polarization
$\vec{J} \cdot (\vec{P}_1 \times \vec{P}_2)$	$\vec{J} \cdot (\vec{P}_1 \times \vec{P}_2)$	$-\vec{J} \cdot (\vec{P}_1 \times \vec{P}_2)$	$\vec{P} =$ transverse polarization

well as the magnetic $\vec{\mu}_m$ dipole moments must be proportional to $\vec{J}$. Thus the interaction of the particle with the electric and magnetic fields has to be of the form:

$$H_{int} = f\vec{J}\cdot\vec{E} \text{ (electric dipole moment)} + g\vec{J}\cdot\vec{B} \text{ (magnetic dipole moment)}. \tag{12.55}$$

with f and g constants. Under the influence of the operator S, S being P or T, $H_{\rm int}$ becomes:

$$SH_{int}S^{-1} = -f\vec{J}\cdot\vec{E} + g\vec{J}\cdot\vec{B}, \quad S = P \text{ or } T\,.$$

Consequently either such a particle is not elementary (like a molecule) or the EM interactions violate P and T. Many experiments have attempted to measure the electric dipole moment of the proton and the neutron. No such quantity has been found.

The current experimental limits are:

$$d_e < 2.7\times 10^{-27} \text{ e-cm} \quad \text{(electron)},$$
$$d_n < 1.1\times 10^{-25} \text{ e-cm} \quad \text{(neutron)}.$$

(more below) Such moments, however, are predicted to exist in many current theories beyond the Standard Model (SM). The SM predicts values which are at least five orders of magnitude smaller. Anyway the search is going on.

12.4.3 *Premature evidence for T violation*

In 1998, the CPLEAR experiment announced that for the first time that a measurement of T violation had been successful. The idea was to measure the difference in the decay rate of $K^0 \to \overline{K}^0$ and its time reversed $\overline{K}^0 \to K^0$, through the so called Kabir asymmetry:

$$A_{\rm T} \equiv -\frac{P_{K\overline{K}} - P_{\overline{K}K}}{P_{K\overline{K}} + P_{\overline{K}K}}\,.$$

This experiment had the capacity to distinguish the initial (K^0 and $\overline{K}^0$), by the observation of the pair $K^{\pm}\pi^{\mp}$ in the disappearance $p\overline{p}$, and the final (K^0 and $\overline{K}^0$), utilizing the semi-leptonic decays ($\bar{K}_0 \to e^+\pi^-\nu,\ K_0 \to e^-\pi^+\bar{\nu}$). The asymmetry was found to be different than zero, which could be interpreted as evidence of T violation. This, however, was not accepted, following the thoughts of Wolfenstein [Wolfenstein (1999)]. Such an asymmetry has recently been obtained, however, by the BaBar collaboration (see below).

12.5 CP, the combined action of P and C symmetries

We have seen that weak interactions violate both C and P invariance. In the situations we have examined, however, the product of these two symmetries CP, i.e. the P operation first and the C operation afterward, is conserved. Indeed the polar diagrams of e^- and e^+ produced in the decay of μ^- and μ^+ respectively exhibit mirror symmetry. So they can be made to coincide after an application of P (reflection with respect to the plane of symmetry). In other words, if in the process:

$$\mu^- \to e^- + \nu_\mu + \bar{\nu}_e,$$

we apply CP we will finally get

$$\mu^+ \to e^+ + \bar{\nu}_\mu + \nu_e,$$

with exactly the same distribution of e^- and e^+. The space inversion changes the helicity. Another example of CP can be found in the behavior of the neutrino. The neutrinos, so long as they can be considered massless, they are characterized by negative helicity. If we act on such a neutrino state with C we will obtain an anti-neutrino with the same helicity, which is not consistent with observation. If, however, we act on it with combined operation of CP we will get an anti-neutrino with positive helicity, consistent with observation. It was the presumed conservation of CP in weak interactions that elevated this symmetry to the highest level of importance.

If CP were an exact symmetry, the laws of Nature would be the same for matter and for antimatter. We observe that most phenomena are C- and P-symmetric, and therefore, also CP-symmetric. In particular, these symmetries are respected by the gravitational, electromagnetic, and strong interactions. The weak interactions, on the other hand, violate C and P in the strongest possible way. For example, the charged W bosons couple to left-handed electrons, e_L^-, and to their CP-conjugate right-handed positrons, e_R^+, but to neither their C-conjugate left-handed positrons, e_L^+, nor their P-conjugate right-handed electrons, e_R^- . While weak interactions violate C and P separately, CP is still preserved in most weak interaction processes. The CP symmetry is, however, violated in certain rare processes, as discovered in neutral K decays in 1964, and observed in recent years in B decays.

There is no experimentally known violation of the CP-symmetry in strong interactions. As there is no known reason for it to be conserved in QCD specifically, this is a "fine tuning" problem known as the strong CP problem, resolved by the Peccei-Quinn involving new scalar particles called axions [Peccei and Quinn (1977)] Clearly QCD (Quantum Chromodynamics) does not violate the CP-symmetry as easily as the electroweak theory; unlike the electroweak theory in which the gauge fields couple to chiral currents, constructed from the fermionic fields, the gluons couple to vector currents. The axion provides a neat solution to this problem as demonstrated in two models, the KSVZ [Kim (1979); Shifman *et al.* (1980)] and DFSZ [Dine *et al.* (1981); Zhitnisky (1980)]. Tis is consistent with experiments, which do not indicate any CP violation in the QCD sector. For example, a generic CP violation in the strongly interacting sector would create the electric dipole moment of the neutron (see table 12.2, under T) which would be comparable to 10^{-14} e-m, while the experimental upper bound is roughly one trillionth that size [Baron *et al.* (2014)]. A better value of 10^{-19} e-m is expected at storage ring experiments [Anastassopoulos *et al.* (2015)].

The universe is made chiefly of matter, rather than consisting of equal parts of matter and antimatter as might be expected. It can be demonstrated that, to create an imbalance in matter and antimatter from an initial condition of balance, the Sakharov conditions must be satisfied, one of which is the existence of CP violation during the extreme conditions of the first seconds after the Big Bang. The Sakharov conditions are i) Baryon number (charge) violating interactions ii) CP violations and iii) absence of equilibrium conditions. It appears that the observed CP violation in the hadronic sector is not adequate to explain the baryon asymmetry in nature.

The third source of of possible CP violation comes from the phase of the Pontecorvo-Maki-Nakagawa-Sakata (PMNS) matrix in the lepton sector. We already know that the neutrino mixing is quite large and the neutrinos are not degenerate in mass. So CP violation can proceed, if the PMNS matrix is not real. Current neutrino experiments are not yet sensitive enough to allow experimental observation of CP violation in the lepton sector, but the NOνA experiment currently under construction could observe some small fraction of possible CP violating phases. Similarly the proposed neutrino experiments Hyper-Kamiokande and LBNE will be sensitive to a relatively large fraction of CP violating phases. Further into the future, a neutrino factory could be sensitive to nearly all possible CP violating phases. Anyway CP violation in the leptonic sector may lead to baryogenesis and explain the baryon asymmetry of the universe. Before going further let us examine the behavior of some simple systems under the action of CP

(1) A system of two charged pions with opposite charge. We have

$$CP|\pi^+\pi^-\rangle = C(-1)^\ell|\pi^+\pi^-\rangle = (-1)^\ell C|\pi^+\pi^-\rangle = (-1)^\ell|\pi^-\pi^+\rangle$$
$$= (-1)^{2\ell}|\pi^+\pi^-\rangle = |\pi^+\pi^-\rangle\,. \qquad (12.56)$$

That is the $|\pi^+\pi^-\rangle$ system has an eigenvalue of CP which is $+1$, regardless of the relative angular momentum of the two pions. Thus we find:

$$CP|\pi^+\pi^-\rangle = \lambda_{CP}|\pi^+\pi^-\rangle, \quad \lambda_{CP} = +1\,. \qquad (12.57)$$

(2) The system of two neutral pions.
Since the wf of $|\pi^0\pi^0\rangle$ consists of two identical bosons, it must be symmetric under their exchange. Thus $(-1)^\ell = 1$, $\ell = 2, 4, \cdots =$ and

$$CP|\pi^0\pi^0\rangle = \lambda_{CP}|\pi^0\pi^0\rangle, \quad \lambda_{CP} = +1\,. \qquad (12.58)$$

(3) The system of three pions with total charge zero, e.g. $|\pi^+\pi^-\pi^0\rangle$.
Now it is necessary to know the relative angular momentum of the two charged pions ℓ_{12} and the angular momentum of the neural pion with respect to the other two, ℓ . Proceeding as above we find:

$$CP|\pi^+\pi^-\pi^0\rangle = C(-1)^{\ell_{12}+\ell}\pi(\pi^0)|\pi^+\pi^-\pi^0\rangle = (-1)^{\ell_{12}+\ell+1}C|\pi^-\pi^+\pi^0\rangle$$
$$= (-1)^{\ell+1}|\pi^+\pi^-\pi^0\rangle \qquad (12.59)$$

(we made use of the fact that $\pi(\pi^0) = -1$). Then

$$CP|\pi^+\pi^-\pi^0\rangle = \lambda_{CP}|\pi^+\pi^-\pi^0\rangle, \quad \lambda_{CP} = \begin{cases} -1, & \text{if } \ell = \text{even} \\ +1, & \text{if } \ell = \text{odd} \end{cases}\,. \qquad (12.60)$$

(4) A system of three neutral pions, $|\pi^0\pi^0\pi^0\rangle$.
Since again this wf must be symmetric under the exchange of any two pions, the total isospin must be odd. Thus it is necessary that

$$CP|\pi^0\pi^0\pi^0\rangle = \lambda_{CP}|\pi^0\pi^0\pi^0\rangle, \quad \lambda_{CP} = -1\,. \qquad (12.61)$$

12.5.1 *The neutral kaon system*

We will now examine in some detail the system of K^0 and $\bar{K}^0$ mesons. Recall that $(S(K^0) = +1, S(\bar{K}^0) = -1)$. The fact that strangeness is not conserved in weak interactions raises some interesting problems. Since the neutral kaons happen to have angular momentum 0, one cannot define a preferred direction in space, which means that P has no effect on them, which means that, in this instance, CP is equivalent to C. Thus

$$CP|K^0\rangle = C|K^0\rangle = |\bar{K}^0\rangle, \tag{12.62}$$

$$CP|\bar{K}^0\rangle = C|\bar{K}^0\rangle = |K^0\rangle. \tag{12.63}$$

Thus the states K^0 and $\bar{K}^0$ are not eigenstates of the CP operaor. We, therefore, define the states:

$$|K_1\rangle = \frac{1}{\sqrt{2}}\left[|K^0\rangle + |\bar{K}^0\rangle\right], \tag{12.64}$$

$$|K_2\rangle = \frac{1}{\sqrt{2}}\left[|K^0\rangle - |\bar{K}^0\rangle\right]. \tag{12.65}$$

Then we find that:

$$CP|K_1\rangle = \ |K_1\rangle, \quad \lambda_{CP} = +1, \tag{12.66}$$

$$CP|K_2\rangle = -|K_2\rangle, \quad \lambda_{CP} = -1. \tag{12.67}$$

Thus in a CP conserving world the state $|K_1\rangle$ can decay into two π-mesons, while the state $|K_2\rangle$ will always decay into three mesons. In principle the state $|K_1\rangle$ can decay into three π-mesons, since three mesons can $\ell = 1$ (see Eq. (12.60). Such a decay, however, is suppressed due to phase space considerations. Furthermore in a CP conserving world $|K_1\rangle$ and $|K_2\rangle$ will be eigenstates of the total Hamiltonian H. Thus for non relativistic energies we have

$$H|K_1\rangle = m_1|K_1\rangle, \quad H|K_2\rangle = m_2|K_2\rangle, \tag{12.68}$$

with m_1, m_2 the corresponding masses. These two masses can be different. Their lifetimes may also be different. Indeed two neutral kaons with such characteristics have been found in nature and have been denoted by K_S and K_L with lifetimes:

$$\begin{aligned} K_S &= 0.89 \times 10^{-10}\text{s (short lived)}, \\ K_L &= 5.2 \times 10^{-8}\text{s (long lived)}. \end{aligned} \tag{12.69}$$

We are thus tempted to identify K_S with K_1 and K_L with K_2. K_S^0 has a mean lifetime $(0.8953 \pm 0.0005) \times 10^{-10}$ s and K_L^0 $(5.116 \pm 0.020) \times 10^{-8}$ s. Thus indeed the particle which decays primarily into three pions has 10^3 times longer lifetime compared to K_S. We note that these particles have masses which are close to each other, i.e.

$$\begin{aligned} \Delta m = m_{K_L} - m_{K_S} &= (0.5349 \pm 0.0022) \times 10^{10}\hbar\text{s}^{-1} \\ &= (3.5208 \pm 0.145) \times 10^{-12}\ \text{MeV}. \end{aligned} \tag{12.70}$$

This difference is tiny compared to their mass m, since

$$\frac{\Delta m}{m} = 1.6 \times 10^{-15}. \tag{12.71}$$

We thus see that in a world in which CP is conserved the mass difference between K_1 and K_2 comes solely from the transitions $K^0 \leftrightarrow \bar{K}^0$, which violate strangeness and, in fact, we have $|\Delta S| = 2$. Such changes cannot be due to the usual weak Lagrangian, but they arise from second order effects.

Since K_1 and K_2 are not stable particles, we can write:

$$E_1 = m_1 + i\frac{\Gamma_1}{2}, \quad E_2 = m_2 + i\frac{\Gamma_2}{2}, \tag{12.72}$$

where Γ_1 and Γ_2 are the widths of these states. Furthermore the wave functions, as functions of time, are given by

$$\psi_j(t) = \psi_j(0)e^{im_j t}e^{-(\Gamma_j/2)t}, \quad j = 1, 2\,, \tag{12.73}$$

with $\psi_j(0)$ the corresponding function of particle j at time $t = 0$. From Eq. (12.73) we find the the probability at time t is

$$|\psi_j(t)|^2 = |\psi_j(0)|^2 e^{-\Gamma_j t}, \tag{12.74}$$

that is $\Gamma_j = 1/\tau_j$, where τ_j is the mean lifetime of particle j (in natural units). So long as CP is conserved the K_1 and K_2 cannot communicate with each other, which means that $\psi_1(t)$ and $\psi_2(t)$ are orthogonal, i.e.

$$|\langle\psi_1(t)|\psi_2(t)\rangle|^2 = 0\,.$$

We should stress that the kaon system exhibits a typical example of quantum mechanical behavior. Which of the four functions $|K^0\rangle, |\bar{K}^0\rangle, |K_S\rangle$, or $|K_L\rangle$ one sees depends on the experiment at hand. If in an experiment one observes only two pions one "sees" $_S$. If, on the other hand, one observes three pions one detects the K_L . If instead, one detects he combinations, $\pi^+e^-\bar{\nu}_e$ or $\pi^-e^+\nu_e$, a sign that the products arise from a particle and its antiparticle, one "sees" the $\bar{K}^0$ and K^0. Sometimes one sees in tables the leptonic channel to be associated with K_L. This is because this channel is long and it does not really affect K_S.

Suppose now that K^0 and $\bar{K}^0$ get admixed via a an interaction that does not conserve strangeness but respects CP. Phenomenologically such a mixing can occur via the virtual transitions $K^0 \leftrightarrow 2\pi \leftrightarrow \bar{K}^0$ shown in Fig. 12.12 (see also fig. 12.16 below).

As we have already mentioned the eigenstates of CP are the K_1 and K_2 (see Eq. (12.65)). If m_1 and m_2 are the masses respectively of these states,

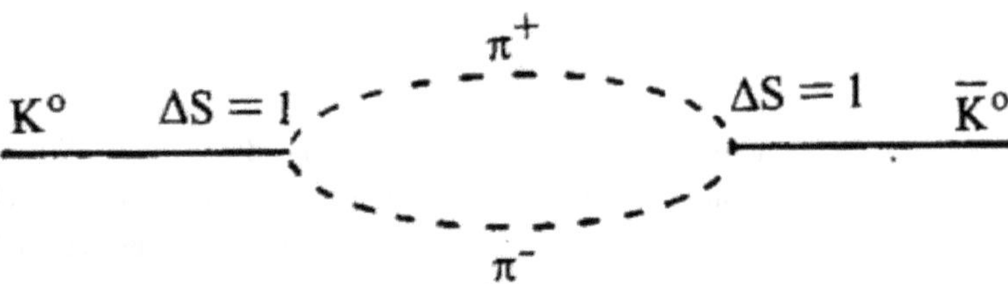

Fig. 12.12: Virtual transitions $K^0 \leftrightarrow 2\pi \leftrightarrow \bar{K}^0$. Such transitions violate strangeness as shown in this figure, but they conserve CP.

Eq. (12.65) yields

$$\begin{aligned}
m_1 &= \langle K_1|H|K_1\rangle = \langle \frac{1}{\sqrt{2}}(K^0+\bar{K}^0)|H|\frac{1}{\sqrt{2}}(K^0+\bar{K}^0)\rangle \\
&= \frac{1}{2}\left[\langle K^0|H|K^0\rangle + \langle \bar{K}^0|H|\bar{K}^0\rangle\right] + \frac{1}{2}\left[\langle K^0|H|\bar{K}^0\rangle + \langle \bar{K}^0|H|K^0\rangle\right], \\
m_2 &= \langle K_2|H|K_2\rangle = \langle \frac{1}{\sqrt{2}}(K^0-\bar{K}^0)|H|\frac{1}{\sqrt{2}}(K^0-\bar{K}^0)\rangle \\
&= \frac{1}{2}\left[\langle K^0|H|K^0\rangle + \langle \bar{K}^0|H|\bar{K}^0\rangle\right] - \frac{1}{2}\left[\langle K^0|H|\bar{K}^0\rangle + \langle \bar{K}^0|H|K^0\rangle\right].
\end{aligned}$$

This leads to:

$$\Delta m = m_1 - m_2 = \langle K^0|H|\bar{K}^0\rangle + \langle \bar{K}^0|H|K^0\rangle. \tag{12.75}$$

Suppose further that a neutral kaon is produced in strong interaction, which conserves strangeness. More specifically consider the reaction

$$\pi^- + p \to \Lambda^0 + K^0,$$

in which a K^0 meson is produced at time $t = 0$ and let us follow its time evolution. At $t = 0$ K^0 can be written as as a linear combination of the eigenstates $|K_1\rangle \equiv |1\rangle$ and $|K_2\rangle \equiv |2\rangle$ of CP operator as follows

$$|K^0(0)\rangle = \frac{1}{\sqrt{2}}\left[|1\rangle + |2\rangle\right].$$

At $t > t_0$, according to the laws of quantum mechanics, we will have:

$$\begin{aligned}
|K^0(t)\rangle &= \frac{1}{\sqrt{2}}e^{iE_1t}|1\rangle + \frac{1}{\sqrt{2}}e^{-iE_2t}|2\rangle \\
&= \frac{e^{-im_1t}}{\sqrt{2}}\left[e^{-\Gamma_1t/2}|1\rangle + e^{-i\Delta mt}e^{-\Gamma_2t/2}|2\rangle\right] \\
&= \frac{e^{im_1t}}{2}\left[(e^{-\Gamma_1t/2} + e^{-i\Delta mt}e^{-\Gamma_2t/2})|K^0\rangle + \right. \\
&\left.(e^{-\Gamma_1t/2} - e^{-i\Delta mt}e^{-\Gamma_2t/2})|\bar{K}^0\rangle\right].
\end{aligned} \tag{12.76}$$

From this relation we see that an initial bunch of K^0. with strangeness has been transformed into one that also contains $\bar{K}^0$. Thus a measurement of strangeness at $t > t_0$ via the strong interaction would provide the probability $P_{K^0\to\bar{K}^0}$ to observe $\bar{K}^0$ with strangeness $S = +1$

$$P_{K^0\to K^0} = |\langle K^0|K^0(t)\rangle|^2 = \frac{1}{4}\left[e^{-\Gamma_1 t} + e^{-\Gamma_2 t} + 2e^{-\Gamma_1 t/2}e^{-\Gamma_2 t/2}\cos(\Delta m t)\right]. \tag{12.77}$$

The first term in the previous equation gives the contribution $P_{K^0\to K^0}$ of K_1, the second that of K_2, while the third the interference of K_1 and K_2. Similarly the probability to observe in the beam a $\bar{K}^0$ particle with strangeness $S = -1$ is

$$P_{K^0\to\bar{K}^0} = |\langle\bar{K}^0|K^0(t)\rangle|^2 = \frac{1}{4}\left[e^{-\Gamma_1 t} + e^{-\Gamma_2 t} - 2e^{-\Gamma_1 t/2}e^{-\Gamma_2 t}\cos(\Delta m t)\right]. \tag{12.78}$$

In neutrino oscillation parlance Eq. (12.77) gives the disappearance probability, while Eq. (12.78) gives the probability of appearance.

12.5.2 *Strangeness oscillations*

From Eqs (12.77) and (12.78) we see that even if the particles K_1 and K_2 were absolutely stable (i.e. if $\Gamma_1 = \Gamma_2 = 0$) we would have a reduction of K^0 mesons and appearance of $\bar{K}^0$. In other words

$$|\langle K^0(t)|K^0(0)\rangle|^2 = \cos^2\left(\frac{\Delta m t}{2}\right), \tag{12.79}$$

$$|\langle\bar{K}^0(t)|K^0(0)\rangle|^2 = \sin^2\left(\frac{\Delta m t}{2}\right). \tag{12.80}$$

This phenomenon, that is the transition $K^0 \leftrightarrow \bar{K}^0$, is called kaon (strangeness) oscillation. The period of the oscillation $T_{\rm osc}$ is found from Eq. (12.80) to be

$$T_{\rm osc} = \frac{2\pi}{\Delta m} = 1.9 \times 10^{10}\text{s}, \tag{12.81}$$

where use of Eq. (12.70) was made. As we see this period is comparable to the life time of K_1.

Such oscillations can be observed in other neutral systems like neutrino oscillations and neutron antineutron, i.e. $n \leftrightarrow \bar{n}$, oscillations, which violate baryon number by two units ($|\Delta B| = 2$). This may happen in GUT theories, which allow the breakdown $B - L$ conservation. Such oscillations have not been observed, since, among other things, the period of oscillation is expected to be very long. Indeed the expected period is

$$T_{\rm osc} > 1.2 \times 10^8\text{s}, \tag{12.82}$$

which corresponds to

$$\Delta m \leq 2.4 \times 10^{-32}\ \text{GeV}\,.$$

It is clear that baryon number violation is much smaller than strangeness non conservation, which is simply a flavor violation, analogous to the neutrino oscillations, $\nu_e \to \nu_\mu$, $\nu_e \to \nu_\tau$ and $\nu_\mu \to \nu_\mu$, which have also been observed and violate lepton flavor (see chapter 11).

12.5.3 *Kaon regeneration*

It is clear from Eqs. (12.77) and (12.78) that if we know the quantities Γ_1 and Γ_2, the mass difference Δm can immediately be obtained by measuring K^0 or $\bar{K}^0$ as a function of time, starting from a pure beam, (e.g. of K^0). The intensity of K^0 and $\bar{K}^0$ can easily be measured, since the interaction of these particles with matter is vastly different. Thus the reaction

$$\bar{K}^0 + p \to \Lambda^0 + \pi^+$$

is open for $\bar{K}^0$ but the final products are not produced, if the K^0 is involved.

We note that, since $\Gamma_1 \gg \Gamma_2$ ($\tau_2 \approx 10^3\tau_1$), the relations (12.77) and (12.78) become:

$$P_{K^0\to K^0} = \frac{1}{4}\left[1 + e^{-\Gamma_1 t/2} + 2\cos^2\left(4\pi\frac{t}{T_{\rm osc}}\right)e^{-\Gamma_1 t/2}\right], \tag{12.83}$$

$$P_{K^0\to \bar{K}^0} = \frac{1}{4}\left[1 + e^{-\Gamma_1 t/2} - 2\cos^2\left(4\pi\frac{t}{T_{\rm osc}}\right)e^{-\Gamma_1 t/2}\right], \tag{12.84}$$

where $T_{\rm osc}$ is given by (12.81). The quantities $P_{K^0\to\bar{K}^0}$ and $P_{K^0\to\bar{K}^0}$ of Eqs. (12.83) and (12.84) are exhibited in Fig. 12.13.

Since the neutral kaons are not visible their measurement is indirect. The main features of the experimental set up is shown in Fig. 12.14. We start with a beam of K^+ in a bubble chamber of deuterium. The K^+ of the beam collide with protons at rest yielding a beam of K^0. These are then transformed into their antiparticles $\bar{K}^0$, which then collide with protons at rest. The following reactions take place chronologically:

$$K^+ + n \to p + K^0,\ \ K^0 \to \bar{K}^0,\ \ \bar{K}^0 + p \to \Lambda^0 + \pi^+,\ \ \Lambda^0 \to \pi^- + p\,.$$

Another measurement of Δm depends on the regeneration of K_1. To this end we consider a beam of K^0. Clearly this contains a component of $K_1(=K_S)$ and a component of $K_2(=K_L)$. After sufficiently long time, so that almost all short lived K_S have vanished, but most of the long lived

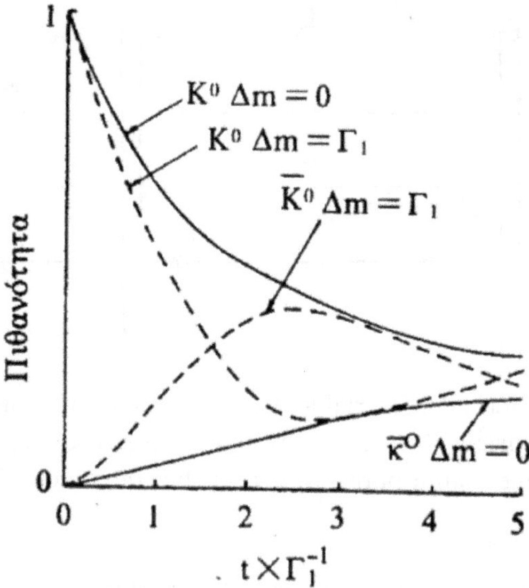

Fig. 12.13: The probabilities $P_{K^0 \to \bar{K}^0}$ and $P_{\bar{K}^0 \to K^0}$ as a function of time t for two extreme values of Δm: *i)* $\Delta m = 0$, and *ii)* $\Delta m = \Gamma_1$.

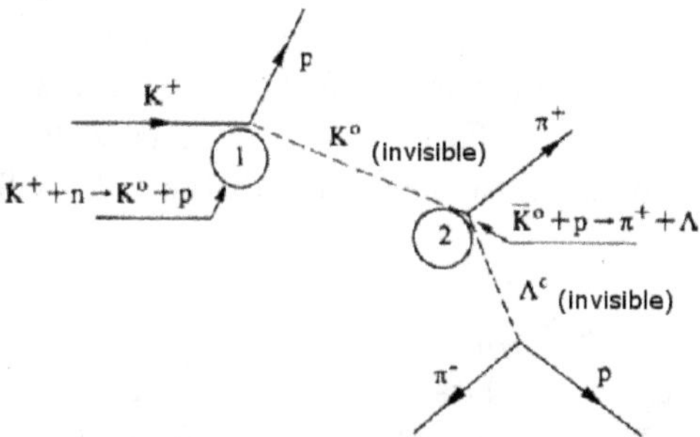

Fig. 12.14: A measurement of Δm employing an initial beam of K^+. In the branch 1 a beam of K^0 is produced, which subsequently is converted to $\bar{K}^0$. The latter in branch 2 interacts with protons yielding π^+ and Λ^0. The Λ^0,s are detected by their products π^- and p.

K_L still exist, i.e. $\Gamma_1 t_0 \gg 1, \Gamma_2 t_0 \ll 1$) the beam contains almost 100% K_2. Then Eq. (12.76) implies

$$K^0(t) = \frac{1}{\sqrt{2}} e^{-i m_2 t_0} e^{-\Gamma_2 t_0} |K_2\rangle = \frac{1}{\sqrt{2}} e^{-i m_2 t_0} e^{-\Gamma_2 t_0} \frac{|K^0\rangle - |\bar{K}^0\rangle}{\sqrt{2}} . \quad (12.85)$$

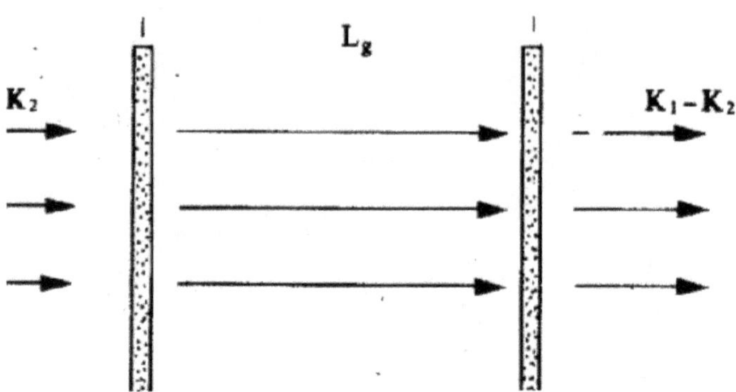

Fig. 12.15: The method of variable gap L_g between the plates, which is used for the measurement of the mass difference Δm of K_1 and K_2. It is shown how starting from a pure $K_2(=K_L)$ the component $K_1(=K_S)$ can be regenerated.

Now if at this point the component $\bar{K}^0$ is absorbed using a suitable absorber (see Fig. 12.15), we get again[7] a pure K^0 beam which now contains 50% K_1 and 50% K_2. We thus have a regeneration of K_1. Their presence is manifested through their decay to the channel $\pi^+\pi^-$.

In experimental set up like that shown in Fig. 12.15), the beam passes through two plates which are at a L_g. The first of these acts as **absorber**, while the second as **regenerator**. Both of them absorb $\bar{K}^0$ to almost 100%. We will assume that the thickness of each plate is negligible. Then, if υ is the velocity of the kaons in the beam, the distance L_g will be covered at a time t_g, given by:

$$t_g = \frac{L_g}{\gamma \upsilon}, \quad \gamma = \left(1 - \frac{\upsilon^2}{c^2}\right)^{-1/2}. \tag{12.87}$$

Suppose that the beam hits the first plate at t_0. At this point it is a pure K_2 beam. Immediately after its exit from the plate, it has been transformed into a pure K^0 beam with intensity $I(t_0)$. After time t_g the number $|K^0(t_g)\rangle$ of kaons K^0, using Eq. (12.76), will be given by

$$|K^0(t_g)\rangle = \frac{e^{im_1 t_g}}{2} A^0(t_0)\big[(e^{-\Gamma_1 t_g/2} + e^{-i\Delta m t_g} e^{-\Gamma_2 t_g/2})|K^0\rangle + (e^{-\Gamma_1 t_g/2} + e^{-i\Delta m t_g} e^{-\Gamma_2 t_g/2})|\bar{K}^0\rangle\big], \tag{12.88}$$

[7] The ratio of absorption n of $\bar{K}^0$ over K^0 is defined via the corresponding cross sections as follows:

$$n = \frac{\sigma(\bar{K}^0)}{\sigma(K^0)}. \tag{12.86}$$

Generally $n > 1$. To simplify the analysis we will assume $n \gg 1$.

where $A^0(t_0)$ is the amplitude of the beam after the first plate. The absorption of $\bar{K}^0$ for a second time in the second plate will yield a beam with amplitude

$$|\psi(t_g)\rangle = \frac{e^{im_1 t_g}}{2} A^0(t_0)(e^{-\Gamma_1 t_g/2} + e^{-i\Delta m t_g} e^{-\Gamma_2 t_g/2})|K^0\rangle,$$

where

$$|K^0\rangle = \frac{1}{\sqrt{2}}(|1\rangle + |2\rangle).$$

Thus the intensity $I(t_g)$ of K_1 of the final beam after the second plate is given by

$$I_{t_g} = |\langle 1|\psi(t_g)\rangle|^2 = \frac{I(t_0)}{8}\left[e^{-\Gamma_1 t_g} + e^{-\Gamma_2 t_g} + 2e^{-(\Gamma_1+\Gamma_2)/2 t_g}\cos(\Delta m t_g)\right]. \tag{12.89}$$

Measuring the intensity of kaons K_1 as a function of the time t_g (or equivalently of the distance L_g of the plates) one can measure $|\Delta m|$. The measurement, however, is not easy for the following reasons.

(1) The relation (12.86) is not fully satisfied. As a result we will not be able to achieve 100% absorption of $\bar{K}^0$.
(2) The quantity Δm is so small that the intensity $I(t_g)$ does not exhibit a second maximum, which, if it were observed, it would have reduced the errors.
(3) CP as we will see below, is not conserved. Thus a phase appears in the argument of cosine in Eq. (12.89).

In spite of all these the result of Eq. (12.70) has been obtained. Other subsequent independent measurements have yielded

$$\Delta m = (0.52860.0028)10^{10}\hbar s^{-1}, \quad \tau_S = (0.89290.0016)^{-10}s.$$

12.5.4 *CP violation*

In 1964 Christenson, Cronin, Fitch and Turlay [Christenson *et al.* (1964)] made a very important discovery, namely they found that the long lived kaons K_L decay with a small branching ratio (2×10^{-3}) to the channel $\pi^+\pi^-$. This demonstrated that the CP symmetry is violated in the kaon system. This implies that the long lived K_L kaon are not pure K_2 but of the form:

$$K_L = \frac{1}{\sqrt{1+|\epsilon|^2}}\left[K_2 + \epsilon K_1\right] \approx K_2 + \epsilon K_1 \tag{12.90}$$

or

$$K_L = \frac{1}{\sqrt{2(1+|\epsilon|^2)}}\left[(1+\epsilon)K_0 - (1-\epsilon)\bar{K}_0\right] \approx \frac{1}{\sqrt{2}}\left[(1+\epsilon)K_0 - (1-\epsilon)\bar{K}_0\right]$$

where ϵ is a small complex quantity ($|\epsilon| < (2-3) \times 10^{-3}$). The short lived K_S will also be of the form:

$$K_S = \frac{1}{\sqrt{1-|\epsilon|^2}}\left[K_1 + \epsilon K_2\right] \approx K_1 - \epsilon K_2 \tag{12.91}$$

or

$$K_S = \frac{1}{\sqrt{2(1+|\epsilon|^2)}}\left[(1+\epsilon)K_0 + (1-\epsilon)\bar{K}_0\right] \approx . \frac{1}{\sqrt{2}}\left[(1-\epsilon)K_0 + (1+\epsilon)\bar{K}_0\right].$$

In Eqs. (12.90) and (12.91) we have, of course, ignored in the final step terms of order ϵ^2 and higher. Since, in general, ϵ is complex K_L and K_S are not necessarily orthogonal (the operator which causes the mixing of K^0 and $\bar{K}^0$ is not necessarily Hermitian (see Eq. (12.72)).

There has been another CP violating process, which has also been observed in the kaon system, namely the decay $K_L \to \pi^+\ell^-\bar{\nu}$ and $K_L \to \pi^-\ell^+\nu$, with ℓ a lepton. The asymmetry is expressed through the parameter δ (asymmetry ratio)

$$\delta = \frac{\Gamma(K_L \to \pi^- l^+ \nu) - \Gamma(K_L \to \pi^+ l^- \bar{\nu})}{\Gamma(K_L \to \pi^- l^+ \nu) + \Gamma(K_L \to \pi^+ l^- \bar{\nu})}. \tag{12.92}$$

If CP were exact, $\delta = 0$. The experimental data show that δ is small, but non zero.

In the analysis of CP violation the parameters $\eta_{\pm}$ and η_{00} are usually employed. They are defined as follows:

$$\eta_{+-} \equiv |\eta^{+-}|e^{i\phi_{+-}} \equiv \frac{\langle \pi^+\pi^-|K_L\rangle}{\langle \pi^+\pi^-|K_S\rangle}, \tag{12.93}$$

$$\eta_{00} \equiv |\eta^{00}|e^{i\phi_{00}} \equiv \frac{\langle \pi^0\pi^0|K_L\rangle}{\langle \pi^0\pi^0|K_S\rangle}, \tag{12.94}$$

where ($\langle \pi^+\pi^-|K_L\rangle$) is the amplitude of the decay of K_L into the channel $\pi^+\pi^-$ ($K_L \to \pi^+\pi^-$), etc. Experimentally it is found that

$$|\eta_{+-}| = (2.281 \pm 0.022) \times 10^{-3}, \ \phi_{+-} = (44.1 \pm 0.09)^0 \approx \frac{\pi}{4}. \tag{12.95}$$

$$\frac{|\eta_{00}|}{|\eta_{+-}|} = (0.9930 \pm 0.020) \times 10^{-3}, \ \phi_{00} - \phi_{+-} = (-1.2 \pm 1)^0 \approx 0, \tag{12.96}$$

i.e. within the errors $|\eta_{+-}| \approx |\eta^{00}|$. Sometimes instead of η_{+-} and η_{00} one uses the quantities ϵ and ϵ' given by

$$\eta^{+-} = \epsilon + \epsilon', \eta^{00} = \epsilon - 2\epsilon' \Rightarrow \epsilon = \frac{1}{3}(2\eta^{+-} + \eta^{00}), \quad \epsilon' = \frac{1}{3}(\eta^{+-} - \eta^{00}). \tag{12.97}$$

Phenomenologically speaking the decay $K_L \to \pi^+\pi^-$ may be attributed to two different mechanisms:

(1) The interaction causing the decay conserves CP.
In this case the pions π^+, π^- originate from the small component ofϵK_1 of K_L, which, as we have seen, can lead to $K_1 \to \pi^+\pi^-$. Then

$$\eta^{+-} = \eta^{00} = \epsilon, \quad \epsilon' = 0. \tag{12.98}$$

We also find that

$$\tan\phi_{+-} = \frac{2(m_L - m_S)}{\Gamma_L - \Gamma_S} \approx 0.96 \quad \text{or} \quad \phi_{+-} \approx 44^0 \tag{12.99}$$

in good agreement with experiment. Furthermore the asymmetry ratio δ of leptonic decays, assuming that only the decays $K_0 \to l^-\pi^+\bar{\nu}$ and $\bar{K}_0 \to l^+\pi^-\nu$ are allowed, becomes:

$$\delta \approx \frac{|1+\epsilon|^2 - |1-\epsilon|^2}{|1+\epsilon|^2 + |1-\epsilon|^2} \approx 2\text{Re}(\epsilon). \tag{12.100}$$

Experimentally it is found that $\delta = (3.87 \pm 0.12) \times 10^{-3}$, in good agreement with the value $\sqrt{2}|\eta^{+-}| \approx 3.2 \times 10^{-3}$ (see Eq. (12.95)).

(2) CP violation is responsible for the mixing of the kaons K_1 and K_2. With steps analogous to those that lead to Eq. (12.75) for Δm we find that, due to CP violation, the states $|1\rangle$ and $|2\rangle$ are no longer eigenstates of H, i.e.

$$i\mu_{21} \equiv \langle 1|H|2\rangle = \frac{1}{2}\left[\langle \bar{K}^0|H|K^0\rangle - \langle K^0|H|\bar{K}^0\rangle\right], \tag{12.101}$$

where μ_{21} is a real quantity. μ_{21} can result from a second order weak interaction as shown in Fig. 12.16. This diagram leads to an effective coupling between K^0 and $\bar{K}^0$ with strength $G_2 = rG_F$, G_F being Fermi's constant and r a complex number. If $r = 0$ such such interactions do not exist and $\Delta m = \mu_{21} = 0$. The real part of r contributes to the difference $m_2 - m_1$, while the imaginary part contributes $\langle 1|H|2\rangle$, that is

$$\frac{\mu_{12}}{m_2 - m_1} = \frac{\text{Im}(r)}{\text{Re}(r)}. \tag{12.102}$$

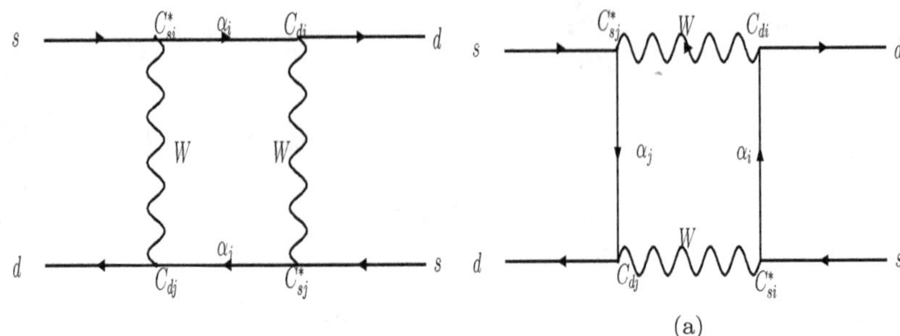

Fig. 12.16: Feynman diagrams leading to $\Delta S = 2$, reactions $K^0 \leftrightarrow \bar{K}^0$. $C_{\gamma,\beta}$, $\gamma = s, d$, $\beta = i, j$ are flavor indices indicating the quark couplings with the intermediate vector bosons (summations over i and j understood). Even though $\sum C_{di} C^*_{si} = 0$ (orthogonality of the corresponding Kobayashi-Maskawa matrix) we obtain a non zero result, since the loop integral depends on the intermediate quark masses. Furthermore the result is zero, if all the elements $C_{\gamma,\beta}$ are real. This diagram contributes to the parameter ϵ, discussed in the text.

Anyway we see that $\mu_{21} \neq 0$, if some of the couplings $C_{\alpha,\beta}$ in Fig. 12.16 are complex. In Fig. 12.16 the only parameters appearing explicitly are the mixing coefficients C_{si}, C_{di} C_{sj} and C_{dj}. The masses of the quarks α_i, $i = u, c, t$, do not appear explicitly, but they come out of the corresponding quark propagators after the loop integral is performed. Note that the relative sign between the two diagrams is $-$, and they are complex of one another. So we get a non zero contribution provided that

$$\Lambda(d, s) = Im \left(\sum_{i,j} C^*_{di} C_{si} C^*_{dj} C_{sj} f(m_i, m_j) \right) \neq 0 \qquad (12.103)$$

where $f(m_i, m_j)$ is a function of the intermediate quark masses arising, as we have mentioned, from the loop integral. Clearly if $f(m_i, m_j)$ is a constant, independent of i and j, we get no CP violation, due to the unitarity of the Kobayashi-Maskawa matrix We also get no CP violation if $Im(C^*_{di} C_{si} C^*_{dj} C_{sj}) = 0$, all i, j.

The same is true for all other neutral mesons made of down type quarks studied for CP violation in recent. The violation being

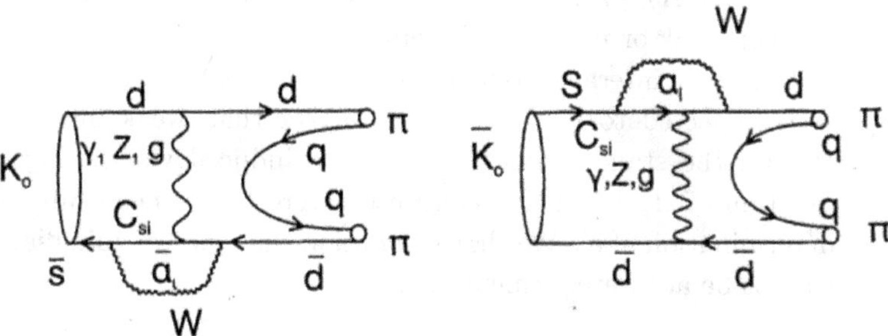

Fig. 12.17: CP violation resulting from $K_L \to \pi\pi$ via $K_2 \to \pi\pi$. This is caused by the change in strangeness $\Delta S = 1$ and is a measure of the parameter ϵ' Note that the relative sign between the two diagrams is $-$. Such diagrams are called of the penguin type. The particles exchanged are, in addition to W, the photon, the Z^0 and the gluons.

proportional to

$$\Lambda(\alpha, \beta) = Im \left(\sum_{i,j} C^*_{\alpha i} C_{\beta i} C^*_{\alpha j} C_{\beta j} f(m_i, m_j) \right),$$

$$(\alpha, \beta) = (d, s), (d, b), (s, b) \tag{12.104}$$

CP violation is commonly associated by the non vanishing of the Jarlskog determinant [Jarlskog (1985)] J:

$$J = \frac{Im\left(C^*_{\alpha i} C_{\beta i} C^*_{\alpha j} C_{\beta j}\right)}{\sum_{\gamma,k} \epsilon_{\alpha\beta\gamma}\epsilon_{ijk}} \tag{12.105}$$

We know that, even though the standard model cannot predict the above couplings, it is possible to have a complex phase in the Kobayashi-Maskawa matrix which may lead to CP violation.

The small mass difference of K_1 and K_2 implies

$$\text{Re}(r) \approx 10^{-6}\text{–}10^{-7}.$$

Furthermore, from the experimental value

$$|\mu_{21}| \approx 3 \times 10^{-3}(m_2 - m_1),$$

it is found that

$$Im(r) \approx 3 \times 10^{-9}\text{–}3 \times 10^{-10}.$$

(3) The decay $K_L \to \pi^+\pi^-$ is caused by the process $K_2 \to \pi^+\pi^-$ via a super-weak or milli-weak interaction.

This is an interaction, which violates CP as well as strangeness. It is expected to be 1000 times weaker than weak interactions. Within the standard model it can be induced via the Feynman diagram of 12.17. This diagram is interesting, since it allows the determination of ϵ'. We should mention that the penguin diagrams cannot be accurately calculated.

In summary the presently accepted values of the CP violating parameters are [Groom *et al.* (2000)]:

$$|\eta_{+-}| = (2.285 \pm 0.019) \times 10^{-3}, \quad \phi_{+-} = (43.5 \pm 0.6)^\circ,$$

$$|\eta_{00}| = (2.275 \pm 0.019) \times 10^{-3}, \quad \phi_{00} = (43.4 \pm 1.0)^\circ.$$

Through the relations,

$$\mathrm{Re}\frac{\epsilon'}{\epsilon} = \frac{1}{6}\left(1 - \frac{\eta_{00}}{\eta_{+-}}\right),$$

it has also been found that:

$$\mathrm{Re}\frac{\epsilon'}{\epsilon} = (2.07 \pm 0.28) \times 10^{-3}, \quad \mathrm{Im}\frac{\epsilon'}{\epsilon} = (1.53 \pm 0.26) \times 10^{-3}.$$

Finally we should mention that the observed CP violation allows us to determine operationally the sign of the electric charge. We define a positive charge as the charge of the lepton mainly produced in the decay of K_L^0 (the production of the negatively charged lepton is much rarer).

12.5.5 *Other CP violating processes involving quarks*

Since r is tiny, the detection of CP violation is indeed hard. In the kaon system a happy coincidence resulted in the fact that CP violation, $r = 3 \times 10^{-3}$, is of the same order with the degeneracy of K_1 and K_2 (favorable resonance). Another possibility is to consider the decays of the neutral mesons composed of up type quarks, like $D_0 = c\bar{u}$ and $\bar{D}_0 = u\bar{c}$ with c, $\bar{c}$ the charmed quark anti-quark respectively.

The existence of $D_0 - \bar{D}_0$ mixing has been established in recent years. In fact the experimental constraints read $x = \frac{\Delta m}{\Gamma} = (0.48 \pm 0.18) \times 10^{-2}$ and $y = \frac{\Delta\Gamma}{2\Gamma} = (0.66 \pm 0.09) \times 10^{-2}$. Thus, the data clearly show that $y \neq 0$, but improved measurements are needed to be sure of the size of x. Long-distance contributions make it difficult to calculate Standard Model predictions for the $D_0 - \bar{D}_0$ mixing parameters. Therefore, the goal of the search for $D_0 - \bar{D}_0$ mixing is not to constrain the CKM parameters, but rather to probe new physics. Here CP violation plays an important role. Within the Standard Model, the CP-violating effects are predicted to be small, since the mixing and the relevant decays are described, to an excellent approximation, by the physics of the first two generations only. The expectation is that the Standard Model size of CP violation in D decays is of order 10^{-3} or less, but theoretical work is ongoing to understand whether QCD effects can significantly enhance it. At present, the most sensitive searches involve the $D_0 \to K^+K^-$, $D_0 \to \pi^+\pi^-$ and $D_0 \to K^{\pm}\pi^{\mp}$ modes.

Extensive searches have been going on for many years in systems involving the b and $\bar{b}$ quarks, like the $B_d = b\bar{d}$ and $\bar{B}_d = d\bar{b}$ and $B_s = b\bar{s}$ and $\bar{B}_s = s\bar{b}$. In these systems CP violation was expected to be much larger than in the kaon system. Other effects like B_d and $\bar{B}_d$ mixing has been measured.

In the $B^0 = d\bar{b}$, with decays into J/ψ, K^0_S, the asymmetry is larger $r = 0.73$, but the difference of the widths is much smaller than the mass difference, which makes the experiment much harder. That is the reason it took so long to be established, but it finally came (see below).

12.5.6 *Evidence for T-violation*

Time-reversal asymmetry in particle physics has finally been seen [Lees *et al.* (2012)] by the BaBar Collaboration, in B-factories, in which B-mesons consisting of a heavy b-quark ($\bar{b}$-quark) and a light anti-quark (quark) are produced. In the SLAC collider the counter-circulating e^- and e^+ had a CM energy was 10.58 GeV and, as result, the $\mathcal{Y}(4s)$ state was produced (this is the $n = 4$, $\ell = 0$ bound state of a $b\bar{b}$ system. This meson immediately decays into pairs of neutral or two opposite charged B-mesons with almost equal probability. The B-mesons are tagged by the charged lepton they decay to. Since the b-quark (anti-quark) cannot decay to the much heavier t-quark (anti-quark), if a positive lepton is produced we know that it is a $B^0 = b\bar{q}$, $q = d, s$. If a negative lepton is produced, we know that the decayed mason was $\bar{B}^0$. Due to their production the two B-mesons are entangled. In other words if one of the mesons, say B^0, is detected we know that its yet undetected partner has to be $\bar{B}^0$. For a better analysis of the data it was decided to use electrons with about three times of the positron energy. Thus the pair is not produced at rest, but with a pair momentum of about 6 GeV/c. Thus a measurement of the distance between the two decays indicates the time difference between the two decays. As in the case of the kaon system we can have oscillations $B^0 \rightleftharpoons \bar{B}^0$. Either of them can also oscillate into the CP eigenstates:

$$B_{\pm} = \frac{1}{\sqrt{2}}\left(B^0 \pm \bar{B}^0\right). \tag{12.106}$$

The main idea was to detect one decay into a CP eigenstate, while its partner revealed a decay into a flavor eigenstate. The latter can be revealed by the tagging mentioned above. The other form of decay reveals itself by the rarer process:

$$\begin{aligned} B_{+/-} &= \psi(1s) + K_{L/S}, \quad K_{L/S} = \text{long/short lived kaon.} \\ \psi(1s) &= J/\psi = \text{neutral charmed meson} \end{aligned} \tag{12.107}$$

Thus the nature of the B-meson is revealed by the decay of the daughter kaon. Le us now suppose that at some time a negative lepton is measured. Then its still surviving partner must be B^0, which after some time Δt, measured by the displacement Δz along the beam direction, decays as, e.g. $B^0 \to B_-$ with a rate Γ_1, verified by the mechanism of Eq. (12.107). This transition rate, Γ_1, is compared with that of its inverse, $B_- \to B^0$, Γ_2. Then an asymmetry is defined as:

$$A_T(\Delta t) = \frac{\Gamma_1 - \Gamma_2}{\Gamma_1 + \Gamma_2}. \tag{12.108}$$

Then if time reversal holds the asymmetry should be zero. A detailed analysis [Lees *et al.* (2012)] showed that this is not the case (see Fig. 12.18). In the figure we indicate the mesons participating in the oscillation, e.g. $B^0 \to B_-(e^+X, c\bar{c}K^0_S)$ means that first $\bar{B} \to e^+$ happens first, the $B^0 \to B_-$ takes place and Δt later the decay $B_- \to c\bar{c}K^0_S$ takes place. Similarly for the other processes of the figure. Thus, for the transition $\bar{B}^0 \to B_-$ in the above notation, the asymmetry parameter is:

$$A_T = \frac{\Gamma^-_{e^-,K^0_L}(\Delta t) - \Gamma^+_{e^+,K^0_s}(\Delta t)}{\Gamma^-_{e^-,K^0_L}(\Delta t) + \Gamma^+_{e^+,K^0_s}(\Delta t)}, \tag{12.109}$$

From Fig. 12.18 it is evident that we have T-violation in this system.

Whole this book was in press a new measurement of CP violation has appeared [Aaij *et al.* (2016)], in $B_0 \to D^+D^-$ decays involved in p-p collisions at center of mass energies 7-8 TeV, by the LHCb collaboration at a significance level of 4.0 standard deviations. This is in agreement with the

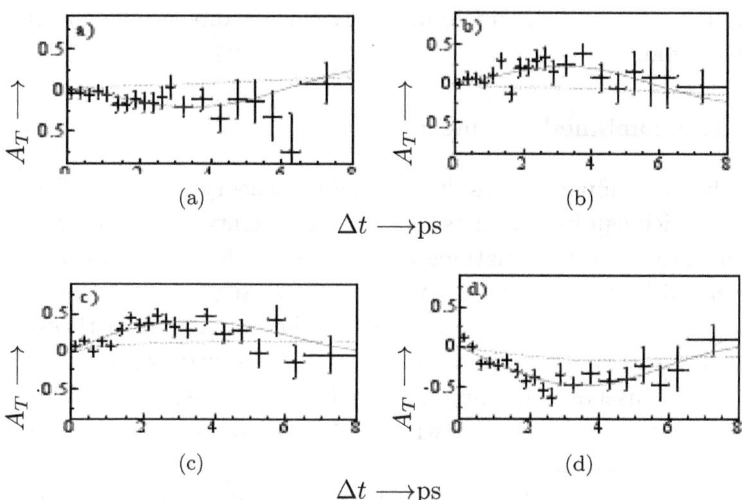

Fig. 12.18: The T-asymmetry parameter as a function of Δt in the BaBar collaboration experiment. (a) $B^0 \to B_-(e^+X, c\bar{c}K_S^0)$(b) $B_+ \to \bar{B}^0(c\bar{c}K_S^0, e^+X)$ (c) $\bar{B}^0 \to B_+(e^+X, J/\psi K_L^0)$ (d)$B_- \to B^0(J/\psi K_L^0, e^+X,)$. The bars indicate the data, while the solid and dashed curves correspond to the best fit with and without T-violation respectively. The final states are time ordered separated by a comma, the earlier time first.

previous BaBar experiment discussed above, with the latter being much more precise. A detailed comparison between the two experiments has not yet been made. The phase shift, however, is constrained to be in the smallest world value of $\Delta\phi = -0.16^{0.19}_{-0.21}$rad, implying that only a small contribution may be coming from higher-order Standard Model contributions.

It is worth noting at this point that the standard model, through Kobayashi-Maskawa matrix with the its complex phase, can describe all the facts regarding CP. It cannot, however, describe the matter-antimatter asymmetry appearing in the Universe by many orders of magnitude. It is perhaps possible that CP violation has been much more pronounced in the early Universe, but this does not come out of what our present models yield as proper physics of that time. For this reason other mechanisms may be required to explain such a violation. Some people believe that the baryon asymmetry may arise from the leptonic sector via the scenario of **leptogenesis**.

As we have already discussed another source CP violation would be the appearance of an electric dipole moment. Such a dipole moment has not yet been found.

12.6 The combined symmetry CPT

One of the most important results of the Field Theory is the so called CPT theorem, which can be stated as follows: Any quantum field theory, which respects Lorentz transformations and the principle of micro causality is characterized by the combined symmetry CPT, regardless of the order of the operations P, C and T. Since most of the available field theories are consistent with this theorem, there appears to be very strong confidence in the CPT conservation. There has yet been no evidence for CPT break down. As a result the breakdown of CP, e.g., in the system of neutral kaons, implies T violation.

Using the relations (12.2), (12.50), and (12.30) for the transformation of various physical observables under the operations P, C, and T, we can easily deduce their transformation properties under the operation of CPT. This way we get

$$\vec{p} \to \vec{p},\ \vec{r} \to -\vec{r},\ \vec{J} \to -\vec{J},\ \vec{Q} \to -\vec{Q},\ \vec{h} \to -\vec{h},$$

$$B \to -B,\ L_i \to -L_i, \quad i = e, \mu, \tau\,. \tag{12.110}$$

We see that, with exception of momentum, all other quantities change sign under the action of CPT.

An important consequence of the CPT theorem is that the reaction

$$A + B \to C + D,$$

as well as its CPT symmetric

$$\bar{C} + \bar{D} \to \bar{A} + \bar{B},$$

must have the same characteristics.

As we have mentioned the electric dipole moment changes sign under the action of P. It can easily be shown that it also changes sign under CP, but it remains unchanged under CPT. Thus if an elementary particle is found to have a non zero electric dipole moment, none of the symmetries C, P, T, and CP is exact.

Chapter 13

Appendix: Some Elementary Aspects of Particle Physics

13.1 The natural system of units

In basic physics we consider a system of three elementary units the length (L), the time (T) and the mass (M). From these the dimensions of the other physical quantities can be determined via definitions ($\upsilon = s/t$ etc.) and the laws of nature, (e.g. $F = ma$). This way any physical quantity dimensionally can take the form:

$$[\Pi] = L^{\lambda} T^{\mu} M^{\nu} \tag{13.1}$$

where λ, μ, ν rational numbers. So if units are defined for these three, the units of all the other quantities can be obtained. Traditionally we select the meter m (m), the second (s) and the kilogram (Kg) as the units of length time and mass respectively. This system is called the International System of Units (IS). In a electromagnetism we introduced one more unit for the charge via Coulomb's law:

$$F = k_1 \frac{q_1 q_2}{r^2}, \quad k_1 = \frac{1}{4\pi\epsilon_0}, \quad \epsilon_0 = \frac{1}{\mu_0} c^2, \quad \mu_0 = 4\pi \times 10^{-7}, \tag{13.2}$$

$$\epsilon_0 = 8.85 \times 10^{-12}, \quad \frac{1}{4\pi\epsilon_0} = 8.99 \times 10^9 m/F \text{ (values used in calculations)},$$

Significant digits 8.854187817 and 8.987551787 respectively (exactly)

and dimensions

$$[k_1] = [\epsilon_0^{-1}] = Q^{-2} L^3 M T^{-2}.$$

The dimensionless quantities have, of courde, the same value in all systems:

$$\alpha = \frac{e^2}{4\pi\epsilon_0} \frac{1}{\hbar c} = \frac{1}{137.035599911} \approx \frac{1}{137}, \text{ (obtained in MKSA)},$$

$$\alpha = \frac{e^2}{4\pi}\frac{1}{\hbar c} = \frac{1}{137.035599911}, \text{ (obtained in Heavyside-Lorentz).}$$

In hindsight it might have been better had we chosen a more a unambiguous system of units like the mass m_0 of an elementary particle (e.g. the mass of the electron or the proton) as the unit of mass, a given wavelength as a unit of length and the period of a simpl microscopic system as the unit of time.

In particle physics we often encounter the constants $\hbar = \frac{h}{2\pi}$, h Planck's constant and c the velocity of light. It would be nice, if we devise a system where these constants do not show up. So we use a system whereby

$$\hbar = 1, c = 1 \text{ and as unit of mass the mass } m_0 \text{ of a particle.} \tag{13.3}$$

This way:

- Length and time have the same dimension.
- The velocity is dimensionless.
- Energy has the dimension of mass.
- length and time have dimension of inverse mass: $E = \hbar\omega = E = \hbar/(2\pi T) \Rightarrow [T] = [E^{-1}] \Rightarrow [T] = [L] = [m_0^{-1}]$.
- For any quantity Π

$$[\Pi] = m_0^{\rho}, \quad \rho = \text{rational}.$$

For the needed conversions we will use:

$$\hbar c = 1.97 \times 10^{-13}\text{MeV - m} = 3.15 \times 10^{-15} \text{ J-m} . \tag{13.4}$$

Thus for the length we get:

$$\begin{aligned}\ell_0 &= \frac{\hbar c}{m_0 c^2} = \frac{\hbar c}{m_p c^2}\frac{m_p}{m_0} = \frac{1.97 \times 10^{-13}\text{MeV - m}}{939\text{MeV}}\frac{m_p}{m_0}\\ &\approx 2.10 \times 10^{-16}\frac{m_p}{m_0} \text{ m},\end{aligned} \tag{13.5}$$

while for the time:

$$t_0 = \frac{\hbar}{m_0 c^2} = \frac{\hbar}{m_p c^2}\frac{m_p}{m_0} = \frac{2.10 \times 10^{-16} \text{ m}}{3.00 \times 10^8 \text{ m/s}} = 7.00 \times 10^{-25}\frac{m_p}{m_0} \text{ s.} \tag{13.6}$$

The left hand side is m_0^{-1}. Thus selecting as unit the proton mass $m_p = 939 \text{ GeV}/c^2$, as a basis (unit) we find:

$$m_p^{-1} = 2.10 \times 10^{-16} \text{ m or 1 m} = 4.76 \times 10^{15} m_p^{-1}, \tag{13.7}$$

$$m_p^{-1} = 7.00 \times 10^{-25} \text{ s or 1s} = 1.43 \times 10^{24} m_p^{-1}. \tag{13.8}$$

Table 13.1: The dimensions of some physical quantities in various systems.

quantity	symbol	MKSA			NSI
		λ	μ	ν	ρ
Action	$\hbar$	2	-1	1	0
velocity	v	1	-1	0	0
mass	m	0	0	1	1
length	ℓ	1	0	0	-1
time	t	0	0	1	-1
momentum	p	1	-1	1	1
Energy	E	2	-2	1	1
Fine structure constant	α	0	0	0	0
Newton's constant	G_N	3	2	-1	-6
Fermi's constant	G_F	5	-2	1	-2

Sometimes we use GeV for a unit of energy. Then

$$\begin{aligned} &\text{GeV}^{-1} = 1.97 \times 10^{-16}\text{m} = 6.57 \times 10^{-25}\text{s},\ 1\text{m} = 5.06 \times 10^{15}\text{GeV}^{-1}, \\ &\quad 1\text{s} = 1.52 \times 10^{24}\text{GeV}^{-1}, \end{aligned} \tag{13.9}$$

$$G_F = 1.16 \times 10^{-5}\hbar c \left(\frac{\hbar}{m_p c}\right)^2 \Rightarrow G_F = 1.16 \times 10^{-5} m_p^{-2}. \tag{13.10}$$

The converse can perhaps be found by some tricks. We will exhibit here a quite general method, which is a bit harder. What we should do is to find a combination of $\hbar, c$ and m_p needed to express G_F in the SI system, if we know that its dimension is energy times length^3, that is

$$G_F = \text{energy} \times (\text{length})^3 = (L/T)^2 M L^3 = L^5 M T^{-2}. \tag{13.11}$$

We try

$$G_F = \hbar^p c^q m_0^r. \tag{13.12}$$

Then from table 13.1 we find:

$$[\hbar] = L^2 T^{-1} M\ ,\ [c] = L T^{-1}\ ,\ [m_0] = M. \tag{13.13}$$

Thus

$$G_F = (L^2 T^{-1} M)^p (L T^{-1})^q M^r = L^{2p+q} T^{-p-q} M^{p+r}. \tag{13.14}$$

Comparing (13.11) and (13.14) we get:

$$2p+q=5\ ,\ -p-q=-2\ ,\ p+r=1\Rightarrow p=3\ ,\ q=-1\ ,\ r=-2, \quad (13.15)$$

$$G_F=\hbar^3c^{-1}m_0^{-2}. \quad (13.16)$$

The numerical value is obtained by comparing it with the expression of G_F given by (13.10)).

Example 1: The life time of muon μ in natural units is given by:

$$\frac{1}{\tau}=\frac{G_F^2}{192\pi^3}m_\mu^5. \quad (13.17)$$

Let us find it in seconds.

$$\frac{1}{\tau}=\frac{(G_Fm_p^2)^2}{192\pi^3}\frac{m_\mu^5}{m_p^5}m_p=\frac{(1.16\times10^{-5})^2}{0.60\times10^4}\left(\frac{0.106}{0.940}\right)^5m_p=3.37\times10^{-19}m_p, \quad (13.18)$$

$$\tau=3.10\times10^{18}m_p^{-1}=3.10\times10^{18}\times7.00\times10^{-25}\text{s}=2.17\times10^{-6}\text{s}. \quad (13.19)$$

Example 2: Electromagnetism.

In the system MKSA the energy of the hydrogen atom is:

$$U=-\frac{e^2}{4\pi\epsilon_0}\frac{1}{r}=-\alpha\frac{\hbar c}{r}, \quad (13.20)$$

where α is the fine structure constant. In particle physics we choose $\epsilon_0=1$, so that it coincides with the Heavyside-Lorentz system. Then the unit of charge is:

$$e=\sqrt{4\pi\alpha}=0.303. \quad (13.21)$$

So if q and q_e are some charge and and the charge of the electron respectively

$$Q=\frac{q}{q_e}\sqrt{4\pi\alpha}\ \text{(natural units)}\ . \quad (13.22)$$

13.2 The Planck natural system of units

Up to now the mass unit has been arbitrary. In an effort to remove this arbitrariness, we write gravitational energy as:

$$U=-G_N\frac{m^2}{r}=-\alpha_G\frac{\hbar c}{r}\ ,\ \alpha_G=\frac{G_Nm^2}{\hbar c}, \quad (13.23)$$

$$m=\sqrt{\frac{\alpha_G\hbar c}{G_N}}. \quad (13.24)$$

Choosing $\alpha_G = 1$ we obtain the so called Planck mass:

$$M_P = \sqrt{\frac{\hbar c}{G_N}}. \tag{13.25}$$

This mass from a particle point of view is tremendous, almost macroscopic. Taking $\hbar c = 1.97 \times 10^{-13} \times 10^6 \times 1.610^{-19}$ Kg m^{3}s^{-2}, $G_N = 6.67 \times 10^{-11}$m^{3}Kg^{-1}s^{-2} we find

$$M_P = \left[\frac{3.15 \times 10^{-26}}{6.67 \times 10^{-11}}\right]^{1/2} \text{Kg} = 2.17 \times 10^{-8}\text{Kg} \approx 0.02 \text{ mgr}. \tag{13.26}$$

This can be written

$$M_P = \frac{2.17 \times 10^{-8}\text{Kg}}{1.67 \times 10^{-27}\text{Kg}} m_p = 1.27 \times 10^{19}\ m_p = 1.20 \times 10^{19}\text{GeV}. \tag{13.27}$$

Note: There exist other definitions of the Planck mass. If we choose $\alpha_G = \alpha = 1/137$, we find $M_P = 1.02 \times 10^{18}$.

Note also that, through the Eq. (13.25) for $M_P = 1$, in the Planck system get

$$G_N = 1\ . \tag{13.28}$$

Furthermore Boltzmann's constant in the Plank scale can be set to unity, $k_B = 1$. Thus a new unit of absolute temperature is defined by

$$T_P = \frac{M_P c^2}{k_B} = 1.42 \times 10^{32}\text{K}. \tag{13.29}$$

In the Plank system we choose ϵ_0 so that

$$\epsilon_0 = \mu_0 = 1/4\pi \Rightarrow e = \sqrt{\alpha}. \tag{13.30}$$

Thus the of charge will be:

$$q_P = \sqrt{4\pi\epsilon_0 \hbar c} = 1.88 \times 10^{-18}\text{C}, \tag{13.31}$$

while the length and time units are:

$$\ell_P = \sqrt{\frac{\hbar G}{c^3}} = 1.62 \times 10^{-35}\text{m}\ ,\ t_P = \sqrt{\frac{\hbar G}{c^5}} = 5.39 \times 10^{-44}\text{s}. \tag{13.32}$$

At energies in the plank scale, gravity will be as strong as electromagnetism. Something that happened in the Universe at the time of the order of the Plank time.

If the Planck system is adopted the school books will define the three basic units as follows:

$$\hbar = 1, \quad c = 1, \quad G_N = 1. \tag{13.33}$$

Then the basic laws of nature will be written as:

$$F = -\frac{M_1 M_2}{r^2}\ (\text{Newton}),\ F = \frac{Q_1 Q_2}{r^2}\ (\text{Coulomb}),\ E = T,\ E = \omega. \tag{13.34}$$

In the every day use one will, of course, have to convert the numerical values to the common units.

13.3 Invariants in kinematics

We will begin with some simple examples.

Consider the elastic collision of two particles A and B with m_A and m_B. We define the invariant quantity:

$$s = (p_A + p_B)^2 = (E_A + E_B)^2 - (\vec{p}_A + \vec{p}_B)^2. \tag{13.35}$$

We will evaluate s in some frames of special interest.

(1) the center of momentum system $\mathbf{p}_A + \mathbf{p}_B = 0$.
In this system the meaning of s will become simpler

$$s = (E_A^* + E_B^*)^2 \to \sqrt{s} = E_A^* + E_B^*.$$

In the other words $\sqrt{s}$ is the energy in the center of momentum frame.

(2) In the laboratory frame in which, let us say, the particle B is at rest.

We have

$$s = (E_A + m_B)^2 - \vec{p}_A{}^2.$$

But

$$\vec{p}_A{}^2 = E_A^2 - m_A^2$$

i.e.

$$s = 2m_B E_A + m_B^2 + m_A^2.$$

Thus

$$\sqrt{s} = E_A^* + E_B^* = \sqrt{2m_B E_A + m_A^2 + m_B^2}.$$

(3) Particles of unequal mass and energies moving in opposite directions, $\mathbf{p}_A = -\mathbf{p}_B$.

An example is the LINAC accelerator HERA accelerating protons and electrons in opposite directions.

$$\begin{aligned} s &= (E_A + E_B)^2 - (p_A - p_B)^2 \\ &= E_A^2 + E_B^2 + 2E_A E_B - p_A^2 - p_B^2 + 2p_A p_B \\ &= m_A^2 + m_B^2 + 2E_A E_B \left(1 + \sqrt{1 - \frac{m_A^2}{E_A^2}} + \sqrt{1 - \frac{m_B^2}{E_B^2}}\right) \\ &\approx 4E_A E_B. \end{aligned} \tag{13.36}$$

The last relation is valid in the limit in which the masses can be neglected in front of the corresponding energies.

Let us return to the fixed target experiment and express $\sqrt{s}$ in terms of kinetic energies.

$$K_{cm} = K_A^* + K_B^* = \sqrt{2m_B E_A + m_A^2 + m_B^2} - m_A - m_B \text{ or}$$
$$K_{cm} = \sqrt{2m_B K_A + (m_A + m_B)^2} - m_A - m_B. \quad (13.37)$$

Thus if A = electron, B = positron ($m_A = m_B = m_e$) at relativistic energies the energy available in the center of momentum frame becomes:

$$K_{cm} \approx \sqrt{2m_c K_A}, \quad (13.38)$$

i.e. it increases as the square root of the energy of the beam. Thus, to obtain an energy of 2 GeV in the Center of momentum system, the beam energy should be:

$$K_A \approx \frac{K_{cm}^2}{2m_e} \approx \frac{4\ \text{GeV}^2}{0.511 \times 2\ \text{MeV}} \approx 4000\ \text{GeV!}$$

Clearly accelerating a particle this way is not a practical method to achieve high energy in the center of momentum.

In reactions the energy in the center of momentum counts, so that we do not lose most of it as kinetic energy of the fragments. Had we accelerated two particles we only needed 1 GeV each to get 2 GeV in the center of momentum.

COLLIDING BEAMS IS THE SECRET TO HIGH ENERGIES

FOR TWO BEAMS OF OPPOSITE CHARGE

ONE MACHINE AND ONE CHANNEL SUFFICE

13.4 Cross sections and luminosities

The cross section in particle physics is expressed in units

$$1\text{b} = 1\text{barn} = 10^{-24}\text{cm}^2 = 100\ \text{fm}^2 \quad (13.39)$$

typical cross sections at energies of 100 GeV are given in table 13.2

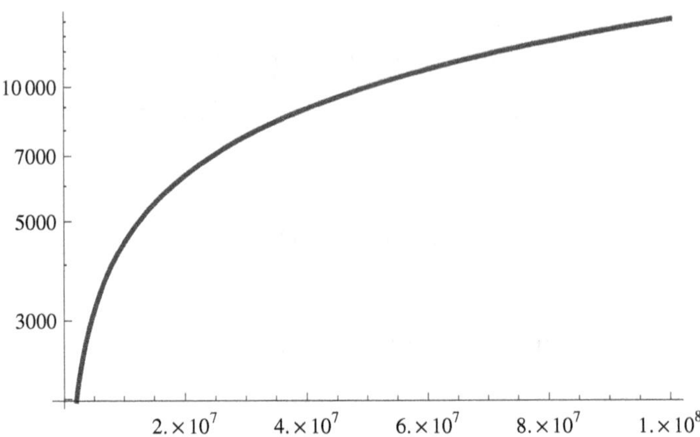

Fig. 13.1: The kinetic energy of two particles of the same mass in the center of momentum frame versus the energy of one of them in a fixed target set up. Both energies are given in units of the rest energy of each particle.

Table 13.2: Typical cross sections for weak, electromagnetic and strong interactions. (μb $= 10^{-6}$b).

	weak	Electromagnetic	strong
σ	$10^{-8}\mu$b	10μb	$10^4\mu$b

13.4.1 *Luminosity in fixed target experiments*

The number of events produced during the collision is proportional to the flux of the particles in the beam (see Fig. 13.2). Indeed, if n is the particle density in volume dV we will have:

$$dN = ndV.$$

But during the time dt only the particles which are vdt way will be able to hit the surface dS. Thus $dV = vdtdS$, i.e. $dN = nvdtdS$ or

$$\text{particle flux } \Phi = \frac{dN}{dSdt} = nv. \tag{13.40}$$

The the number of particles going through an area dS is going to be:

$$\frac{dN}{dt} = nvdS$$

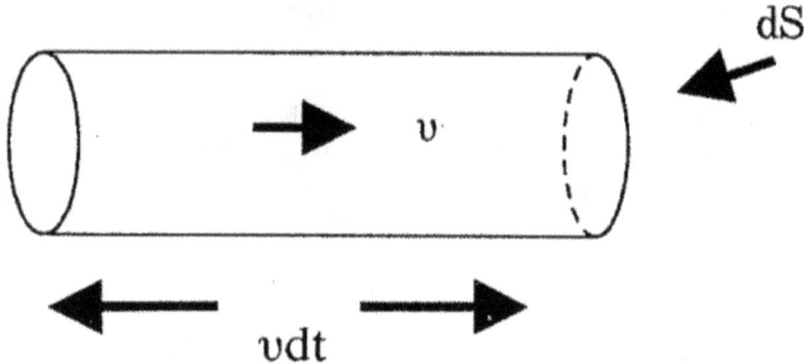

Fig. 13.2: The geometry relevant for obtaining the particle flux, $\Phi = dN/dtdS = n\upsilon$, where $n =$ is the number of particles per unit volume.

For the process $A + B \to C + D$, (A beam, B target) we have $S = N_B\sigma$, i.e.

$$\frac{dN}{dt} = n_A\upsilon_A N_B\sigma = L\sigma, \quad \sigma = \sigma(A + B \to C + D). \tag{13.41}$$

The quantity L given by

$$L = n_A\upsilon_A N_B \tag{13.42}$$

is called **fixed target luminosity** of the accelerator and is measured in $\text{cm}^{-2}\ \text{s}^{-1}$.

13.4.2 *Luminosity in colliding beam experiments*

For colliding beams (C.B.):

$$n_A = \frac{N_A}{V_A} = \frac{N_A}{\ell S}, \tag{13.43}$$

where ℓ is the length of the accelerator and S cross sectional area of the beam. Now the quantity of interest is the relative velocity, $\upsilon \to \upsilon_{AB} = |\upsilon_A - \upsilon_B|$. Thus for a colliding beam accelerator

$$L = \frac{N_A N_B}{S}\frac{\upsilon_{AB}}{\ell} \quad \text{(C.B. Luminosity)}. \tag{13.44}$$

In the case of a circular C.B. accelerator we have

$$\ell = 2\pi R, \quad \upsilon_{AB} = 2\upsilon_A = 2\omega R = 4\pi f R$$

that is

$$L = \frac{N_A N_B}{S} f. \tag{13.45}$$

If we have k bunches accelerated simultaneously

$$L = 2k\frac{N_A N_B}{S}f. \tag{13.46}$$

Usually instead of N_A and N_B one uses the beam intensities I_A and I_B, i.e the number of particles going through the unit area in unit time. Thus in the interval of one period $= 1/f$ we have:

$$N_A = I_A ST, \quad N_B = I_B ST,$$

$$L = 2kI_A I_B \frac{S}{f}. \tag{13.47}$$

Usually $I_A = I_B$ (we do not want any particles not to find partners). Then

$$L = 2kI^2\frac{S}{f}, \quad I = I_A = I_B.$$

Unfortunately the luminosities that can be achieved in colliding beam accelerators typically are:

$$L \approx 10^{31}\ cm^{-2}s^{-1}(e^+, e^-), \quad L \approx 10^{32}\ cm^{-2}s^{-1}(\bar{p}, p), \tag{13.48}$$

i.e. they relatively small compared to those of a fixed target accelerator. Indeed for fixed target:

$$n_A v_B = \frac{1}{S}\frac{dN_A}{dt}.$$

Furthermore if the target has mass M_B, $M_B = N_B A_B m_p$ with A_B the molecular weight of the substance of B and m_p the proton mass. Thus

$$n_B = \frac{N_B}{V_B} = \frac{N_B}{(M_B/\rho_B)} = \frac{N_B \rho_B}{m_p A_p},$$

$$L = \frac{dN_A}{dt}\left(\frac{\rho_B}{m_p A_B}\right)\ell_B. \tag{13.49}$$

As an example consider a synchrotron of fixed target so that 2×10^{12} protons fall per second on liquid hydrogen with thickness of 1 m and density $\rho = 0.084$ gr/cm^3. Then

$$A_B = 1, \ m_p = 1.7 \times 10^{-24}\text{gr},$$

The achieved luminosity is:

$$L = 2 \times 10^{12}\text{s}^{-1}\left(\frac{0.065\text{s gr/cm}^3}{1.7 \times 10^{-24}\ \text{gr}}\right) \times 10^2\text{cm} == 7.73 \times 10^{36}\text{cm}^{-2}\text{s}^{-1}.$$

Much larger than that of a colliding beam accelerator.

The small luminosity is the price one has to pay to achieve high energies.

13.5 Kinematics of particle decay

The simplest case is the decay of one particle into two:

$$A \to B + C.$$

Four momentum conservation implies

$$p_A = p_B + p_C.$$

In the rest frame of the decaying particle we have

$$p_A = (m_A, 0), \quad p_B = (E_B, \vec{p}_B), \quad p_C = (E_C, \vec{p}_C), \, \vec{p}_C = -\vec{p}_B.$$

Combining them with the relations:

$$\begin{aligned} E_B^2 &= \vec{p}_B^{\,2} + m_B^2 = \vec{p}^{\,2} + m_B^2, \\ E_C^2 &= \vec{p}_C^{\,2} + m_C^2 = \vec{p}^{\,2} + m_C^2, \\ m_A &= E_B + E_C, \end{aligned}$$

we find

$$m_A = \sqrt{p^2 + m_B^2} + \sqrt{p^2 + m_C^2}, \quad p = p_B = p_C. \tag{13.50}$$

It thus follows:

$$p = \frac{\sqrt{\lambda(m_A^2, m_B^2, m_C^2)}}{2m_A}, \tag{13.51}$$

where

$$\lambda(x, y, z) = x^2 + y^2 + x^2 - 2xy - 2xz - 2yz$$

is the triangular function. Then it is easy to show:

$$\begin{aligned} E_B &= \frac{m_A^2 + m_B^2 - m_C^2}{2m_A}, \\ E_C &= \frac{m_A^2 - m_B^2 + m_C^2}{2m_A}. \end{aligned}$$

Example 3: We will investigate the kinematics for resonance production in pion proton scattering.

$$\pi + p \to \pi + N^*$$

with N^* a particle with mass $m_{N^*} > m_p > m_\pi$. In the center of momentum

$$\vec{p}_\pi^{\,*} + \vec{p}_p^{\,*} = 0. \tag{13.52}$$

The energy of the resonance is known if the pion energy is known. The minimum required energy in the center of momentum is:

$$E^*_{\text{tot}} = m_\pi + m_{N^*}.$$

Obviously the initial particles must have energies such that:

$$E^*_\pi + E^*_p = m_\pi + m_{N^*}.$$

The question is: What is the kinetic energy of the π particles in the lab so that the resonance is produced? We have

$$(E^*_\pi + E^*_p)^2 = (E_\pi + E_p)^2 - (\vec{p}_\pi + \vec{p}_p)^2 = m_p^2 + E_\pi^2 + 2E_\pi m_p - E_\pi^2 + m_\pi^2,$$

that is

$$E_\pi = \frac{(E^*_{\text{tot}})^2 - (m_\pi^2 + m_p^2)}{2m_p}.$$

The threshold pion energy must be such that $E^*_{\text{tot}} = m_\pi + m_{N^*}$, i.e.

$$E_\pi^{th} = \frac{(m_\pi + m_{N^*})^2 - (m_\pi^2 + m_p^2)}{2m_p} = \frac{2m_\pi m_{N^*} + m_{N^*}^2}{2m_p} - \frac{m_p}{2}. \tag{13.53}$$

The π kinetic energy is $K_\pi = E_\pi - m_\pi$. We thus see that the mass of the resonance can be computed if E_π^{th} is known. Indeed from Eq. (13.53) we obtain:

$$m_{N^*} = \sqrt{2m_p E_\pi^{th} + m_p^2 + m_\pi^2} - m_\pi.$$

13.6 Invariants in the scattering of two particles

Invariants are quantities which are the same in all frames. Consider the reaction

$$A + B \to C + D$$

If for the moment we ignore the spin and the internal quantum numbers the cross section will depend on p_A, p_B, p_C and p_D. The possible invariants arising from these momenta are of the form:

$$p_i^\mu p_{j,\mu} \quad i,j = A, B, C, D.$$

But

$$p_A^2 = m_A^2,\ p_B^2 = m_B^2,\ p_C^2 = m_C^2,\ p_D^2 = m_D^2,$$

in other words 4 of these quantities are trivial. We are thus left with the six quantities:

$$p_A \cdot p_B, \quad p_A \cdot p_C, \quad p_A \cdot p_D, \quad p_B \cdot p_C, \quad p_B \cdot p_D, \quad p_C \cdot p_D. \tag{13.54}$$

They are not all independent, however, since we have four relations between them arising from energy-momentum conservation:

$$p_A^\mu + p_B^\mu = p_C^\mu + p_D^\mu \quad \mu = 0, 1, 2, 3.$$

The two independent variables are any of the following three:

$$s = (p_A + p_B)^2 = (p_C + p_D)^2 \tag{13.55}$$

$$t = (p_A - p_C)^2 = (p_D - p_B)^2 \tag{13.56}$$

$$u = (p_A - p_D)^2 = (p_C - p_B)^2, \tag{13.57}$$

where $\sqrt{s}$ is the center of mass energy and $\sqrt{-t}$ 4-momentum transfer. The quantities s, t and u are called Mandelstam variables. They correspond to the square of the propagator in the three possible topologies (channels) shown in Fig. 13.3. One can easily show that:

$$s + t + u = m_A^2 + m_B^2 + m_C^2 + m_D^2. \tag{13.58}$$

In a scattering experiment we can express the kinematical variables in terms of s, t and u. Consider, e.g., the reaction

$$A + B \to C + D$$

in the center of momentum system. The sum of the three momenta before and after the reaction is zero. Thus:

$$\vec{p}_A^{\,*} + \vec{p}_B^{\,*} = \vec{p}_C^{\,*} + \vec{p}_D^{\,*} = 0. \tag{13.59}$$

The four momenta are given by

$$p_A = (E_A^*, \vec{p}), \quad p_B = (E_B^*, -\vec{p}), \tag{13.60}$$

$$p_C = (E_C^*, \vec{p}'), \quad p_D = (E_D^*, -\vec{p}'), \tag{13.61}$$

with

$$\vec{p} = \vec{p}_A^{\,*} = -\vec{p}_B^{\,*}, \quad \vec{p}^{\,'} = \vec{p}_C^{\,*} = -\vec{p}_D^{\,*}.$$

from these we can find s, t and u. Thus, if θ is the scattering angle, i.e. the angle between $\vec{p}$ and $\vec{p}'$ we have

$$s = (E_A^* + E_B^*)^2 = \left[\sqrt{m_A^2 + \vec{p}^2} + \sqrt{m_B^2 + \vec{p}^2}\right]^2$$

$$= (E_C^* + E_D^*)^2 = \left[\sqrt{m_C^2 + \vec{p}^{\,'2}} + \sqrt{m_D^2 + \vec{p}^{\,'2}}\right]^2,$$

$$t = E_A^* - E_C^*)^2 - (\vec{p} - \vec{p})'^2 = m_A^2 + m_C^2 - 2E_A^* E_C^* + 2|\vec{p}||\vec{p}^{\,'}| \cos\theta,$$

$$u = E_A^* - E_D^*)^2 - (\vec{p} + \vec{p})'^2 = m_A^2 + m_D^2 - 2E_A^* E_D^* - 2|\vec{p}||\vec{p}^{\,'}| \cos\theta.$$

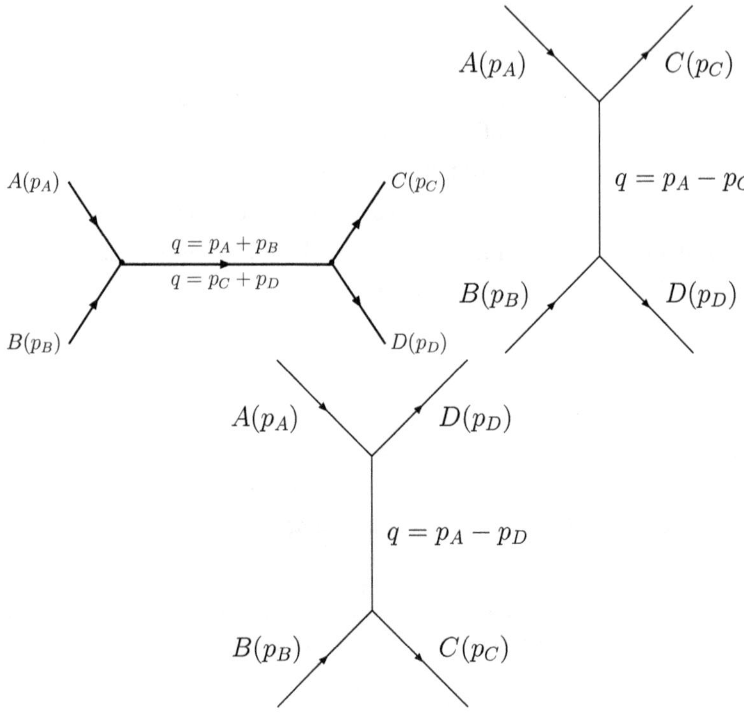

Fig. 13.3: The three available channels of the reaction $A + B \to C + D$. The arrows indicate the oncoming and outgoing particles.

Solving them in terms of p, p' and $\cos\theta$ we find:

$$|\vec{p}|^2 = \frac{1}{4s}\left[s - (m_A + m_B)^2\right]\left[s - (m_A - m_B)^2\right],$$

$$|\vec{p}\,'|^2 = \frac{1}{4s}\left[s - (m_C + m_D)^2\right]\left[s - (m_C - m_D)^2\right]. \tag{13.62}$$

$$\cos\theta = \frac{s(t-u) + (m_A^2 - m_B^2)(m_C^2 - m_D^2)}{[(s - (m_A^2 + m_B^2))^2 - 4m_A^2 m_B^2]^{1/2}[(s - (m_C^2 + m_D^2))^2 - 4m_C^2 m_D^2]^{1/2}}. \tag{13.63}$$

Solving them for the energy we get:

$$E_A^* = \frac{1}{2\sqrt{s}}(s + m_A^2 - m_B^2), \tag{13.64}$$

$$E_B^* = \frac{1}{2\sqrt{s}}(s + m_B^2 - m_A^2), \tag{13.65}$$

$$E_C^* = \frac{1}{2\sqrt{s}}(s + m_C^2 - m_D^2), \tag{13.66}$$

$$E_D^* = \frac{1}{2\sqrt{s}}(s + m_D^2 - m_C^2). \tag{13.67}$$

In the special case of elastic scattering:

$$A + B \to C + D,$$

we have

$$|\vec{p}| = |\vec{p}\,'|, \quad \cos\theta = 1 + \frac{t}{2p^2}.$$

If we consider the laboratory frame with particle B as target we get the picture of Fig. 13.4.

Fig. 13.4: The scattering $A + B \to C + D$ in a frame in which B is at rest (laboratory frame).

The 4-momenta in this system are given by

$$p_A = (E_{AL}, \vec{p}_L) \quad p_B = (m_B, \vec{0}),$$
$$p_C = (E_{CL}, \vec{p}\,'_L) \quad p_D = (E_{DL}, \vec{p}_L - \vec{p}\,'_L).$$

Obviously

$$E_{DL} = E_{AL} + m_B - E_{CL}.$$

The s, t and u are given as follows:

$$s = (E_{AL} + m_B)^2 - p_L^2 = m_A^2 + m_B^2 + 2E_{AL}m_B, \tag{13.68}$$

$$t = 2p_L p'_L \cos\theta_L + m_A^2 + m_C^2 - 2E_{AL}E_{CL}, \tag{13.69}$$

$$u = m_A^2 + m_B^2 + m_C^2 + m_D^2 - s - t. \tag{13.70}$$

We can follow the same procedure in the center of momentum frame. We find:

$$E_{AL} = \frac{1}{2m_B}(s - m_A^2 - m_B^2), \tag{13.71}$$

$$|\vec{p}_L| = \frac{1}{2m_B}(s^2 + m_A^4 + m_B^4 - 2s + m_A^4 - 2s + m_B^4 - 2 + m_A^2 + m_B^2)^{1/2}, \tag{13.72}$$

$$\begin{aligned}\cos\theta_L &= \cos\theta_{AC} = \\ &= \frac{(s - m_A^2 - m_B^2)(s - m_C^2 - m_D^2) + t(s + m_B^2 - m_A^2) - 2m_B^2(m_A^2 + m_C^2)}{\left[(s + m_A^4 + m_B^4 - 2sm_A^2 - 2sm_B^2 - 2m_A^2 m_B^2)\right]^{1/2}} \times \\ &\quad \frac{1}{\left[(u + m_B^4 + m_C^4 - 2um_B^2 - 2um_C^2 - 2m_B^2 m_C^2)\right]^{1/2}}.\end{aligned} \tag{13.73}$$

All the above above can be generalized in more complicated processes.

13.7 Detection of resonances

Suppose we consider the reaction:

$$p + \pi^- \to n + \pi^- + \pi^+ \tag{13.74}$$

This could have proceeded directly producing the above three particles or indirectly via a very short lived particle indicated by ρ^0, i.e.

$$p + \pi^- \to n + \rho^0, \quad \rho^0 \to \pi^- + \pi^+ \tag{13.75}$$

Could one tell?

13.7.1 *The signature of a resonance*

Suppose we measure the invariant mass of the two produced pions.

$$s = (E_1 + E_2)^2 - (\mathbf{p}_1 + \mathbf{p}_2)^2 \tag{13.76}$$

This quantity can easily be measured by measuring the energies and momenta of the two produced pions. If the reaction proceeds as in Eq. (13.75) s must be very sharp and be equal to the mass of the intermediate particle. If, on the other hand, it proceeds as in Eq. (13.74) this quantity must take continuous values between a minimum and a maximum determined by the energy and momenta of the two initial particles and the energy and

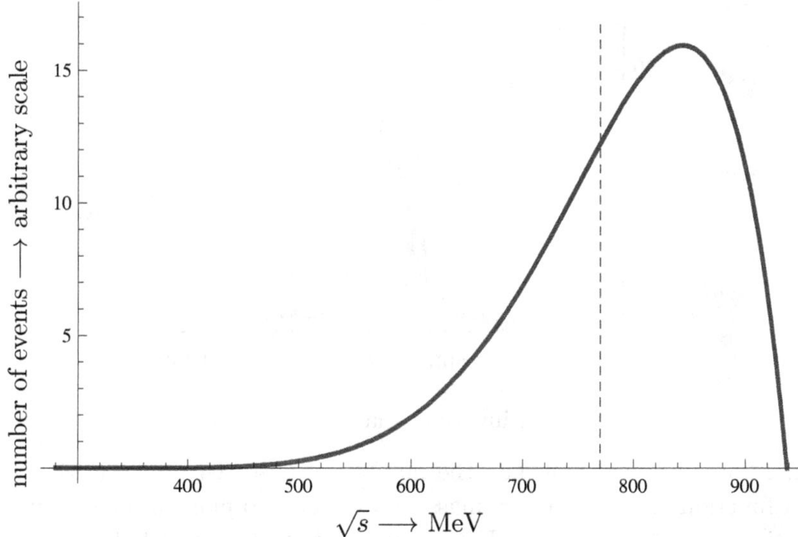

Fig. 13.5: The spectrum of invariant mass of the final particles π^+ and π^-. If the reaction proceeds through the the decay of an intermediate very short lived particle ρ^0, one expects a single line (the dotted in the figure). If the two pions are produced directly the spectrum is continuous as expected from of three particles in the final state. The scale for the number of events is arbitrary. The energy is in MeV.

momentum of the produced neutron. So given enough energy one expects the spectrum indicated in Fig. 13.5. The experimental situation is given in Fig. 13.6.

The question is how does one find the mass of the resonance? Clearly in the center of momentum we have

$$\sqrt{s} = m_n + m_\rho = E^*_\pi + E^*_p = \sqrt{(E_\pi + E_p)^2 - (\mathbf{p}_\pi + \mathbf{p}_p)^2}$$

Since the proton is at rest we find

$$s = m_\pi^2 + m_p^2 + 2E_\pi m_p \rightarrow m_\pi^2 + m_p^2 + 2E_\pi m_p = \left(E^*_\pi + E^*_p\right)^2$$

We thus get:

$$E_\pi = \frac{\left(E^*_\pi + E^*_p\right)^2 - m_\pi^2 - m_p^2}{2m_p}$$

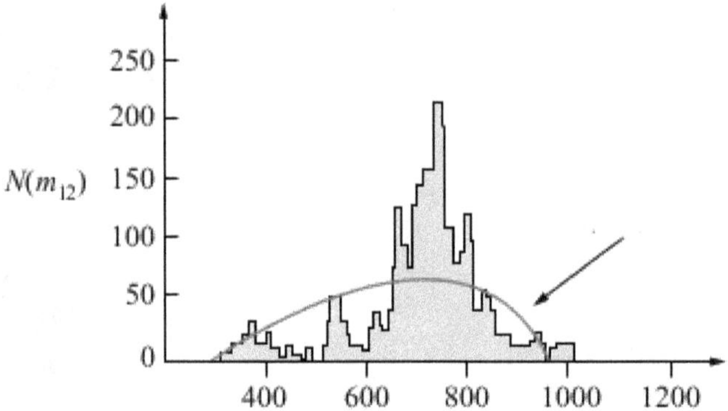

The two pion invariant mass $m_{12} \longrightarrow$ MeV/c^2

Fig. 13.6: The spectrum, i.e. the number of observed events (104 total) as a function of the invariant mass m_{12} of the two pions produced in the reaction $p + \pi^- \rightarrow n\pi^+\pi^-$. The arrow points to the expected continuous spectrum in the absence of the resonance. The resonance is clearly seen.

The minimum (threshold) pion energy obviously happens when $E^*_\pi + E^*_p = m_\rho + m_n$. We thus get

$$(E_\pi)_{th} = \frac{(m_\rho + m_n)^2 - m_\pi^2 - m_p^2}{2m_p} \rightarrow m_\rho = \sqrt{2m_p (E_\pi)_{th} + m_\pi^2 + m_p^2} - m_n \tag{13.77}$$

Experimentally it is found the pion threshold energy is 1068 MeV. This leads to 765 MeV for the ρ^0 mass, which is in agreement with Fig. 13.6. From this figure one can obtain the resonance width, which is $\Gamma = 125$MeV.

13.7.2 *The notion of a resonance*

It is known from elementary quantum theory that the number of unstable particles decaying in time dt is given by:

$$dN(t) = -\lambda N(t)dt. \tag{13.78}$$

Integrating this we obtain

$$N = N_0 e^{-\lambda t}. \tag{13.79}$$

The quantity $\tau \equiv 1/\lambda$ is known as the average life time since:

$$\tau = \langle t \rangle = \frac{N_0 \int_0^\infty t e^{-\lambda t} dt}{N_0 \int_0^\infty e^{-\lambda t} dt} = \frac{\frac{1}{\lambda^2}}{\frac{1}{\lambda}} = \frac{1}{\lambda}. \tag{13.80}$$

The half life $T_{1/2}$ is defined so that $N(T_{1/2}) = \frac{N_0}{2}$

$$\frac{N_0}{2} = N_0 e^{-\lambda T_{1/2}},$$

that is

$$T_{1/2} = \frac{1}{\lambda} \ln 2 = \tau \ln 2. \tag{13.81}$$

How are all the above related to the wave function describing the particle? In its rest frame ($\mathbf{p} = 0$) it takes the form:

$$\Psi(t) = \Psi(0) e^{-i\frac{E}{\hbar}t}.$$

Thus if the energy is real the probability of finding the particle is a constant independent of time:

$$|\Psi(t)|^2 = |\Psi(0)|^2 \text{ independent of } t, \text{ if } E \text{ is real}$$

In other words such a particle does not decay, it lives for ever. In the above formalism a decay occurs only if the energy has an imaginary part, that is,

$$E = E_0 - i\frac{\Gamma}{2}, \quad E_0, \Gamma = \text{real}.$$

(the factor of $\frac{1}{2}$ was introduced for later convenience). In this case we get

$$|\Psi(t)|^2 = |\Psi(0)|^2 e^{-\frac{\Gamma}{\hbar}t}. \tag{13.82}$$

In other words it is in agreement with Eq. (13.79), if

$$\frac{\Gamma}{\hbar} = \lambda \text{ or } \Gamma\tau = \hbar, \tag{13.83}$$

The wave function $\Psi(t)$ takes the form

$$\Psi(t) = \begin{bmatrix} \Psi(0) e^{-i\frac{E_0}{\hbar}t - \frac{\Gamma}{2m}t} & , t > 0 \\ 0 & , t < 0 \end{bmatrix} \tag{13.84}$$

The appearance of a complex energy is at first sight disturbing, since we learned in non relativistic quantum mechanics that the eigenvalues of a Hermitian operator are real. Hermiticity, however, is linked with the boundary conditions. In quantum mechanics we assumed that the wave function vanishes at infinity. This, however, does not include the decay, since then there is an outgoing spherical way to infinity. Complex boundary conditions yield complex eigenvalues. An analogous situation occur in relativistic field theory [Majore (2015)]. Now the Feynman propagators can have complex energies.

The above discussion aside, what is the meaning of the parameter Γ? Before we give an answer to this let us ask ourselves what is the probability for the particle to have energy between E and $E + dE$? To this end we must find the wave function in terms of the frequency ω, with $E = \hbar\omega$. In other words we must find the Fourier transform of Eq. (13.84). This way we find:

$$\begin{aligned}\tilde{\Psi}(\omega) &= \frac{1}{\sqrt{2\pi}} \int_{-\infty}^{+\infty} \Psi(t) e^{i\omega t} dt = \\ &= \frac{\Psi(0)}{\sqrt{2\pi}} \int_0^{\infty} e^{-i\frac{E_0}{\hbar}t - \frac{\Gamma}{2\hbar}t} e^{i\omega t} dt = \\ &= \frac{\Psi(0)}{\sqrt{2\pi}} \frac{e^{i\left(\omega - \frac{E_0}{\hbar}\right)t - \frac{\Gamma}{2\hbar}t}}{\left(\omega - \frac{E_0}{\hbar}\right)i - \frac{\Gamma}{2\hbar}} \Bigg|_0^{\infty}\end{aligned}$$

or

$$\tilde{\Psi}(\omega) = \frac{\Psi(0)}{\sqrt{2\pi}} \frac{i\hbar}{(\hbar\omega - E_0) + i\frac{\Gamma}{2}} \tag{13.85}$$

Recalling that $E = \hbar\omega$, we get

$$\tilde{\Psi}(E) = \frac{\Psi(0)}{\sqrt{2\pi}} \frac{i\hbar}{(E - E_0) + i\frac{\Gamma}{2}}$$

or

$$|\tilde{\Psi}(E)|^2 = \frac{\hbar^2}{2\pi} \frac{|\Psi(0)|^2}{(E - E_0)^2 + \frac{\Gamma^2}{4}}.$$

The condition

$$\int_{-\infty}^{+\infty} |\tilde{\Psi}(E)|^2 dE = 1$$

yields

$$|\tilde{\Psi}(0)|^2 = \frac{\Gamma}{\hbar^2} \tag{13.86}$$

((13.86) can be obtained from the normalization condition of Eq. (13.84)). Thus we finally get:

$$\rho(E) = |\tilde{\Psi}(E)|^2 = \frac{\Gamma}{2\pi} \frac{1}{(E - E_0)^2 + \left(\frac{\Gamma}{2}\right)^2} \tag{13.87}$$

as the probability of finding the particle with energy between E and $E+dE$. In other words the parameter Γ widens the state of the particle, so that it no longer has definite energy and this leads to its destruction. The width of a state that arises from the decay of the particle is called natural width. The shape of Fig. 13.7b is called Lorentzian or Breit-Wigner. Γ is the width at half the maximum value of $\rho(E)$ (see Fig. 13.7b).

From the above we see that

$$\tau\Gamma = \hbar.$$

Thus returning to Fig. 13.6 we find that even from this first experiment $\Gamma = 125$ MeV. This leads to a life time measurement

$$\tau = 6 \times 10^{-24}\text{s},$$

which very close to the presently accepted value mean value of 4.5×10^{-24}s Yes, with ingenious ideas, experimental physicists can measure so small life times! It is through resonances that the life times of particles like the W-bosons etc, listed periodically by the particle data group (PDG) [Groom *et al.* (2000)] have been measured.

Furthermore from the relation $\Delta E \cdot \Delta t \approx \hbar$ we see that, in order to measure the energy with accuracy $\Delta E = \Gamma$ the uncertainty in time must be $\Delta t = \tau$.

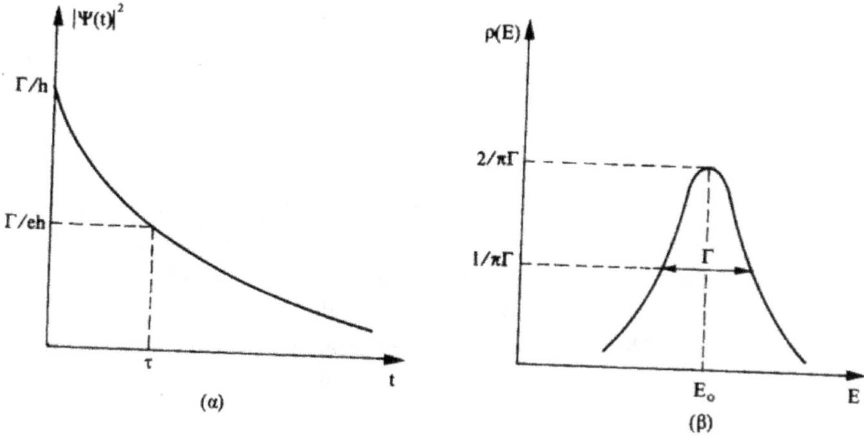

Fig. 13.7: The probability $|\Psi(t)|^2$ and $|\Psi_E|^2$ (b). The wider the bell given by $\rho(E)$ the smaller is τ, i.e. the function $|\tilde{\Psi}_E|^2$ falls faster.

13.8 Transformation properties of non invariant quantities

One very useful quantity is the cross section σ. It has the meaning of an effective surface of the target particles. Since the lengths, which are perpendicular to the velocity, do not Lorentz contract, the cross section is the same in two inertial frames. Consider, e.g., the process:

$$A + B \to C + D,$$

then $\sigma = \sigma(s,t)$.

What do we do, if a quantity is not invariant? Such a quantity is, e.g., the differential cross section $d\sigma/d\Omega$, Ω =the solid angle. How can we transform it from one system to the next? We can multiply such a quantity with some suitable factor to get an invariant. One can show, e.g. that the quantity $(d^3\vec{p})/E$ is invariant. Thus for two inertial frames we have:

$$\frac{E}{p^2 dp}\frac{d\sigma}{d\Omega} = \frac{E'}{(p')^2 dp'}\frac{d\sigma'}{d\Omega'} \quad \text{or} \quad \frac{d\sigma}{pdEd\Omega} = \frac{d\sigma'}{p'dE'd\Omega'} \tag{13.88}$$

(we used the relation $pdp = EdE$).

From the known relation between the two momenta and after integrating Eq. (13.88) over the momenta we find the desired relation.

13.9 Problems

(1) The HERA accelerator in DESY uses colliding beams of protons and electrons. The proton energy is 820 GeV and the electron energy 30 GeV.

- Find the total energy in the center of momentum system.
- What energy would be required to achieve this in fixed target experiments i) if the proton is the target and ii) if the electron is the target.
- Find the energy of each beam in a standard electron-positron collider to achieve the same goal.
- Do the same for a proton-proton colliding beam accelerator.
- Comment on the feasibility of each proposal.

(2) In pion-nucleon scattering many fermion resonances were discovered, e.g. the $I = 3/2$, $s = 3/2$ resonances with a mass 1250 GeV indicated by Δ^{++}, Δ^{+}, Δ^{0} and Δ^{-}. The following reactions were considered:

$$\pi^+ + p \to \pi^+ + p + \pi^0$$

$$\pi^- + p \to \pi^+ + n + \pi^-$$

- Indicate which resonance was found in which reaction. Are there any resonances observed in both reactions.

- Find the threshold pion kinetic energy required for the production of each resonance (the initial proton is at rest).
- Obtain the life time for each resonance, if their width is 140 MeV.
- Give the quark content of each resonance. Does it appear reasonable in the case of Δ^{++}?

Data: proton mass 938 MeV, neutron mass 940 MeV, charged pion mas 140 MeV, neutral pion mass 135 MeV.

Bibliography

Aaij, R. *et al.* (2016). *Phys. Rev. Lett.* **117**, p. 261801, the LHCb collaboration; arXiv:1608.06420 (astro-ph.CO).

Ade, P. *et al.* (2015). Planck 2015 results. XIII. Cosmological parameters, arXiv:1502.01589 (astro-ph.CO).

Ahlen, S. P., III, F. A., Brodzinski, R. L., Druckier, A. K., Gelmini, G., and Spergel, D. N. (1987). *Phys. Lett. B* **195**, p. 603.

An, F. J. *et al.* (2012). *Phys. Rev. Lett.* **108**, p. 171803, arXiv:1203.1669 [hep-ex].

Anastassopoulos, V. *et al.* (2015). ArXiv:1502.04317 Accelerator Physics (physics.acc-ph); High Energy Physics - Experiment (hep-ex).

Aprile, E. *et al.* (2011). *Phys. Rev. Lett.* **107**, p. 131302, arXiv:1104.2549v3 [astro-ph.CO].

Araki, T. *et al.* (2005). *Phys. Rev. Lett.* **94**, p. 081801.

Asztalos, S. J. *et al.* (2010). *Phys. Rev. Lett.* **104**, p. 041301, the ADMX collaboration; arXiv:0910.5914 (astro-ph.CO).

Baron, J. *et al.* (2014). *Science* **343**, p. 269, dOI: 10.1126/science.1248213.

Baru, A. and Bronzin, G. L. (1971). *J. Math. Phys.* **12**, p. 841.

Battistony, G. *et al.* (1983). *Phys. Lett. B* **133**, p. 454.

Bionta, R. *et al.* (1983). *Phys. Rev. Lett.* **51**, p. 27.

Bjorken, J. D. and Drell, S. D. (1964). *Relativistic Quantum Mechanics* (McGraw-Hill).

Capozzi, F., Fogli, G., Lisi, E., Marrone, A., Montanino, D., and Palazzo, A. (2014). *Phys. Rev. D* **89**, p. 093018, arXiv:1312.2878 [hep-ph].

Castano, D. J., Piard, E., and Rammond, P. (1994). *Phys. Rev. D* **49**, p. 49.

Chodos, A., Jaffe, R., K.Johnson, Kiskis, J., and Weisskoph, V. F. (1974). *Phys. Rev. D* , p. 3471.

Christenson, J. H., Cronin, J. W., Fitch, V. L., and Turlay, R. (1964). *Phys. Rev. Lett.* **13**, p. 380.

Compton, A. H. (1923). *Phys. Rev.* **21**, p. 483.

Dine, M., Fischler, W., and Strdnicki, M. (1981). *Phys. Lett. B* **104**, p. 199.

Eichten, E. *et al.* (1980). *Phys. Rev. D* **21**, p. 203.

Ellis, J. (1980) The phenomenology of Unified Theories, TH. 3124, CERN, CH.

Frauenfelder, H. and Henley, E. M. (1974). *Subatomic Physics* (Prentice Hall).

Fritsch, H. and Minkowski, P. (1975). *Ann. Phys.* **93**, p. 193.

Fukuda, Y. *et al.* (1998a). *Phys. Rev. Lett.* **81**, p. 1562, super-Kamiokande Collaboration; (hep-ex/9807003).

Fukuda, Y. *et al.* (1998b). *Phys. Lett. B* **436**, p. 44, super-Kamiokande Collaboration; (hep-ex/9805006).

Georgi, H. and Dimpopoulos, S. (1981). *Nucl. Phys. B* **193**, p. 150.

Georgi, H. and Georgi, S. L. (1974). *Phys.Rev. Lett.* **32**, p. 438.

Goldhaber, M., Grodzins, L., and Sunyar, A. W. (1958). *Phys. Rev.* **109**, p. 1015, helicity of Neutrinos; doi:10.1103-PhysRev.109.1015.

Griffiths, D. (1987). *Introduction to Elementary Particles* (John Wiley).

Groom, D. E. *et al.* (2000). *Eur. Phys. J* **C 25**, p. 1, [Particle Data Group].

Haber, A. H. and Kane, G. L. (1985). *Phys. Rep.* **114**, p. 75.

Hakaya, T. (2005). *Nucl. Phys. B (Proc. Suppl.)* **138**, p. 376.

Hayato, Y. *et al.* (2005). (Super Kamiokande Collaboration) hep-ph/9904020.

Henley, E. M., Oka, T., and Vergados, J. D. (1986). *Phys. Lett.* **166 B**, p. 274.

Hinshaw, G. *et al.* (2008). For WMAP, arXiv:0803.0732; B. Gold *et al.*, arXiv:0803.0715; M. R. Nolta *et al.*, arXiv:0803.0593; J. Dunkley *et al.*, arXiv:0803.0586; E. L. Wright *et al.*, arXiv:0803.0577, R. S. Hill *et al.*, arXiv:0803.0570; E. Komatsu *et al.*, arXiv:0803.0547.

Jarlskog, C. (1985). *Phys. Rev. Lett.* **55**, p. 1039.

Kim, J. E. (1979). *Phys. Rev. Lett.* **43**, p. 137.

Klein, O. and Nishima, Y. (1929). *Z. Physik.* **52**, p. 853.

Langacker, P. (1981). *Phys. Rep.* **72**, p. 185.

Lees, J. *et al.* (2012). *Phys. Rev. Lett.* **109**, p. 211801, b. Schwarzschild, Physics today, November 2012, p. 16.;.(BaBar collaboration); arXiv:1207.5832 (hep-ex).

Majore, M. (2015). *A Modern Introduction to Quantum Field Theory* (Oxford University Press).

Massey, R., Kitching, T., and Richard, J. (2010). *Rep. Prog. Phys.* **73**, p. 086901, arXiv:1001.1739 (astro-ph).

Peccei, R. and Quinn, H. (1977). *Phys. Rev. Lett.* **38**, p. 440.

Reines, F., Cowan, C., and Goldhaber, M. (1954). *Phys. Rev.* **96**, p. 1154.

Ross, G. G. (1982). *Grand Unified Theories* (Perseus Books Group), chapter 9.

Sakharov, A. D. (1991). *Sov. Phys. Usp.* **34 (5)**, p. 392, delayed publication, submitted 23 September 1966.

Salaris, M. and Cassisi, S. (2005). *Evolution of stars and stellar populations* (John Wiley and Sons), iSBN 0-470-09220-3.

Seiden, A. (2005). *Particle Physics* (Addison Wesley).

Shifman, M. A., Vainshtein, A., and Zakharov, V. I. (1980). *Nuc. Phys. B* **166**, p. 493.

Shirman, Y. (2009). TASI 2008 Lectures: Introduction to Supersymmetry and Supersymmetry Breaking; arXiv:0907.0039 [hep-ph].

Slansky, R. (1981). *Phys. Rep.* **79**, p. 1.

Spergel, D. *et al.* (2003). *Astrophys. J. Suppl* **148(1)**, pp. 175–94, first-year Wilkinson Microwave Anisotropy Probe (WMAP) observations: Determination of cosmological parameters.

Spergel, D. N. *et al.* (2007). *ApJS* **170**, p. 377, the Wilkinson Microwave Anisotropy Probe (WMAP) Three Year Results: Implications for Cosmology; (astro-ph) arXiv:astro-ph/0603449.

Tegmark, M. *et al.* (2004). *Phys. Rev. D* **69(10)**, p. 10351, cosmological parameters from SDSS and WMAP.

Vergados, J. D. (2016). *Group and Representation Theory* (World Scientific).

Vergados, J. D. (1986). *Phys. Rep.* **133**, p. 1.

Vergados, J. D. (2007). *Lect. Notes Phys.* **720**, p. 69, lectures given at the Third Aegean Summer School *The Invisible Universe: Dark Matter and Dark Energy,* 26 September-1 October 2005, Karfas, Island of Chios, Greece; arXiv:hep-ph/0601064.

Vergados, J. D., Ejiri, H., and Simkovic, F. (2012). *Rep. Prog. Phys.* **75**, p. 106301.

Vergados, J. D., Ejiri, H., and Simkovic, F. (2016). *IJMPE.* **25(11)**, p. 10630007, World Scientific Publishing Company DOI: 10.1142/S0218301316300071.

Vergados, J. D., Giomataris, Y., and Novikov, Y. N. (2010). *J. Phys. Conf. Ser.* **259**, p. 012100, arXiv:1010.4388 (hep-ph).

Vergados, J. D. and Semertzidis, Y. (2016). *Nucl. Phys. B* **897**, p. 821, center for Axion and Precision Physics Research, IBS, Daejeon, Republic of Korea; arXiv:1601.04765 (hep-ph).

Weinberg, S. (1972.). *Gravitation and Cosmology* (Wiley).

Weinberg, S. (1995.). *The Quantum Theory of Fields, Volume 1: Foundations* (Cambridge University Press).

Weinberg, S. (1996.). *The Quantum Theory of Fields, Volume 2: Modern applications* (Cambridge University Press).

Weinberg, S. (2008.). *Cosmology* (Oxford University Press, Oxford, UK).

Wolfenstein, L. (1999). *Phys. Rev. Lett.* **83**, p. 911.

Wu, C. S. *et al.* (1957). *Phys. Rev.* **105**, p. 1413.

Wybourne, B. G. (1974). *Classical Groups for Physicists* (Wiley NY).

Yang, C. . (1962). *Elementary Particles* (Princeton University Press, Princeton N.J.).

Zhitnisky, A. (1980). *Sov. J. Nuc. Phys.* **31**, p. 260, in Russian.

Zioutas, K. *et al.* (2017). Search for axions in streaming dark matter.

Index

Aharonov-Bohm effect, 97
fundamental representation, 24
inhomogeneous Maxwell equations, 100
representations,irreducible, 26

Abelian group, 2
Abelian transformation, 47
Abelian gauge symmetry, 94
Abelian subalgebra, 22
adjoined, 24
adjoined representation, 23, 56, 105
algebra, 58
allowed parameter space, 220
angles, 11
anti-screening, 160
antihermitian, 7
antiparticle, 80, 91
antisymmetric tensor, 101
antiunitary operator, 377
apparent magnitude, 309
asymmetry baryon-antbaryion, 297
asymptotic freedom, 159
atmospheric neutrino oscillations, 331
atmospheric neutrinos, 331
axial gauge, 96
axion, 102, 306, 381

background microwave radiation, 248
baryon asymmetry, 235
baryon number, 88
baryon to phoon ratio, 240
baryon-antbaryion asymmetry, 297
baryons, 88
BBN, 254
Big Bang, 244
big bang, 244
big bang nucleosynthesis, 254
black body radiation, 249
bolometric candles, 308
boosts, 43
boson decay widths, 190
bosons, 87
bullet cluster, 302

C for composite systems, 369
C in Dirac theory, 368
C violation, 373
Cabbibo theory, 147
Cambell-Hausdorff, 7
candles (prototype), 245
Casimir operator, 13, 15
CBMR, 248
CDM-Λ, 316
cepheids, 245
Chandrashekar limit, 308
charge conjugation, 79
charge conjugation C, 366
charge operator, 127
charged boson matrix, 126
charged currents, 147
charged gauge bosons, 129
chiral, 77
chiral basis, 137

chiral multiplets, 89
chiral states, 80
chiral transformation, 84
chiral transformations, 118
chirality, 77, 78
CHOOZE, 322
chromodynamics, 109
closed universe, 260
co-factor, 3
collaboration PLANCK, 316
colliding beams, 407
color group SU(3), 89
color r,g,b, 90
color sextet, 162
combined symmetry CPT, 400
commutative, 1
commutator rules, 44
complex representation, 53
Compton scattering, 208
confinement, 159
confining potential, 169–171
conjugate representation, 16
conservation law, 44, 63, 64
conserved current, 46
contact interaction, 150
contragredient representation, 16
cosmic background microwave radiation, 248
cosmological constant, 266
cosmological neutrino bound, 344
cosmological priciples, 243
Cosmology, 243
Coulomb gauge, 96, 97
covariant derivative, 93, 105
covariant form, 65
CP invariance, 102
CP of simple systems, 382
CP symmetry, 381
CP violation, 391
CP-violation, 235
cplor, 88
CPT, 353
cross sections, 173, 178
currents under C, 324
curvature, 260

dark energy, 266
dark matter, 223, 297, 304
Daya Bay, 322
decay widths, 179
degeneracy, 277
degeneracy n(T), 278, 279
degrees of freedom, 133
detailed balance, 378
determinant, 3
diagonalization (non Hermitian matrices), 142
Dirac equation, 49, 65
Dirac matrices, 66
Dirac-Pauli representation, 72
direct detection, 307
discrete group, 2
Doppler effect, 246
Dynkin lbels, 227

Eddington, 252
Einstein equation, 260
elecroweak Lagrangian, 111
electric dipole moment, 380
Electromagnetic tensor, 99
electron-proton scattering, 198
electron-proton scattering (inelastic), 202
electroweak Lagrangian, 127
EM current, 147
EM field tensor, 99
EM invariants, 102
energy flux, 245
energy fraction, 261
energy momentum tensor, 260
energy negative, 70
Energy negative interpretation, 73
energy positive, 68
energy production, 252
equation (Einstein), 260
equilibrium condition, 238
equivalence of observers, 248
evolution (classical prelude), 257
evolution of the universe, 256
expanding universe, 244
experiments satellite, 246
exponential matrix, 6

Fermi interaction, 151
Fermi theory, 319
Fermion currents, 137
Fermion field, 94
Fermion field Lagrangian, 110
Fermion fields, 49
Fermion mass matrix, 140
Fermion masses, 137
Fermion masses (SUSY), 218
Fermion SM Lagrangian, 111
Fermions, 87
Feynman diagrams, 173
Feynman rules, 173, 177
field theory, 49
flat universe, 260
flavor, 90
flavor SU(3), 50
flux, 245
form factors, 202
fraction Ω, 261
Friedman equations, 258, 261
fundamental, 17
fundamental representation, 17, 20
fundamental representation, 22
fusion, 254

G-prity, 371
gauge boson mass, 124
gauge field invariant Lagrangians, 108
gauge field transformation, 104
gauge invariance, 93
gauge invariant Lagrangians, 107
gauge non Abelian transformation, 103
gauge particles, 213
gauge SU(2) transformation, 106
gauge SU(3) transformation, 107
gauge SU(5) boson masses, 234
gauge SU(5) bosons, 228
gauge SU(5) couplings, 235
gauge symmetry, 88
gauge transformation, 95
gaugino, 211
gaugino-higgsino mixing, 221
Gell-Mann matrices, 157
generalized derivative, 93
generations, 90
generator, 46
generators, 21, 58, 89
Glashow, 87
global transformations, 44
global symmetry, 88
gluon exchange, 157, 165
gluon exchange potential, 171
gluon self-couplings, 158
gluons, 129
Goldstone boson, 116
Goldstone bosons, 121
Goldstone bosons eaten up, 129
Grand unified theories, 225
gravitational lensing, 301
group, 1
GSW, 365
GUT, 225
GUT SU(5), 226

hadron jets, 161
hadronic channel, 190
hadrons, 88
harmonic oscillator, 169
helicity, 73, 76, 88, 353
helicity of neutrino, 361
helium abundance, 251
hermitian, 3
Hermitian conjugate, 3
Higgs, 88
Higgs (physical), 125
Higgs CP-odd, 219
Higgs discovery, 152
Higgs isodoublet, 91
Higgs Mechanism, 123
Higgs mechanism (gauged), 123
Higgs mechanism (SUSY), 214
Higgs potential, 113
Higgs SU(5), 231
Higgs SU(5) mechanism, 231
Higgs SUSY, 213
higgsino, 211
Hipparchus, 309
homomorphism, 32
horizon, 272, 275
Hubble law, 246

hypecharge, 89
hyperchage, 90

identity matrix, 3
independent parameters, 4, 9
infinitesimal operators, 44
inflation, 288
inflaton, 293
invariant amplitude, 179, 180
invariants EM, 102
invariants in scattering, 412
inverted hierarchy, 340
irreducible representations, 26
irreducible SU(3) representations, 162
isentropic, 277
isodoublet, 90
isomorphism, 32
isosinglet, 91
isospin (weak) goup SU(2), 89
isotriplet, 34, 132

jets, 161

Kaluza-Klein, 306
KamLAND, 336
kaon oscillation, 385
kaon regeneration, 388
KCM, 147
KCM matrix, 12
kinematical invariants, 406
kinematics in particle decay, 411
kinetic energy part, 169
Klein Gordon equation, 63
Kobayashi-Maskawa, 147, 393
Kobayashi-Maskawa phase, 237
Kronecker product, 26

Lagrangian density, 46
left handed, 90
left handed currents, 144
left-right, 48
leptogenesis, 399
lepton flavor, 88
lepton flavor conservation, 320
lepton number, 88
leptonic channel, 190
leptonic charged currents, 148
leptons, 88
Lie algebra, 13, 15
life time (proton), 235
lightest supersymmetric particle, 223
local symmetry, 88, 91
longitudinal component, 132
Lorentz gauge, 95
Lorentz scalars, 137
Lorentz transformation, 43
low energy approximation, 168
LSP, 223, 306
luminocity, 245
luminosity colliding beams, 409
luminosity fixed target, 408

Madelstam invariants, 182
magnitude apparent, 309
Majorana mass, 323
Majorana representation, 73
mass eigenstates, 140
massless gauge boson, 129
matrix group, 2
matter and curviture, 269
Maxwell equations, 94
metric (Robertson-Walker), 259
metric tensor, 14
minimum condition, 114
Moeller velocity, 180
MSSM, 212, 214
multicomponent, 269
multiplicative QN, 355
muon-antimuon production, 193

natural system of units, 401
negative energy, 65
neutral boson matrix, 126
neutral currents, 148
neutral kaons, 383
neutralinos, 222
neutrino appearance, 328
neutrino channel, 190
neutrino disappearance, 328
neutrino helicity, 361
neutrino helicity determination, 363
neutrino mass, 322

neutrino mass scale, 341
neutrino mixing, 322
neutrino mixing matrix, 339
neutrino oscillation data, 340
neutrino oscillations, 319, 321, 327
neutrino probes, 346
neutrino-electron scattering, 205
neutrinoless double beta decay, 341
Noether theorem, 44
non invariant, 421
normal hierarchy, 340
NOSTOS, 337
nucleosynthesis, 250

observations, 307
observer equivalence, 248
Ω_i(energy fraction), 261, 262
open universe, 260
operator charge conjugation, 80
orbital part, 168
orthogonal, 6
orthogonal group, 19
oscillation length, 328

parity, 75, 355
parity conservation, 357
parity in Dirac theory, 356
parity violation, 88, 358
particle history (universe), 278
particle spectrum (SUSY), 220
particle SU(5) content, 227
Paschos-Wolfenstein, 152
Pauli matrices, 7, 26, 50, 66
Peccei-Quinn, 381
Penzias, 249
phase Kobayashi-Maskawa, 237
phase space integral, 180, 189
phases, 11
photon, 127
photon polarization, 98
physical Higgs, 129
PLANCK, 246, 316
PLANCK collaboration, 316
plane wave, 64
PMNS matrix, 12, 339
Poisson equation, 97
polarization, 88
polarizations, 129
Pontecorvo, 321
preserved EM symmetry, 128
Primakoff effect, 307
primodial, 257
primordial nucleosynthesis, 250
problems of SCM, 284
projection operator, 72
projection operators, 188
proper length, 272
proper length (for candles), 274
proper orthogonal O(4), 35
proton life time, 235
prototype candles, 245
pseudoscalar, 102
pseudoscalar current, 81
pseudovector current, 81

QCD, 157
quantum chromodynamics, 157
quantum fluctuation bubbles, 296
quantum fluctuations, 295
quark confinement, 159
quark interactions, 164
quark slavery, 159
quarks, 88
quintesense, 297

R(3), 20
R-parity, 220
radiation dominance, 264, 277
rates, 173, 178
reactor neutrino oscillations, 334
receding of galaxies, 245
regular, 57
reionization, 283
relative parity, 356
relic neutrinos, 278
representation, 89
representations, 16
resonance, 418
resonance detection, 416
resonance signature, 416
resonance width, 421
Ricci tensor, 260

right handed, 90
right handed currents, 146
Robertson-Walker metric, 259
rotation, 47
rotational velocities, 298
rotations, 43
Runge-Lenz, 38
Rutherford scattering, 182

s-fermions, 213
s-lepton, 211
s-quark, 211
Saha effect, 249
satellite experiments, 246
scalar current, 81
scalar field, 93
scalar field (complex), 113
scalar mass, 116
scalar mass matrix, 116
scalar particles, 88
scalar potential, 94, 113, 293
scalar SM doublet, 118
scalar SM Lagrangian, 111
scale factor, 262
scattering Compton, 208
scattering electron-proton, 202
scattering invariants, 412
scattering neutrino-electron, 205
scattering proton-electron, 198
Schroedinger equation, 63
screening, 160
see-saw mechanism, 325
semisimple, 14
semisimple group, 2
sigma model, 117
simple group, 2
SM currents, 144
SM parameter determination, 151
SM symmetry, 89, 155
SNO, 322
SO(2,1), 58
SO(3), 20
SO(4), 15, 38
SO(3)x SO(3), 38
soft supersymmetry breaking, 220
solar neutrino oscillations, 334
space inversion, 74
space inversion P, 353
special unitary group, 5
spinors, 68
spontaneous symmetry breaking, 88, 116, 158
SSB, 123
standard candles, 307
standard cosmological model, 272
Standard model, 87
static universe, 266
strangeness oscillations, 387
strong SM Lagrangian, 112
structure constants, 9, 12, 13, 23
structure functions, 9
SU(2), 22
SU(3), 8, 9, 15, 22
SU(5), 226
SU(5) gauge boson masses, 234
SU(5) gauge couplings, 235
SU(5) Higgs, 231
SU(5) Higgs mechanism, 231
subalgebra, 22
subgroup, 1
SU(n), 5
sun energy output, 252
SuperKamiokande, 322, 332
supernova Ia, 308
supersymmetry, 211
SUSY, 211
SUSY breaking, 220
SUSY Fermion masses, 218
SUSY Higgs, 213
SUSY Higgs mechanism, 214
SUSY particle spectrum, 220
symmetry restoration, 122

T in classical physics, 374
T in quantum mechanics, 376
T violation, 397
T2k, 322
temperature, 277
tensor current, 81
three generation oscillations, 329
time reversal T, 374
trace techniques, 185

triton decay, 344
two real scalars, 117

U(n), 22
unbroken operator, 128
unimodular, 8
unimodular SU(n), 22
unitary group, 3, 5
unitary matrix, 3
unitary gauge, 123, 131
universe evolution, 256

vacuum expectation value, 119
vacuum polarization, 159
vector boson decay, 186
vector boson discovery, 191
vector boson self-coupling, 153
vector current, 81
vector potential, 94
volometric, 245

W-boson mass, 126
Weak angle, 126
weak interaction-handedness, 364
Weinberg angle, 126
Weyl representation, 73, 78
width of resonance, 421
Wilson, 249
WIMP, 305
WMAP, 246

Yukawa, 87
Yukawa couplings, 112
Yukawa interactions, 138
Yukawa Lagrangian, 112

Z-boson mass, 126

www.ingramcontent.com/pod-product-compliance
Lightning Source LLC
Chambersburg PA
CBHW072256210326
41458CB00074B/1783